AF572630

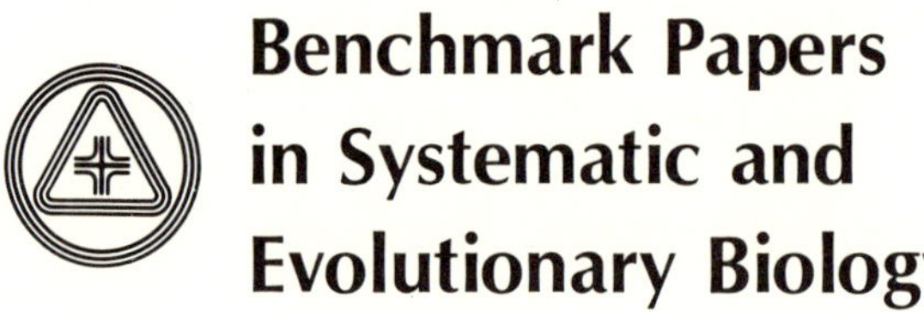

Benchmark Papers in Systematic and Evolutionary Biology

Series Editor: Howell V. Daly
University of California, Berkeley

PUBLISHED VOLUMES AND VOLUMES IN PREPARATION

MULTIVARIATE STATISTICAL METHODS: Among-Groups Covariation
William R. Atchley and Edwin H. Bryant
MULTIVARIATE STATISTICAL METHODS: Within-Groups Covariation
Edwin H. Bryant and William R. Atchley
NUMERICAL TAXONOMY
M. W. Moss
ZOOGEOGRAPHY
E. I. Schlinger
CONCEPTS OF SPECIES
C. N. Slobodchicoff

Benchmark Papers in Systematic and Evolutionary Biology / 2

A BENCHMARK® Books Series

MULTIVARIATE STATISTICAL METHODS
Within-Groups Covariation

Edited by

EDWIN H. BRYANT
University of Houston

and **WILLIAM R. ATCHLEY**
Texas Tech University

Dowden, Hutchinson & Ross, Inc.
Stroudsburg, Pennsylvania

Distributed by
HALSTED PRESS *A Division of John Wiley & Sons, Inc.*

Benchmark Papers in Systematic and Evolutionary Biology, Volume 2
Library of Congress Catalog Card Number: 75–9891
ISBN: 0–470–11471–1

77 76 75 1 2 3 4 5
Manufactured in the United States of America.

LIBRARY OF CONGRESS CATALOGING IN PUBLICATION DATA
Main entry under title:

Multivariate statistical methods, within-groups covariation.

(Benchmark papers in systematic and evolutionary biology ; 2)
1. Biometry--Addresses, essays, lectures.
2. Multivariate analysis--Addresses, essays, lectures. 3. Variation (Biology)--Mathematical models--Addresses, essays, lectures. I. Bryant, Edwin H. II. Atchley, William R.
QH323.5.M84 519.5'3 75-9891
ISBN 0-470-11471-1

Exclusive Distributor: **Halsted Press**
A Division of John Wiley & Sons, Inc.

ACKNOWLEDGMENTS AND PERMISSIONS

ACKNOWLEDGMENTS

AMERICAN PSYCHOLOGICAL ASSOCIATION
Journal of Educational Psychology
Analysis of a Complex of Statistical Variables into Principal Components
The Most Predictable Criterion
Psychological Review
Multiple Factor Analysis

AMERICAN METEOROLOGICAL SOCIETY—*Journal of Atmospheric Science*
Canonical Correlation and its Relationship to Discriminant Analysis and Multiple Regression

GROWTH PUBLISHING COMPANY, INC.—*Growth*
Size and Shape Variation in the Painted Turtle, A Principal Component Analysis

THE SOCIETY OF SYSTEMATIC ZOOLOGY—*Systematic Zoology*
An Application of Factor Analysis to the Systematics of *Kalotermes*
Evolutionary Implications of Wing and Size Variation in the Red-Winged Blackbird in Relation to Geographic and Climatic Factors: A Multiple Regression Analysis
Factor Analysis in Morphometric Traits of the House Mouse

TAYLOR & FRANCIS LTD—*The Philosophical Magazine*
On Lines and Planes of Closest Fit to Systems of Points in Space

PERMISSIONS

The following papers have been reprinted with the permission of the authors and copyright holders.

AMERICAN SOCIETY OF ICHTHYOLOGISTS AND HERPETOLOGISTS—*Copeia*
Chronological Correlation Between Change in Weather and Change in Morphology of the Pacific Tree Frog in Southern California

THE BIOMETRIC SOCIETY—*Biometrics*
Factor Analysis: An Introduction to Essentials: I. The Purpose and Underlying Models

BLACKWELL SCIENTIFIC PUBLICATIONS LTD (FOR THE BRITISH ECOLOGICAL SOCIETY)
Ecological Aspects of the Mineral Nutrition of Plants
The Application of Ordination Techniques
Journal of Animal Ecology
Multivariate Analytical Treatment of Quantitative Species Associations: An Example From Palaeoecology

Acknowledgments and Permissions

THE BRITISH PSYCHOLOGICAL SOCIETY—*The British Journal of Mathematical and Statistical Psychology*

The Theoretical Foundations of Principal Factor Analysis, Canonical Factor Analysis and Alpha Factor Analysis

FREDERICK KUDER—*Educational and Psychological Measurement*

The "Maxplane" Program for Factor Rotation to Oblique Simple Structure

INDIAN STATISTICAL INSTITUTE—*Sankhya*

The Use and Interpretation of Principal Component Analysis in Applied Research

STICHTING INTERNATIONAL BUREAU FOR PLANT TAXONOMY AND NOMENCLATURE, UTRECHT—*Taxon*

Perspectives on the Application of Multivariate Statistics to Taxonomy

PSYCHOMETRIC SOCIETY—*Psychometrika*

A Comparison of Four Methods of Constructing Factor Scores
Estimation and Tests of Significance in Factor Analysis
Rotation for Simple Loadings
The Varimax Criterion for Analytic Rotation in Factor Analysis

THE ROYAL SOCIETY OF EDINBURGH—*Proceedings of the Royal Society of Edinburgh*

On the Theory of Statistical Regression

THE UNIVERSITY OF CHICAGO PRESS

American Naturalist

Components of Sexual Dimorphism in *Chironomus* Larvae (Diptera: Chironomidae)

The Journal of Geology

The Theory and Application of Canonical Trend Surfaces

SERIES EDITOR'S PREFACE

The volumes of the Benchmark Series in Systematics and Evolutionary Biology will make accessible to students and scientists alike a broad selection of topics from these interrelated fields. While other series will be primarily ecological, genetic, or behavioral, this series will concern the classification of organisms, techniques useful in systematics, and those processes or aspects of biological relationships that are essentially evolutionary in nature. Within the past decade or two, entirely new areas of research have emerged as techniques were developed and used and concepts altered by discoveries. Many exciting developments have sprung from the juncture of systematic and evolutionary biology with other disciplines: biogeography and the geological drift of continents; molecular evolution and immunological cross-reactions; genetics of natural populations and the biochemical electrophoresis of proteins; evolutionary strategies and mathematical models. In these examples and others, we are witnessing the convergence of investigations on the central questions of evolution by scientists of disparate backgrounds. It is not an easy task for the undergraduate or researcher to acquire information drawn from so many sources. This Benchmark series will not only provide reproductions of selected original papers, but will assist in interpreting and indexing current, interdisciplinary contributions.

Other volumes will be mainly historical, because systematists retain a strong interest in the early attempts to cope with classifying organisms and to understand their lineages. The development of the theory of evolution, nature of species, and biological nomenclatures are among these retrospective topics planned for the series. Facsimile reproductions and translations will make classic papers readily available, some for the first time.

The first two volumes in this series were jointly prepared by William R. Atchley and Edwin H. Bryant. With the advent of electronic computers, methods of multivariate statistics are now widely employed in the solution of biological problems. New applications in systematics, ecology, and genetics are being recognized as more biologists become familiar with both the techniques and the machines. Although young, Atchley and Bryant are nevertheless seasoned investigators and carry with them the excellent training of the University of Kansas, which has been a center for biometrical instruction and research. They have divided the subject and senior authorship as follows: The first volume, by Atchley and Bryant, deals with covariation of measurements among groups; the second volume, by Bryant and Atchley, concerns covariation within groups. Consequently, systematists who wish to compare the geographic variation of populations, phenetic similarity of taxa, or devise identification procedures will find Part I especially useful. Where the subgroups are not yet known or the interrelationships of characters are of major concern, the sections of Part II on principal-component, factor, and regression analysis will be of interest. Both volumes should be at hand because the methods can be used in a variety of contexts.

HOWELL V. DALY

PREFACE

This collection of "benchmark" papers in multivariate statistics is an attempt to assemble a set of classical papers that deal with the analysis of within-groups covariation. To properly relate these classical papers to current research in quantitative biology, we have integrated them with additional examples of recent applications from systematics and ecology. This volume of papers is concerned with principal-component analysis, factor analysis, and multivariate regression and correlation. A companion volume, published separately in the series, deals with techniques primarily orientated toward the analysis of among-groups covariation.

Papers were selected for inclusion in this volume which we thought would be of widest interest to the researcher as well as the advanced graduate student interested in quantitative methodology in biology. In general, theoretical papers were selected which we believe represent either the most important historical developments or the most lucid summaries of a particular aspect of that type of analysis. Papers that deal with applications were included if they seemed to discuss clearly and in some detail the methodology in question and its application to a specific biological problem.

The papers that deal with biological applications tend to be from as broad a spectrum of the disciplines of systematics and ecology as was possible from the literature available. These biological papers are of much more recent vintage because electronic computers, which are necessary to analyze complex biological data, have been widely available only for the last 15 years or so. We would also like to point out that we selected papers because of the statistical applications; no attempt was made to evaluate the biological validity of the conclusions that were reached.

The greatest obstacle in this endeavor, other than the practical one of being able to include only a limited number of papers, was to surmount the problem that a large segment of the biologists in systematics and ecology (including the authors) have limited preparation in mathematics. Thus, if we included only the classical papers in mathematical statistics, the volumes would have very little utility to the average biologists. However, we feel that the same would be true had we included only papers on the applications of multivariate methods in biology. It is our opinion that too many biologists analyze quantitative data by imitation of previous work and without any real understanding of the underlying assumptions of the procedures being used. This is an unfortunate drawback of the wide availability of computers and easily used multivariate statistics programs.

By way of organization, each topic is briefly introduced, a set of theoretical papers given, and then papers with biological applications are presented. In each case we have tried to justify the inclusion of a paper, and in some cases explain why others were not included. Finally, a short bibliography of additional references on theory and applications has been included.

We have not provided a formal theoretical introduction to each major statistical technique but rather have used a historical approach in introducing the subject matter. Since this is a volume of "classical" papers, we felt that this approach was justified. For those readers who might be unfamiliar with the methods of analysis, we have provided a brief description of the primary aims of each technique in the form of a "taxonomic key" as well as reprinting a recent survey paper by F. James Rohlf on multivariate analysis. Further, in each part there is generally at least one reprinted paper that contains a lucid theoretical introduction to that particular type of analysis.

We would like to express our deep appreciation to our colleague, F. James Rohlf, who read a draft of this volume and offered many valuable suggestions. Any omissions or errors, however, remain our own.

EDWIN H. BRYANT
WILLIAM R. ATCHLEY

CONTENTS

Acknowledgments and Permissions v
Series Editor's Preface vii
Preface ix
Contents by Author xv

Introduction 1

1 ROHLF, F. J.: Perspectives on the Application of Multivariate Statistics to Taxonomy 5
Taxon, **20**(1), 85–90 (1971)

PART I: PRINCIPAL COMPONENTS ANALYSIS

Editors' Comments on Papers 2 Through 6 12

2 PEARSON, K.: On Lines and Planes of Closest Fit to Systems of Points in Space 17
Phil. Mag., Ser. 6, **2**(11), 559–572 (1901)

3 HOTELLING, H.: Analysis of a Complex of Statistical Variables into Principal Components 31
J. Educ. Psychol., **24**(6), 417–441 (1933)

4 RAO, C. R.: The Use and Interpretation of Principal Component Analysis in Applied Research 56
Sankhya, Ser. A, **26,** 329–358 (1964)

5 JOLICOEUR, P., and J. E. MOSIMANN: Size and Shape Variation in the Painted Turtle. A Principal Component Analysis 86
Growth, **24**(4), 339–354 (1960)

6 GITTINS, R.: The Application of Ordination Techniques 102
Ecological Aspects of the Mineral Nutrition of Plants (British Ecol. Soc. Symp. 9), I. H. Rosison (ed.), Blackwell Scientific Publications, 1969, pp. 37–66

PART II: FACTOR ANALYSIS

Editors' Comments on Papers 7 Through 18 134

7 **THURSTONE, L. L.:** Multiple Factor Analysis — **144**
Psychol. Rev., **38**, 406–427 (1931)

8 **CATTELL, R. B.:** Factor Analysis: An Introduction to Essentials: I. The Purpose and Underlying Models — **166**
Biometrics, **21**(1), 190–215 (1965)

9 **RAO, C. R.:** Estimation and Tests of Significance in Factor Analysis — **191**
Psychometrika, **20**(2), 93–111 (1955)

10 **McDONALD, R. P.:** The Theoretical Foundations of Principal Factor Analysis, Canonical Factor Analysis and Alpha Factor Analysis — **210**
British J. Math. Stat. Psychol., **23**, Pt. 1, 1–21 (May 1970)

11 **KAISER, H. F.:** The Varimax Criterion for Analytic Rotation in Factor Analysis — **231**
Psychometrika, **23**(3), 187–200 (1958)

12 **CATTELL, R. B., and J. L. MUERLE:** The "Maxplane" Program for Factor Rotation to Oblique Simple Structure — **245**
Educ. Psychol. Measurement, **20**(3), 569–590 (1960)

13 **JENNRICH, R. I., and P. F. SAMPSON:** Rotation for Simple Loadings — **267**
Psychometrika, **31**(3), 313–323 (1966)

14 **McDONALD, R. P., and E. J. BURR:** A Comparison of Four Methods of Constructing Factor Scores — **278**
Psychometrika, **32**(4), 381–401 (1967)

15 **STROUD, C. P.:** An Application of Factor Analysis to the Systematics of *Kalotermes* — **299**
Syst. Zool., **2**(2), 76–92 (1953)

16 **REYMENT, R. A.:** Multivariate Analytical Treatment of Quantitative Species Associations: An Example from Palaeoecology — **316**
J. Animal Ecol., **32**, 535–547 (1963)

17 **WALLACE, J. T., and R. S. BADER:** Factor Analysis in Morphometric Traits of the House Mouse — **329**
Syst. Zool, **16**(2), 144–148 (1967)

18 **ATCHLEY, W. R.:** Components of Sexual Dimorphism in *Chironomus* Larvae (Diptera: Chironomidae) — **334**
Amer. Naturalist, **105**(945), 455–466 (1971)

PART III: MULTIVARIATE REGRESSION AND CORRELATION

Editors' Comments on Papers 19 Through 24 — **348**

19 **BARTLETT, M. S.:** On the Theory of Statistical Regression — **355**
Proc. Roy. Soc. Edinburgh, **53**, 260–283 (1933)

20 **HOTELLING, H.:** The Most Predictable Criterion — **379**
J. Educ. Psychol, **26**, 139–142 (1935)

21 GLAHN, H. R.: Canonical Correlation and Its Relationship to Discriminant Analysis and Multiple Regression **383**
J. Atmospheric Sci., **25**(1), 23–31 (1968)

22 POWER, D. M.: Evolutionary Implications of Wing and Size Variation in the Red-Winged Blackbird in Relation to Geographic and Climatic Factors: A Multiple Regression Analysis **392**
Syst. Zool., **18**(4), 363–373 (1969)

23 VOGT, T., and D. L. JAMESON: Chronological Correlation Between Change in Weather and Change in Morphology of the Pacific Tree Frog in Southern California **403**
Copeia, No. 1, 135–144 (1970)

24 LEE, P. J.: The Theory and Application of Canonical Trend Surfaces **413**
J. Geol., **77**, 303–318 (1969)

Author Citation Index **429**
Subject Index **433**

CONTENTS BY AUTHOR

Atchley, W. R., 334
Bader, R. S., 329
Bartlett, M. S., 355
Burr, E. J., 278
Cattell, R. B., 166, 245
Gittins, R., 102
Glahn, H. R., 383
Hotelling, H., 31, 379
Jameson, D. L., 403
Jennrich, R. I., 267
Jolicoeur, P., 86
Kaiser, H. F., 231
Lee, P. J., 413
McDonald, R. P., 210, 278
Mosimann, J. E., 86
Muerle, J. L., 245
Pearson, K., 17
Power, D. M., 392
Rao, C. R., 56, 191
Reyment, R. A., 316
Rohlf, F. J., 5
Sampson, P. F., 267
Stroud, C. P., 299
Thurstone, L. L., 144
Vogt, T., 403
Wallace, J. T., 329

INTRODUCTION

Ecologists and systematists are utilizing multivariate statistical methods with increasing frequency, as any perusal of the current literature in the field with affirm. Like their univariate counterparts, multivariate techniques allow one to evaluate, by means of sample observations, hypotheses regarding statistical populations. They are not merely a more sophisticated replacement for numerous univariate analyses, however. Since biological variables are usually intercorrelated to varying degrees, separate single-variable analyses can misleadingly overestimate the true dimensionality of divergence; hence the correct approach must involve appropriate multivariate analyses of such covariation.

In general, multivariate procedures are called for when several variables have been assessed for each object under study. However, there can be considerable flexibility in the interpretation of both variables and objects. In systematics, for example, the objects may be specimens from one or more collection sites and the variables the characters recorded on these specimens, while in ecology the objects may be the collection sites themselves and the variables, the species residing there. Whatever variables or objects are under study, the basic data matrix displays the observed values of the variables on the objects and thus consists on N rows representing these objects and p columns representing the variables.

An important distinction in multivariate analysis is whether the N objects represent samples from a single population or from several populations. For example, a basic data matrix in systematics may contain measurements of several variables on specimens from a single species and locality or on specimens from several localities of the same or different species. In the first case one might be interested in size and shape relationships among individuals within a population; in the second, in describing the pattern of adaptive divergence among populations. A companion volume (Atchley and Bryant, 1975) concerns those analyses appropriate to among-group covariation analyses; this book concentrates on analyses of within-group covariation.

In spite of this apparent dichotomy of multivariate procedures, it can still be confusing to select an appropriate methodology for one's particular hypothesis and type of data. To aid in this decision we have provided here and in Atchley and Bryant (1975) a brief key to the analyses covered in both books and in addition we have reprinted a short paper by F. J. Rohlf (Paper 1), in which he gives a lucid nonmathematical introduction to these same topics. For the reader interested in a more in-depth introduction to these techniques, we have provided a list of general treatises on multivariate analysis.

Even though multivariate techniques are powerful tools for biologists, one should be cautious in applying them in "cookbook" fashion, without understanding the underlying theory. Under such circumstances, one can easily abuse a particular methodology by violating its assumptions. Although considerable mathematical expertise is necessary to understand some of the theoretical developments in multivariate statistics, many can be appreciated with a good working knowledge of matrix algebra. We therefore encourage readers with a limited background in this area to consult a basic text on matrix algebra, such as Searle (1966), Horst (1963), or Graybill (1969), before attempting to read many of the more theoretical papers which we have included in this volume.

KEY TO MULTIVARIATE ANALYSES

1.	The objects were sampled from more than one population	**2**
	The objects were sampled from a single population	**3**
2.	The main purpose of the analysis is to determine if the samples could have been drawn from a single statistical population; i.e., are the mean vectors of the populations equal?	*multivariate analysis of variance*

	The main purpose of the analysis is to find linear combinations of the variables that maximize differences among preexisting populations; i.e., one wishes to sort the objects into their appropriate populations with minimal error	*discriminant analysis*
	The main purpose of the analysis is to sort a previously unpartitioned heterogeneous collection of objects into a series of sets; i.e., one wishes to identify sets and allocate objects to these sets simultaneously	*cluster analysis*
	The main purpose of the analysis is to arrange the objects graphically in few dimensions, while retaining maximal fidelity to the original interobject relationships	*nonmetric scaling*
3.	The variables can be logically divided into two (or more) sets and one wishes to establish maximal linear relationships among these sets	*multiple regression and correlation*
	The variables logically belong to a homogeneous set	**4**
4.	The main purpose of the analysis is to describe parsimoniously the total variance in a sample in few dimensions; i.e., one wishes to reduce the dimensionality of the original data while minimizing loss of information. These few dimensions are the linear combinations of the original variables that successively account for the major independent patterns of variation in the sample	*principal components*
	The main purpose of the analysis is to resolve the intercorrelations among variables into their putative underlying causes; i.e., one wishes to reproduce only the intercorrelations rather than the total variances	*factor analysis*

BIBLIOGRAPHY

Anderson, T. W. 1958. *An Introduction to Multivariate Statistical Analysis*. John Wiley & Sons, Inc., New York. 374 pp.

Atchley, W. R., and E. H. Bryant. 1975. *Multivariate Statistical Methods: Among-Groups Covariation*. Dowden, Hutchinson & Ross, Inc., Stroudsburg, Pa.

Bartlett, M. S. 1965. Multivariate statistics, in *Theoretical and Mathematical Biology* (T. H. Waterman and H. J. Morowitz, eds.). Xerox College Publishing, Lexington, Mass. pp. 201–224.

Blackith, R. E., and R. A. Reyment. 1971. *Multivariate Morphometrics*. Academic Press, New York. 412 pp.

Cooley, W. W., and P. R. Lohnes. 1962. *Multivariate Procedures for the Behavioral Sciences*. John Wiley & Sons, Inc., New York, 211 pp.

———, and P. R. Lohnes. 1971. *Multivariate Data Analysis*. John Wiley & Sons, Inc., New York. 364 pp.

Cramer, E. M., and R. D. Bock. 1966. Multivariate analysis. *Rev. Educ. Res. 36:*604–614.

Dempster, A. P. 1969. *Elements of Continuous Multivariate Analysis*. Addison-Wesley, Menlo Park, Calif. 388 pp.

———. 1971. An overview of multivariate data analysis. *J. Multivariate Anal. 1:*316–346.

DuBois, P. H. 1957. *Multivariate Correlation Analysis*. Harper & Row, Inc., New York. 202 pp.

Gower, J. C. 1967. Multivariate analysis and multidimensional geometry. *The Statistician 17:*13–28.

Graybill, F. A. 1969. *Introduction to Matrices with Applications in Statistics*. Wadsworth Publishing Co., Inc., Belmont, Calif. 372 pp.

Horst, P. 1963. *Matrix Algebra for Social Scientists*. Holt, Rinehart and Winston, Inc., New York. 517 pp.

Kendall, M. G. 1969. *A Course in Multivariate Analysis*. Hofman, New York. 185 pp.

Kshirsagar, A. M. 1972. *Multivariate Analysis*. Marcel Dekker, Inc., New York. 534 pp.

Lee, P. J. 1970. Multivariate analysis for the fisheries biology. *Fish. Res. Board Canada Tech. Rept. 244.* 182 pp. (mimeo).

Morrison, D. F. 1967. *Multivariate Statistical Methods*. McGraw-Hill Book Company, New York. 338 pp.

Overall, J. E., and C. J. Klett. 1972. *Applied Multivariate Analysis*. McGraw-Hill Book Company, New York. 500 pp.

Rao, C. R. 1952. *Advanced Statistical Methods in Biometric Research*. John Wiley & Sons, Inc., New York. 390 pp.

———. 1960. Multivariate analysis: an indispensable statistical aid in applied research. *Sankhya 22:*317–338.

———. 1965. *Linear Statistical Inference and Its Applications*. John Wiley & Sons, Inc., New York. 522 pp.

———. 1972. Recent trends of research work in multivariate analysis. *Biometrics 28:*3–22.

Reyment, R. S. 1969. Biometrical techniques in systematics, in *Systematic Biology*. National Academy of Science, Washington, D.C. pp. 542–587.

Seal, H. 1964. *Multivariate Statistical Analysis for Biologists*. Methuen & Company Ltd., London. 209 pp.

Searle, S. R. 1966. *Matrix Algebra for the Biological Sciences*. John Wiley & Sons, Inc., New York. 296 pp.

Sokal, R. R. 1965. Statistical methods in systematics. *Biol. Rev. 40:*337–391.

Tatsuoka, M. M. 1971. *Multivariate Analysis: Techniques for Education and Psychological Research*. John Wiley & Sons, Inc., New York. 310 pp.

Van de Geer, J. P. 1971. *Introduction to Multivariate Analysis for the Social Sciences*. W. H. Freeman and Company, San Francisco. 293 pp.

1

Reprinted from *Taxon,* **20**(1), 85–90 (1971)

PERSPECTIVES ON THE APPLICATION OF MULTIVARIATE STATISTICS TO TAXONOMY *

F. James Rohlf *

Summary

A brief outline is given of the principal types of multivariate statistical techniques which have found use in taxonomy. Techniques such as correlation, principal components, canonical correlation, and factor analyses are described for problems dealing with analysis of covariation within a single sample. Techniques such as canonical variate, cluster, multidimensional scaling, and network analyses are described for dealing with analyses of among sample variation. The purpose of this account is to give an intuitive understanding of what the various techniques have to offer to research in taxonomy.

Introduction

Since taxonomy is concerned with the classification of organisms based upon relationships (both cladistic and phenetic) inferred from characteristics of the *whole* organism, statistical analysis in this field must take into consideration the simultaneous covariation of many characters of the organism as possible. Thus taxonomy differs in an important way from fields such as physiology or biochemistry where investigations often are concerned with the effect of a certain combination of treatments upon a single variable of particular interest. In taxonomy there is often no special interest in the particular characters used. They are a means to an end, needed in order to compare samples of organisms taken from different localities or from what are believed to be different taxa. For these reasons, the techniques of multivariate statistics are of particular importance in taxonomy.

In the account given below I have outlined a variety of techniques which have found use in taxonomy. The account is purposely nonmathematical. Its intention is to give one a general intuitive feeling for the types of questions which can be answered using presently available multivariate techniques and to introduce some of the jargon of the field so that one can communicate the type of analysis desired to someone who can arrange for the actual computations to be performed (since most of the analysis require an enormous amount of arithmetic, the actual numerical computations will almost always have to be done on a highspeed digital computer).

Several texts are available dealing with the applications of multivariate statistics, e.g., Morrison (1967), Seal (1964), Cooley and Lohnes (1962), and Rao (1952). While these texts all have brief introductions to matrix algebra, the books Searle (1966) and Graybill (1969) should be consulted for a more complete understanding.

The account given below is divided into two main sections. The first one discusses techniques which analyze patterns of covariation found within a single sample and the second is concerned with analysis of variation between samples.

* Contribution number 29 from the Program in Ecology and Evolution, State University of New York at Stony Brook.

** Department of Ecology and Evolution, State University of New York at Stony Brook, Stony Brook, New York 11790.

There are a number of ways in which the patterns of variation and covariation within a single sample can be described. If 3 or fewer characters have been used, then frequency distributions and scatter diagrams can give one a useful intuitive appreciation of the variation found. If many characters have been used, then one can still plot scatter diagrams for various combinations of the characters taken 2 or 3 at a time, but it is usually difficult to fully appreciate complex patterns of covariation. A conventional statistical description of the sample would involve the computation of the mean for each variable and the variance-covariance matrix (a symmetrical table containing the variances of each character down the main diagonal and covariances in the off diagonal cells). If the sample represents a random sample from a multivariate normal distribution then such statistics contain sufficient information for estimating various properties of the population from which the sample was drawn. If the population was not normally distributed, then other statistics must be computed. In univariate statistics, one can compute higher moments such as g_1 and g_2 to measure skewness and kurtosis (Sokal and Rohlf, 1969). The analogous matrices in multivariate statistics are difficult to interpret. For this reason most workers resort to graphical techniques in such situations.

It is difficult to test for goodness of fit of an observed sample to a multivariate normal distribution. One can test whether each character taken separately fits a univariate normal distribution. If even one character does not fit a normal distribution, then the entire suite of characters does not fit a multivariate normal distribution. However, even if they all fit it does not guarantee that the entire suite of characters is consistant with a multivariate normal, since there can be a variety of complex interactions among the characters. If one has very large samples, the p-dimensional space can be partitioned into a series of regions and compare the frequency of observations in each region with that which would be expected based on a multivariate normal distribution (using the sample means and covariances). With samples of the size usually employed in taxonomy, this is not practical unless only a very few characters are used. The only alternative is to perform some sort of multidimensional scaling analysis (ordination) which will enable one to reduce the dimensionality of the system which needs to be considered. That is, to construct a few axes which contain most of the information about the covariation among the observations found in the original characters. If 3 or fewer axes are sufficient, then one can examine scatter diagrams constructed by projecting the specimens onto these axes and then plotting them against one another. Techniques such as non-metric multidimensional scaling (Kruskal, 1964 and Rohlf, 1970) and principal components analysis (Seal, 1964; Rohlf, 1970; Jolicoeur and Mosiman, 1960) have been used in taxonomy.

If one is satisfied that the data are consistant with the multivariate normal distribution, then principal components analysis can serve as a particularly compact means to describe the variation found in ones sample. The first principal component indicates the direction in hyperspace in which the observations differ most (the relative magnitude of the first eigenvalue indicates the extent to which the observations vary in this direction. This direction often corresponds to variation in overall size of the specimens (but it can sometimes represent the directions in which polymorphs vary if the sample is heterogeneous). The other principal components are often more difficult to interpret but they usually correspond to various shape differences between specimens. These differences are usually expressed as contrasts (high positive coefficients for some characters and high negative coefficients for other characters). The particular contrasts which result from the analysis are a consequence of the structure of the correlations between the characters. The information given by a principal components analysis can also be used to construct equal frequency ellipses which enclose regions expected to enclose $(1-\alpha)$ 100% of the observations. An example of this construction for the 2-dimensional case is given in Sokal and Rohlf (1969).

If a major purpose of the analysis is the investigation of patterns of intercorrelations among the characters, it is often useful to perform a factor analysis with rotation to simple structure (Harmon, 1967). This type of analysis expresses basically the same information but displays the correlation structure present in a much simpler form. Here each axis (or factor) corresponds to a group of characters as indicated by high (in absolute value) correlations between each factor and a set of characters. Characters not belonging to a set should have correlations near zero. Examples of the use of factor analysis in systematics are: Rohlf and Sokal (1958), Gould and Garwood (1969). Some other examples are listed in Seal (1964).

When the suite of characters can be logically divided into two sets and the relationships (if any) between the two sets is of interest, one can employ canonical correlation analysis. This technique obtains that a linear combination of the characters from each set of characters is such that these two linear functions have the highest possible correlation. This largest correlation is called the canonical correlation and measures the extent to which relationships in one set of characters can be predicted by a knowledge of the other set of characters. For example, one could use this type of analysis to locate those features in the adult stage which can be predicted based upon a knowledge of the larval stage. Morrison (1967) gives an outline of the necessary computations.

Description of variation among samples

There are several approaches to the study of variation among samples. The "proper" approach depends upon the statistical model and the purposes of the analysis (i.e., the questions being asked).

The question most commonly asked is: "Are the samples homogeneous?" If each sample can be assumed to have been drawn from a multivariate normal distribution then we can use a generalization of Bartlet's test to test for homogeneity of the variance-covariance matrices (Seal, 1964; Reyment, 1969). If they are homogeneous then we can use the techniques of multivariate analysis of variance to test whether the means of the samples are significantly heterogeneous. Of course, we must remember that if the samples were drawn from different geographic regions of a species or from different species, then we *know* that the true means (and probably also the variances and covariances) are different in different statistical populations. What we are testing is whether we have sufficient evidence to demonstrate that such differences exist and to set confidence intervals on the magnitude of the differences. If the test of significance yields a significant result, then it usually will be of interest to isolate those characters whose differences between various samples were most important in contributing to the overall significance test (just as we would turn to either *a priori* or to multiple comparison tests in a similar situation in univariate anova). However, it is difficult to know how to fully break down the overall multivariate test in the most meaningful way. If one has designed the sampling so that one can test a variety of *a priori* hypotheses, then one is relatively well off. One can then partition the overall test into a series of tests reflecting differences due to time of year vs. locality vs. sex vs. food plant, etc. If, however, one simply has those samples which are available one must use some kind of *a posteriori* test. Several multivariate multiple comparisons tests have been devised. Gabriel and Sokal (1969) described a test which can be used for this purpose, but it has the disadvantage that it produces "too many" answers. The voluminous output of this procedure reflects the fact that there are a very large number of ways in which multivariate samples can differ from one another. The results of this type of test are usually expressed in terms of so-called maximum nonsignificant subsets. These sets have the property that the addition of any other sample or variable to the set would cause it to be significantly heterogeneous.

The description of the patterns of variation among samples in terms of sets (which may be partially overlapping) is usually not very convenient. Other techniques (which have less statistical rigor) have been developed to more conveniently express statistical relationships among the variables over the samples.

There are three main classes of techniques which are used to reveal the relationships among the samples in the p-dimensional space: multidimensional scaling, cluster analysis, and network analysis. These techniques all come under the heading of multivariate data analysis since their main purpose is to give insight into ones data and to place less stress on tests of significance. In biology these techniques are associated with the field of numerical taxonomy where they have been found very useful in elucidating taxonomic relationships.

Multidimensional scaling is used when one wishes to express relationships among the sample means in terms of their coordinates on a few specially constructed coordinate axes. The goal is to preserve as much of the information about interpoint distances as possible while reducing the number of variables to be considered from p down to k (where k is 1, 2, 3, or perhaps 4 at the most). If the variation within all of the samples is homogeneous (or at least the orientation of the scatter ellipsoids are similar) then one can validly compute a pooled within group variance-covariance matrix and then perform a canonical variates analysis (see Jolicoeur, 1959; Seal, 1964). In this type of analysis relationships among the samples are expressed relative to the average covariation found within the samples.

This type of analysis is also sometimes called a generalized discriminant analysis since in the special case where there are only two samples the canonical variable is the discriminant function. When there are more than two samples, the canonical variables constitute a set of linear combinations of the variables which best discriminate between the groups. They can be used to form a probabilistic identification scheme (Cooley and Lohnes, 1962).

If the within group variation is not homogeneous (particularly if the orientation of the scatter ellipsoids differ) then it is difficult to make use of the within sample information and one must base ones analyses on the among sample variation. For example, one can perform a principal components analysis on the among sample correlation matrix to obtain vectors which indicate the major trends of variation among groups. One can then project the standardized sample means onto these axes in order to be able to prepare a scatter diagram depicting the among group variation relative to the total amount of variation found among the samples (since the correlation matrix and standardized data were used).

When the samples correspond to higher taxa then one expects the within sample covariation to be heterogeneous. This is one reason why there is seldom any attempt to take within sample covariation into account in numerical taxonomy. Often the taxa being sampled are sufficiently distinct that only a few specimens are used to represent each taxon. This is a valid shortcut whenever the among sample variation is much larger than the within sample variation as would be expected, for example, when the samples correspond to different species sampled throughout a family. In such cases there seems little point in worrying about tests of significance — the species are obviously different from one another. What is uncertain is their relative degrees of overall similarity and the way in which this can be most simply expressed. Another alternative (which sometimes is capable of expressing the relationships in fewer dimensions) is non-metric multidimensional scaling (see Kruskal, 1964; Rohlf, 1970). If a sufficient amount of the among group variation can be expressed in k-dimensions ($k << p$) then one can visually look for patterns in the differences among the samples (results are mostly intuitive, few tests of significance are possible here, but one often gains considerable insight into ones data).

Cluster analysis sorts the samples into a series of sets. These sets may be mutually exclusive, hierarchic, or partially overlapping in various ways. Hierarchical clustering schemes have been used most commonly in taxonomy. Typically these techniques begin with a matrix of distances between sample means (computed in various ways) and a search for other points which are relatively close together and separated by gaps from other such groups. The distance coefficient can be computed in such a way as to take the within sample covariation into consideration if this is desired. The generalized distance D is one way in which this can be done (Rohlf, 1970; Seal, 1964; Rao, 1952). For data in which the relationships among the samples are hierarchic cluster analyses works rather efficiently. They tend to be somewhat unsatisfactory on data in which the distribution of points in the p-dimensional space form very elongated clusters or where there are many points which are intermediate between clusters. These techniques also do not reveal the fact that some clusters may be in between other clusters (see Rohlf, 1970 for a general discussion).

Network analyses express relationships in terms of a graph (in the sense of graph theory, Ore, 1963) which consists of vertices (corresponding to the samples) and edges (which are connections between vertices). The existence of an edge implies that the two vertices so connected share some relation between them (e.g., they are nearest neighbors in the p-dimensional space). The shortest simply connected network has been found to be useful in numerical taxonomy since it indicates in a convenient fashion the closest neighbor of each point. Kruskal (1956) and Prim (1957) give algorithms for constructing such networks. Jardine and Sibson (1968) have suggested the use of networks which are more than simply connected and thus more capable of summarizing multivariate relationships (and hence more complex to understand). An example of the use of a shortest connection network in taxonomy is given in Rohlf (1970).

Comprehension of multivariate relationships is difficult. This difficulty is not helped by the fact that classical multivariate statistical techniques tend to result in a single number which is used for tests of significance. Such statistics are often difficult to interpret in terms of the particular samples and variables under investigation. For this reason more emphasis has been placed in the last few years upon a variety of graphical techniques which allow one to visualize many parameters of the sample simultaneously. The account given above is an attempt to give one a brief intuitive introduction to the types of techniques which are apt to be found useful in taxonomy. A number of workers are attempting to develop new mathematical tools which will allow a simple but efficient graphical summarization of multivariate relationships. Such developments, if successful, could have a large impact upon taxonomic methodology.

References

COOLEY, W. W. and P. R. LOHNES 1962 – Multivariate procedures for the behavioral sciences. Wiley: New York 211 pp.

GABRIEL, K. R. and R. R. SOKAL 1969 – A new statistical approach to geographic variation analysis. Systematic Zool. 18: 259–278.

GRAYBILL, F. A. 1969 – Introduction to matrices with applications in Statistics. Wadsworth: Belmont, Calif. 372 pp.

GOULD, S. J. and R. A. GARWOOD 1969 – Levels of integration in mammalian dentitions: An analysis of correlations in *Nesophontes micrus* (Insectivora) and *Oryzomys couesi* (Rodentia). Evolution, 23: 276–300.

HARMON, H. H. 1967 – Modern factor analysis. Chicago, 470 pp.

JARDINE, N. and R. SIBSON 1968 – The construction of hierarchic and non-hierarchic classifications. Computer Jour. 11: 177–184.

JOLICOEUR, P. 1959 – Multivariate geographical variation in the wolf, *Canis Lupus* L. Evolution, 13: 283–299.

JOLICOEUR, P. and J. E. MOSIMANN 1960 – Size and shape variation in the painted turtle. A principal component analysis. Growth, 24: 339–354.

KRUSKAL, J. B. 1956 – On the shortest spanning subtree of a graph and the traveling salesman problem. Proc. Amer. Math. Soc., 7: 48–50.

KRUSKAL, J. B. 1964 – Non-metric multidimensional scaling. Psychometrica, 29: 1–27.

MORRISON, D. F. 1967 – Multivariate statistical methods. McGraw-Hill: New York, 338 pp.

ORE, O. 1963 – Graphs and their uses. Random House: New York, 131 pp.

PRIM, R. C. 1957 – Shortest connection networks and some generalizations. Bell System Technical Jour. 36: 1389–1401.

RAO, C. R. 1952 – Advanced statistical methods in biometrical research. Wiley: New York, 390 pp.

REYMENT, R. A. 1969 – Biometrical techniques in systematics. In Systematic Biology. Publ. 1692 National Academy of Sciences. pp. 541–594.

ROHLF, F. J. 1970 – Adaptive hierarchical clustering schemes. Systematic Zool., 19: 58–82.

ROHLF, F. J. and R. R. SOKAL 1958 – The description of taxonomic relationships by factor analysis. Systematic Zool., 11: 1–16.

SEAL, H. 1964 – Multivariate statistical analysis for biologists. Wiley: New York, 207 pp.

SEARLE, S. R. 1966 – Matrix algebra for the biological sciences. Wiley: New York, 296 pp.

SOKAL, R. R. and F. J. ROHLF 1969 – Biometry. Freeman: San Francisco, 776 pp.

Part I

PRINCIPAL-COMPONENTS ANALYSIS

Editors' Comments on Papers 2 Through 6

2 **PEARSON**
On Lines and Planes of Closest Fit to Systems of Points in Space

3 **HOTELLING**
Analysis of a Complex of Statistical Variables into Principal Components

4 **RAO**
The Use and Interpretation of Principal Component Analysis in Applied Research

5 **JOLICOEUR and MOSIMANN**
Size and Shape Variation in the Painted Turtle. A Principal Component Analysis

6 **GITTINS**
The Application of Ordination Techniques

The aim of principal-components analysis is to resolve the total variation of a set of variables into linearly independent composite variables which successively account for the maximal variability in the data. The unique linear combinations of the original variables that satisfy these successive variance maximization criteria are the eigenvectors or principal axes of the covariance or correlation matrix among variables, weighted by the square root of their respective eigenvalues. Each principal component is then proportional to the variance it explains and the eigenvalues are the variances of the composite variables along their principal-component axis. In matrix notation, the fundamental equation of principal components is $Y = AZ$, where A is the matrix of scaled eigenvectors, Z is the original data matrix, and Y contains the principal components.

Geometrically, the first principal axis is that line which has a minimum sum of squared perpendicular projections of all variables onto it; the second, orthogonal to the first, minimizes the sum of perpendicular distances remaining; and so on for the succeeding axes. These axes, along with their eigenvalues, thus describe the shape of the distribution of variables in a multivariate scatter diagram. More specifically, if the data follow a multivariate normal distribution, the principal-component axes define a hyperellipsoid for a $(1 - \alpha)\%$ contour of the multivariate normal density function.

Sample variables are generally intercorrelated to varying degrees (that is, the data contain some redundancy), and hence not all principal components are needed to summarize the data adequately. In practice, then, one usually retains only the first few components that account for the major patterns of variation. Because each successive component accounts for a smaller amount of variance in the sample, one is assured that the total variance accounted for by these first few axes is maximal for the chosen dimensionality. Thus principal-components analysis is widely utilized as a method for summarizing data in few dimensions while retaining most of the essential information in the sample. One drawback of the method is the operational ambiguity in deciding upon the number of axes to retain in a reduced solution: the various empirical tests that have been suggested for determining appropriate dimensionality (see Part II, Factor Analysis, for relevant references) may result in different decisions by different investigators. On the other hand, if the distribution is multivariate normal, Bartlett (1950, 1951) and Lawley (1956) have discussed some statistical criteria that may be applied to this question.

Although references to the principal axes of a bivariate ellipsoid can be found in the literature prior to 1900, Karl Pearson (Paper 2) provided the mathematical basis for principal-components methodology by first describing the line of best fit through an ellipsoidal cluster of points. Later, in a two-part paper published in 1933, Harold Hotelling formulated the rigorous modern definition of principal components as the axes which successively account for maximal variability in a sample. We reproduce only the first of these papers here (Paper 3), because it substantially outlines the methodology; however, the reader is encouraged to consult the sequel paper for Hotelling's entire treatment. The papers by Pearson and Hotelling comprise the core of principal-components theory, but because they antedated the development of most other multivariate techniques, a discussion of the important interrelationships of principal components to these other methods is lacking. However, an excellent comparative treatment of the uses of principal components in applied research was supplied by C. R. Rao in 1964 (Paper 4).

While principal-components analysis resolves variation among variables, situations arise where one desires instead to assess relationships among objects. The projections or scores of these objects can be computed using principal-components methodology (e.g., Kaiser, 1962; Harman, 1967). In an alternative attempt to ordinate these objects directly, some investigators have computed the principal components of the *Q*-mode matrix of correlations among objects. However, Gower (1966) has criticized this approach and recommends that when such direct ordinations are desired, one should compute the principal axes of suitable distance measures among objects. He refers to this method as "principal-coordinates analysis" to distinguish it from standard principal components. Principal coordinates have been utilized on biological data by Sokal and Rohlf (1970) and Rohlf (1972). Although *Biometrika* would not give us permission to reprint Gower's work here, the paper by Rohlf (1972), which provides a comparative empirical assessment of principal components, principal coordinates, and nonmetric scaling, is reproduced in Part IV, Nonmetric Scaling, of Atchley and Bryant (1975).

The applications of principal-components analysis in biology have been largely confined to systematics, morphometrics, and phytosociology, although sporadic applications can be found in many other fields, including genetics. A paper by Jolicoeur and Mosimann (Paper 5) represents one of the earliest applications of principal components to morphometrics. This paper is especially appropriate here because the authors in introducing a new methodology to the field, presented a refreshingly elementary and detailed account of the technique utilizing a small tractable set of data.

By far the most profuse application of principal-components analysis has been in ordination studies in phytosociology, where the method was introduced by D. W. Goodall in 1954. Similar utilization of principal components occurs in systematics, where it provides an ordination among taxa (e.g., Powers and Rohlf, 1972). Although some authors have recently questioned its utility in these circumstances (e.g., Gauch and Whittaker, 1973), Gittins (Paper 6) has claimed that it "probably represents the most successful ordination technique currently employed in studies of vegetation." We reproduce the latter paper here because it is a particularly lucid account of the use of principal components in plant ordination studies, and also because of its comprehensive but elementary treatment of the methodology, it can serve as an adequate introduction to this analysis for those readers desiring a less mathematical approach. It is unfortunate that when Goodall introduced principal-components analysis to phytosociology, he referred to it as factor analysis, a confusion that has persisted in this literature. We caution the reader to be aware of erroneous references to these two very different methodologies.

In many instances biologists have rotated the principal-component axes to new positions corresponding to the "simple structure" criteria of factor analysis (see Part II, Factor Analysis, for a discussion of rotation). The results have been variously referred to as principal components or factor analysis; however, these new axes are no longer principal components nor the factors of factor analysis. Nevertheless, these rotated components often provide a more ecologically meaningful summarization of the data than the original axes. Hence, such rotations will undoubtedly continue to appear in biological literature and we felt that they should be mentioned here. Among the papers utilizing rotated components are those by Ivimey-Cook and Proctor (1967) in phytosociology and by Gould (1969) and Power (1971) in morphometrics.

BIBLIOGRAPHY

Anderson, T. W. 1963. Asymptotic theory for principal components analysis. *Ann. Math. Stat. 34:*122–148.

Atchley, W. R., and E. H. Bryant. 1975. *Multivariate Statistical Methods: Among-Groups Covariation.* Dowden, Hutchinson & Ross, Inc., Stroudsburg, Pa.

Bailey, D. W. 1956. A comparison of genetic and environmental principal components of morphogenesis in mice. *Growth 20:*63–74.

Bartlett, M. S. 1950. Tests of significance in factor analysis. *British J. Psychol. 3:*77–85.

———. 1951. A further note on tests of significance in factor analysis. *British J. Psychol. 4:*1–2.

Gauch, H. G., and R. H. Whittaker. 1973. Comparison of ordination techniques. *Ecology 53:*868–875.

Goodall, D. W. 1954. Objective methods for the classification of vegetation. III. An essay in the use of factor analysis. *Austral. J. Bot. 2:*304–324.

Gould, S. J. 1969. Character variation in two land snails from the Dutch Leeward Islands: geography, environment, and evolution. *Syst. Zool. 18:*185–200.

Gower, J. C. 1966. Some distance properties of latent root and vector methods used in multivariate analysis. *Biometrika 53:*325–338.

Harman, H. H. 1967. *Modern Factor Analysis.* University of Chicago Press, Chicago. 474 pp.

Hotelling, H. 1933. Analysis of a complex of statistical variates into principal components. *J. Educ. Psychol. 24:*498–520.

Ivimey-Cook, R. B., and M. C. F. Proctor. 1967. Factor analysis of data from an East Devon heath: a comparison of principal component and rotated solutions. *J. Ecol. 55:*405–413.

Jeffers, J. N. R. 1962. Principal component analysis of designed experiments. *Statistics 12:*230–242.

Jolicoeur, P. 1963. Bilateral symmetry and asymmetry in limb bones of *Martes americana* and man. *Rev. Canadian Biol. 22:*409–432.

Kaiser, H. F. 1962. Formulas for component scores. *Psychometrika 27:*33–37.

Lawley, D. N. 1956. Tests of significance for latent roots of covariance and correlation matrices. *Biometrika 43:*128–136.

Power, D. M. 1971. Statistical analysis of character correlations in Brewer's blackbirds. *Syst. Zool. 20:*186–203.

Powers, D. A., and F. J. Rohlf. 1972. A numerical taxonomic study of Caribbean and Hawaiian reef corals. *Syst. Zool. 21:*53–64.

Rohlf, F. J. 1972. An empirical comparison of three ordination techniques in numerical taxonomy. *Syst. Zool. 27:*271–280.

Sinha, R. N., H. A. Wallace, and F. S. Chebib. 1969. Principal component analysis of interrelationships among fungi, mites, and insects, in grain bulk ecosystems. *Ecology 50:*536–547.

Sokal, R. R., and F. J. Rohlf. 1970. The intelligent ignoramous, an experiment in numerical taxonomy. *Taxon 19:*305–319.

2

Reprinted from *Phil. Mag.*, Ser. 6, **2**(11), 559–572 (1901)

On Lines and Planes of Closest Fit to Systems of Points in Space. *By* KARL PEARSON, *F.R.S.*, *University College, London* *.

(1) IN many physical, statistical, and biological investigations it is desirable to represent a system of points in plane, three, or higher dimensioned space by the "best-fitting" straight line or plane. Analytically this consists in taking

$$y = a_0 + a_1 x, \quad \text{or} \quad z = a_0 + a_1 x + b_1 y,$$

$$\text{or} \quad z = a_0 + a_1 x_1 + a_2 x_2 + a_3 x_3 + \ldots + a_n x_n,$$

where $y, x, z, x_1, x_2, \ldots x_n$ are variables, and determining the "best" values for the constants $a_0, a_1, b_1, a_0, a_1, a_2, a_3, \ldots a_n$ in relation to the observed corresponding values of the variables. In nearly all the cases dealt with in the text-books of least squares, the variables on the right of our equations are treated as the independent, those on the left as the dependent variables. The result of this treatment is that we get one straight line or plane if we treat some one variable as independent, and a quite different one if we treat another variable as the independent variable. There is no paradox about this; it is, in fact, an easily understood and most important feature of the theory of a system of correlated variables. The most probable value of y for a given value of x, say, is not given by the same relation as the most probable value of x for a given value of y. Or, to take a concrete example, the most probable stature of a man with a given length of leg l being s, the most probable length of leg for a man of stature s will not be l. The "best-fitting" lines and planes for the cases of 2 up to n variables for a correlated system are given in my memoir on regression †. They depend upon a determination of the means, standard-deviations, and correlation-coefficients of the system. In such cases the values of the independent variables are supposed to be accurately known, and the probable value of the dependent variable is ascertained.

(2) In many cases of physics and biology, however, the "independent" variable is subject to just as much deviation or error as the "dependent" variable. We do not, for example, know x accurately and then proceed to find y, but both x and y are found by experiment or observation. We observe x and y and seek for a unique functional relation between them. Men of given stature may have a variety

* Communicated by the Author.

† Phil. Trans. vol. clxxxvii. A, pp. 301 *et seq.*

of leg-lengths; but a point at a given time will have one position only, although our observations of *both* time and position may be in error, and vary from experiment to experiment. In the case we are about to deal with, we suppose the observed variables—all subject to error—to be plotted in plane, three-dimensioned or higher space, and we endeavour to take a line (or plane) which will be the "best fit" to such a system of points.

Of course the term "best fit" is really arbitrary; but a good fit will clearly be obtained if we make the sum of the squares of the perpendiculars from the system of points upon the line or plane a minimum.

For example:—Let $P_1, P_2, \ldots P_n$ be the system of points with coordinates x_1, y_1; x_2, y_2; $\ldots x_n\ y_n$, and perpendicular distances $p_1, p_2, \ldots p_n$ from a line A B. Then we shall make

$$U = S(p^2) = \text{a minimum}.$$

If y were the dependent variable, we should have made

$$S(y' - y)^2 = \text{a minimum}$$

(y' being the ordinate of the theoretical line at the point x which corresponds to y), had we wanted to determine the best-fitting line in the usual manner.

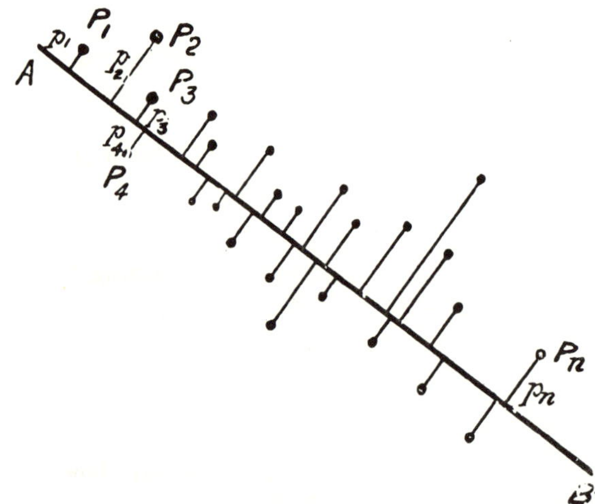

Now clearly $U = S(p^2)$ is the moment of momentum, the second moment of the system of points, supposed equally loaded, about the line A B. But the second moment of a system about a series of parallel lines is always least for the line going through the centroid. Hence: *The best-fitting straight line for a system of points in a space of any order goes through the centroid of the system.*

Now let there be n points each fixed by q variables $x_1, x_2 \ldots x_q$, and let

$$\bar{x}_1 = S(x_1)/n, \quad \bar{x}_2 = S(x_2)/n \ldots \bar{x}_q = S(x_q)/n. \quad . \quad \text{(i.)}$$

fix the centroid, or the *mean values* of the variables;

$$\sigma^2 x_1 = S(x_1^2)/n - \bar{x}_1^2, \quad \sigma^2 x_2 = S(x_2^2)/n - \bar{x}_2^2, \ldots$$
$$\sigma^2 x_q = S(x_q^2)/n - \bar{x}_q^2, \quad . \quad \text{(ii.)}$$

fix the *standard-deviations* (errors of mean square), or indirectly the moments of inertia or second-moments about the axes of coordinates, through the centroid parallel to the axes of the variables $x_1, x_2 \ldots x_q$. And, lastly, let

$$r_{x_u x_v} = \frac{S(x_u x_v) - n\bar{x}_u\bar{x}_v}{n\sigma x_u \sigma x_v} \quad . \quad . \quad . \quad . \quad . \quad \text{(iii.)}$$

for all pairs of values of u and v from 1, 2, 3, ... q, fix the *correlations* of the variables, or indirectly the products of inertia or product-moments about the axes.

Now let $l_1, l_2, l_3 \ldots l_q$ be the generalized direction-cosines of a plane at perpendicular distance p from the origin. We shall have

$$l_1^2 + l_2^2 + l_3^2 + \ldots + l_q^2 = 1. \quad . \quad . \quad . \quad . \quad \text{(iv.)}$$

Further, if U be the sum of the squares of the perpendicular distances of the system of n points from the plane

$$l_1 x_1 + l_2 x_2 + l_3 x_3 + \ldots + l_q x_q = p, \quad . \quad . \quad . \quad . \quad \text{(v.)}$$

we require to make a minimum of

$$U = S(l_1 x_1 + l_2 x_2 + l_3 x_3 + \ldots + l_q x_q - p)^2, \quad . \quad . \quad \text{(vi.)}$$

by variation of $l_1, l_2, \ldots l_q, p$ subject to (iv.). Differentiate first with regard to p and we have

$$l_1 S(x_1) + l_2 S(x_2) + l_3 S(x_3) + \ldots + l_q S(x_q) - np = 0;$$
$$\therefore \quad p = l_1 \bar{x}_1 + l_2 \bar{x}_2 + \ldots + l_q \bar{x}_n, \quad . \quad . \quad . \quad . \quad \text{(vii.)}$$

which shows us from (v.) that: *the best-fitting plane passes through the centroid of the system.*

Now vary (vi.) and add to it Q times the variation of (iv.), Q being an undetermined multiplier. We have, by equating to zero the coefficient of dl_u,

$$l_1 S(x_1 x_u) + l_2 S(x_2 x_u) + \ldots + l_u S(x_u^2) + \ldots + l_q S(x_q x_u)$$
$$- pS(x_u) + Ql_u = 0.$$

Or, substituting for p from (vii.) and using (ii.) and (iii.):

$$l_1\sigma_{x_1}\sigma_{x_u}r_{x_1x_u}+l_2\sigma_{x_2}\sigma_{x_u}r_{x_2x_u}+\ldots l_u\sigma_{x_u}^2+\ldots+l_q\sigma_{x_q}\sigma_{x_u}r_{x_ux_q}+\frac{Q}{n}l_u=0, \quad . \quad . \quad . \quad \text{(viii.)}$$

is the type equation.

Now (vi.) may be written :—

$$U=n\{l_1^2\sigma^2_{x_1}+l_2^2\sigma^2_{x_2}+\ldots+l_q^2\sigma^2_{x_q}+2l_1l_2\sigma_{x_1}\sigma_{x_2}r_{x_1x_2}+\ldots+2l_{q-1}l_q\sigma_{x_{q-1}}\sigma_{x_q}r_{x_{q-1}x_q}\} \quad . \quad \text{(viii.) } bis$$

Multiplying each type equation by its corresponding l_u, adding together and remembering (iv.), we find

$$\frac{U_m}{n}+\frac{Q}{n}=0, \quad \text{or} \quad Q=-U_m,$$

where U_m is the minimum value of U.

Now let Σ^2 be the mean square of the residuals, or

$$\Sigma^2=\frac{S(l_1x_1+l_2x_2+\ldots+l_qx_q-p)^2}{n}.$$

Then

$$\frac{Q}{n}=-\Sigma^2,$$

and a physical meaning has been given to Q, $\sqrt{-Q/n}$ is the "mean square residual,"—*i. e.*, the quantity, the square of which is the mean square of the residuals.

The type equation (viii.) may now be written :

$$l_1\sigma_{x_1}\sigma_{x_u}r_{x_1x_u}+l_2\sigma_{x_2}\sigma_{x_u}r_{x_2x_u}+\ldots+l_u(\sigma^2_{x_u}-\Sigma^2)+l_q\sigma_{x_q}\sigma_{x_u}r_{x_qx_u}=0. \quad . \quad \text{(ix.)}$$

We can eliminate the l's and dividing out row and column of resulting determinant by the corresponding σ, we have:

$$\begin{vmatrix} 1-\dfrac{\Sigma^2}{\sigma^2_{x_1}} & r_{x_1x_2}, & r_{x_1x_3},\ldots r_{x_1x_q} \\ r_{x_2x_1}, & 1-\dfrac{\Sigma^2}{\sigma^2_{x_2}}, & r_{x_2x_3},\ldots r_{x_2x_q} \\ \cdots & \cdots & \cdots \\ \cdots & \cdots & \cdots \\ r_{x_qx_1}, & r_{x_qx_2}, & r_{x_qx_3}, \quad 1-\dfrac{\Sigma^2}{\sigma^2_{x_q}} \end{vmatrix}=0, \quad . \quad . \quad \text{(x.)}$$

as a determinantal equation to find Σ^2. We must choose the least root of this equation, for the mean square residual must be as small as possible. Substitute this value of Σ^2 in the type equations (viii.), and we find the required values of $l_1, l_2 \ldots l_q$, using (iv.).

This is the complete analytical solution of the problem of drawing the best-fitting plane through n non-coplanar points. We see that it depends only on a knowledge of the means, standard-deviations, and correlations of the q variables.

Whenever we may suppose that variation is due to errors of observation or measurement,—*i. e.*, is not organic, but there exists a unique functional relation between the true values of the variables,—then, assuming it of the first degree, we may determine the best values of the constants in the manner given above.

(3) A geometrical interpretation is of course to be found from (viii.) *bis*. Consider the quadric

$$\sigma^2_{x_1}x_1^2 + \sigma^2_{x_2}x_2^2 + \ldots + \sigma^2_{x_q}x_q^2 + 2\sigma_{x_1}\sigma_{x_2}r_{x_1x_2}x_1x_2$$
$$+ \ldots + 2\sigma_{x_{q-1}}\sigma_{x_q}r_{x_{q-1}x_q}x_{q-1}x_q = \epsilon^4, \quad . \quad . \quad \text{(xi.)}$$

where ϵ is any line. Then this quadric will be "ellipsoidal" since the coefficients of $x_1^2 \ldots x_q^2$ are all positive quantities. Let R be its radius-vector measured in the direction $l_1, l_2, \ldots l_q$, or perpendicular to the plane from which we are measuring the residuals; then clearly:

$$U = n\epsilon^4/R^2,$$

or

$$\Sigma^2 = \epsilon^4/R^2. \quad . \quad . \quad . \quad . \quad . \quad . \quad . \quad \text{(xii.)}$$

Thus the inverse square of the radius of this "ellipsoid" measures the square of the mean square residual. We shall speak of the ellipsoid as the *ellipsoid of residuals*. Since Σ is to be a minimum, R must be a maximum; or we conclude: *that the best-fitting plane is perpendicular to the greatest axis of the ellipsoid of residuals and the minimum mean square residual varies inversely as the length of this axis.*

A case of failure can only arise if the ellipsoid of residuals degenerates into an "oblate spheroid," *i.e.*, when every plane through its shorter axis is one of "best fit," or into a sphere, when every plane through the centroid of the system of points is an equally good fit. This sphericity of distribution of points in space involves the vanishing of all the correlations between the variables and the equality of all their standard-deviations. It corresponds to isotropic inertia in the theory of moments in dynamics.

(4) The theory of the best-fitting straight line need not

detain us long. Let its equation be

$$\frac{x_1 - x_1'}{l_1} = \frac{x_2 - x_2'}{l_2} = \frac{x_3 - x_3'}{l_3} = \ldots \frac{x_q - x_q'}{l_q} = \rho. \quad . \quad \text{(xiii.)}$$

Draw the plane perpendicular to this line through x_1', x_2', $x_3' \ldots x_q'$; i. e.,

$$l_1 x_1 + l_2 x_2 + l_3 x_3 + \ldots + l_q x_q = \mathrm{H},$$

where $\mathrm{H} = l_1 x_1' + l_2 x_2' + l_3 x_3' + \ldots + l_q x_q'$.

Then if p be the perpendicular from any point in space on the line (xiii.):

$$p^2 = (x_1 - x_1')^2 + (x_2 - x_2')^2 + \ldots + (x_q - x_q')^2$$
$$- \{ l_1(x_1 - x_1') + l_2(x_2 - x_2') + l_3(x_3 - x_3') + \ldots + l_q(x_q - x_q') \}^2.$$

Now x_1', $x_2' \ldots x_q'$ and l_1, $l_2 \ldots l_q$, subject to the relation $l_1^2 + l_2^2 + l_3^2 + \ldots + l_q^2 = 1$, are the constants at our disposal. Sum p^2 and differentiate to find when $\mathrm{U} = \mathrm{S}(p^2)$ is a minimum. We have for type equation

$$\mathrm{S}(x_u - x_u') - l_u [\mathrm{S}\{ l_1(x_1 - x_1') + l_2(x_2 - x_2') + \ldots + l_q(x_q - x_q') \}] = 0,$$

whence we see:

$$\frac{\mathrm{S}(x_u) - n x_u'}{l_u} = \text{symmetrical function of } x\text{'s}.$$

Or, we must have

$$\frac{\bar{x}_1 - x_1'}{l_1} = \frac{\bar{x}_2 - x_2'}{l_2} = \ldots = \frac{\bar{x}_q - x_q'}{l_q},$$

which show us that the straight line passes (as we have already noted) through the centroid of the system. We can accordingly take $x_1', x_2' \ldots x_q'$ to be that centroid, and we find:

$$\Sigma'^2 = \frac{\mathrm{U}}{n} = \frac{\mathrm{S}(p^2)}{n} = \sigma^2_{x_1} + \sigma^2_{x_2} + \ldots + \sigma^2_{x_q}$$
$$- [l_1^2 \sigma^2_{x_1} + l_2^2 \sigma^2_{x_2} + \ldots + l^2_q \sigma^2_{x_q}$$
$$+ 2 l_1 l_2 \sigma_{x_1} \sigma_{x_2} r_{x_1 x_2} + \ldots + 2 l_{q-1} l_q \sigma_{x_{q-1}} \sigma_{x_q} r_{x_{q-1} x_q}].$$

But the expression in square brackets is precisely the square of the mean square residual with regard to the plane,

$$l_1(x_1 - \bar{x}_1) + l_2(x_2 - \bar{x}_2) + \ldots + l_q(x_q - \bar{x}_q) = 0,$$

or Σ^2. Thus we have:

$$\Sigma'^2 = \sigma^2_{x_1} + \sigma^2_{x_2} + \ldots + \sigma^2_{x_q} - \Sigma^2.$$

Now clearly $\sigma^2_{x_1} + \sigma^2_{x_2} + \ldots + \sigma^2_{x_q}$ is a constant. Hence Σ'^2 will be a minimum when Σ^2 is a maximum, or when the plane perpendicular to the best-fitting line is perpendicular to the least axis of the ellipsoid of residuals. Thus we find: *That the line which fits best a system of n points in q-fold space passes through the centroid of the system and coincides in direction with the least axis of the ellipsoid of residuals.*

The mean square residual (which measures of course the closeness of the fit) is given by

$$\Sigma' = \sqrt{\sigma^2_{x_1} + \sigma^2_{x_2} + \ldots + \sigma^2_{x_q} - \frac{\epsilon^4}{R^2}}, \quad . \quad . \quad \text{(xiv.)}$$

where **R** is the least radius of the ellipsoid of residuals. The direction-cosines of the line can be found from (ix.) by giving Σ^2 the least value among the roots of (x.).

Clearly the plane of best fit passes through the line of best fit, and is further perpendicular to the greatest radius, the maximum axis of the ellipsoid of residuals.

(5) While the geometry of lines and planes of best fit is thus seen to be very simple from the standpoint of inertia ellipsoids,—particularly from the consideration of the surface which, for the theory of errors, I have termed the ellipsoid of residuals,—they most frequently occur, perhaps, in the case of correlated variations or errors, and it is thus of interest to consider them in relation to the ellipses and ellipsoids which arise as "contours" in correlation surfaces.

Now take the case of two variables x and y only, the type-ellipse of the contours of the correlation surface is, when referred to its centroid as origin:

$$\frac{x^2}{\sigma^2_x} + \frac{y^2}{\sigma^2_y} - \frac{2r_{xy}\,xy}{\sigma_x\,\sigma_y} = 1.$$

Compare this with the ellipse of residuals

$$\sigma^2_x x'^2 + \sigma^2_y y'^2 + 2\sigma_x \sigma_y r_{xy} x'y' = \epsilon^4.$$

Clearly if we take $x' = y$, $y' = -x$, and $\epsilon^4 = \sigma^2_x \sigma^2_y$ the ellipse of residuals becomes the correlation type-ellipse. Further, $x'^2 + y'^2 = x^2 + y^2$, or the two ellipses have equal rays, but they are at right-angles to each other. Thus the best-fitting straight line for the system of points coincides in direction with the major axis of the correlation ellipse, and the mean square residual for this line

$$= \frac{\text{product of standard deviations}}{\text{semi-major axis of correlation ellipse}}.$$

The geometry of these results is indicated in the accompanying diagram :—

EE′ is found by making $S(y'-y)^2$ a minimum,
FF′ „ „ $S(x'-x)^2$ „
AA′ „ „ $S(p^2)$ „

The equation to EE′ referred to C is $y=\frac{r_{xy}\sigma_y}{\sigma_x}x$,

„ „ FF′ „ „ $x=\frac{r_{xy}\sigma_x}{\sigma_y}y$.

The angle θ which AA′ makes with Ox is determined by

$$\tan 2\theta=\frac{2r_{xy}\sigma_x\sigma_y}{\sigma^2_x-\sigma^2_y}.$$

Further:

$$(\text{Mean sq. residual})^2=\sigma^2_x\sigma^2_y/\cot^2\theta$$
$$=\tfrac{1}{2}(\sigma^2_x+\sigma^2_y)-\tfrac{1}{2}\sqrt{(\sigma^2_x-\sigma^2_y)^2+4r^2_{xy}\sigma^2_x\sigma^2_y}.$$

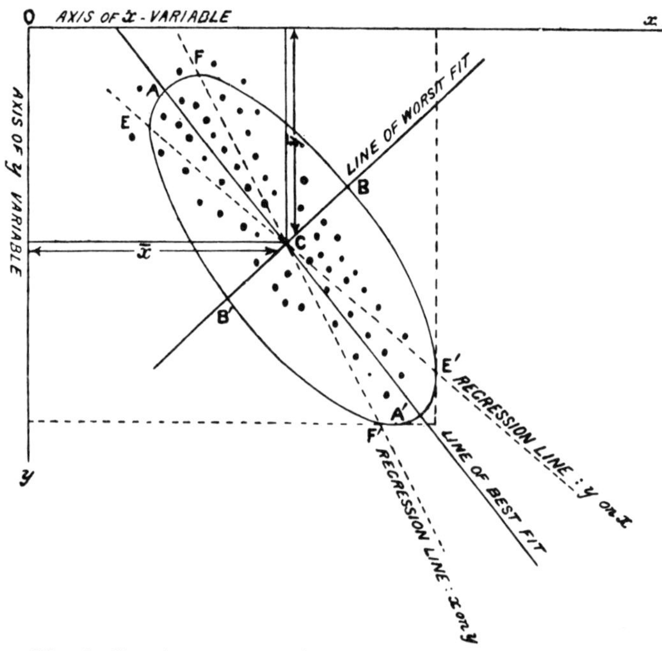

Physically the axes of the correlation type-ellipse are the directions of independent or uncorrelated variation. Hence the line of best fit is a direction of uncorrelated variation.

(6) We turn to the correlation type-"ellipsoid" for q variables. It is*:

$$\Delta_{11}\frac{x_1^2}{\sigma^2_{x_1}}+\Delta_{22}\frac{x_2^2}{\sigma^2_{x_2}}+\ldots\Delta_{qq}\frac{x_q^2}{\sigma^2_{x_q}}+2\Delta_{12}\frac{x_1x_2}{\sigma_{x_1}\sigma_{x_2}}$$

$$+\ldots\ldots+2\Delta_{q-1q}\frac{x_{q-1}x_q}{\sigma_{x_{q-1}}\sigma_{x_q}}=1, \quad . \quad . \quad \text{(xv.)}$$

where $\Delta_{11}, \Delta_{22}, \Delta_{12}\ldots\Delta_{q-1\,q}, \Delta_{qq}$ are the minors corresponding to the constituents marked by the same subscripts of the determinant:

$$\Delta=\begin{vmatrix} 1 & r_{12} & r_{13} \ldots\ldots r_{1q} \\ r_{21} & 1 & r_{23} \ldots\ldots r_{2q} \\ . \quad . & . \quad . & . \quad . \quad . \quad . \\ r_{q1} & r_{q2} & r_{q3} \ldots\ldots 1 \end{vmatrix} \quad . \quad . \quad . \quad . \quad . \quad . \quad \text{(xvi.)},$$

Now let us find the directions and magnitudes of the principal axes of this ellipsoid. We must make

$$u^2=x_1^2+x_2^2+\ldots+x_q^2$$

a maximum. Or if Q be an indeterminate multiplier, we have:

$$(\Delta_{11}+Q\sigma^2_{x_1})\frac{x_1}{\sigma_{x_1}}+\Delta_{12}\frac{x_2}{\sigma_{x_2}}+\Delta_{13}\frac{x_3}{\sigma_{x_3}}+\ldots+\Delta_{1q}\frac{x_q}{\sigma_{x_q}}=0,$$

$$\Delta_{1q}\frac{x_1}{\sigma_{x_1}}+(\Delta_{22}+Q\sigma^2_{x_2})\frac{x_2}{\sigma_{x_2}}+\Delta_{23}\frac{x_3}{\sigma_{x_3}}+\ldots+\Delta_{2q}\frac{x_q}{\sigma_{x_q}}=0,$$

. . . .

. . . .

$$\Delta_{1q}\frac{x_1}{\sigma_{x_1}}+\Delta_{2q}\frac{x_2}{\sigma_{x_2}}+\Delta_{3q}\frac{x_3}{\sigma_{x_3}}+\ldots+(\Delta_{qq}+Q\sigma^2_{x_q})\frac{x_q}{\sigma_{x_q}}=0 \text{(xvii.)}$$

Multiply the last $q-1$ of these equations by $r_{12}, r_{13}, \ldots r_{1q}$ respectively and add them to the first, then we know that:

$$\Delta_{11}+r_{12}\Delta_{12}+r_{13}\Delta_{13}+\ldots+r_{1q}\Delta_{1q}=\Delta,$$

and if u be not 1:

$$\Delta_{1u}+r_{12}\Delta_{2u}+r_{13}\Delta_{3u}+\ldots+r_{1q}\Delta_{qu}=0.$$

* Phil. Trans. vol. clxxxvii. A, p. 302.

Hence:

$$(\Delta+Q\sigma^2_x)\frac{x_1}{\sigma x_1}+r_{12}Q\sigma^2_{x_2}\frac{x_2}{\sigma x_2}+r_{13}Q\sigma^2_{x_3}\frac{x_3}{\sigma x_3}+\ldots+r_{1q}Q\sigma^2_{x_q}\frac{x_q}{\sigma x_q}=0,$$

which may be written:

$$\left(\frac{\Delta}{Q\sigma^2_{x_1}}+1\right)x_1\sigma_{x_1}+r_{12}x_2\sigma_{x_2}+r_{13}x_3\sigma_{x_3}+\ldots+r_{1q}x_q\sigma_{x_q}=0.$$

Or system (xvii.) may be replaced by:

$$x_1\sigma^2_{x_1}\left(1-\frac{\Delta R^2}{\sigma^2_{x_1}}\right)+r_{12}x_2\sigma_{x_2}\sigma_{x_1}+r_{13}x_3\sigma_{x_3}\sigma_{x_1}+\ldots+r_{1q}x_q\sigma_{x_q}\sigma_{x_1}=0,$$

$$r_{12}x_1\sigma_{x_1}\sigma_{x_1}+x_2{}^2\sigma^2_{x_2}\left(1-\frac{\Delta R^2}{\sigma^2_{x_2}}\right)+r_{23}x_3\sigma_{x_3}\sigma_{x_2}+\ldots+r_{2q}x_q\sigma_{x_q}\sigma_{x_2}=0,$$

. . . .

$$r_{1q}x_1\sigma_{x_q}\sigma_{x_1}+r_{2q}x_2\sigma_{x_2}\sigma_{x_q}+r_{3q}x_3\sigma_{x_3}\sigma_x+\ldots+x_q{}^2\sigma^2_x\left(1-\frac{\Delta R}{\sigma^2_{x_q}}\right)^2=0$$

. . . (xviii.)

For multiplying (xvii.) by x_1, x_2, ... x_q respectively and adding, we find

$$1+Qu^2=0;$$

or if R be a maximum or minimum value of u, $Q=-1/R^2$.

Now compare these equations with those we obtain for the directions and magnitudes of axes of the ellipsoid of residuals:

$$\sigma^2_{x_1}x_1'^2+\sigma^2_{x_2}x_2'^2+\ldots+\sigma^2_{x_q}x_q'^2+2r_{12}\sigma_{x_1}\sigma_{x_2}x_1'x_2'+\ldots$$
$$+2r_{q-1q}\sigma x_{q-1}\sigma_{x_q}x'_{q-1}x_q'=\epsilon^4. \text{ (xix.)}$$

These are:

$$x_1'\sigma^2_{x_1}\left(1-\frac{\epsilon^4}{R'^2\sigma^2_{x_1}}\right)+r_{12}x_2'\sigma_{x_2}\sigma_{x_1}+\ldots+r_{1q}x_q'\sigma_{x_q}\sigma_{x_1}=0,$$

$$r_{12}x_1'\sigma_{x_1}\sigma_{x_2}+x_2'\sigma^2_{x_2}\left(1-\frac{\epsilon^4}{R'^2\sigma^2_{x_2}}\right)+\ldots+r_{2q}x_q'\sigma_{x_q}\sigma_{x_2}=0,$$

. . . .

$$r_{1q}x_1'\sigma_{x_1}\sigma_{x_q}+r_{2q}x_2'\sigma_{x_2}\sigma_{x_q}+\ldots+x_q'\sigma^2_{x_q}\left(1-\frac{\epsilon^4}{R_1{}^2\sigma^2_{x_q}}\right)=0$$

. . . (xx.)

Now eliminate the x's from (xviii.) and the x''s from (xx.) and we have precisely the same determinant to find

ΔR^2 and ϵ^4/R'^2. Hence for the semi-axes or max.-min. values:

$$\Delta R^2 = \epsilon^4/R'^2 \quad . \quad . \quad . \quad . \quad . \quad . \quad . \quad \text{(xxi.)}$$

Equations (xviii.) and (xx.) will now give the same values for the ratios of the x's and of the x''s, or for any axis:

$$x_1/x_1' = x_2/x_2' = x_3/x_3' = \ldots = x_q/x_q' \quad . \quad . \quad \text{(xxii.)}$$

But (xxii.) combined with (xxi.) gives us:

$$x_1 = \frac{\epsilon^2}{\sqrt{\Delta}}\frac{x_1'}{R'^2}, \quad x_2 = \frac{\epsilon^2}{\sqrt{\Delta}}\frac{x_2'}{R'^2}, \ldots, x_q = \frac{\epsilon^2}{\sqrt{\Delta}}\frac{x_q'}{R'^2} \quad \text{(xxiii.)}$$

In other words, if we define points given by (xxiii.) to be corresponding points,—*i. e.*, if corresponding points lie on the same line at distances inversely as each other from the origin,—then the ends of the principal axes of the two ellipsoids are corresponding points. Thus the principal axes of the correlation ellipsoid coincide with those of the ellipsoid of residuals in direction, and a minimum axis of the one is a maximum axis of the other and *vice versa*. We therefore conclude:

(i.) That the best fitting plane to a system of points is perpendicular to the least axis of the correlation ellipsoid, and that if $2\,R_{\text{min.}}$ be the length of this axis the mean square residual $= \sqrt{\Delta} \times R_{\text{min.}}$ where Δ is the well-known determinant of the correlation coefficients.

(ii.) The best-fitting straight line to a system of points coincides in direction with the maximum axis of the correlation ellipsoid, and the mean square residual

$$= \sqrt{\sigma^2_{x_1} + \sigma^2_{x_2} + \sigma^2_{x_3} + \ldots + \sigma^2_{x_q} - \Delta.\, R^2_{\text{max.}}}, \quad . \quad \text{(xxiv.)}$$

where $2\,R_{\text{max.}}$ is the length of the maximum axis.

We have thus the properties of the best-fitting plane and line in terms of the correlation ellipsoid, which is the one generally adopted for variation problems. At the same time our investigation shows us that the q directions of independent variation and the standard-deviations of the independent variables may be found from the ellipsoid of residuals, which will usually be a process involving much simpler arithmetic.

(7) *Numerical Illustrations.*

Case (i.). Find the best fitting straight line to the following system of points supposed of equal weight:

$x = 0$	$y = 5{\cdot}9$	$x = 4{\cdot}4$	$y = 3{\cdot}7$
$x = {\cdot}9$	$y = 5{\cdot}4$	$x = 5{\cdot}2$	$y = 2{\cdot}8$
$x = 1{\cdot}8$	$y = 4{\cdot}4$	$x = 6{\cdot}1$	$y = 2{\cdot}8$
$x = 2{\cdot}6$	$y = 4{\cdot}6$	$x = 6{\cdot}5$	$y = 2{\cdot}4$
$x = 3{\cdot}3$	$y = 3{\cdot}5$	$x = 7{\cdot}4$	$y = 1{\cdot}5$

We have at once:

$$\bar{x}=3{\cdot}82 \qquad \bar{y}=3{\cdot}70$$
$$\sigma_x=2{\cdot}3748 \qquad \sigma_y=1{\cdot}31225$$
$$r_{xy}=-{\cdot}9765$$
$$\tan 2\theta=2r_{xy}\sigma_x\sigma_y/(\sigma_x^2-\sigma_y^2)=-1\ 5535.$$

Hence: $\tan\theta=-{\cdot}54556$, or the best-fitting line passes through the point 3·82, 3·70 at a slope of $-{\cdot}546$. This line is shown in the accompanying diagram by AB. The mean

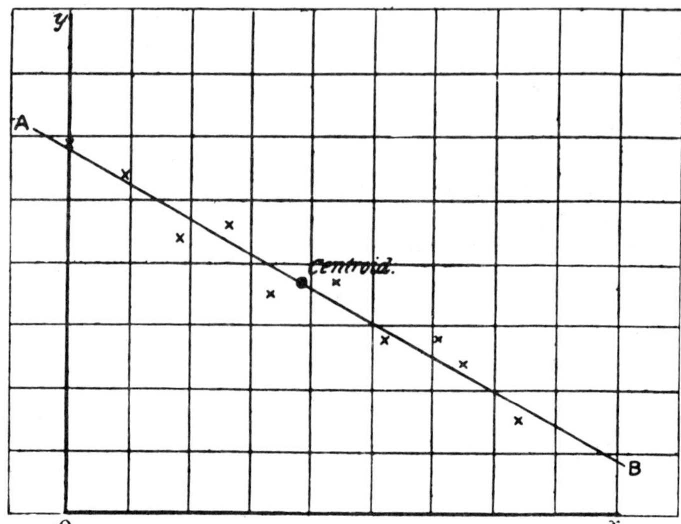

square residual is ·2484. Had we made $S(y-y')^2$ a minimum, the slope of the "best-fitting" line would have been $-{\cdot}5396$ and the mean square residual ·2828; had we made $S(x-x')^2$ a minimum, the slope of the "best-fitting" line would have been $-{\cdot}5659$, and the mean square residual ·5118. These lines are of course the regression lines of slopes $r_{xy}\sigma_y/\sigma_x$ and $\sigma_y/(r_{xy}\sigma_x)$ to the horizontal and mean square vertical and horizontal residuals of $\sigma_y\sqrt{1-r^2}$ and $\sigma_x\sqrt{1-r^2}$ respectively.

Illustration (2).—The following system gives four values of a certain function z:

	$x=2$.	$x=4$.
$y=16$	219	127
$y=26$	261	231

Let us find the best-fitting plane, treating these as four points in three-dimensioned space. We have at once

$$\bar{x}=3, \qquad \bar{y}=21, \qquad \bar{z}=209{\cdot}5$$

$$\sigma_x=1, \qquad \sigma_y=5, \qquad \sigma_z=50{\cdot}0275$$

$$r_{zy}\sigma_y\sigma_z=182{\cdot}5, \quad r_{xz}\sigma_x\sigma_z=-30{\cdot}5, \quad r_{xy}\sigma_x\sigma_y=0.$$

Thus the ellipsoid of residuals is:

$$x^2+25y^2+2502{\cdot}75z^2+365yz-61xz=\epsilon^4.$$

The equations to find the direction-cosines are:

$$\left(2+2\frac{Q}{n}\right)l_1+0\,.\,l_2-61\,.\,l_3=0,$$

$$0\,.\,l_1+\left(50+2\frac{Q}{n}\right)l_2+365\,l_3=0$$

$$-61l_1+365l_2+\left(5005{\cdot}5+2\frac{Q}{n}\right)l_3=0.$$

Whence writing $2\frac{Q}{n}=\chi$ (n=number of points=4) the cubic for χ is:

$$C=\chi^3+5057{\cdot}5\chi^2+123{,}440\chi+48{,}050=0.$$

We want the *least* root:

$\chi=0$, C= +; $\chi=-{\cdot}5$, C= −; $\chi=-100$, C= +; $\chi=-\infty$, C= −. Thus the required root lies between 0 and $-{\cdot}5$. It is easily found to be

$$\chi_1=-{\cdot}395{,}660.$$

Thus $-\frac{Q}{n}={\cdot}197830$, and the mean square residual

$$=\sqrt{-\frac{Q}{n}}={\cdot}4448.$$

We easily deduce:

$$\frac{l_1}{38{\cdot}02187}=-\frac{l_2}{7{\cdot}35823}=\frac{l_3}{1}.$$

Thus the best-fitting plane is:

$$38{\cdot}02187\,(x-3)-7{\cdot}35823\,(y-21)+z-209{\cdot}5=0,$$

or:

$$z+38{\cdot}02187x-7{\cdot}35823y-169{\cdot}03778=0. \quad . \quad \text{(xxv.)}$$

If we find the values for z for given x and y, say those of the four points, which are

211·7, 135·7, 283·3, 207·3,

we should not be impressed by the goodness of the fit. But

the small value of the mean square residual shows how close to each of the points the plane really goes when we measure its distance from a point not by the vertical intercept, but by the perpendicular from the point on the plane. Thus the *vertical* distance from $x=4$, $y=16$, $z=127$, to the plane is 8·7, but the *perpendicular* distance is only ·1988.

If χ_2 and χ_3 be the other two roots of the cubic C we easily find:

$$\chi_2\chi_3=121{,}442{\cdot}65, \quad \chi_2+\chi_3=-5057{\cdot}10434.$$

Thus we have the quadratic to find χ_2 and χ_3

$$\chi^2+5057{\cdot}10434\chi+121{,}442{\cdot}65=0,$$

or:

$$\chi_2=-24{\cdot}12895, \quad \chi_3=-5032{\cdot}97445.$$

χ_3 gives the least axis of the ellipsoid of residuals; hence the direction-cosines of this axis are given by

$$\frac{l_1}{-{\cdot}012{,}125}=\frac{l_2}{{\cdot}073{,}249}=\frac{l_3}{1{\cdot}}$$

We have accordingly for the equation of the best-fitting straight line to the four points:

$$\frac{x-3}{-12{\cdot}125}=\frac{y-21}{73{\cdot}249}=\frac{z-209{\cdot}5}{1000} \quad . \quad . \quad . \quad \text{(xxvi.)}$$

The mean square residual for this line Σ' is given by (xiv.)

$$\begin{aligned}\Sigma' &= \sqrt{\sigma^2_x+\sigma^2_y+\sigma^2_z+\tfrac{1}{2}\chi_3}\\ &= \sqrt{2528{\cdot}75-2516{\cdot}487225}\\ &= 3{\cdot}5018.\end{aligned}$$

This again is remarkably small, considering how far our four points are from being co-linear.

The reader will easily prove directly that the best line (xxvi.) really lies in the best plane (xxv.).

These two illustrations may suffice to show that the methods of this paper can be easily applied to numerical problems; the labour is not largely increased if we have a considerable number of points. It becomes more cumbersome if we have four, five, or more variables or characters which involve the determination of the least (or greatest) root (as the case may be) or an equation of the fourth, fifth, or higher order. Still, the coefficients being numerical and all the roots real and negative, it is not very difficult to localize them.

3

Reprinted from *J. Educ. Psychol.*, **24**(6), 417–441 (1933)

ANALYSIS OF A COMPLEX OF STATISTICAL VARIABLES INTO PRINCIPAL COMPONENTS[1]

HAROLD HOTELLING

Columbia University

1. INTRODUCTION

Consider n variables attaching to each individual of a population. These statistical variables $x_1, x_2, \ldots, x_n$ might for example be scores made by school children in tests of speed and skill in solving arithmetical problems or in reading; or they might be various physical properties of telephone poles, or the rates of exchange among various currencies. The x's will ordinarily be correlated. It is natural to ask whether some more fundamental set of independent variables exists, perhaps fewer in number than the x's, which determine the values the x's will take. If $\gamma_1, \gamma_2, \ldots$ are such variables, we shall then have a set of relations of the form

$$x_i = f_i(\gamma_1, \gamma_2, \ldots) \qquad (i = 1, 2, \ldots, n) \tag{1}$$

Quantities such as the γ's have been called mental factors in recent psychological literature. However in view of the prospect of application of these ideas outside of psychology, and the conflicting usage attaching to the word "factor" in mathematics, it will be better simply to call the γ's *components* of the complex depicted by the tests.

We shall consider only normally distributed systems of components having zero correlations and unit variances. If we use the symbol E to denote the expectation, or mean value in the population, of the quantity following it, the condition that the means shall be zero is expressed by

$$E\gamma_i = 0.$$

The assumptions of unit variances and zero correlations may be combined in the statement

[1] A study made in part under the auspices of the Unitary Traits Committee and the Carnegie Corporation.

The author is indebted to Professor Truman L. Kelley, who was responsible for the initiation of this study and the propounding of many of the questions to which answers are here attempted; also to Professors L. L. Thurstone, Clark V. Hull, C. Spearman, and E. L. Thorndike, who raised some of the further questions treated.

$$E\gamma_i\gamma_j = \delta_{ij} \tag{2}$$

where δ_{ij}, the so-called Kronecker delta, equals unity if i equals j, zero if they are unequal.

If, following the notation of T. L. Kelley, we express the x's in "standard measures," by taking the deviation of each from its mean value and dividing by its standard deviation, we obtain a set of quantities $z_1, z_2, \ldots, z_n$ for which our formulas will be simpler. Confining ourselves to the case in which the functions f_i are linear, the equations (1) then take the form

$$z_i = \Sigma_j a_{ij}\gamma_j, \tag{3}$$

constant terms disappearing because both the z's and γ's have zero means. The summation will be taken from 1 to n; this will include as special cases situations in which there are fewer components than tests, since some of the a_{ij}'s may be zero. However we shall assume in what immediately follows that this is not the case, and that the determinant a of the a_{ij}'s is not zero.

We shall make use of the tensor analysis convention that the repetition of a literal subscript in a term shall, unless otherwise explicitly indicated, denote summation with respect to that subscript from 1 to n. This not only saves writing a large number of summation signs, but has a mnemonic value in helping to indicate what to do next. According to this convention we write (3) in the form:

$$z_i = a_{ij}\gamma_j. \tag{3}$$

Let A_{ij} denote the cofactor of a_{ij} in a, divided by a. Then by the elementary theory of determinants,

$$a_{ij}A_{ik} = \delta_{jk}, \qquad a_{ij}A_{kj} = \delta_{ik}. \tag{4}$$

We may solve (3) for the γ's by multiplying both sides by A_{ik}, summing with respect to i from 1 to n, and using (4). Since $\delta_{jk}\gamma_j$ is a sum consisting of terms which all vanish except γ_k, this gives:

$$\gamma_k = A_{ik}z_i. \tag{5}$$

Let r_{ik} be the correlation between x_i and x_k, equal to unity if $i = k$. This is the same as the correlation between z_i and z_k; and

$$r_{ik} = Ez_iz_k.$$

Here substitute the value for z_i given by (3), and for z_k an expression obtained from (3) by replacing i by k, and j by l. With the help of (2) we then obtain:

$$r_{ik} = a_{ij}a_{kl}E\gamma_j\gamma_l = a_{ij}a_{kl}\delta_{jl}$$
$$= a_{ij}a_{kj}. \tag{6}$$

Since $r_{ik} = r_{ki}$ the number of equations (6) is only $\frac{1}{2}n(n+1)$. They are therefore insufficient for determining the n^2 quantities a_{ij} when the correlations between the tests are known. Thus systems of uncorrelated components γ may be chosen, consistently with the observed correlations, in $\infty^{\frac{1}{2}n(n-1)}$ ways. This variety of choices of components corresponds to the $\frac{1}{2}n(n-1)$ degrees of freedom of a rigid rotation in a space of n dimensions.

It might be thought that additional equations for determining the a_{ij} could be obtained with the moments of higher order, or of other parameters of the population. But if we retain our assumption that the x's are linearly compounded of normally distributed components, this is not the case. Indeed, the x's then have a multivariate normal distribution; and every parameter of such a distribution is a function of the means, variances, and covariances, whose available information is fully embodied in the equations (6) and in the assumption of standard measures. If for example we multiply together four such equations as (3) and take the mean value of each side so as to get an equation in the a_{ij}'s, this equation will, with the help of (6) and the expressions for the fourth moments of a multiple normal distribution, reduce to an identity.

Various modes of escape from the indeterminateness have been considered. The number n^2 of unknowns a_{ij} may be reduced by supposing that there are fewer than n components, which amounts to setting some of the a_{ij} equal to zero. If carried far enough, this results in fewer equations than unknowns, so that consistency conditions upon r_{ij} may be obtained. A similar situation arises from other arbitrary specializations of the a_{ij}, the number of components possibly even exceeding the number of tests. Thus Spearman, putting (in different notation)

$$z_i = a_{io}\gamma_o + a_{ii}\gamma_i, \tag{7}$$

obtained as consistency conditions the famous tetrad equations,

$$r_{ij}r_{kl} - r_{ik}r_{jl} = 0. \tag{8}$$

When these are satisfied for every set of different values of the subscripts, the a_{ij} appearing in (7) are determined uniquely except for sign. Systems involving less specialization and leading to different and fewer consistency conditions have been considered by Truman L. Kelley in "Crossroads in the Mind of Man."[1]

The consistency conditions are of course never satisfied exactly in a sample. Whether the extent of their non-fulfillment in a sample of given size is sufficient to render incredible their fulfillment in the population depends not only on the standard of credibility adopted, but also on the solution of mathematical problems whose study is still incomplete. Apart from this question of sampling, it may well be argued that it is unlikely that the conditions should be fulfilled exactly in the population, and that for sufficiently large samples tetrads such as the left member of (8) may confidently be expected to exceed any assigned multiple of their probable errors, just as the correlation between any two mental or physical measurements is not likely to be exactly zero. This argument is not necessarily conclusive, since small tetrads, even in the infinite population, may like very small correlations be treated as negligible, for economy of thought. It does, however, bring out the special character of the assumption that the number of components is less than the number of tests, as well as of other simplifying particularisations of the a_{ij}.

In order to go as far as may reasonably be possible in a given case in expressing the test scores x_i in terms of a smaller number of components, an orderly procedure is required for selecting the components in the order of the definiteness of their existence, or of their importance for our purposes, and rejecting any which prove to be of little importance, or which are not clearly defined by the data. An analogous situation arises in fitting empirical curves. A series of the form

$$y = a + bx + cx^2 + \cdots$$

may be fitted, the number of terms used being limited by the increasing probable errors of the coefficients of higher order, and also by the diminishing contributions to the total variance of y by these higher order terms. If the series is modified so as to consist of orthogonal functions, the successive coefficients have zero intercorrelations. Only those terms should be retained which are significant. Another analogy is the use of regression equations involving more and more variables $x_1, x_2, x_3, \ldots$ to explain or predict y, these being chosen in the order of their contributions to the variance of y.

These analogies suggest that, in choosing among the infinity of possible modes of resolution of our variables into components, we

[1] Stanford University Press, 1928.

begin with a component γ_1 whose contributions to the variances of the x_1 have as great a total as possible; that we next take a component γ_2, independent of γ_1, whose contribution to the residual variance is as great as possible; and that we proceed in this way to determine the components, not exceeding n in number, and perhaps neglecting those whose contributions to the total variance are small. This we shall call *the method of principal components.* Its technique will be considered in the subsequent sections.

If $z_1, z_2, \ldots z_n$ be taken as rectangular coordinates in n dimensions, each point represents a possible individual. If, as we assume, the population is normally distributed, the loci of uniform density are concentric, similar, and similarly placed ellipsoids. The method of principal components, we shall see, is equivalent to choosing a set of coordinate axes coinciding with the principal axes of these ellipsoids.

Now since the set of x_i is capable of transformations such as changes of units and other linear transformations, the ellipsoids may be squeezed and stretched in any way. The method of principal components can therefore be applied only if for each x_i there exists a unit of measure of unique importance, and if, furthermore, linear transformations, or at least those which do not correspond to rotations of axes, are unimportant. In other words, a *metric*—a definition of *distance*—must be assumed in the n-dimensional space, and not simply a set of axes; we must use Euclidean, not affine geometry, if the principal axes of the ellipsoids are to possess significance. For various purposes it might well happen that different metrics would be suitable. For example the assumption that all the tests, and all the sets of components to be considered, shall have their chance errors independent of those of the others in the set and of equal variance, provides a unique metric. Other possible metrics might be derived from economic considerations, as by requiring that the component traits shall be of equal market value per unit and must not compete with or complement each other. The particular metric implied by the method of principal components is based on the assumption that the unweighted sum of the variances, where the total variance of each test is taken as unity, is the essential quantity to be analyzed.

Weights may in effect be introduced by changing the units of measure, so as to make the standard deviations of the tests no longer unity. The correlations which appear in our subsequent work would in that case be replaced by covariances, and the 1's in the diagonal of the determinant by the variances. However we shall not treat this obvious generalization, excepting to discuss in Section 11 a possible criterion for suitable weighting. Analysis of the unweighted sum of variances has somewhat the same sort of validity as the use of an unweighted mean of observations when we do not know what the weights should be.

A question bound to arise is whether the ensuing analysis should be applied to the "raw" correlations or to those corrected for attenuation. This is equivalent to the question whether the unit of measure in the n-space is to be the standard deviation of the true or of the observed scores. If the true scores' standard deviations are to be used as units, the analysis must be based on the corrected correlations, with 1's in the diagonal. This seems for some purposes a reasonable procedure, and is exemplified in Section 5, p. 432, to which the reader may now pass directly if he is interested in learning the method rather than in its theory. If on the other hand the standard deviations of the inexact *observed* scores are taken as units, the analysis must be performed upon a matrix having the reliability coefficients in the principal diagonal, with the raw correlations elsewhere. An advantage of this last method is that the relative influence of the more reliable tests upon the results is in general enhanced.

An easily verified property of the method is that the first of our principal components has a greater mean square correlation with the tests than does any other variable; and that among all variables uncorrelated with the first $q - 1$ principal components ($q = 2, 3, \cdots, n$), that having the greatest mean square correlation with the tests is the qth principal component. The argument is similar to that of the next section, and will not be given explicitly.

2. DERIVATION OF THE METHOD

Upon squaring each side of (3) and taking the mean value, it is evident that the variance of z_i may be written

$$a^2_{i1} + a^2_{i2} + \cdots + a^2_{in},$$

and that the first term is correctly described as the contribution of γ_1 to the variance of z_i. The sum of the contributions of γ_1 to the variances of all the z's is

$$S = a^2_{11} + a^2_{21} + \cdots + a^2_{n1},$$

which in our abbreviated notation may be written

$$S = a_{i1}a_{i1}. \tag{9}$$

Subject to (6), which we rewrite

$$a_{ij}a_{kj} = r_{ik}, \tag{6}$$

our present object is to choose the coefficients a_{ij} so as to make S a maximum. To this end we write

$$2T = S - \lambda_{ih}a_{ij}a_{hj},$$

where the $\lambda_{ih}(= \lambda_{hi})$ are Lagrange multipliers. We put

$$\frac{\partial T}{\partial a_{i1}} = a_{i1} - \lambda_{ih}a_{h1} = 0, \tag{10}$$

$$\frac{\partial T}{\partial a_{ij}} = -\lambda_{ih}a_{hj} = 0 \qquad (j \neq 1) \tag{11}$$

These two sets of equations may be combined in the single form

$$\frac{\partial T}{\partial a_{ij}} = \delta_{1j}a_{i1} - \lambda_{ih}a_{hj} = 0 \tag{12}$$

According to (11), the linear equations

$$\lambda_{ih}x_h = 0 \tag{13}$$

have the $n - 1$ solutions

$$\begin{array}{l} a_{12},\ a_{22},\ \ldots,\ a_{n2} \\ a_{13},\ a_{23},\ \ldots,\ a_{n3} \\ \ldots\ldots\ldots\ldots\ldots \\ a_{1n},\ a_{2n},\ \ldots,\ a_{nn} \end{array}$$

These must all be linearly independent, for otherwise the determinant a would vanish, contrary to hypothesis. Hence the rank of the system (13) is 1. Therefore quantities $\alpha_1, \alpha_2, \ldots, \alpha_n, \beta_1, \ldots \beta_n$, can be found such that $\lambda_{ih} = \alpha_i\beta_h$. But since $\lambda_{ij} = \lambda_{ji}$, it follows that $\alpha_i\beta_h = \alpha_h\beta_i$, so that the β's are proportional to the α's. Hence we put $\beta_i = \epsilon\alpha_i$, and

$$\lambda_{ih} = \epsilon\alpha_i\alpha_h \tag{14}$$

Consequently (10) may be written in the form

$$a_{i1} = \epsilon\alpha_i\alpha_h a_{h1};$$

or, putting

$$\alpha_h a_{h1} = \sqrt{\frac{k}{\epsilon}},$$

in the form

$$\alpha_i = \frac{a_{i1}}{\sqrt{k\epsilon}}.$$

Substituting this in (14) and the result in (12) gives

$$a_{i1}(k\delta_{1j} - a_{h1}a_{hj}) = 0.$$

Since the quantities a_{i1} are not all to be zero, this leads to

$$a_{h1}a_{hj} - k\delta_{1j} = 0 \tag{15}$$

Multiply by a_{mj}, sum for j, and use (6). The result is:

$$r_{hm}a_{h1} - ka_{m1} = 0. \tag{16}$$

Writing out these equations explicitly for $m = 1, 2, \cdots, n$, and dropping the subscript 1 from the unknowns a_{h1}:

$$\begin{aligned} (1 - k)a_1 + r_{12}a_2 + \cdots + r_{1n}a_n &= 0 \\ r_{12}a_1 + (1 - k)a_2 + \cdots + r_{2n}a_n &= 0 \\ \cdots\cdots\cdots\cdots\cdots\cdots\cdots\cdots \\ r_{1n}a_1 + r_{2n}a_2 + \cdots + (1 - k)a_n &= 0 \end{aligned} \tag{17}$$

In order that these equations have solutions in which not all unknowns are zero, it is necessary and sufficient that

$$f(k) = \begin{vmatrix} 1 - k & r_{12} & r_{13} & \cdots & r_{1n} \\ r_{21} & 1 - k & r_{23} & \cdots & r_{2n} \\ \cdots & \cdots & \cdots & \cdots & \cdots \\ r_{n1} & r_{n2} & r_{n3} & \cdots & 1 - k \end{vmatrix} = 0 \tag{18}$$

Equations such as (18) were first studied in connection with the perturbations of the planets, and are known as *characteristic equations.* The general theory of such equations[1] shows that all the roots are real, and that if there is a q-fold multiple root, the determinant in (18) has rank $n - q$ when this root is substituted for k.

If we substitute a simple (not a multiple) root of the characteristic equation for k in (17), we obtain a set of linear homogeneous equations of rank $n - 1$. These will have a family of solutions, all of which are proportional to one solution. The factor of proportionality is found by putting $j = 1$ in (15); this shows that the sum of squares, S, which we are trying to maximize, is just equal to one of the roots of the characteristic equation. Since it is to be a maximum, we must obviously use the greatest root.

The problem of finding a component γ_1 which will account for as large as possible a part of the total variance is thus solved by finding

[1] *Cf.* for example G. Kowalewski, Einführung in die Determinantentheorie, 1st ed., Leipzig, 1909, pp. 126 and 274. The proof of the reality of the roots was given by Cauchy in the Philosophical Magazine for 1852, and may be found in the treatment of quadric surfaces in books on solid analytic geometry.

the largest root k_1 of the characteristic equation (18), substituting in (17), finding any solution $a_1, a_2, \ldots, a_n$ of these linear equations, and dividing these last values by the sum of their squares and multiplying by $\sqrt{k_1}$. The resulting quantities are the coefficients of γ_1 in the expression (3) giving the test scores in terms of the independent components. A simplified numerical method is given in §4 below.

If the q largest roots of the characteristic equation are equal to k_1, the determination of $a_{11}, a_{21}, \ldots, a_{n1}$ is not unique, since the rank of the linear equations (17) is then only n-q. In this case q linearly independent solutions,

$$\begin{array}{l} a_{11}, a_{21}, \ldots, a_{n1} \\ a_{12}, a_{22}, \ldots, a_{n2} \\ \cdots\cdots\cdots\cdots\cdots \\ a_{1q}, a_{2q}, \ldots, a_{nq}, \end{array}$$

of (17), can be found. Moreover these solutions may be so chosen as to be "orthogonal" to each other, in the sense that

$$a_{ij}a_{im} = \delta_{jm}k_1.$$

They may then be taken as the coefficients of q independent components $\gamma_1, \ldots, \gamma_q$, all of which contribute equally to the total variance.

When the coefficients of the component γ_1 which makes the largest contribution to the total variance, or of the q components which equally make maximum contributions, have been determined, the next problem is to find a component making a maximum contribution to the residual portion of the variance. The argument and procedure are virtually the same as before. If just one component has previously been determined, the subscript i in (10), (11), (12), and (13) takes the values $2, 3, \ldots, n$, and the subscript 1 is replaced by 2. Proceeding in this way we determine the coefficients of $\gamma_1, \gamma_2, \gamma_3, \ldots, \gamma_n$, in the order of the contributions of these components to the sum of the variances of the z's.

The orthogonality among principal components which we have postulated for multiple roots holds also for simple roots. This follows from (15), which, for $j = 2$, gives $a_{h1}a_{h2} = 0$; and similarly we have for all unequal values of i and j, $a_{hi}a_{hj} = 0$.

Having thus derived the equations for the z's in terms of the γ's, it is desirable to solve these, so as to be able to assign a value to each of these principal components γ in terms of the test scores. This will make it possible to assign a value for each principal component to any individual upon whom the tests had been made.

The solution is remarkably simple.

In (15), k is the same as k_1. For the ith component we replace the subscript 1 in (15) by i, and k by k_i. This gives

$$a_{hi}a_{hj} = k_i\delta_{ij}, \qquad \text{(not summed for } i\text{)} \tag{19}$$

which simply means that the sum of the products of corresponding elements of two different columns of the determinant a is zero, and that the sum of the squares of the elements of a column is the root of the characteristic equation corresponding to this column.

Recalling that we have denoted the ratio of the cofactor of a_{mj} to a by A_{mj}, we multiply both sides of (19) by A_{mj} and sum for j. With the help of (4) this gives

$$a_{mi} = k_iA_{mi} \qquad \text{(not summed for } i\text{)} \tag{20}$$

Since (5) gives by a mere change of indices,

$$\gamma_i = A_{mi}z_m,$$

it follows that

$$\gamma_i = \frac{a_{mi}z_m}{k_i} \qquad \text{(not summed for } i\text{).} \tag{21}$$

3. GEOMETRICAL MEANING

Geometrically the foregoing procedure corresponds to rotating the rectangular axes of $z_1, z_2, \ldots, z_n$ so that the new coordinate axes lie along the principal axes of the ellipsoids of uniform density. The squares of the lengths of the principal axes of one of these ellipsoids are proportional to the k's. These facts are not immediately obvious, but may be easily proved as follows. Let ω be the determinant of the correlation coefficients r_{ij}, and let

$$R_{ji} = R_{ij} = \frac{\text{cofactor of } r_{ij} \text{ in } \omega}{\omega},$$

so that

$$r_{ij}R_{ik} = \delta_{jk}. \tag{22}$$

From the theory of multiple normal distribution, the ellipsoids of uniform density are given by

$$R_{ij}z_iz_j = \text{constant.} \tag{23}$$

The procedure developed in treatises on solid analytic geometry is to solve the equation

$$\begin{vmatrix} R_{11}-\lambda & R_{12} & \cdots & R_{1n} \\ R_{21} & R_{22}-\lambda & \cdots & R_{2n} \\ \cdots & \cdots & \cdots & \cdots \\ R_{n1} & R_{n2} & & R_{nn}-\lambda \end{vmatrix} = 0, \qquad (24)$$

to substitute the roots $\lambda_1, \lambda_2, \ldots, \lambda_n$ in the homogeneous linear equations

$$(R_{ij} - \lambda\delta_{ij})l_j = 0, \qquad (25)$$

and for each root to solve these equations. Calling the solution corresponding to λ_1

$$l_{11}, l_{21}, \ldots, l_{n1},$$

that corresponding to λ_2

$$l_{12}, l_{22}, \ldots, l_{n2},$$

and so on, the equations of rotation to new rectangular axes $y_1, \ldots, y_n$ are

$$z_i = l_{ij}y_j; \qquad (26)$$

the solution of these equations for the y_j has the same coefficients in transposed order, and runs:

$$y_j = l_{ij}z_i.$$

The equation of the ellipsoid in the new coordinates is

$$\lambda_1 y_1^2 + \lambda_2 y_2^2 + \cdots + \lambda_n y_n^2 = \text{constant} \qquad (27)$$

Now multiply both sides of (24) by

$$\omega = \begin{vmatrix} 1 & r_{12} & r_{13} & \cdots & r_{1n} \\ r_{21} & 1 & r_{23} & \cdots & r_{2n} \\ \cdots & \cdots & \cdots & \cdots & \cdots \\ r_{n1} & r_{n2} & r_{n3} & \cdots & 1 \end{vmatrix}$$

By (22), the result is

$$\begin{vmatrix} 1-\lambda & -\lambda r_{12} & \cdots & -\lambda r_{1n} \\ -\lambda r_{21} & 1-\lambda & \cdots & -\lambda r_{2n} \\ \cdots & \cdots & \cdots & \cdots \\ -\lambda r_{n1} & -\lambda r_{n2} & \cdots & 1-\lambda \end{vmatrix} = 0$$

Upon dividing each row by $-\lambda$, and setting $k = 1/\lambda$, this reduces to the characteristic equation (18). If we also set $k = 1/\lambda$ in (25), multiply by r_{ih}, and sum for i, the resulting equations have the same coefficients as (17). For simple roots the solutions are therefore the same, apart from a common factor. For multiple roots the solutions have in both cases the same type of indeterminateness. Since $\lambda_i = 1/k_i$, (27) becomes

$$\frac{y_1^2}{k_1} + \frac{y_2^2}{k_2} + \cdots + \frac{y_n^2}{k_n} = \text{constant}, \tag{28}$$

which shows that the squares of the lengths of the axes are proportional to $k_1, k_2, \ldots, k_n$.

If, instead of the y's or z's, the γ's or other independent and equally variable quantities be taken as rectangular coordinates, the ellipsoids are squeezed and stretched into spheres. Each test is represented by a line through the origin. The correlation between two tests is the cosine of the angle between their lines. L. L. Thurstone has used coordinates equivalent to these.[1]

Not only must the roots of the characteristic equation be real; they must all be positive. If there were a negative root, (28) would represent, not an ellipsoid, but a hyperboloid extending to infinity. Since the density of probability is to be uniform over this locus, the probability of a sample deviating in certain tests from the mean by more than any given amount would be infinite, which is absurd.

According to the original Spearman theory of the mind, one important general factor accounts for the bulk of the variance of mental tests, other components being of minor importance. If this is true, one of the roots of the characteristic equation should be much larger than any of the others. The ellipsoids should be needle-shaped. But if two or more of the roots are equal, there will be a corresponding number of independent components which contribute equally to the variance. In this case the ellipsoids will be figures of revolution. If $n = 3$ and the two largest roots are equal, while the third is very small, the ellipsoids will be thin discs.

In distinguishing among such theories it is of course not the absolute values of the roots that is important, but their ratios, and particularly the ratios among the largest of the roots. The sum of

---

[1] Multiple Factor Analysis. *Psychological Review*, Vol. XXXVIII, 1931, pp. 406–427.

Since this was written Professor Thurstone has kindly sent me a pamphlet he has prepared for class use, in which he uses the same geometric interpretation as in the present section, and discusses the problem from essentially the same standpoint as that taken in §1. His iterative procedure appears to have no relation to that of §4. In June, 1932, Professor Thurstone presented at the Syracuse meeting of the American Association for the Advancement of Science certain of the considerations which have served as a point of departure for this paper.

the roots always equals the number n of the tests, as appears from the form of (18); hence the fraction of the total variance contributed by the ith component is k_i/n.

Developing (18) we have

$$f(k) = (-1)^n(k^n - nk^{n-1} + S_2k^{n-2} - S_3k^{n-3} + \cdots + S_n) = 0, \quad (29)$$

where S_2 is the sum of the two-rowed principal minors in the determinant ω of the correlations, S_3 the sum of the three-rowed principal minors, and so on.

4. ITERATIVE SOLUTION

The explicit calculation of the determinant ω and its principal minors, and the solution of the characteristic equation and of the homogeneous linear equations (17), would be a laborious computation. A vast saving of arithmetical effort is effected by the following iterative method, which yields simultaneously a root and the corresponding coefficients, the roots appearing in order of magnitude, the greatest first. This makes it possible to stop whenever it is evident that all the important principal components have been obtained. Since the sum of the roots is n, the fraction of the total accounted for at any stage is always in evidence. During the calculation of each principal component, an error at any stage is rectified in the next. The risk of serious numerical error is therefore negligible, especially if all the roots are calculated and their sum compared with n.

If numbers $a_1, a_2, \ldots a_n$ proportional to the direction cosines of any line through the origin be substituted in the equations

$$a_i' = r_{ij}a_j \qquad (i = 1, 2, \cdots, n) \quad (30)$$

the quantities on the left will be proportional to the direction cosines of a new line through the origin into which we shall consider the original line to have moved. Under this transformation, the invariant lines will be those for which the quantities k exist such that $a_i' = ka_i$. In this case, (30) reduces to (17). Thus, for each invariant line, the direction cosines are proportional to a solution of (17), while k is a root of the characteristic equation. It follows that the invariant lines are the principal axes. Hence, if numbers $a_1, \ldots, a_n$ can be found which, substituted in the right-hand members of (30), give $ka_1, \ldots, ka_n$, these numbers are proportional to the direction cosines of one of the principal axes and to the coefficients of one of the principal components γ in the expressions for the test scores z_i, while k is the sum of the contributions of this component to the variances of the test scores.

If two or more roots of the characteristic equation are equal, the ellipsoids are figures of revolution, with two or more equal axes. Every line in the plane or hyperplane of the equal axes is invariant under the transformation; the axes themselves may be taken as arbitrary perpendicular lines in this plane or hyperplane.

If with respect to new coordinate axes coinciding with the principal axes of the ellipsoids the direction cosines of a line are proportional to $b_1, b_2, \ldots, b_n$, the transformation

$$b_1' = k_1 b_1,\ b_2' = k_2 b_2,\ \cdots,\ b_n' = k_n b_n, \tag{31}$$

where $k_1, k_2, \ldots, k_n$ are as before the roots of the characteristic equation, will geometrically be the same as (30), since the invariant lines are the principal axes, together perhaps with the lines in the plane or hyperplane determined by two or more equal axes. Algebraically, this amounts to putting.

$$a_i' = a_{ih} b_h', \qquad a_j = a_{jm} b_m.$$

Substituting these expressions in (30) and in the result setting

$$r_{ij} a_{jl} = k_l a_{il} \qquad \text{(not summed for } l\text{)},$$

a relation which is the generalization of (16), we multiply by A_{im}, sum for i, and use (4). The result is (31).

Let the notation be so arranged that

$$k_1 \geqslant k_2 \geqslant \cdots \geqslant k_n.$$

If k_1 is greater than k_2, and if $b_1 \neq 0$, we then have from (31) that each of the ratios $b_2'/b_1', \ldots, b_n'/b_1'$ is numerically less than the corresponding ratios $b_2/b_1, \ldots, b_n/b_1$. When the transformation is repeated, the absolute values of these ratios are further diminished, and with further repetitions approach zero in geometrical progressions. If, however, the q greatest roots are equal, the ratios among the first q direction cosines remain unchanged under the transformation, while the remaining $p - q$ direction cosines approach zero. Thus if we start with any line which does not lie in the hyperplane of the $p - 1$ axes perpendicular to the longest axis, this line will, under iteration of the transformation, approach the longest axis if the greatest root is unique, or, if there are several equal roots greater than the rest, some position which may be taken as the longest axis.

From this it is evident that the coefficients of the greatest principal component of the tests may be obtained with any required accuracy by inserting an arbitrary set of numbers $a_1, \ldots, a_n$ in the right members of (30), multiplying or dividing the resulting numbers a_1', $\ldots, a_n'$ by any constant, again substituting in the right members of (30), and repeating the process until, to the required degree of accuracy, the quantities obtained are multiples of the preceding quantities by a constant k_1. This constant will be the greatest root of the characteristic equation. The process fails only in the infinitely improbable case of the initial values being linearly dependent upon principal components other than the first; the greatest of these components is then approached.

The coefficients $a_{11}, a_{21}, \ldots, a_{n1}$ of the first principal component are found by multiplying each of the quantities $a_1, a_2, \ldots, a_n$ by $\sqrt{k_1/\Sigma a_j^2}$.

After the coefficients of γ_1 have been determined, those of γ_2 are sought. This might be done by starting with arbitrary values orthogonal to the coefficients of γ_1, which would give the desired result if the calculations were carried out exactly. But on account of the practical necessity of working only to a limited number of decimal places, the transformed line will at each stage deviate from the plane perpendicular to the first axis, and will drift off toward this first axis. The tendency could be offset by applying corrections to restore orthogonality each time, but this is excessively laborious. Instead, we revise the matrix of correlations by subtracting $a_{i1}a_{j1}$ from the element in the ith row and jth column. This produces the matrix of covariances of the quantities $z_i - a_{i1}\gamma_1$, which are uncorrelated with γ_1. The iterative process applied to the reduced matrix yields k_2 and the coefficients of γ_2. To obtain γ_3 we apply the iterative process to the further reduced matrix in which the element in the ith row and jth column is $r_{ij} - a_{i1}a_{j1} - a_{i2}a_{j2}$; and so on. When this method is used, it is not even essential to take the initial trial values of the coefficients at any of the later stages precisely orthogonal to the coefficients already determined; round numbers may be chosen at the beginning, and the process will converge to the correct values anyhow.

A convenient procedure is to divide each of the trial values of any set of coefficients by a fixed one of them. The next value obtained for this coefficient will then be an approximation to the appropriate characteristic number k. The process should be started with trial values of one digit each, the largest of these values corresponding to variates which are on the whole most highly correlated with the rest, as judged by inspection. Each digit should be accurately determined, by repetition until stationary values are reached, before the calculations are carried to another place.

The labor is sometimes reduced if the deviations of a set of trial a's from the preceding set are multiplied by the rows of the correlation matrix and the results added to the last trial a's to get the next set, instead of multiplying the trial values themselves by the correlations. However, this method does not automatically correct errors unless the values obtained finally are multiplied by the correlations and added.

5. EXAMPLE

Truman L. Kelly (*op. cit.*, p. 100) gives the correlations found in a sample of 140 seventh-grade children among numerous tests. We select the correlations, corrected for attenuation, among: (1) Reading speed, (2) reading power, (3) arithmetic speed, (4) arithmetic power. Curtailed to three places, these are, in the natural order:

$$\begin{vmatrix} 1. & .698 & .264 & .081 \\ .698 & 1. & -.061 & .092 \\ .264 & -.061 & 1. & .594 \\ .081 & .092 & .594 & 1. \end{vmatrix}$$

The correlations being slightly higher on the whole for the first than for the later tests, we take as trial values

$$1, \quad .9, \quad .8, \quad .7.$$

Multiplying by the rows of the matrix of correlations, we have, to two place, 1.90, 1.61, 1.42, 1.33. If we divide each of these values by the first in order to make them comparable with the initial quantities, we obtain as our second approximation:

$$1, \quad .85, \quad .75, \quad .70.$$

Multiplying these by the correlations, we obtain 1.85, 1.57, 1.37, and 1.30. Dividing these by 1.85 gives

$$1, \quad .85, \quad .74, \quad .70.$$

This is close enough to warrant carrying the calculations to one more decimal place. We next obtain 1.846, 1.567, 1.368, 1.304; and upon division by 1.846,

$$1, \quad .849, \quad .741, \quad .707$$

The next trial gives, after division by 1.846,

1, .849, .743, .706,

values which we adopt after noting that the differences,

0 0 .002, −.001,

multiplied by any row of the correlation matrix, produce a total which does not affect the third decimal place. We divide $k_1 = 1.846$ by the sum of the squares of the final trial values, extract the square root, and multiply by the trial values. This gives

$$\begin{aligned} a_{11} &= .816 \qquad k_1 = 1.846 \\ a_{21} &= .693 \\ a_{31} &= .606 \\ a_{41} &= .576 \end{aligned}$$

Subtracting the products $a_{i1}a_{j1}$ from the elements of the correlation matrix we obtain

$$\begin{Vmatrix} .330 & .129 & -.233 & -.392 \\ .129 & .517 & -.484 & -.310 \\ -.233 & -.484 & .631 & .243 \\ -.392 & -.310 & .243 & .666 \end{Vmatrix}$$

Starting from the trial values −.6, −1, 1, 1, we obtain for the second principal component:

$$\begin{aligned} a_{12} &= -.438 \qquad k_2 = 1.465 \\ a_{22} &= -.620 \\ a_{32} &= .674 \\ a_{42} &= .660 \end{aligned}$$

Subtracting the products of these numbers from the reduced matrix we find:

$$\begin{Vmatrix} .138 & -.142 & .062 & -.103 \\ -.142 & .133 & -.066 & .099 \\ .062 & -.066 & .177 & -.202 \\ -.103 & .099 & -.202 & .230 \end{Vmatrix}$$

This time we take the trial values −1, 1, −1, 1.3, and obtain another root, another set of coefficients, and another reduced matrix. A fourth application of the iterative process completes the resolution of the tests into their principal components. The results are combined in the table below, in which each column corresponds to a principal component, and each of the last four rows to a test. The entries in the last four rows are the coefficients of the γ's in the expressions for the z's; they are at the same time the correlations of the γ's with the z's.

					Totals
Root	1.846	1.465	.521	.167	3.999
Percentage of total variance	46½	36½	13	4	100
Reading speed	.818	−.438	−.292	.240	
Reading power	.695	−.620	.288	−.229	
Arithmetic speed	.608	.674	−.376	−.193	
Arithmetic power	.578	.660	.459	.143	

The chief component seems to measure general ability; the second, a difference between arithmetical and verbal ability. These two account for eighty-three per cent of the variance. An additional thirteen per cent seems to be largely a matter of speed vs. deliberation. The remaining variance is trivial.

6. SAMPLING ERRORS

The exact distribution of the roots of the characteristic equation, or of their ratio, can be found at once when $n = 2$. Indeed, we have in this case,

$$k_1 = 1 + r, \qquad k_2 = 1 - r;$$

and the distribution of r in samples of N from a normally correlated population is fully known.

For the case of zero correlation in the population, the distribution of r reduces to

$$\frac{1}{\sqrt{\pi}} \frac{\Gamma[\frac{1}{2}(N-1)]}{\Gamma[\frac{1}{2}(N-2)]} (1 - r^2)^{\frac{N-4}{2}} dr.$$

If we put

$$u = \frac{k_1}{k_2} = \frac{1+r}{1-r}, \text{ so that } r = \frac{k_1 - k_2}{k_1 + k_2},$$

this is transformed into

$$\frac{2}{\sqrt{\pi}} \frac{\Gamma[\frac{1}{2}(N-1)]}{\Gamma[\frac{1}{2}(N-2)]} \frac{(4u)^{\frac{N-4}{2}}}{(u+1)^{N-2}} du.$$

For $n > 2$, this same distribution may be used as a close approximation for differentiating between any two roots.

Thus, to determine whether k_1 is significantly greater than k_2 in the example worked out in the last section, we compute

$$r = \frac{1.846 - 1.465}{1.846 + 1.465} = .115$$

Since the sample consists of one hundred forty individuals, we may treat this value of r as a sample from a normal distribution of zero mean and standard deviation $1/\sqrt{139} = .085$. Since r is only 1.35 times its standard error, the probability of a greater discrepancy is about .18, and the two roots cannot be called significantly different. The fact that a single component accounts for so much as 46½ per cent of the variance of the four tests tends to support the idea introduced by Spearman that one general factor enters into all tests to a dominating extent; but this argument is considerably weakened by the fact that γ_2 contributes nearly as much variance as γ_1, the magnitudes of the two contributions being indeed not clearly distinguishable in a sample of this size. The contribution of γ_3 is however definitely less than that of γ_2; for

$$r = \frac{1.465 - .521}{1.465 + .521} = .474$$

is some 5.58 times the standard error .085. The third characteristic root exceeds the fourth even more definitely, the ratio of the corresponding value of r to its standard error being 6.1.

Apart from the question of equality of any two roots, it may sometimes be desired to find upper and lower limits such that, corresponding to any given degree of probability, the ratio of the roots may be said to lie between these limits. Such limits may be deduced from the corresponding ones for the correlation coefficient. To obtain these, we may utilize R. A. Fisher's transformation to $z = \tanh^{-1} r$, whose distribution is nearly normal with a variance, $1/(N - 3)$, independent of the population value.[1]

As an example, let us find upper and lower fiduciary limits corresponding to the probability .05 for the ratio k_2/k_3, which has just been seen to differ clearly from unity. From Fisher's table, we find that the value $r = .474$ corresponds to $z = .515$. The table of the normal probability integral shows that a quantity deviates from its mean more than 1.96 times its standard deviation with a probability .05. The deviation

$$1.96\sigma_z = \frac{1.96}{\sqrt{137}} = .117$$

[1] Statistical Methods for Research Workers, Oliver and Boyd, Chap. VI.

is therefore to be added to and subtracted from the sample value .515, giving .398 and .632 as the fiduciary limits for z. Again referring to the table of hyperbolic tangents, we find .378 and .559 as the corresponding values of r. Inserting each of these limits for r in the expression defining u, we have 2.21 and 3.53 as the extreme values of the ratio which can plausibly be assumed, corresponding to the probability .05.

This use of the sampling distribution of r is subject to two qualifications. In the first place, it applies only in comparing two components which are definitely identified, otherwise than by their having a particular order among the entire set of n components determined, such as being the two greatest, or the greatest and least. This situation is common to all problems in which a number of observations are made on a quantity, and two of these observations are examined for the significance of their difference from each other. If the two are selected because of their positions relatively to the others in the sample, they cannot be compared accurately by means of the standard error of the distribution as if they were not so selected. In many examples, however, including the foregoing, this consideration does not materially affect the conclusions to be drawn.

It must also be remembered that the directions as well as the magnitudes of the principal components are subject to sampling errors, and that these errors in determination of the directions will interfere with the accuracy of the foregoing treatment of the k's by transformation into correlation coefficients. That the inaccuracy introduced in this way is very slight is suggested by the geometry. Suppose that a principal plane of an ellipsoid derived from a sample makes a small angle θ with the corresponding principal plane of the population ellipsoid. Then the section of the population ellipsoid made by the sample plane will be an ellipse whose principal axes bear to the corresponding principal axes of the population ellipsoid ratios differing from unity by quantities of order θ^2. Hence if the standard error of θ, like most standard errors, is of order $1/\sqrt{N}$, those of the semiaxes, and consequently those of the k's and of the r's calculated from them, will be of order $1/N$. This suggests that a suitable correction for this kind of error could be made in the foregoing example by changing the standard error used, namely .085, by something of the order of $1/140 = .007$.

Essentially the same results may be reached in another way. The k's are, in the population, variances of independent variates. The ratio u of two independent estimates of variance, each based on $N - 2$

degrees of freedom, as for example in independent samples of $N - 1$, has exactly the distribution of u given above. This way of looking at the matter has the further advantage of showing that, for very large samples, log k may be treated approximately as a normally distributed variate with variance $1/(2N - 4)$. The n values of log k may then be treated as independent samples from a normal distribution having this variance.

7. MINIMUM NUMBER OF INDEPENDENT COMPONENTS. RELIABILITY COEFFICIENTS

But of still greater importance is the fact that by treating the k's as estimates of variance in n orthogonal directions we may compare them, not only with each other, but also with the variance to be expected on account of the inaccuracy of the tests as revealed by their self-correlations, or reliability coefficients. This is of major interest, for if some of the principal components found contribute so little to the variance that their reality is in doubt, it is possible to assume that the number of independent components is less than the number of variables measured. The following tests may therefore serve as substitutes both for Spearman's use of tetrads and for the more elaborate criteria developed by Kelley in "Crossroads in the Mind of Man."

Upon administering a test twice to the same individuals a measure of its accuracy may be obtained from the correlation of the two sets of scores. This correlation is known as the reliability coefficient; for the ith test we shall denote it by r_I. If the test score is thought of as made up of two parts, a true score, whose variance we shall take as unity, and a random error of variance σ_i^2, it is easy to see that

$$r_I = \frac{1}{1 + \sigma_i^2},$$

whence σ_i^2 may be determined from the data. The random errors are supposed to be independent in the several tests. For the four tests which we have been using as an example the reliability coefficients given by Kelley are, in order

$$.9197, \quad .8942, \quad 9083, \quad .5639.$$

Hence we find

$$\sigma_1^2 = .0873 \qquad \sigma_2^2 = .1183 \qquad \sigma_3^2 = .1010 \qquad \sigma_4^2 = .7734.$$

Since the covariance of the ith and jth tests is the same whether based on the true or the total scores, the correlations will differ in the two cases, that based on the true score being larger in the ratio $\sqrt{(1 + \sigma_i^2)(1 + \sigma_j^2)}$. If r'_{ij} is the correlation between the observed scores, the correlation between true scores is estimated as

$$r_{ij} = r'_{ij}\sqrt{r_I r_J} = r'_{ij}\sqrt{(1 + \sigma^2{}_i)(1 + \sigma^2{}_j)} \qquad (i \neq j;\ \text{not summed for } i \text{ or } j), \quad (32)$$

and is known as a correlation corrected for attenuation. The exact sampling distribution of this quantity has never been determined, but for sufficiently large samples it may be used with confidence, subject to the validity of the assumption that the errors of measurement in the several tests are uncorrelated with the test scores and with each other. Since we wish to deal with real quantities as far as possible, we have based our analysis into principal components upon these corrected coefficients r_{ij}.

If the number of independent components of the n true scores is less than n, the scatter diagram of the true scores will lie in a flat space of smaller dimensionality immersed in the n-dimensional space. The scatter diagram of the observed scores will however be n-dimensional in character, since the scatter diagram of the errors of measurement is n-dimensional. The scatter diagram corresponding to the correlations corrected for attenuation would in this case be of the smaller dimensionality if calculated from the whole population; but on account of the fluctuations of the errors of measurement this would not in general be true for samples. We are therefore interested in comparing the variances of the principal components we find, and particularly the least of these variances, with those to be expected on the basis of the reliability coefficients.

Now by (21), we have upon omitting the subscript i,

$$\gamma = \frac{a_1 z_1 + a_2 z_2 + \cdots + a_n z_n}{k}. \qquad (33)$$

The variance of this expression which results from errors of measurement is

$$\sigma'^2 = \frac{a_1^2\sigma_1^2 + a_2^2\sigma_2^2 + \cdots + a_n^2\sigma_n^2}{k^2}.$$

On the hypothesis that γ has no real existence, but arises purely from errors of measurement, its variance, which we have taken as unity, should differ from σ'^2 only on account of fluctuations in these errors. On this hypothesis the equation

$$k = \bar{k}$$

where

$$\bar{k} = \sqrt{a_1^2\sigma_1^2 + a_2^2\sigma_2^2 + \cdots + a_n^2\sigma_n^2},$$

should fail to be satisfied only in so far as k and $\bar{k}$ are affected by random sampling errors. These two quantities appear to be uncorrelated with each other; for although the k's are derived from correlation coefficients corrected for attenuation with the help of the data from which the $\bar{k}$'s are deduced, still this process is analogous to that in Fisher's analysis of variance of subtracting from the total variance the intraclass variance before comparing with the interclass variance, which is independent of this difference. If we rely upon this analogy, we may compare k with $\bar{k}$ simply by adding their variances, provided the samples are large enough to allow the difference to be treated as normally distributed.

Instead of comparing k with $\bar{k}$ directly, we might use any function of k and the corresponding function of $\bar{k}$. In choosing among such functions as k, k^2, $\sqrt{k}$, and log k, it is to be recalled that k is of the nature of a variance, as pointed out at the end of the last section. In comparing estimates of variance, the logarithm is used by R. A. Fisher because its standard error is independent of the value of the variance in the population, and depends only on the number of cases in the sample—or, more generally, upon the number of degrees of freedom on which the estimate of variance is based. Against this advantage must be set the fact that the square root of an estimate of variance has a more nearly normal distribution than either the estimate of variance itself or its logarithm, and the use of the standard error presupposes a normal distribution. Further, log $\bar{k}$ has a standard error which may be found approximately from the definition of $\bar{k}$ above, and which lacks the property of depending only on the number of cases, the a's and σ's being involved in it also. The advantage of the logarithm is thus lost when a comparison is made with $\bar{k}$, but the gain in accuracy in the use of $\sqrt{k}$ persists, though the higher moments of the various functions of $\bar{k}$ have not yet been investigated. Consequently we shall make our comparisons in terms of the square roots.

Since

$$\sqrt{\bar{k}} = (\Sigma a_j^2\sigma_j^2)^{1/4},$$

we have, apart from terms of higher order, the following relation between deviations of sample from population values:

$$\delta\sqrt{\bar{k}} = \tfrac{1}{2}(\Sigma a_j^2\sigma_j^2)^{-3/4}(\Sigma a_j^2\sigma_j\delta\sigma_j).$$

The mean value of this expression is zero, provided the estimate of σ_j^2 is without bias. This will be the case if σ_j^2 is calculated as the ratio of the sum of the squares of the differences between test and retest to $2N$; this method of calculation appears to be approximately equivalent to that from reliability coefficients. Neglecting any bias, then, the variance of $\sqrt{\bar{k}}$ will be the mathematical expectation of the square of the above expression. Since the errors of measurement of the tests are independent of each other,

$$E\delta(\sigma_i)\delta(\sigma_j) = 0, \text{ if } i \neq j,$$

while the usual formula for the variance of the standard deviation gives

$$E(\delta\sigma_j)^2 = \frac{\sigma_j^2}{2N}.$$

Making these substitutions after squaring $\delta\sqrt{\bar{k}}$, we obtain finally,

$$\sigma^2_{\sqrt{\bar{k}}} = \frac{\Sigma a_j^4 \sigma_j^4}{8N\bar{k}^3}.$$

To this we add in each case

$$\sigma^2_{\sqrt{k}} = \frac{k}{2N}$$

to obtain the variance of $\sqrt{k} - \sqrt{\bar{k}}$. The results for the four tests based on one hundred forty cases which we have been considering are given in the following table. The variances found for $\sqrt{k}$ were all considerably higher than those for $\sqrt{\bar{k}}$, the ratios to the latter ranging from 2½ to sixty-seven.

Principal component	k	$\sqrt{k}$	$\sqrt{\bar{k}}$	Ratio of difference to standard error
1	1.846	1.359	.801	6.72
2	1.465	1.210	.814	5.28
3	.521	.722	.665	1.31
4	.167	.406	.413	−.28

The value for the fourth component is actually less than the value to be expected on the basis of errors of measurement, while that for the third component does not significantly exceed the expected value. For the other two, however, the excess is decidedly significant.

We conclude that the true scores, if we could find them, would display a scatter diagram of at least two dimensions, *i.e.* that there are at least two genuine independent components; but this experimental material supplies no evidence for more than two independent components. Thus we can definitely affirm the existence of two such components, though we cannot distinguish definitely between them with one hundred forty individuals. In the scatter diagram, we can fix with some definiteness the plane of these two leading components, but on account of their nearly equal contributions to the variance we cannot be at all sure of their directions within the plane. The ellipses of the scatter diagram are too nearly circular for this. It is possible, but far from certain on this evidence, that they are really three-dimensional, close in form to oblate spheroids.

(*To be concluded in October issue.*)

4

Reprinted from *Sanhkya*, Ser. A, **26**, 329–358 (1964)

THE USE AND INTERPRETATION OF PRINCIPAL COMPONENT ANALYSIS IN APPLIED RESEARCH*

By C. RADHAKRISHNA RAO

Indian Statistical Institute

Visiting at Stanford University

SUMMARY. The paper provides various interpretations of principal components in the analysis of multiple measurements. A number of generalizations of principal components have been made specially in the study of a set of variables in relation to another set known as instrumental variables. The use of generalized principal components in applied research has been indicated. Finally, the difference between factor analysis and principal component analysis has been explained.

1. Introduction

During the last few years, I had the opportunity of surveying some of the multivariate statistical techniques and evaluating their usefulness in practical research work (Rao, 1960, 1961). I undertook to do this mainly because the superiority of the multivariate methods over the simpler univariate analysis has been questioned by some statisticians. It was said that an examination of individual measurements by simpler univariate methods is an essential step which can be supplemented by more elaborate multivariate analysis when this appears feasible and necessary. It was also thought that multivariate methods have only provided new outlets for mathematical theory without materially assisting scientific research.

It is, indeed, necessary to criticise uncritical and inappropriate applications, which are perhaps natural when new techniques are put forward. But a criticism of multivariate methods as such is not justifiable. I have considered a number of practical examples specially in one of the papers (Rao, 1961) to demonstrate the use of multivariate analysis and to emphasize the need for exploring the potentialities of the existing techniques and for further research.

The purpose of the present paper is to examine, in some detail, the role of principal component analysis in applied research. When a large number of measurements are available, it is natural to enquire whether they could be replaced by a fewer number of the measurements or of their functions, *without loss of much information*, for convenience in the analysis and in the interpretation of data. Principal components, which are linear functions of the measurements, are suggested for this purpose. It is, therefore, relevant to examine in what sense principal components provide a reduction of the data without much loss of *information we are seeking from the data.*

The following are some typical statements found in the literature on applications of principal component analysis. 'The transformed variables (the first few principal components) are useful in connection with preliminary investigation of a large number of samples of species from different localities.' 'For an orientatory

---

* Technical Report No. 9. Prepared Under Contract OE 2-10-065 with U.S. Office of Education.

type of investigation, it would probably suffice with the variable provided by the first principal component (out of four) for this represents 3/4 of the total variation.' 'At first principal components are most likely to be regarded by the nonmathematicians as a highly arbitrary set of manipulations. Such a reaction should be dismissed, as soon as the geometrical meaning is considered : principal component analysis merely leads to new angles of viewing data, analysis best suited to disclose the nature of size and shape variation.' In no case are the statements made substantiated by statistical analysis to show that the information neglected does not lead to misleading conclusions. Often, the principal component analysis is first undertaken without any clear objective and then an attempt is made to interpret the derived results.

The discussion in the present paper has the following aims : (a) To provide various interpretations of principal components. (b) To examine the situations where the principal component analysis can be undertaken with a definite purpose or in an exploratory way in the earlier stages of investigation of a research problem. (c) To indicate the difference between the principal component analysis and the factor analysis. (d) To generalize the principal component analysis in a number of directions useful in applied research.

2. Eigen values and vectors of matrices

For a theoretical development of the principal component analysis and its interpretation it is necessary to use some results on the canonical reduction of matrices, which are summarized in this section for use in the later sections.

(i) *Eigen values and vectors of a matrix.* Let $\mathbf{\Sigma}$ be a non-negative (i.e., positive definite or positive semi-definite) matrix of order $p \times p$. The roots of the determinantal equation

$$|\mathbf{\Sigma}-\lambda \boldsymbol{I}| = 0 \qquad \dots \quad (2.1)$$

are called the eigen values of $\mathbf{\Sigma}$. The equation (2.1) has p real non-negative roots, $\lambda_1 \geqslant \lambda_2 \geqslant \dots \geqslant \lambda_p$. Corresponding to each root λ_i of (2.1), there exists a column vector $\boldsymbol{P}_i$, such that

$$\mathbf{\Sigma}\, \boldsymbol{P}_i = \lambda_i \boldsymbol{P}_i, \qquad \dots \quad (2.2)$$

which is called an eigen vector. Then the following results hold :

(a) $\boldsymbol{P}_1, \dots, \boldsymbol{P}_p$ can be chosen to be orthonormal whether $\lambda_1, \dots, \lambda_p$ are distinct or not.

(b)
$$\mathbf{\Sigma} = \lambda_1 \boldsymbol{P}_1 \boldsymbol{P}_1' + \dots + \lambda_p \boldsymbol{P}_p \boldsymbol{P}_p'.$$
$$\boldsymbol{I} = \boldsymbol{P}_1 \boldsymbol{P}_1' + \dots + \boldsymbol{P}_p \boldsymbol{P}_p' \qquad \dots \quad (2.3)$$

(c) Let $\boldsymbol{L}_1, \dots, \boldsymbol{L}_q$ be any set of orthonormal vectors. Then

$$\sum_{i=1}^{q} \boldsymbol{L}_i' \mathbf{\Sigma} \boldsymbol{L}_i \leqslant \sum_{1}^{q} \boldsymbol{P}_i' \mathbf{\Sigma} \boldsymbol{P}_i = \lambda_1 + \dots + \lambda_q \qquad \dots \quad (2.4)$$
$$q = 1, \dots p,$$

and the maximum is attained at $\boldsymbol{L}_i = \boldsymbol{P}_i$.

(d) Let $\boldsymbol{B}$ be any matrix of rank q. Then

$$\min_{\boldsymbol{B}} \|\boldsymbol{\Sigma}-\boldsymbol{B}\| = (\lambda_{q+1}^2+\ldots+\lambda_p^2)^{\frac{1}{2}} \qquad \ldots \ (2.5)$$

and the minimum is attained when

$$\boldsymbol{B} = \lambda_1 \boldsymbol{P}_1\boldsymbol{P}_1'+\ldots+\lambda_q\boldsymbol{P}_q\boldsymbol{P}_q'. \qquad \ldots \ (2.6)$$

In (2.5), the symbol $\|\boldsymbol{A}\|$ denotes the Euclidean norm, which is the square root of the sum of squares of the elements of $\boldsymbol{A}$. Or in other words, the choice of $\boldsymbol{B}$ as in (2.6) is the best fitting matrix of given rank q to $\boldsymbol{\Sigma}$.

(ii) *Eigen values and vectors associated with a pair of matrices.* Let $\boldsymbol{\Sigma}$ and $\boldsymbol{\Gamma}$ be two symmetric matrices of order $(p\times p)$ such that $\boldsymbol{\Gamma}$ is positive definite. Consider the determinantal equation

$$|\boldsymbol{\Sigma}-\lambda\boldsymbol{\Gamma}| = 0. \qquad \ldots \ (2.7)$$

The equation (2.7) has p roots, $\lambda_1 \geqslant \lambda_2 \geqslant \ldots \geqslant \lambda_p$ which may be called eigen values of $\boldsymbol{\Sigma}$ with respect to $\boldsymbol{\Gamma}$. Corresponding to each root λ_i there is a vector $\boldsymbol{P}_i$ such that

$$\boldsymbol{\Sigma}\,\boldsymbol{P}_i = \lambda_i\,\boldsymbol{\Gamma}\,\boldsymbol{P}_i,$$

which is called an eigen vector. Then the following results hold:

(a) $\boldsymbol{P}_1, \ldots, \boldsymbol{P}_p$ can be chosen such that

$$\boldsymbol{P}_i'\boldsymbol{\Gamma}\,\boldsymbol{P}_i = 1,\ \boldsymbol{P}_i'\boldsymbol{\Gamma}\,\boldsymbol{P}_j = 0,\ \ i \neq j \qquad \ldots \ (2.8)$$
$$i, j = 1, \ldots, p.$$

(b)
$$\boldsymbol{\Gamma}^{-1}\boldsymbol{\Sigma}\boldsymbol{\Gamma}^{-1} = \lambda_1\boldsymbol{P}_1\boldsymbol{P}_1'+\ldots+\lambda_p\boldsymbol{P}_p\boldsymbol{P}_p'$$
$$\boldsymbol{\Gamma}^{-1} = \boldsymbol{P}_1\boldsymbol{P}_1'+\ldots+\boldsymbol{P}_p\boldsymbol{P}_p'. \qquad \ldots \ (2.9)$$

(c) Let $\boldsymbol{L}_1, \ldots, \boldsymbol{L}_q$ be any set of vectors satisfying the same conditions as $\boldsymbol{P}_i$ in (2.8). Then

$$\sum_{i=1}^{q} \boldsymbol{L}_i'\boldsymbol{\Sigma}\boldsymbol{L}_i \leqslant \sum_{i=1}^{q} \boldsymbol{P}_i'\boldsymbol{\Sigma}\boldsymbol{P}_i = \lambda_1+\ldots+\lambda_q \qquad \ldots \ (2.10)$$
$$q = 1, \ldots, p.$$

(iii) *The Eigen values and vectors under restrictions.* Consider a symmetric $p\times p$ matrix $\boldsymbol{\Sigma}$ and a $p\times k$ matrix $\boldsymbol{C}$ of rank k. Let $\boldsymbol{L}_1, \ldots, \boldsymbol{L}_q$ be p dimensional vectors satisfying the conditions

$$\left.\begin{array}{ll}\text{(a)} & \boldsymbol{L}_i'\boldsymbol{L}_i = 1,\ \boldsymbol{L}_i'\boldsymbol{L}_j = 0 \quad \text{for } i \neq j \\ \text{(b)} & \boldsymbol{L}_i'\boldsymbol{C} = 0, \quad i = 1, \ldots, q.\end{array}\right\} \qquad \ldots \ (2.11)$$

Then the maximum of

$$\boldsymbol{L}_1'\boldsymbol{\Sigma}\,\boldsymbol{L}_1+\ldots+\boldsymbol{L}_q'\boldsymbol{\Sigma}\,\boldsymbol{L}_q \qquad \ldots \ (2.12)$$

is attained when $\boldsymbol{L}_i = \boldsymbol{R}_i$, the i-th eigen vector of the matrix $(\boldsymbol{I}-\boldsymbol{C}(\boldsymbol{C}'\boldsymbol{C})^{-1}\boldsymbol{C}')\boldsymbol{\Sigma}$. The maximum value of (2.12) is then $\nu_1+\ldots+\nu_q$ where ν_i is the i-th eigen value of the matrix $(\boldsymbol{I}-\boldsymbol{C}(\boldsymbol{C}'\boldsymbol{C})^{-1}\boldsymbol{C}')\boldsymbol{\Sigma}$. It is easy to see that if ν_i, $\boldsymbol{R}_i$ are the eigen values and vectors of $(\boldsymbol{I}-\boldsymbol{C}(\boldsymbol{C}'\boldsymbol{C})^{-1}\boldsymbol{C}')\boldsymbol{\Sigma}$ and μ_i, $\boldsymbol{M}_i$ are the eigen values and vectors of the symmetric matrix $\boldsymbol{\Sigma}^{1/2}(\boldsymbol{I}-\boldsymbol{C}(\boldsymbol{C}'\boldsymbol{C})^{-1}\boldsymbol{C})\boldsymbol{\Sigma}^{1/2}$, then $\lambda_i = \mu_i$ and $\boldsymbol{R}_i = \gamma_i\ \boldsymbol{\Sigma}^{-1/2}\boldsymbol{M}_i$, where $\gamma_i = \boldsymbol{M}_i'\ \boldsymbol{\Sigma}^{-1}\boldsymbol{M}_i$. Thus the eigen vectors $\boldsymbol{R}_i$ are obtained from the eigen vectors of a symmetric matrix.

(iv) *Relation between the eigen values and vectors of the matrices* $\boldsymbol{AA}'$ *and* $\boldsymbol{A}'\boldsymbol{A}$. Let $\boldsymbol{A}$ be a matrix of order $n\times p$, and rank r. Then $\boldsymbol{AA}'$ and $\boldsymbol{A}'\boldsymbol{A}$ are both symmetric and have the same non-zero eigen values $\lambda_1, \ldots, \lambda_r$. The multiplicity of the zero root is $n-r$ for $\boldsymbol{AA}'$ and $p-r$ for $\boldsymbol{A}'\boldsymbol{A}$. Further, let $\boldsymbol{P}_i$ be the eigen vector of $\boldsymbol{AA}'$ and $\boldsymbol{Q}_i$ be the eigen vector of $\boldsymbol{A}'\boldsymbol{A}$ corresponding to the same non-zero eigen value λ_i. Then $\boldsymbol{Q}_i = \lambda_i^{-1/2}\ \boldsymbol{A}'\boldsymbol{P}_i$ and $\boldsymbol{P}_i = \lambda_i^{-1/2}\ \boldsymbol{A}\boldsymbol{Q}_i$, so that $\boldsymbol{P}_i$ can be obtained from $\boldsymbol{Q}_i$ and vice versa.

(v) Let $\boldsymbol{A}$ be $n\times p$ matrix of rank r. To determine a matrix $\boldsymbol{B}$ of order $n\times p$ and of rank $k \leqslant r$ such that $\|\boldsymbol{A}-\boldsymbol{B}\|$ is a minimum.

Let $\boldsymbol{B} = \boldsymbol{CD}$ where $\boldsymbol{C}$ is $n\times k$ matrix with orthonormal columns. For given $\boldsymbol{C}$, it is easily shown that $\|\boldsymbol{A}-\boldsymbol{B}\|$ is a minimum when $\boldsymbol{D} = \boldsymbol{C}'\boldsymbol{A}$. With such a choice of $\boldsymbol{D}$

$$\begin{aligned}\|\boldsymbol{A}-\boldsymbol{B}\|^2 &= \|(\boldsymbol{I}-\boldsymbol{CC}')\boldsymbol{A}\|^2\\ &= \text{trace } [(\boldsymbol{I}-\boldsymbol{CC}')\boldsymbol{A}\ \boldsymbol{A}'(\boldsymbol{I}-\boldsymbol{CC}')]\\ &= \text{trace } [\boldsymbol{AA}'(\boldsymbol{I}-\boldsymbol{CC}')]\\ &= \text{trace } \boldsymbol{AA}'-\text{trace } \boldsymbol{AA}'\ \boldsymbol{CC}'\\ &= \text{trace } \boldsymbol{AA}'-\text{trace } \boldsymbol{C}'\boldsymbol{A}\ \boldsymbol{A}'\boldsymbol{C}.\end{aligned}$$

We need choose $\boldsymbol{C}$ such that trace $\boldsymbol{C}'\boldsymbol{A}\ \boldsymbol{A}'\boldsymbol{C}$ is a maximum. Then using the result (2.4), the columns of $\boldsymbol{C}$ are the first k eigen vectors of $\boldsymbol{AA}'$.

(vi) Let $\boldsymbol{L}_1, \ldots, \boldsymbol{L}_q$ be p dimensional vectors satisfying the conditions

(a) $\boldsymbol{L}_i'\ \boldsymbol{\Lambda}\ \boldsymbol{L}_i = 1, \qquad \boldsymbol{L}_i'\ \boldsymbol{\Lambda}\ \boldsymbol{L}_j = 0 \quad \text{for} \quad i \neq j$

(b) $\boldsymbol{L}_i'\ \boldsymbol{C} = 0, \qquad i = 1,\ldots, q$

where $\boldsymbol{\Lambda}$ is a positive definite matrix and $\boldsymbol{C}$ is $p\times k$ matrix of rank k. Then the maximum of

$$\boldsymbol{L}_1'\ \boldsymbol{\Sigma}\ \boldsymbol{L}_1+\ldots+\boldsymbol{L}_q'\ \boldsymbol{\Sigma}\ \boldsymbol{L}_q$$

is attained when $\boldsymbol{L}_i = \boldsymbol{R}_i$, the i-th eigen vector of $\boldsymbol{\Sigma}-\boldsymbol{C}(\boldsymbol{C}'\boldsymbol{\Lambda}^{-1}\ \boldsymbol{C})^{-1}\ \boldsymbol{C}'\boldsymbol{\Lambda}^{-1}\ \boldsymbol{\Sigma}$ with respect to $\boldsymbol{\Lambda}$.

The problem is reduced to that of (iii) by considering the vectors $\boldsymbol{M}_i = \boldsymbol{G}'\boldsymbol{L}_i$ where $\boldsymbol{GG}' = \boldsymbol{\Lambda}$, $i = 1, \ldots, q$. In terms of $\boldsymbol{M}_i$, the problem is the same as in (iii).

3. Fitting a subspace to a set of points in a higher dimensional space

Principal component analysis involving the application of eigen values and vectors of a matrix was first encountered by Karl Pearson (1901) and Frisch (1929) in the problem of fitting a line, a plane or in general a subspace to a scatter of points in a higher dimensional space. Let

$$\mathscr{X} = \begin{pmatrix} X_{11} \dots X_{1n} \\ \cdot \ \cdot \ \cdot \ \cdot \ \cdot \\ X_{p1} \dots X_{pn} \end{pmatrix} \qquad \dots \quad (3.1)$$

represent n points (each column vector representing a point) in a p dimensional space. Any typical point is represented by $\boldsymbol{X}$. The i-th point is $\boldsymbol{X}_i$ and the centre of gravity of the points is $\overline{\boldsymbol{X}}$, with the s-th coordinate

$$\overline{\boldsymbol{X}}_s = \sum_{j=1}^{n} X_{sj} \div n. \qquad \dots \quad (3.2)$$

The points $\boldsymbol{X}_i$ as measured from the centre of gravity are

$$\mathscr{X}_d = \begin{pmatrix} X_{11}-\overline{X}_1 \dots X_{1n}-\overline{X}_1 \\ \cdot \quad \dots \quad \cdot \\ X_{p1}-\overline{X}_p \dots X_{pn}-\overline{X}_p \end{pmatrix}. \qquad \dots \quad (3.3)$$

Let us define $$\boldsymbol{\Sigma} = \mathscr{X}_d \mathscr{X}_d' = (\sigma_{ij}) \qquad \dots \quad (3.4)$$

as the total dispersion matrix (or scatter) of the points. In (3.4) the element σ_{ij} is computed by the formula

$$\sigma_{ij} = \sum_{r=1}^{n} (X_{ir} - \overline{X}_i)(X_{jr} - \overline{X}_j) = \sum_{r=1}^{n} X_{ir} X_{jr} - n \overline{X}_i \overline{X}_j.$$

Now a q dimensional subspace is specified by a point (origin) and q orthogonal axes (each axis specified by its direction cosines) passing through it. Pearson defined a best fitting subspace as that for which the sum of squares of the perpendiculars from the points to the subspace is a minimum. It is easily shown that such a subspace passes through the centre of gravity of the points. Further, the sum of squares of the perpendiculars from the points to the subspace defined by the centre of gravity and q orthogonal vectors $\boldsymbol{L}_1, \dots, \boldsymbol{L}_q$ is

$$\sum_{i=1}^{p} \sigma_{ii} - \sum_{i=1}^{q} \boldsymbol{L}_i' \boldsymbol{\Sigma} \boldsymbol{L}_i. \qquad \dots \quad (3.5)$$

Minimizing (3.5) is the same as maximizing $\Sigma\ \boldsymbol{L}_i' \boldsymbol{\Sigma} \boldsymbol{L}_i$ and an application of the result (2.4) shows that the maximum is attained when

$$\boldsymbol{L}_i = \boldsymbol{P}_i, \quad i=1, \dots, q \qquad \dots \quad (3.6)$$

where $\boldsymbol{P}_1, \dots, \boldsymbol{P}_q$ are the first q eigen vectors of the matrix $\boldsymbol{\Sigma}$.

What we need in practice is an actual representation of points in the best fitting lower dimensional subspace. This is simply done by computing for each $\boldsymbol{X}$ the q coordinates

$$\boldsymbol{P}_1' \boldsymbol{X}, \dots, \boldsymbol{P}_q' \boldsymbol{X}. \qquad \dots \quad (3.7)$$

Thus the points in the projected space referred to q orthogonal axes are

$$\mathscr{Y} = \begin{pmatrix} \boldsymbol{P}_1'\boldsymbol{X}_1 \dots \boldsymbol{P}_1'\boldsymbol{X}_n \\ \cdot \quad \dots \quad \cdot \\ \boldsymbol{P}_q'\boldsymbol{X}_1 \dots \boldsymbol{P}_q'\boldsymbol{X}_n \end{pmatrix}. \qquad \dots \quad (3.8)$$

The first row of $\mathscr{Y}$ gives the best one dimensional representation, the first two rows of $\mathscr{Y}$ give the best two dimensional representation, and so on.

In practice, it is also necessary to know how good the representation of points in a lower dimensional space is. A suitable criterion for this purpose is the sum of squares of the distances between the original and the projected points. For the best q space, the criterion has the value

$$\lambda_{q+1}+\dots+\lambda_p \qquad \dots \quad (3.9)$$

using the result (2.4). Instead of the absolute value (3.9), we may choose as a measure of goodness of fit the ratio

$$\frac{\lambda_{q+1}+\dots+\lambda_p}{\lambda_1+\dots+\lambda_p}. \qquad \dots \quad (3.10)$$

The choice of q, then depends on the smallness of (3.10) or the largeness of the ratio

$$\frac{\lambda_1+\dots+\lambda_q}{\lambda_1+\dots+\lambda_p}. \qquad \dots \quad (3.11)$$

It may happen that the overall measure (3.11) has a small value but the configuration of the points as a whole is distorted in the projected space due to some isolated points being far away from the best fitting space. The examination of such isolated points is of interest in practical work and may provide an interpretation of the nature of heterogeneity in the data. For this purpose it is necessary to compute the length of the perpendicular of each point on the best fitting space. The square of the perpendicular from $\boldsymbol{X}_i$, the i-th point is

$$d_i^2 = (\boldsymbol{X}_i-\bar{\boldsymbol{X}})'(\boldsymbol{X}_i-\bar{\boldsymbol{X}})-[\boldsymbol{P}_1'(\boldsymbol{X}_i-\bar{\boldsymbol{X}})]^2-\dots-[\boldsymbol{P}_q'(\boldsymbol{X}_i-\bar{\boldsymbol{X}})]^2$$
$$i = 1, \dots, n.$$

An examination of the values $d_1^2, \dots, d_n^2$ will enable us to find out the outliers if any.

4. Other interpretations of the transformation (3.8)

The problem of representing the points (3.1) in a lower dimensional space may be posed as the determination of a transformation matrix $\boldsymbol{T}$ of order $p \times q$, and rank q transforming a p-vector $\boldsymbol{X}$ into a q-vector $\boldsymbol{Y}, \boldsymbol{Y} = \boldsymbol{T}'\boldsymbol{X}$, and satisfying some optimum properties. The transformation from $\boldsymbol{Y}$ to $\boldsymbol{X}$ is not one to one when $q < p$ and consequently certain properties of the configuration of the points in the original space (which is invariant when $q = p$) are altered. Our aim is to choose $\boldsymbol{T}$ which preserves the configuration of the points to the maximum possible extent. We shall show that some intuitive measures of closeness between the configurations in the original space and the transformed subspace, when minimized with respect to the transformation matrix $\boldsymbol{T}$ lead to the transformation (3.8) as the optimum.

Let us observe that the columns of $\boldsymbol{T}$ can be chosen to be orthonormal, without loss of generality.

Maximizing the sum of squares of the distances. The sum of squares of the distances between all possible pairs of points in the p-space is

$$\sum_{i,j=1}^{n}\sum (\boldsymbol{X}_i-\boldsymbol{X}_j)'(\boldsymbol{X}_i-\boldsymbol{X}_j) = n \text{ trace } \boldsymbol{\Sigma}, \quad \ldots \quad (4.1)$$

while the corresponding expresion in the q-space is

$$\sum_{i,j=1}^{n}\sum (\boldsymbol{Y}_i-\boldsymbol{Y}_j)'(\boldsymbol{Y}_i-\boldsymbol{Y}_j) = n \text{ trace } \boldsymbol{T}' \boldsymbol{\Sigma} \boldsymbol{T}$$

$$= n(\boldsymbol{T}_1' \boldsymbol{\Sigma} \boldsymbol{T}_1+\ldots+\boldsymbol{T}_q' \boldsymbol{\Sigma} \boldsymbol{T}_q) \quad \ldots \quad (4.2)$$

where $\boldsymbol{T}_1, \ldots, \boldsymbol{T}_q$ are the columns of $\boldsymbol{T}$. It is seen that (4.2) $\leqslant$ (4.1) and (4.2) = (4.1) when (not necessarily only when) $p = q$. Let us choose $\boldsymbol{T}$ such that (4.2) is a maximum. Applying the result (2.4), we find that the maximum is attained when $\boldsymbol{T}_i^* = \boldsymbol{P}_i$, $i = 1, \ldots, q$, which is the solution obtained in Section 3.

Closest fit to the distances and the angles of the lines joining the points to the centre of gravity. In the p-space the vectors joining the points to their centre of gravity are represented by the matrix $\mathscr{X}_d$ as defined in (3.3). It may be seen that in the matrix

$$\boldsymbol{\Lambda} = \mathscr{X}_d' \mathscr{X}_d \quad \ldots \quad (4.3)$$

the i-th diagonal entry is the square of the distance of the i-th point from the centre of gravity and the (i, j) entry divided by the square root of the i-th and j-th entries is the cosine of the angle between the lines joining the i-th and j-th points to the centre of gravity. Thus the matrix $\boldsymbol{\Lambda}$ represents the distances and the angles between the lines joining the points to their centre of gravity. In the q-space, under the transformation $\boldsymbol{Y} = \boldsymbol{T}'\boldsymbol{X}$ the matrix corresponding to (4.3) is

$$\boldsymbol{B} = \mathscr{Y}_d' \mathscr{Y}_d = \mathscr{X}_d' \boldsymbol{T} \boldsymbol{T}' \mathscr{X}_d. \quad \ldots \quad (4.4)$$

We wish to determine $\boldsymbol{T}$ such that the matrix $\boldsymbol{B}$ is as close as possible to $\boldsymbol{\Lambda}$. We may measure the closeness of $\boldsymbol{B}$ and $\boldsymbol{\Lambda}$ by the Euclidean norm $||\boldsymbol{\Lambda}-\boldsymbol{B}||$, which is the square root of the sum of squares of the elements in $(\boldsymbol{\Lambda}-\boldsymbol{B})$. Then the problem is one choosing $\boldsymbol{B}$ such that $||\boldsymbol{\Lambda}-\boldsymbol{B}||$ is a minimum. An application of the result (2.6) gives the optimum choice of $\boldsymbol{B}$ as

$$\mathscr{X}_d'\boldsymbol{T} \boldsymbol{T}'\mathscr{X}_d = \lambda_1\boldsymbol{Q}_1\boldsymbol{Q}_1'+\ldots+\lambda_q \boldsymbol{Q}_q \boldsymbol{Q}_q' \quad \ldots \quad (4.5)$$

where $\boldsymbol{Q}_1, \ldots, \boldsymbol{Q}_q$ are the eigen vectors of $\boldsymbol{\Lambda} = \mathscr{X}_d'\mathscr{X}_d$. The optimum choice of $\boldsymbol{T}$ is obtained from (4.5) as

$$\mathscr{X}_d \boldsymbol{T}^* = (\sqrt{\lambda_1}\boldsymbol{Q}_1 \ldots \sqrt{\lambda_q} \boldsymbol{Q}_q)$$

$$= (\boldsymbol{X}_d' \boldsymbol{P}_1 \ldots \boldsymbol{X}_d' \boldsymbol{P}_q) \quad \ldots \quad (4.6)$$

using the result (iv) of Section 2. The choice

$$\boldsymbol{T}^* = (\boldsymbol{P}_1 \ldots \boldsymbol{P}_q) \quad \ldots \quad (4.7)$$

satisfies the equation (4.6). Thus $\boldsymbol{T}_i^* = \boldsymbol{P}_i$, where $\boldsymbol{T}_i^*$ is the i-th column of $\boldsymbol{T}^*$, which is the solution obtained in Section 3. This means that the representation of the points in a q-dimensional space as in (3.8) preserves the distances and angles of the configuration to a large extent.

5. Reduction in the number of points

In Sections 3 and 4, a reduction in the number of dimensions of the space is secured through a transformation of the type $\boldsymbol{Y} = \boldsymbol{T}'\boldsymbol{X}$ by choosing an optimum $\boldsymbol{T}$. We now consider the dual problem of reducing the number of points to a few *typical points* keeping the number of dimensions the same. For this purpose we consider the matrix $\mathscr{X}_d$ which represents the points with the origin at the centre of gravity. The reduction in the number of points is secured by a transformation of the type $\mathscr{Y}_d = \mathscr{X}_d \boldsymbol{M}$ where $\boldsymbol{M}$ is $n \times q$ matrix.

Reduction providing the closest fit to the total dispersion matrix. The total dispersion matrix based on the reduced points is

$$\mathscr{Y}_d \mathscr{Y}_d' = \mathscr{X}_d \boldsymbol{M}\boldsymbol{M}' \mathscr{X}_d' \quad \ldots \quad (5.1)$$

while that based on all points is $\boldsymbol{\Sigma}$. We wish to choose $\mathscr{Y}_d$ in such a way that

$$\| \boldsymbol{\Sigma} - \mathscr{Y}_d \mathscr{Y}_d' \| \quad \ldots \quad (5.2)$$

is a minimum. Applying the result (2.6), the minimum of (5.2) is attained when

$$\mathscr{Y}_d \mathscr{Y}_d' = \lambda_1 \boldsymbol{P}_1 \boldsymbol{P}_1' + \ldots + \lambda_q \boldsymbol{P}_q \boldsymbol{P}_q'$$

or when

$$\mathscr{Y}_d = (\sqrt{\lambda_1}\boldsymbol{P}_1 \ldots \sqrt{\lambda_q}\boldsymbol{P}_q). \quad \ldots \quad (5.3)$$

Thus the typical points summarizing the deviations from the centre of gravity of the original points are the eigen vectors scaled by the square roots of the corresponding eigen values.

To determine the transformation matrix (if necessary) let us observe that, using result (iv) of Section 2,

$$\begin{aligned} \mathscr{Y}_d &= (\sqrt{\lambda_1}\boldsymbol{P}_1 \ldots \sqrt{\lambda_q}\boldsymbol{P}_q) \\ &= (\mathscr{X}_d \boldsymbol{Q}_1 \ldots \mathscr{X}_d \boldsymbol{Q}_q), \end{aligned} \quad \ldots \quad (5.4)$$

where $\boldsymbol{Q}_1, \ldots, \boldsymbol{Q}_q$ are the first q eigen vectors of the matrix $\mathscr{X}_d' \mathscr{X}_d$. But (5.4) is the same as $\mathscr{X}_d \boldsymbol{M}^*$ where $\boldsymbol{M}^*$ is the matrix with $\boldsymbol{Q}_1, \ldots, \boldsymbol{Q}_q$ as its columns.

An interpretation of the typical points. In Section 3, it is shown that the best fitting q-space is specified by $\bar{\boldsymbol{X}}$, the centre of gravity and the eigen vectors $\boldsymbol{P}_1, \ldots, \boldsymbol{P}_q$. Any point in this space referred to the p *original axes* can be represented by a vector of the form

$$\bar{\boldsymbol{X}} + a_1 \boldsymbol{P}_1 + \ldots + a_q \boldsymbol{P}_q \quad \ldots \quad (5.5)$$

where $a_1, \ldots, a_q$ are arbitrary. By a change in scale, (5.5) can be written as

$$\bar{\boldsymbol{X}} + b_1 \sqrt{\lambda_1}\boldsymbol{P}_1 + \ldots + b_q \sqrt{\lambda_q}\boldsymbol{P}_q \quad \ldots \quad (5.6)$$

involving the typical points $\sqrt{\lambda_1}\boldsymbol{P}_1, \ldots, \sqrt{\lambda_q}\boldsymbol{P}_q$. Let $\boldsymbol{X}_i^{(q)}$ be the projection of $\boldsymbol{X}_i$ on this subspace. Then there exist constants $b_{1i}, \ldots, b_{qi}$ such that

$$\boldsymbol{X}_i^{(q)} = \bar{\boldsymbol{X}} + b_{1i} \sqrt{\lambda_1}\boldsymbol{P}_1 + \ldots + b_{qi} \sqrt{\lambda_q}\boldsymbol{P}_q. \quad \ldots \quad (5.7)$$

Multiplying both sides of (5.7) by $\boldsymbol{P}_j'$ and observing that $\boldsymbol{P}_j'\boldsymbol{X}_i^{(g)} = \boldsymbol{P}_j'\boldsymbol{X}_i$ we obtain the value of b_{ji} as

$$b_{ji} = \lambda^{-1/2}\boldsymbol{P}_j'(\boldsymbol{X}_i-\overline{\boldsymbol{X}}). \qquad \dots \quad (5.8)$$

With such a choice of b_{ji}

$$\boldsymbol{X}_i^{(g)} = \overline{\boldsymbol{X}}+b_{1i}\sqrt{\lambda_1}\boldsymbol{P}_1+\dots+b_{qi}\sqrt{\lambda_q}\boldsymbol{P}_q \qquad \dots \quad (5.9)$$

$$i = 1, \dots, n.$$

The point $\boldsymbol{X}_i^{(g)}$ so determined may be called the graduated or the fitted point of $\boldsymbol{X}_i$. If $\boldsymbol{X}_i-\boldsymbol{X}_i^{(g)}$ is small for each i, the result (5.9) shows that the average point and the typical points

$$\overline{\boldsymbol{X}},\ \sqrt{\lambda_1}\,\boldsymbol{P}_1, \dots, \sqrt{\lambda_q}\boldsymbol{P}_q$$

form an *approximate basis* of the n observed points

$$\boldsymbol{X}_i \sim \overline{\boldsymbol{X}}+b_{1i}\sqrt{\lambda_1}\boldsymbol{P}_1+\dots+b_{qi}\sqrt{\lambda_q}\boldsymbol{P}_q \qquad \dots \quad (5.10)$$

$$i = 1, \dots, n.$$

The representation (5.10) admits a statistical interpretation under the following model for the observed points $\boldsymbol{X}_i$:

$$\boldsymbol{X}_i = \boldsymbol{\mu}+\beta_{1i}\boldsymbol{\pi}_1+\dots+\beta_{qi}\boldsymbol{\pi}_q+\boldsymbol{\epsilon}_i \qquad \dots \quad (5.11)$$

where $\boldsymbol{\mu}, \boldsymbol{\pi}_1, \dots, \boldsymbol{\pi}_q$ are fixed vectors, $\beta_{1i}, \dots, \beta_{qi}$ are constants specific to the i-th point and $\boldsymbol{\epsilon}_i$ is a vector of random errors. The model (5.11) implies that any observed point when corrected for errors of measurement can be expressed as linear combination of $(q+1)$ basic vectors only. It can be shown that, when the dispersion matrix of $\boldsymbol{\epsilon}_i$ is of the form $\sigma^2\boldsymbol{I}$, consistent estimators of $\boldsymbol{\mu}$, $\boldsymbol{\pi}_i$ and β_{ji} are provided by $\overline{\boldsymbol{X}}$, $\sqrt{\lambda_i}\,\boldsymbol{P}_i$ and b_{ji} as defined in (5.10). Under the additional condition that the p components of $\boldsymbol{\epsilon}_i$ have independent normal distributions, $\boldsymbol{X}$, $\sqrt{\lambda_i}\,\boldsymbol{P}_i$ and b_{ji} can be shown to be maximum likelihood estimators. Such a statistical interpretation is not true when the dispersion matrix of $\boldsymbol{\epsilon}_i$ is not of the form $\sigma^2\boldsymbol{I}$.

The typical points $\sqrt{\lambda_1}\,\boldsymbol{P}_1, \dots, \sqrt{\lambda_q}\,\boldsymbol{P}_q$ used in the representation (5.10) may not admit any physical interpretation. But in special situations it may be possible to characterize each typical point as indicating some aspect of the measurements as a whole (see Simonds, 1964).

6. FITTING A SUBSPACE WHEN THE POINTS ARE IN AN OBLIQUE SPACE

In the analysis of Section 3, it is implicitly assumed that the original points are represented in a p-dimensional Euclidean space with orthogonal axes. There are, however, situations where oblique axes are chosen to represent the coordinates. In such a case, the distance between any two points $\boldsymbol{X}_i$, $\boldsymbol{X}_j$ is a quadratic form

$$(\boldsymbol{X}_i-\boldsymbol{X}_j)'\boldsymbol{\Gamma}^{-1}(\boldsymbol{X}_i-\boldsymbol{X}_j) \quad \dots \quad (6.1)$$

using a positive definite matrix $\boldsymbol{\Gamma}$. The scatter of points analogous to (3.10) is

$$\sum_{i,j=1}^{n}\sum (\boldsymbol{X}_i-\boldsymbol{X}_j)'\boldsymbol{\Gamma}^{-1}(\boldsymbol{X}_i-\boldsymbol{X}_j). \quad \dots \quad (6.2)$$

Let us transform $\boldsymbol{X}$ to $\boldsymbol{Y}$ with q coordinates

$$\boldsymbol{Y}' = (\boldsymbol{L}_1'\boldsymbol{X}, \dots, \boldsymbol{L}_q'\boldsymbol{X})$$

where $\boldsymbol{L}_i'\boldsymbol{\Gamma}\boldsymbol{L}_i = 1$ and $\boldsymbol{L}_i'\boldsymbol{\Gamma}\boldsymbol{L}_j = 0$, $i \neq j$. The scatter of points in the q-space with $\boldsymbol{L}_1, \dots, \boldsymbol{L}_q$ as orthogonal axes is

$$\Sigma\Sigma(\boldsymbol{Y}_i-\boldsymbol{Y}_j)'(\boldsymbol{Y}_i-\boldsymbol{Y}_j) = n(\boldsymbol{L}_1'\boldsymbol{\Sigma}\boldsymbol{L}_1+\dots+\boldsymbol{L}_q'\boldsymbol{\Sigma}\boldsymbol{L}_q). \quad \dots \quad (6.3)$$

An application of the result (2.10) shows that (6.3) is a maximum when $\boldsymbol{L}_i = \boldsymbol{P}_i$, $i = 1, \dots, q$, where $\boldsymbol{P}_1, \dots, \boldsymbol{P}_q$ are the first q eigen vectors associated with the determinantal equation $|\boldsymbol{\Sigma}-\boldsymbol{\Gamma}| = 0$. The adequacy of a q dimensional fit is judged by the largeness of the ratio

$$\frac{\lambda_1+\dots+\lambda_q}{\lambda_1+\dots+\lambda_p} \quad \dots \quad (6.4)$$

where $\lambda_1, \dots, \lambda_p$ are the roots of $|\boldsymbol{\Sigma}-\lambda\boldsymbol{\Gamma}| = 0$.

7. PRINCIPAL COMPONENTS OF A VECTOR RANDOM VARIABLE

The concept of principal component analysis as applied to a random variable is due to Hotelling (1933, 1935, 1936a).

Let us consider a p dimensional random variable $\boldsymbol{X}$ with mean $\boldsymbol{\mu}$ and dispersion matrix $\boldsymbol{\Sigma}$. Consider a transformation, $\boldsymbol{Y} = \boldsymbol{T}'\boldsymbol{X}$, where $\boldsymbol{T}$ is a $p\times q$ matrix, so that $\boldsymbol{Y}$ is q dimensional. When $q < p$, there is loss of information in replacing $\boldsymbol{X}$ by $\boldsymbol{Y}$. For a given q, we wish to determine $\boldsymbol{T}$ such that there is minimum loss.

In defining a loss function we may be guided by the question as to what extent we can predict $\boldsymbol{X}$ knowing $\boldsymbol{Y}$. The pedictive efficiency of $\boldsymbol{Y}$ for $\boldsymbol{X}$ depends on the residual dispersion matrix of $\boldsymbol{X}$ after subtracting its best linear predictor in terms of $\boldsymbol{Y}$. Now the joint dispersion matrix of $\boldsymbol{X}$ and $\boldsymbol{Y}$ is

$$\begin{pmatrix} \boldsymbol{\Sigma} & \boldsymbol{\Sigma T} \\ \boldsymbol{T}'\boldsymbol{\Sigma} & \boldsymbol{T}'\boldsymbol{\Sigma T} \end{pmatrix} \quad \dots \quad (7.1)$$

and the residual dispersion matrix is

$$\boldsymbol{\Sigma}-\boldsymbol{\Sigma T}(\boldsymbol{T}'\boldsymbol{\Sigma T})^{-1}\boldsymbol{T}'\boldsymbol{\Sigma}. \quad \dots \quad (7.2)$$

The smaller the values of the elements in (7.2), the greater is the predictive efficiency.

We consider two overall measures, the trace and the Euclidean norms of (7.2),

$$\text{(a)} \quad \text{trace } (\boldsymbol{\Sigma}-\boldsymbol{\Sigma}\, \boldsymbol{T}(\boldsymbol{T}'\, \boldsymbol{\Sigma}\, \boldsymbol{T})^{-1}\boldsymbol{T}'\boldsymbol{\Sigma}) \qquad \ldots \quad (7.3)$$

$$\text{(b)} \quad \|\boldsymbol{\Sigma}-\boldsymbol{\Sigma}\, \boldsymbol{T}(\boldsymbol{T}'\, \boldsymbol{\Sigma}\boldsymbol{T})^{-1}\boldsymbol{T}'\boldsymbol{\Sigma}\|. \qquad \ldots \quad (7.4)$$

We shall show that any one of these measures when minimized leads to the same choice of $\boldsymbol{T}$ as the matrix of the first q eigen vectors of $\boldsymbol{\Sigma}$.

Let us observe that the column vectors $\boldsymbol{T}_1, \ldots, \boldsymbol{T}_q$ of $\boldsymbol{T}$ can be chosen, without loss of generality, to satisfy the conditions

$$\boldsymbol{T}_i'\, \boldsymbol{\Sigma}\, \boldsymbol{T}_j = 0, \quad i \neq j \qquad \ldots \quad (7.5)$$

which imply that the components of the transformed variable are uncorrelated.

Consider the measure (7.3)

$$\begin{aligned} &\text{trace } (\boldsymbol{\Sigma}-\boldsymbol{\Sigma}\, \boldsymbol{T}(\boldsymbol{T}'\, \boldsymbol{\Sigma}\, \boldsymbol{T})^{-1}\boldsymbol{T}'\boldsymbol{\Sigma}) \\ &= \text{trace } \boldsymbol{\Sigma} - \text{trace } (\boldsymbol{T}'\, \boldsymbol{\Sigma}\, \boldsymbol{T})^{-1}\boldsymbol{T}'\boldsymbol{\Sigma}\, \boldsymbol{\Sigma}\, \boldsymbol{T}) \\ &= \text{trace } \boldsymbol{\Sigma} - \left(\frac{\boldsymbol{T}_1'\boldsymbol{\Sigma}\, \boldsymbol{\Sigma}\, \boldsymbol{T}_1}{\boldsymbol{T}_1'\boldsymbol{\Sigma}\, \boldsymbol{T}_1} + \ldots + \frac{\boldsymbol{T}_q'\boldsymbol{\Sigma}\, \boldsymbol{\Sigma}\, \boldsymbol{T}_q}{\boldsymbol{T}_q'\, \boldsymbol{\Sigma}\, \boldsymbol{T}_q}\right) \end{aligned} \qquad \ldots \quad (7.6)$$

using the conditions (7.5). Minimizing (7.6) is the same as maximizing

$$\frac{\boldsymbol{T}_1'\boldsymbol{\Sigma}\, \boldsymbol{\Sigma}\, \boldsymbol{T}_1}{\boldsymbol{T}_1'\boldsymbol{\Sigma}\, \boldsymbol{T}_1} + \ldots + \frac{\boldsymbol{T}_q'\boldsymbol{\Sigma}\, \boldsymbol{\Sigma}\, \boldsymbol{T}_q}{\boldsymbol{T}_q'\boldsymbol{\Sigma}\, \boldsymbol{T}_q}.$$

Applying the result (2.10) the optimum choice of $\boldsymbol{T}_i$ is the i-th eigen vector $\boldsymbol{\Sigma}\boldsymbol{\Sigma}$ with respect to $\boldsymbol{\Sigma}$, which is the same as $\boldsymbol{P}_i$ the i-th eigen vector of $\boldsymbol{\Sigma}$. The minimum value of (7.6) is then

$$\lambda_{q+1} + \ldots + \lambda_p \qquad \ldots \quad (7.7)$$

the sum of the smallest $p-q$ eigen values of $\boldsymbol{\Sigma}$.

Consider the problem of minimizing

$$\|\boldsymbol{\Sigma}-\boldsymbol{\Sigma}\boldsymbol{T}(\boldsymbol{T}'\, \boldsymbol{\Sigma}\, \boldsymbol{T})^{-1}\boldsymbol{T}'\, \boldsymbol{\Sigma}\|. \qquad \ldots \quad (7.8)$$

The result (2.6) shows that the minimum of (7.8) is attained when

$$\boldsymbol{\Sigma}\boldsymbol{T}(\boldsymbol{T}'\, \boldsymbol{\Sigma}\, \boldsymbol{T})^{-1}\boldsymbol{T}\, \boldsymbol{\Sigma} = \lambda_1\boldsymbol{P}_1\boldsymbol{P}_1' + \ldots + \lambda_q\boldsymbol{P}_q\boldsymbol{P}_q' \qquad \ldots \quad (7.9)$$

and it is easy to verify that (7.9) holds when $\boldsymbol{T}_i^* = \boldsymbol{P}_i$. The minimum value of (7.8) is

$$\lambda_{q+1}^2 \cdots + \lambda_p^2 \qquad \ldots \quad (7.10)$$

where $\lambda_{q+1}, \ldots, \lambda_p$ are as defined in (7.7).

The transformed variables $\boldsymbol{P}'_1\boldsymbol{X}, \ldots, \boldsymbol{P}_q'\boldsymbol{X}$ are called the first q principal components of the random variable $\boldsymbol{X}$ and are interpreted variously. The usual interpretation is as follows. Under a complete orthogonal transformation $\boldsymbol{Y} = \boldsymbol{0}'\boldsymbol{X}$, i.e., from a p dimensional variable to another p dimensional variable, the trace of the dispersion matrix, which is the sum of the variances of the variables, remains invariant,

Thus

$$\text{trace } \mathbf{\Sigma} = \text{trace } \mathbf{0}'\mathbf{\Sigma}\,\mathbf{0} = \mathbf{0}_1'\mathbf{\Sigma}\,\mathbf{0}_1+\ldots+\mathbf{0}_p'\,\mathbf{\Sigma}\,\mathbf{0}_p \qquad \ldots \quad (7.11)$$

where $\mathbf{0}'\,\mathbf{\Sigma}\,\mathbf{0}$ is the dispersion matrix of $\boldsymbol{Y}$ and $\mathbf{0}_i$ is the i-th column vector of $\mathbf{0}$. The total of the variances of the first q variables in $\boldsymbol{Y}$ is

$$\mathbf{0}_1'\,\mathbf{\Sigma}\,\mathbf{0}_1+\ldots+\mathbf{0}_q'\,\mathbf{\Sigma}\,\mathbf{0}_q \leqslant \text{trace } \mathbf{\Sigma}. \qquad \ldots \quad (7.12)$$

It is seen from (2.6) that the optimum choice of $\mathbf{0}_i$ is $\boldsymbol{P}_i$ for (7.12) to be a maximum, i.e., the first q transformed variables are $\boldsymbol{P}_1'\boldsymbol{X}, \ldots, \boldsymbol{P}_q'\boldsymbol{X}$, the q principal components of $\boldsymbol{X}$.

The maximum value of (7.12) is $\lambda_1+\ldots+\lambda_q$ while the total of the variances of the variables is $\lambda_1+\ldots+\lambda_p$, in which case the first q principal components are said to explain $100(\lambda_1+\ldots+\lambda_q)/(\lambda_1+\ldots+\lambda_p)$ percent of the total variance.

The interpretation of the principal components as the best predictors of $\boldsymbol{X}$ suggests extensions of the principal component analysis in many directions, which are considered in the later sections.

8. Principal components of instrumental variables

Let $\boldsymbol{X}$ be the vector of p main variables, and $\boldsymbol{Z}$ the vector of m instrumental variables. In theory $\boldsymbol{Z}$ may include some or all the elements of $\boldsymbol{X}$. Denote the joint dispersion matrix of $(\boldsymbol{X}, \boldsymbol{Z})$ by

$$\begin{pmatrix} \mathbf{\Sigma} & \mathbf{\Theta} \\ \mathbf{\Theta}' & \mathbf{\Gamma} \end{pmatrix}. \qquad \ldots \quad (8.1)$$

We wish to replace $\boldsymbol{Z}$ by a q dimensional random variable $\boldsymbol{Y} = \boldsymbol{M}'\boldsymbol{Z}$ in such a way that the predictive efficiency of $\boldsymbol{Y}$ for $\boldsymbol{X}$ is a maximum, (Rao, 1962a). The dispersion matrix of $(\boldsymbol{X}, \boldsymbol{Y})$ is

$$\begin{pmatrix} \mathbf{\Sigma} & \mathbf{\Theta}\,\boldsymbol{M} \\ \boldsymbol{M}'\mathbf{\Theta}' & \boldsymbol{M}'\mathbf{\Gamma}\boldsymbol{M} \end{pmatrix} \qquad \ldots \quad (8.2)$$

and the residual dispersion matrix of $\boldsymbol{X}$ subtracting its best linear predictor in terms of $\boldsymbol{Y}$ is

$$\mathbf{\Sigma}-\mathbf{\Theta}\boldsymbol{M}(\boldsymbol{M}'\mathbf{\Gamma}\,\boldsymbol{M})^{-1}\boldsymbol{M}'\mathbf{\Theta}. \qquad \ldots \quad (8.3)$$

As in Section 7, we may consider the two measures of predictive efficiency of $\boldsymbol{Y}$

$$\text{(a)} \quad \text{trace } (\mathbf{\Sigma}-\mathbf{\Theta}\,\boldsymbol{M}\,(\boldsymbol{M}'\,\mathbf{\Gamma}\,\boldsymbol{M})^{-1}\,\boldsymbol{M}'\mathbf{\Theta}') \qquad \ldots \quad (8.4)$$

$$\text{(b)} \quad \|\mathbf{\Sigma}-\mathbf{\Theta}\,\boldsymbol{M}(\boldsymbol{M}'\,\mathbf{\Gamma}\,\boldsymbol{M})^{-1}\boldsymbol{M}'\mathbf{\Theta}'\|. \qquad \ldots \quad (8.5)$$

Unfortunately, the solution seems to depend on which measure is chosen for minimization.

Minimizing (8.4) is the same as maximizing

$$\begin{aligned} &\text{trace } \mathbf{\Theta}\,\boldsymbol{M}(\boldsymbol{M}'\mathbf{\Gamma}\,\boldsymbol{M})^{-1}\boldsymbol{M}'\mathbf{\Theta}' \\ &= \text{trace } (\boldsymbol{M}'\,\mathbf{\Gamma}\,\boldsymbol{M})^{-1}\boldsymbol{M}'\mathbf{\Theta}'\,\mathbf{\Theta}\,\boldsymbol{M} \\ &= \frac{\boldsymbol{M}_1'\mathbf{\Theta}'\,\mathbf{\Theta}\boldsymbol{M}_1}{\boldsymbol{M}_1'\,\mathbf{\Gamma}\,\boldsymbol{M}_1}+\ldots+\frac{\boldsymbol{M}_q'\,\mathbf{\Theta}'\mathbf{\Theta}\,\boldsymbol{M}_q}{\boldsymbol{M}_q'\,\mathbf{\Gamma}\,\boldsymbol{M}_q} \end{aligned} \qquad \ldots \quad (8.6)$$

assuming that $\boldsymbol{M}_i\,\mathbf{\Gamma}\,\boldsymbol{M}_j = 0, i \neq j$, without loss of generality. Applying the result (2.10), the best choice of $\boldsymbol{M}_1, \ldots, \boldsymbol{M}_q$ is the set of the first q eigen vectors of the matrix $\mathbf{\Theta}'\,\mathbf{\Theta}$ with respect to Γ, i.e. associated with the determinantal equation

$$|\mathbf{\Theta}'\mathbf{\Theta}-\lambda\mathbf{\Gamma}| = 0. \qquad \ldots \quad (8.7)$$

It may be obsrved that when $\boldsymbol{X}$ is a proper subset of $\boldsymbol{Z}$, the optimum choice of $\boldsymbol{Y}$ consists of the first q principal components of $\boldsymbol{X}$ alone.

The present analysis is different from that of canonical correlations and canonical variates as developed by Hotelling (1936b) for studying the association between two vector random variables.

There seems to be no simple method of minimizing the measure (8.5). The solution would be, however, different from that obtained by minimizing (8.4). The relative merits of the two solutions are worth exploring.

9. Weighted principal components

The measures of predictive efficiency (7.3), (7.4), (8.4) and (8.5) are not invariant under change of scales of the elements of $\boldsymbol{X}$. This is an undesirable feature as different solutions to principal components can be found by choosing different scales. The difficulty can be overcome by defining a loss function associated with the prediction of each element of $\boldsymbol{X}$ by any chosen set of q linear functions $\boldsymbol{Y}$ of the instrumental variable $\boldsymbol{Z}$. A simple loss function for the prediction of $\boldsymbol{X}_i$, the i-th element of $\boldsymbol{X}$, may be chosen as w_i^2 times the residual variance, where w_i^2, $i = 1, \ldots, p$, are assigned quantities. Denoting by $\boldsymbol{W}$, the diagonal matrix of the elements $w_1, \ldots, w_p$ the total loss in the prediction of $\boldsymbol{X}$ is the weighted trace

$$\text{trace}\, \boldsymbol{W\Sigma W} - \text{trace}\ \boldsymbol{W \Theta M}(\boldsymbol{M}' \boldsymbol{\Gamma M})^{-1}\boldsymbol{M}' \boldsymbol{\Theta}' \boldsymbol{W}. \quad \ldots \quad (9.1)$$

The problem is one of maximizing

$$\text{trace}\ \boldsymbol{W \Theta M M}' \boldsymbol{\Theta}' \boldsymbol{W} = \text{trace}\ \boldsymbol{M}' \boldsymbol{\Theta}' \boldsymbol{W}^2 \boldsymbol{\Theta M}, \quad \ldots \quad (9.2)$$

choosing $\boldsymbol{M}' \boldsymbol{\Gamma M} = \boldsymbol{I}$, without loss of generality. The best choice of $\boldsymbol{M}$ is the matrix of the first q eigen vectors associated with the determinantal equation

$$|\boldsymbol{\Theta}' \boldsymbol{W}^2 \boldsymbol{\Theta} - \lambda \boldsymbol{\Gamma}| = 0. \quad \ldots \quad (9.3)$$

In the special case when the instrumental variable is the same as (or contains) the main variable, the equation (9.3) reduces to

$$|\boldsymbol{\Sigma W}^2 \boldsymbol{\Sigma} = \lambda \boldsymbol{\Sigma}\ | = 0 \quad \ldots \quad (9.4)$$

which is equivalent to

$$|\boldsymbol{\Sigma} - \lambda \boldsymbol{W}^{-2}| = 0. \quad \ldots \quad (9.5)$$

10. Principal components with restrictions on individual variances

In the previous sections the predictive efficiency of a random variable $\boldsymbol{Y}$ for predicting $\boldsymbol{X}$ is judged by the sum (or weighted sum) of residual variances for the different components of $\boldsymbol{X}$. An optimum choice of $\boldsymbol{Y}$ (in a given class) was made by maximizing the predictive efficiency, i.e., by minimizing the sum of residual variances. But it may so happen that for such an optimum choice of $\boldsymbol{Y}$, the residual variances for some elements of $\boldsymbol{X}$ are above certain desired levels. In such a case it may be necessary to impose some restrictions on the individual variances and maximize the overall predictive efficiency.

Let $\boldsymbol{Y}$ be the vector of q linear functions, $\boldsymbol{L}_1'\boldsymbol{X}, \ldots, \boldsymbol{L}_q'\boldsymbol{X}$, of $\boldsymbol{X}$, where $\boldsymbol{L}_i'\boldsymbol{\Sigma}\boldsymbol{L}_i = 1$, $\boldsymbol{L}_i'\boldsymbol{\Sigma}\boldsymbol{L}_j = 0$, $i \neq j$. The residual variance in predicting X_i, the i-th element of $\boldsymbol{X}$, by $\boldsymbol{Y}$ is

$$\sigma_{ii} - (\boldsymbol{L}_1'\boldsymbol{\Sigma}_i)^2 - \ldots - (\boldsymbol{L}_q'\boldsymbol{\Sigma}_i)^2 \quad \ldots \quad (10.1)$$

where $\boldsymbol{\Sigma}_i$ is the i-th column vector of $\boldsymbol{\Sigma}$. We impose the conditions

$$\sigma_{ii} - (\boldsymbol{L}_1'\boldsymbol{\Sigma}_i)^2 - \ldots - (\boldsymbol{L}_q'\boldsymbol{\Sigma}_i)^2 \leqslant e_i^2 \text{ (given)}$$

or

$$(\boldsymbol{L}_1'\boldsymbol{\Sigma}_i)^2 + \ldots + (\boldsymbol{L}'\boldsymbol{\Sigma}_i)^2 \geqslant \sigma_{ii} - e_i^2 = \delta_i^2 \text{ (say)} \quad \ldots \quad (10.2)$$

$$i = 1, \ldots, p.$$

The sum of squares of residual variances is

$$\Sigma\, \sigma_{ii} - (\boldsymbol{L}_1'\boldsymbol{\Sigma}\boldsymbol{\Sigma}\boldsymbol{L}_1 + \ldots + \boldsymbol{L}_q'\boldsymbol{\Sigma}\boldsymbol{\Sigma}\boldsymbol{L}_q). \quad \ldots \quad (10.3)$$

The problem is one of maximizing

$$\boldsymbol{L}_1'\boldsymbol{\Sigma}\boldsymbol{\Sigma}\boldsymbol{L}_1 + \ldots + \boldsymbol{L}_q'\boldsymbol{\Sigma}\boldsymbol{\Sigma}\boldsymbol{L}_q \quad \ldots \quad (10.4)$$

subject to the conditions

$$\left.\begin{array}{l} \boldsymbol{L}_i'\boldsymbol{\Sigma}\boldsymbol{L}_i = 1, \quad i = 1, \ldots, q \\ \boldsymbol{L}_i'\boldsymbol{\Sigma}\boldsymbol{L}_j = 0, \quad i \neq j, \\ (\boldsymbol{L}_1'\boldsymbol{\Sigma}_i)^2 + \ldots + (\boldsymbol{L}_q'\boldsymbol{\Sigma}_i)^2 \geqslant \delta_i^2, \; i = 1, \ldots, p. \end{array}\right\} \quad \ldots \quad (10.5)$$

We are thus led to a very complicated problem in non-linear programming.

When $q = 1$, there is only a single linear function $\boldsymbol{L}_1'\boldsymbol{X}$ to be determined. The expression to be maximized is

$$\boldsymbol{L}_1'\boldsymbol{\Sigma}\boldsymbol{\Sigma}\boldsymbol{L}_1 \quad \ldots \quad (10.6)$$

subject to the conditions

$$\boldsymbol{L}_1'\boldsymbol{\Sigma}\boldsymbol{L}_1 = 1 \text{ and } (\boldsymbol{L}_1'\boldsymbol{\Sigma}_i)^2 \geqslant \delta_i^2, \quad i = 1, \ldots, p. \quad \ldots \quad (10.7)$$

Making the substitution $\boldsymbol{\Sigma}\boldsymbol{L}_1 = \boldsymbol{U}$ or $\boldsymbol{L}_1 = \boldsymbol{\Sigma}^{-1}\boldsymbol{U}$, we need only maximize

$$\boldsymbol{U}'\boldsymbol{U} \quad \ldots \quad (10.8)$$

subject to the conditions

$$\boldsymbol{U}'\boldsymbol{\Sigma}^{-1}\boldsymbol{U} = 1 \text{ and } U_i^2 \geqslant \delta_i^2, \quad i = 1, \ldots, p. \quad \ldots \quad (10.9)$$

The solution is complicated even in this simple case. It has been shown elsewhere (Rao, 1962b, 1964) how to obtain a solution for small values of p. The general problem awaits solution.

11. Principal components of $\boldsymbol{X}$ uncorrelated with the instrumental variable $\boldsymbol{Z}$

In Section 8, the instrumental variable $\boldsymbol{Z}$ is analyzed into principal components on the basis of their predictive efficiency for elements of a variable $\boldsymbol{X}$. In some problems, it is of interest to determine the principal components of $\boldsymbol{X}$ in the class of linear functions of $\boldsymbol{X}$ uncorrelated with instrumental variables. The problem may be stated as one of determining q linear functions $\boldsymbol{L}_1'\boldsymbol{X}, \ldots, \boldsymbol{L}_q'\boldsymbol{X}$ such that

$$V(\boldsymbol{L}_1'\boldsymbol{X}) + \ldots + V(\boldsymbol{L}_q'\boldsymbol{X}) = \boldsymbol{L}_1'\boldsymbol{\Sigma}\boldsymbol{L}_1 + \ldots + \boldsymbol{L}_q'\boldsymbol{\Sigma}\boldsymbol{L}_q \quad \ldots \quad (11.1)$$

is a maximum subject to the conditions

$$\left.\begin{aligned} &\boldsymbol{L}_i' \boldsymbol{L}_i = 1, \quad \boldsymbol{L}_i' \boldsymbol{L}_j = 0 \\ &\operatorname{cov}(\boldsymbol{L}_i' \boldsymbol{X}, \boldsymbol{Z}) = 0, \ i = 1, \ldots, q. \end{aligned}\right\} \quad \ldots \quad (11.2)$$

As before, let the dispersion matrix of $(\boldsymbol{X}, \boldsymbol{Z})$ be

$$\begin{pmatrix} \boldsymbol{\Sigma} & \boldsymbol{\Theta} \\ \boldsymbol{\Theta}' & \boldsymbol{\Gamma} \end{pmatrix}. \quad \ldots \quad (11.3)$$

In terms of the elements of the matrix (11.3), the problem is one of maximizing

$$\boldsymbol{L}_1' \boldsymbol{\Sigma} \boldsymbol{L}_1 + \ldots + \boldsymbol{L}_q' \boldsymbol{\Sigma} \boldsymbol{L}_q \quad \ldots \quad (11.4)$$

subject to the conditions

$$\left.\begin{aligned} &\boldsymbol{L}_i' \boldsymbol{L}_i = 1, \ \boldsymbol{L}_i' \boldsymbol{L}_j = 0, \quad i \neq j \\ &\boldsymbol{L}_i' \boldsymbol{\Theta} = 0, \\ &\qquad i, j = 1, \ldots, q. \end{aligned}\right\} \quad \ldots \quad (11.5)$$

An application of the result (2.12) shows that the maximum of (11.4) is attained when $\boldsymbol{L}_1, \ldots, \boldsymbol{L}_q$ are the first q right eigen vectors of the matrix

$$(\boldsymbol{I} - \boldsymbol{\Theta} (\boldsymbol{\Theta}'\boldsymbol{\Theta})^{-1} \boldsymbol{\Theta}')\boldsymbol{\Sigma}. \quad \ldots \quad (11.6)$$

The principal components so determined may be useful in the following situation.

Suppose we have a vector variable $\boldsymbol{X}$ representing p 'economic transactions' measured in terms of a common unit (like the cash value). The observed values of $\boldsymbol{X}$ over time constitute a multiple time series. Some years ago, Stone (1947) considered the problem of isolating linear functions of $\boldsymbol{X}$ which have an intrinsic economic significance from those which represent trend with time and those which measure random errors. For this purpose, he computed the dispersion matrix of the p-variables considering the observations at different points of time as repeated values of $\boldsymbol{X}$ and extracted the principal components of $\boldsymbol{X}$ from the estimated dispersion matrix, without any reference to the time factor. The problem was then posed as that of interpreting the dominant principal components which accounted for a high percentage of the total variance. The first principal component with the largest variance was interpreted as representing linear trend and the rest were interpreted in economic terms.

It is not clear why the trend, even if it is linear, should be reflected in the first principal component only. Then there is the difficulty of choosing more than one principal component to explain the trend, if it is non-linear in time.

To apply the present analysis, we consider an instrumental variable $\boldsymbol{Z}$ consisting of orthogonal functions of time, $\phi_1(t)$, $\phi_2(t)$, ..., representing the first, second,... degree polynomial trend with respect to time. Then we obtain the principal components of $\boldsymbol{X}$ in the class of linear functions not showing any trend, which may then be interpreted in economic terms.

12. Size and shape factors

Biologists use functions of measurements representing size and shape of organisms in studying differences within and between groups. Size and shape are not, however, well-defined concepts and their measurement will be arbitrary to some extent. Let us examine some functions proposed for this purpose.

Penrose (1947) defines size as the linear function

$$\frac{X_1}{\sigma_1}+\ldots+\frac{X_p}{\sigma_p} \qquad \ldots \quad (12.1)$$

where σ_i is the standard deviation of X_i and shape as any linear function

$$c_1\frac{X_1}{\sigma_1}+\ldots+c_p\frac{X_p}{\sigma_p} \qquad \ldots \quad (12.2)$$

with $c_1+\ldots+c_p = 0$. In situations where X_1, X_2, X_3 correspond, in some sense, to length, width and height of an organism, Mosimann (1950) proposed the product

$$X_1X_2X_3 \qquad \ldots \quad (12.3)$$

as a volumetric or a ponderal definition of size. It may be noted that a function of the type

$$X_1^{\beta_1}\, X_2^{\beta_2}\, X_3^{\beta_3} \qquad \ldots \quad (12.4)$$

studied by Rao and Shaw (1948) provides a better index of volume for a suitable choice of the exponents $\beta_1, \beta_2, \beta_3$. In terms of the logarithms of the measurements the size function (12.4) has the linear form

$$\beta_1 \log X_1+\beta_2 \log X_2+\beta_3 \log X_3. \qquad \ldots \quad (12.5)$$

It has been suggested by Jolicoeur and Mosimann (1960) that the first principal component, which has maximum variation, may be taken as size factor provided all the coefficients are positive and other principal components with positive and negative coefficients as shape factors. A justification for such an interpretation of principal components is as follows.

Consider the i-th element X_i of $\boldsymbol{X}$ and the j-th principal component $\boldsymbol{P}_j'\boldsymbol{X}$ of $\boldsymbol{X}$. The regression of X_i on $\boldsymbol{P}_j\boldsymbol{X}$ is simply P_{ji}, the i-th element in $\boldsymbol{P}_j$. Now, a unit increase in $\boldsymbol{P}_j'\boldsymbol{X}$ produces on the average an increase of P_{ji} in X_i. If all the coefficients $P_{j1}, \ldots, P_{jp}$ are positive, a unit increase in $\boldsymbol{P}_j'\boldsymbol{X}$ increases the value of each of the measurements, in which case $\boldsymbol{P}_j'\boldsymbol{X}$ may be called a size factor. If some of the coefficients are positive and others negative, then an increase in $\boldsymbol{P}_j'\boldsymbol{X}$, increases the values of some of the measurements and decreases the values of the others, in which case $\boldsymbol{P}_j'\boldsymbol{X}$ may be called a shape factor. Penrose's size and shape functions have similar properties. In either case, there seems to be no suitable interpretation of the relative magnitudes of changes in the different measurements or of the distribution of positive and negative coefficients in the case of a shape factor. There seem to be too many arbitrary elements involved as the principal components are not explicitly derived to represent size and shape factors in any well-defined manner.

We shall first generalize the problem by attempting to measure size and shape variation with respect to a given set of measurements $X_1, \ldots, X_p$ in terms of instrumental variables $Z_1, \ldots, Z_m$, where the latter may include some or all of the former.

Let the dispersion matrix of $(\boldsymbol{X}, \boldsymbol{Z})$ where $\boldsymbol{X}' = (X_1, \ldots, X_p)$ and $\boldsymbol{Z}' = (Z_1, \ldots, Z_m)$, be

$$\begin{pmatrix} \boldsymbol{\Sigma} & \boldsymbol{\Theta} \\ \boldsymbol{\Theta}' & \boldsymbol{\Gamma} \end{pmatrix} \qquad \ldots \quad (12.6)$$

and consider a linear function of $\boldsymbol{Z}$, with unit variance,

$$\boldsymbol{B}'\boldsymbol{Z} = B_1 Z_1 + \ldots + B_m Z_m, \quad \boldsymbol{B}'\Gamma\, \boldsymbol{B} = 1. \qquad \ldots \quad (12.7)$$

The regression coefficients of $X_1, \ldots, X_p$ on $\boldsymbol{B}'\boldsymbol{Z}$ are the components of the vector $\boldsymbol{\Theta}\, \boldsymbol{B}$. Let us specify the ratios of the regression coefficients, taking into account the signs, as the elements of a vector $\boldsymbol{R}$, so that

$$\boldsymbol{\Theta B} = \rho \boldsymbol{R} \qquad \ldots \quad (12.8)$$

where ρ is a constant. If all the components of $\boldsymbol{R}$ are chosen as positive, then any solution for $\boldsymbol{B}$ of the equation (12.8) provides a size function. If some components of $\boldsymbol{R}$ are positive and others negative, then a solution for $\boldsymbol{B}$ provides a shape function. In the case when $\boldsymbol{\Theta}$ is a square matrix of rank m, there is a unique $\boldsymbol{B}$ and ρ for a given $\boldsymbol{R}$, satisfying the equations

$$\boldsymbol{\Theta B} = \rho \boldsymbol{R}, \quad \boldsymbol{B}'\boldsymbol{\Gamma B} = 1. \qquad \ldots \quad (12.9)$$

Otherwise there may be a multiplicity of solutions. In such a case $\boldsymbol{B}$ may be chosen to maximize ρ.

Introducing Lagrangian multipliers $\boldsymbol{C}_1$ (a p-dimensional vector) and c_2 (a constant), the function to be differentiated is

$$\rho + \boldsymbol{C}_1'(\boldsymbol{\Theta B} - \rho \boldsymbol{R}) + c_2(\boldsymbol{B}'\boldsymbol{\Gamma}\, \boldsymbol{B} - 1). \qquad \ldots \quad (12.10)$$

Differentiating with respect to $\boldsymbol{B}$, $\boldsymbol{C}_1$, ρ and c_2 we obtain the equations

$$\left.\begin{aligned} \boldsymbol{\Theta}'\, \boldsymbol{C}_1 + c_2 \boldsymbol{\Gamma}\, \boldsymbol{B} &= 0 \\ \boldsymbol{\Theta B} &= \rho R \\ \boldsymbol{R}'\boldsymbol{C}_1 &= 1 \\ \boldsymbol{B}'\boldsymbol{\Gamma}\, \boldsymbol{B} &= 1. \end{aligned}\right\} \qquad \ldots \quad (12.11)$$

The first two equations can be written as

$$\left.\begin{aligned} \boldsymbol{\Theta}'\boldsymbol{C} + \boldsymbol{\Gamma}\, \rho^{-1}\boldsymbol{B} &= 0 \\ \boldsymbol{\Theta}\, \rho^{-1}\boldsymbol{B} &= \boldsymbol{R} \end{aligned}\right\} \qquad \ldots \quad (12.12)$$

where $C = (c_2 \rho)^{-1} C_1$. Solving the equations (12.12) we have a solution for $\rho^{-1}\boldsymbol{B}$, i.e., the size or shape function is determined apart from a constant multiplier. If necessary, the function can be standardized by reducing its variance to unity. To obtain an explicit representation for $\rho^{-1}\boldsymbol{B}$, let

$$\begin{pmatrix} \boldsymbol{\Theta}' & \boldsymbol{\Gamma} \\ \boldsymbol{0} & \boldsymbol{\Theta} \end{pmatrix}^{-1} = \begin{pmatrix} \boldsymbol{A} & \boldsymbol{E} \\ \boldsymbol{E} & \boldsymbol{D} \end{pmatrix}. \qquad \ldots \quad (12.13)$$

Then $\rho^{-1}\boldsymbol{B} = \boldsymbol{DR}$. The size and shape functions so determined have greater flexibility, although there will still be some arbitrariness in the choice of the vector of ratios, $\boldsymbol{R}$, when it cannot be specified by other considerations.

For instance, anthropologists use the ratio of head breadth to head length, called the cephalic index, to measure the shape of the head. The present approach suggests that the shape function can be built out of a *number* of length and breadth measurements on the head, by choosing $\boldsymbol{R}$ such that the ratios for length measurements have a positive sign and those for breadth measurements have a negative sign. The shape function so determined has the property that an increase in its value increases the lengths and decreases the breadths. Further, other measurements on the body such as stature, chest girth etc., can be brought in to obtain a better measure of head shape. Examples of such functions are given in Rao (1961, 1962a). It has also been found that the choice of absolute values of the elements of $\boldsymbol{R}$ in proportion to the standard deviations $\sigma_1, \ldots, \sigma_p$ of the measurements $X_1, \ldots, X_p$ leads to reasonable results.

The size and shape functions determined by the above method may not be uncorrelated as in the case of the principal components, although this can be achieved by choosing the vectors $\boldsymbol{R}$ for size and shape functions suitably. But lack of correlation is not an important property if size and shape variations are examined individually. However, if size and shape functions are used jointly in any study, their correlation should be taken into account in the statistical analysis. For instance, if a chart representing the configuration of a set of groups with respect to mean size and shape is desired, one could use 'size' and 'shape corrected for size' (by regression) and represent them on orthogonal axes.

13. Principal component and factor analyses

Some authors do not make a distinction between these two analyses and this has, no doubt, caused some amount of confusion. It is, therefore, of interest to examine the differences in the nature of information provided by these two techniques. It is shown that they provide distinct answers to two different enquiries concerning a hypothetical structure of the elements of a vector random variable.

Let us suppose that there exist hypothetical uncorrelated factors (variables), $F_1, F_2, \ldots$, presumably infinite in number, such that

$$\left.\begin{array}{l} X_1 = a_{11}F_1 + a_{12}F_2 + \ldots \\ \cdot \quad \cdot \quad \cdot \quad \cdot \quad \ldots \\ X_p = a_{p1}F_1 + a_{p2}F_2 + \ldots \end{array}\right\} \qquad \ldots \quad (13.1)$$

or in matrix notation

$$\boldsymbol{X} = \boldsymbol{A}\,\boldsymbol{F}. \quad \dots \quad (13.2)$$

It may be seen that the factor structure with common and specific factors, assumed in psychological work is a special (and perhaps an unrealistic !) case of (13.1). The representation (13.1) is not unique for by an orthogonal transformation on the hypothetical variable $\boldsymbol{F}$, the structural equation (13.2) can be written as

$$\boldsymbol{X} = \boldsymbol{B}\boldsymbol{G} \quad \dots \quad (13.3)$$

where $\boldsymbol{G}$ represents a new (transformed) factor variable.

The problem we wish to investigate is one of *approximating* $X_1, \dots, X_p$ by linear functions of as few factors, which may be called dominant, as possible. An approximation to $\boldsymbol{X}$ in terms of q factors, to be denoted by $\boldsymbol{X}^{(q)}$ is obtained by considering a representation of $\boldsymbol{X}$, such as (13.1) and truncating the right hand side at the q-th term. In matrix notation

$$\boldsymbol{X}^{(q)} = \boldsymbol{A}_q\,\boldsymbol{F}^{(q)} \quad \dots \quad (13.4)$$

where $\boldsymbol{A}_q$ is the matrix of the first q columns of $\boldsymbol{A}$ and $\boldsymbol{F}^{(q)}$ is the vector of the variables $F_1, \dots, F_q$. The different choices of $\boldsymbol{A}_q$ are obtained by making orthogonal transformations on $\boldsymbol{F}$ (the entire set of hypothetical factors), leading to different representations of $\boldsymbol{X}$ and applying the truncation procedure.

We have not yet specified the nature of the approximation we are seeking. It is seen that the dispersion matrix of $\boldsymbol{X}^{(q)}$ is $\boldsymbol{A}_q\boldsymbol{A}_q'$ (assuming without loss of generality that all F_i have unit variance), while the dispersion matrix of $\boldsymbol{X}$ is $\boldsymbol{\Sigma}$. Let us suppose that the approximation aims at the closeness of $\boldsymbol{\Sigma}$ and $\boldsymbol{A}_q\boldsymbol{A}_q'$. In such a case, we may determine $\boldsymbol{A}_q$ by minimizing the Euclidean norm of the residual dispersion matrix $||\boldsymbol{\Sigma}-\boldsymbol{A}_q\boldsymbol{A}_q'||$. Applying the result (2.6), the answer is

$$\boldsymbol{A}_q' = (\sqrt{\lambda_1}\,\boldsymbol{P}_1\ \sqrt{\lambda_2}\,\boldsymbol{P}_2 \dots \sqrt{\lambda_q}\,\boldsymbol{P}_q) \quad \dots \quad (13.5)$$

i.e., the i-th column of $\boldsymbol{A}_q$ is $\sqrt{\lambda_i}\,\boldsymbol{P}_i$ where λ_i is the i-th eigen value and $\boldsymbol{P}_i$ is the i-th eigen vector of $\boldsymbol{\Sigma}$. Then it may be shown that the best estimates of the factors are

$$F_i^q = \lambda_i^{-1/2}\boldsymbol{P}_i'\boldsymbol{X}, \quad i = 1, \dots, q \quad \dots \quad (13.6)$$

where $\boldsymbol{P}_i'\boldsymbol{X}$ is the i-th principal component of $\boldsymbol{X}$. We are thus led to principal component analysis as a definite answer to the problem posed.

Using estimated F_i^q, the q-th approximation to $\boldsymbol{X}$ as in (13.6), the residual is

$$\boldsymbol{X}^q-\boldsymbol{X} = (\boldsymbol{P}_{q+1}\boldsymbol{P}_{q+1}'+\dots+\boldsymbol{P}_p\boldsymbol{P}_p')\boldsymbol{X}. \quad \dots \quad (13.7)$$

The dispersion matrix of the residual is

$$(\boldsymbol{P}_{q+1}\boldsymbol{P}'_{+1}+\dots+\boldsymbol{P}_p\boldsymbol{P}_p')\boldsymbol{\Sigma}(\boldsymbol{P}_{q+1}\boldsymbol{P}_{q+1}'+\dots+\boldsymbol{P}_p\boldsymbol{P}_p)'.$$
$$=\lambda_{q+1}\boldsymbol{P}_{q+1}\boldsymbol{P}_{q+1}'+\dots+\lambda_p\boldsymbol{P}_p\boldsymbol{P}_p' \quad \dots \quad (13.8)$$

which is small if $\lambda_{q+1}, \dots, \lambda_p$ are small. In any problem we can examine the actual residuals (13.7) in judging the adequacy of a given order of approximation.

We have seen how the criterion of minimizing $||\boldsymbol{\Sigma}-\boldsymbol{A}_q\boldsymbol{A}_q'||$ led to the principal component analysis. We shall show that a different criterion leads to factor analysis.

The amount of correlation between X_i, X_j explained by the q factors $F_1, \ldots, F_q$ is

$$\begin{aligned} &\text{cov}\,(X_i^q, X_j^q) \div \sqrt{V(X_i)V(X_j)} \\ &= (a_{i1}a_{j1}+\ldots+a_{iq}a_{jq})/\sigma_i\sigma_j \\ &= b_{i1}b_{j1}+\ldots+b_{iq}b_{jq} \end{aligned}$$

where σ_i is the standard deviation of X_i and $b_{ik} = a_{ik}/\sigma_i$. The off diagonal elements of $\boldsymbol{B}_q\boldsymbol{B}_q'$ where $\boldsymbol{B}_q$ is the matrix with its i-th row as $(b_{i1}, \ldots, b_{iq})$ give all possible correlations explained by the q factors. Let $\boldsymbol{\Lambda}$ be the actual correlation matrix of the variables $X_1, \ldots, X_p$. We wish to choose $\boldsymbol{B}_q$ such that the off diagonal elements of $\boldsymbol{\Lambda}-\boldsymbol{B}_q\boldsymbol{B}_q'$ are as small as possible. A single measure of difference is the sum of squares of the off diagonal elements of $\boldsymbol{\Lambda}-\boldsymbol{B}_q\boldsymbol{B}_q'$ and the mathematical problem is one of minimizing this measure subject to the condition that the diagonal entries of $\boldsymbol{\Lambda}-\boldsymbol{B}_q\boldsymbol{B}_q'$ are non-negative. The factor structure provided by the optimum $\boldsymbol{B}_q$ is known as *factor analysis*; the attempt in such a case is to explain the correlations among the measurements rather than the variances of the individual measurements. The problem of an optimum choice of $\boldsymbol{B}_q$ is not easy to solve, and no definite solution can be obtained. An iterative method is suggested by Rao (1955) but the convergence of the procedure has not been adequately studied. For further literature see Whittle (1953), and the references in Maxwell and Lawley (1963).

14. An application to a problem in multidimensional scaling

The data consist of $n(n-1)/2$ independent determinations (possibly subject to errors) of distances between n stimuli and the problem is to determine the smallest Euclidean space in which the stimuli can be represented as points (Torgerson, 1958). Let $\boldsymbol{X}_1, \ldots, \boldsymbol{X}_n$ be points representing the n stimuli in a Euclidean space, and $\bar{\boldsymbol{X}}$ be the centre of gravity of the points. Then

$$\begin{aligned} (\boldsymbol{X}_i-\bar{\boldsymbol{X}})'(\boldsymbol{X}_j-\bar{\boldsymbol{X}}) &= n^{-2}[(\boldsymbol{X}_i-\boldsymbol{X}_1)+\ldots+(\boldsymbol{X}_i-\boldsymbol{X}_n)]'[(\boldsymbol{X}_j-\boldsymbol{X}_1)+\ldots+(\boldsymbol{X}_j-\boldsymbol{X}_n)] \\ &= n^{-2}\sum_k\sum_m [(\boldsymbol{X}_j-\boldsymbol{X}_k)'(\boldsymbol{X}_j-\boldsymbol{X}_k)+(\boldsymbol{X}_i-\boldsymbol{X}_m)'(\boldsymbol{X}_i-\boldsymbol{X}_m) \\ &\qquad -(\boldsymbol{X}_i-\boldsymbol{X}_j)'(\boldsymbol{X}_i-\boldsymbol{X}_j)-(\boldsymbol{X}_k-\boldsymbol{X}_m)'(\boldsymbol{X}_k-\boldsymbol{X}_m)] \\ &= \frac{1}{2n^2}\sum_k\sum_m (\delta_{jk}+\delta_{im}-\delta_{ij}-\delta_{mk}) \end{aligned} \qquad \ldots (14.1)$$

where δ_{rs} denotes the square of the true distance between the points $\boldsymbol{X}_r$ and $\boldsymbol{X}_s$. Using the notation $\sum_m \delta_{im} = \delta_{i\cdot}$ and $\sum_k\sum_m \delta_{km} = \delta_{\cdot\cdot}$ the expression (14.1) can be written

$$\frac{1}{2}\left(-\delta_{ij}+\frac{\delta_{i\cdot}+\delta_{\cdot j}}{n}-\frac{\delta_{\cdot\cdot}}{n^2}\right). \qquad \ldots (14.2)$$

Thus, given the mutual distances we can express the configuration of the points by the matrix $\mathscr{X}_d'\mathscr{X}_d$ whose (i, j) entry is given by (14.2), where $\mathscr{X}_d$ is as defined in (3.3). The matrix $\mathscr{X}_d'\mathscr{X}_d$, as introduced in Section 4, represents the lengths of the lines joining the points to the centre of gravity and the angles between them. The best representation of the points in a Euclidean space of q dimensions is provided by the $q \times m$ matrix whose i-th row is $\sqrt{\lambda_i}\, Q_i$ where λ_i and Q_i are the i-th eigen value and vector of $\mathscr{X}_d'\mathscr{X}_d$.

However, such an analysis is not directly applicable when the exact distances δ_{ij} are not known, but only estimates d_{ij} of δ_{ij} are available. Substituting d_{ij} for δ_{ij} in (14.2) we have an estimate of $\mathscr{X}_d'\mathscr{X}_d$ which we may denote by $B = (b_{ij})$. The matrix $\boldsymbol{B}$ need not be non-negative definite. We may still compute the eigen values and vectors of $\boldsymbol{B}$. If $\hat{\lambda}_i$ and $\hat{Q}_i$ denote the i-th eigen value and vector of $\boldsymbol{B}$, an estimate of the best q dimensional representation is given by the $q \times m$ matrix whose i-th row is $\sqrt{\hat{\lambda}_i}\, \hat{Q}_i$, provided $\hat{\lambda}_1, \ldots, \hat{\lambda}_q$ are positive.

One drawback of the solution derived from $\boldsymbol{B}$ is that the estimated points may not have the origin as the centre of gravity. This may be secured by determining the eigen vectors of $\boldsymbol{B}$ with the restriction that the sum of the elements of each eigen vector is zero. Let $\boldsymbol{U}$ represent a column vector of n unities. Then it can be shown, using result (iii) of Section 2, that the eigen vectors of the matrix

$$\left(\boldsymbol{I} - \frac{\boldsymbol{U}\boldsymbol{U}'}{n}\right) \boldsymbol{B} \qquad \ldots \quad (14.3)$$

satisfy the required conditions. Let $\mu_1, \mu_2, \ldots$ be the eigen values and $\boldsymbol{B}_1, \boldsymbol{B}_2, \ldots$ the corresponding eigen vectors of the matrix (14.3). Then the rows of the estimated matrix of points (columns representing the points) in a q dimensional space are $\sqrt{\mu_1}\, \boldsymbol{B}_1', \ldots, \sqrt{\mu_q}\boldsymbol{B}_q'$.

15. Estimation of structural relationship, heterogeneity of data, etc.

A recent application by Wernimont (1963) demonstrates the use of principal component analysis in detecting heterogeneity of data. In his study there were 9 spectrophotometers and each instrument was used to obtain a series of six absorbance curves (as a discrete set of measurements at 20 different wave lengths) of solutions at three different concentrations (30, 60 and 90) and on two different days. The data consisted of 54 sets of 20 measurements and in the notation of the present paper $n = 54$ and $p = 20$. According to Beer's law, if the same wave lengths have been used for all the instruments, the 54 points in the 20 dimensional space should ideally be on a line through the origin when there are no errors in the absorbance readings. The direction cosines of such a line would provide the constants of Beer's law for a given set of wave lengths. If, indeed, the wave length settings differed from one instrument to another the configuration of points would not be confined to a straight line. The object of investigation was to examine whether the instruments had systematic errors in the wave length setting. If some of the instruments did not allow a proper setting of the desired wave lengths, the data would be heterogeneous ; this could be detected by asking the question whether the observed points could be regarded as clustering round a straight line.

This is exactly what the principal components can detect provided the errors of absorbance measurements at different wave lengths are independent and have nearly the same distribution.

According to Beer's law, the vector $\boldsymbol{X}$ of measurements at 20 different wave lengths can be written as

$$\boldsymbol{X} = c\boldsymbol{\beta}+\boldsymbol{\epsilon} \quad \dots \quad (15.1)$$

where $\boldsymbol{\beta}$ depends on the wave lengths, c is the concentration of the solution used and $\boldsymbol{\epsilon}$ represents the errors of absorbance readings. Thus for different values of c, the true locus of the points $\boldsymbol{X}$ is a straight line through the origin, with direction cosines proportional to $\boldsymbol{\beta}$. But if $(\boldsymbol{\beta}+\boldsymbol{\xi}_{ij})$ is the vector appropriate to the i-th spectrophotometer on the j-th day, where ξ_{ij} represents the systematic error, then for the measurements with the i-th spectrophotometer on the j-th day at concentration c

$$\boldsymbol{X} = (\boldsymbol{\beta}+\boldsymbol{\xi}_{ij})c+\boldsymbol{\epsilon} \quad \dots \quad (15.2)$$

so that Beer's law holds for a given instrument on a given day. In such a case we should expect the 54 points to cluster round a maximum of 18 different lines corresponding to the nine instruments and two days. The 18 lines may lie in a space of as many as 18 dimensions depending on the complexity of the systematic errors. Let us examine this problem by the principal component analysis by pooling the data from all the instruments. The 54 vectors provide a 20×20 dispersion matrix. (Actually in a problem with the model such as (15.2), the principal component analysis could be carried out on the uncorrected dispersion matrix. The computations referred to are on the corrected dispersion matrix as reported by Wernimont.) The first two eigen values of this matrix have been found to be (0.119590×10^7) and (0.356424×10^3) which explain 99.94 and 0.03 percent of total variation respectively. The rest of the eigen values are much smaller indicating that the configuration of the points can be examined by one or two dimensional representations.

To find the coordinates of the projected points on the best two dimensional plane we have to determine the first two eigen vectors $\boldsymbol{P}_1$ and $\boldsymbol{P}_2$ and compute the coordinates

$$\boldsymbol{P}_1'\boldsymbol{X},\ \boldsymbol{P}_2'\boldsymbol{X} \quad \dots \quad (15.3)$$

corresponding to each original point $\boldsymbol{X}$. Transferring the origin to $\boldsymbol{P}_1'\bar{\boldsymbol{X}}$, $\boldsymbol{P}_2'\bar{\boldsymbol{X}}$, the coordinates are

$$\boldsymbol{P}_1'(\boldsymbol{X}-\bar{\boldsymbol{X}}),\ \boldsymbol{P}_2'(\boldsymbol{X}-\bar{\boldsymbol{X}}). \quad \dots \quad (15.4)$$

For each point, it is also necessary to compute the length of the perpendicular on the best fitting plane to examine whether the representation is adequate with respect to that particular point. The square of the perpendicular from the point $\boldsymbol{X}$ is

$$R^2 = (\boldsymbol{X}-\bar{\boldsymbol{X}})'(\boldsymbol{X}-\bar{\boldsymbol{X}})-[\boldsymbol{P}_1'(\boldsymbol{X}-\bar{\boldsymbol{X}})]^2-[\boldsymbol{P}_2'(\boldsymbol{X}-\bar{\boldsymbol{X}})]^2. \quad \dots \quad (15.5)$$

The following table gives the values of (15.4) scaled by the reciprocal of the square root of the eigen values (which is not necessary for our purpose, but adopted from the computations provided by Wernimont) and R^2 for two sets of six absorbance curves of S. P. Meters 1 and 7. Similar values are available for all the 54 absorbance curves.

no.	S.P.M.	concentration	day	$P_1'(X-\bar{X})/\sqrt{\lambda_1}$	$P_2'(X-\bar{X})/\sqrt{\lambda_2}$	R^2
1.1	1	30	1	−1.220	0.614	54.17
1.2		60	1	−0.016	1.419	254.51
1.3		90	1	1.173	1.875	443.93
1.4		30	2	−1.200	0.574	122.15
1.5		60	2	0.001	1.518	372.83
1.6		90	2	1.200	1.877	334.46
7.1	7	30	1	−1.235	0.590	715.20
7.2		60	1	−0.065	0.140	1166.08
7.3		90	1	1.148	−0.266	714.84
7.4		30	2	−1.210	0.649	405.33
7.5		60	2	−0.061	0.261	610.58
7.6		90	2	1.134	0.074	655.72

Using the coordinates $(P_1'(X-\bar{X})/\sqrt{\lambda_1},\ P_2'(X-\bar{X}/\sqrt{\lambda_2})$, the 54 points are plotted on two charts (to avoid over crowding). In the charts (A.1, A.2, A.3) and (A.4, A.5, A.6) represent the absorbance curves at three levels of concentration obtained with S_A (the spectrphotometer A) on the first and second days respectively. The following conclusions emerge. (a) Generally, the points (A.1, A.2, A.3) are close to a line and so also the points (A.4, A.5, A.6) indicating the validity of Beer's law. (b) The points (A.1, A.2, A.3) and (A.4, A.5, A.6) lie on the same line except for S_6 and S_9, indicating errors specific to the instruments. (c) The lines for different instruments are differently oriented indicating systematic errors.

16. Determination of clusters of points

Let $X_1, \ldots, X_n$ be points in a p-dimensional space. In some problems it is of interest to examine whether the n points can be classified into groups or clusters such that the points within a cluster are close together, while the clusters themselves are far apart. Thus if the p coordinates of each point represent the production of p agricultural commodities in a region and different points represent different regions we may like to group together regions which are similar with respect to over all agricultural production (Kendall, 1939). Or if the p coordinates of each point represent the mean anthropological characters of a population and different points represent different populations we may like to examine whether the populations can be grouped into distinct clusters on the basis of similarity of the characters (Mahalanobis, 1936;

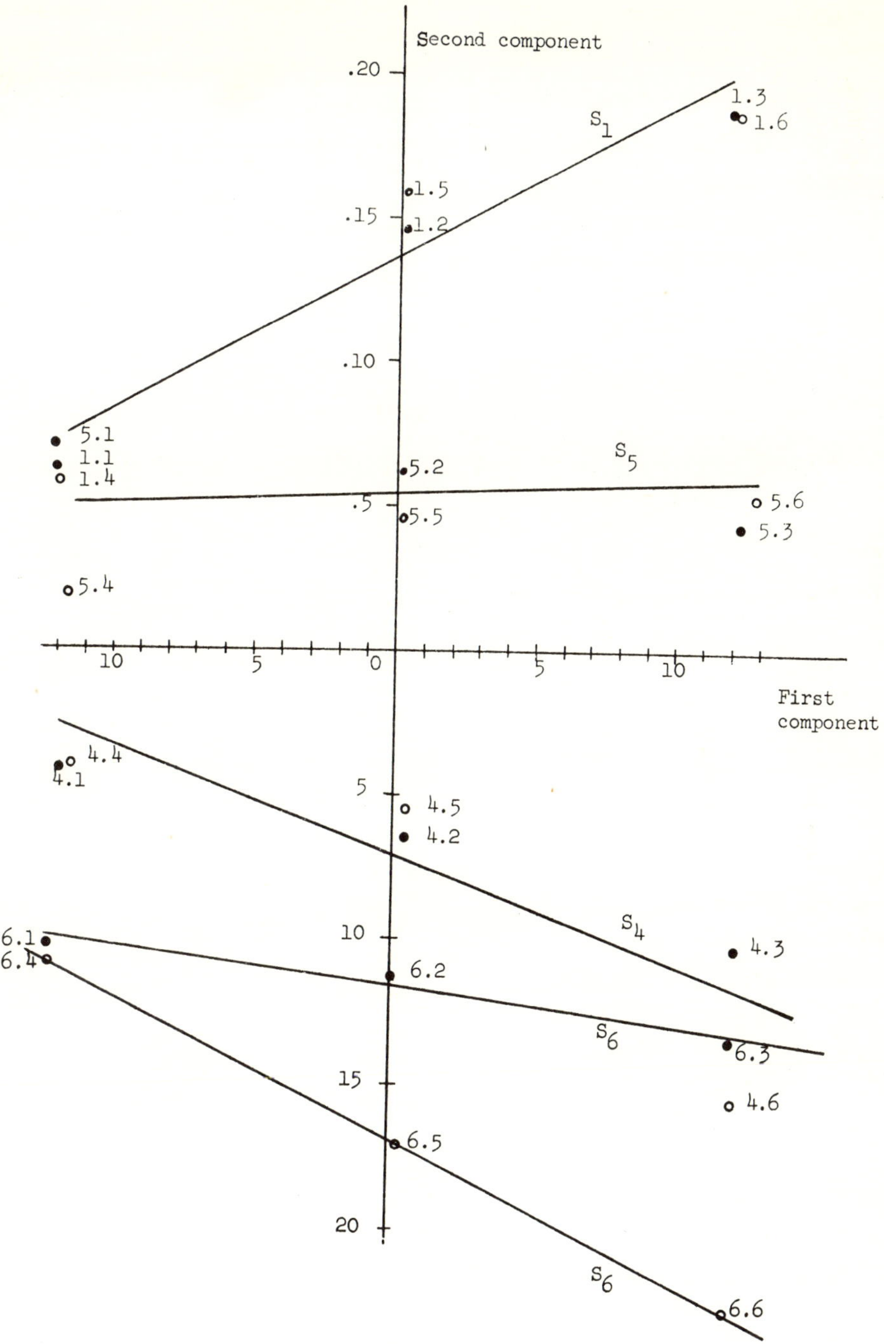

Chart 1. Configuration of the points with respect to the first two Principal Components. The points A.1, A.2, A.3 represent the measurements at three concentrations with spectrophotometer S_A and the points A.4, A.5, A.6 the corresponding measurements on a subsequent day.

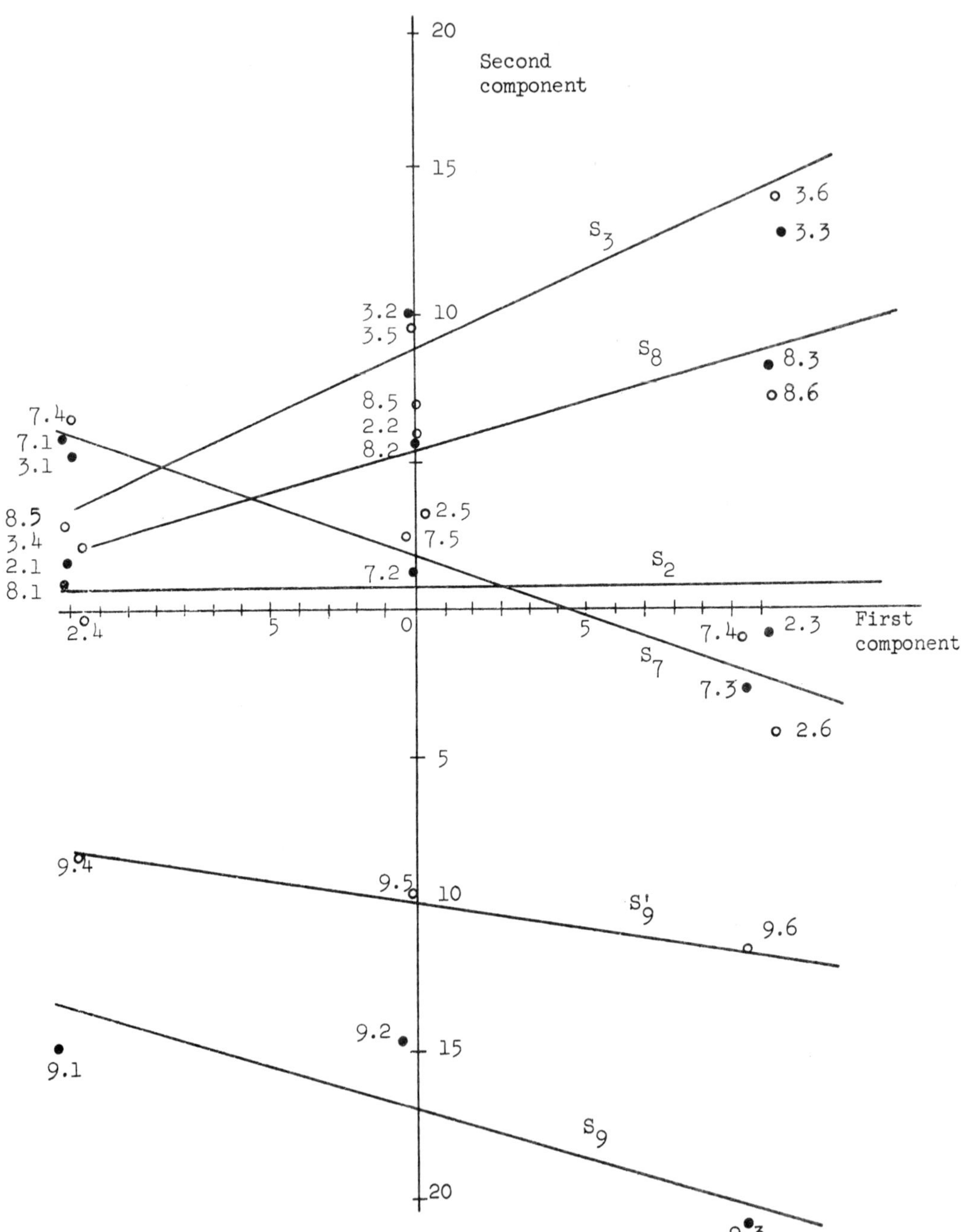

Chart 2. Configuration of the points with respect to the first two Principal Components etc.

Mahalanobis, Majumdar and Rao, 1949 etc.). In all such situations we need to have a measure of distance between two points. One may choose as the distance between $\boldsymbol{X}_i$ and $\boldsymbol{X}_j$

$$d_{ij} = (\boldsymbol{X}_i - \boldsymbol{X}_j)'(\boldsymbol{X}_i - \boldsymbol{X}_j) \qquad \ldots \quad (16.1)$$

representing the points in a Euclidean space with orthogonal axes or

$$d_{ij} = (\boldsymbol{X}_i - \boldsymbol{X}_j)\boldsymbol{\Lambda}^{-1}(\boldsymbol{X}_i - \boldsymbol{X}_j) \qquad \ldots \quad (16.2)$$

representing the points in a Euclidean space with oblique axes. When $\boldsymbol{\Lambda}$ is the co-variance matrix of the characters within a population (or region) the distance (16.2) is known as Mahalanobis distance.

Whatever distance function is chosen, we have a $n \times n$ matrix $D = (d_{ij})$ of all possible distances. We, then, look for clusters of points, by fixing a few pairs of points which are close together, but the pairs themselves being more distant from each other, and building clusters around them by adjoining neighbouring points. In practice such a programme can be successfully carried out although considerable difficulties are involved (see Mahalanobis, Majumdar and Rao, 1949; Majumdar and Rao, 1958).

When $p = 1$, we have one dimensional ordering of points and no visual representation of points on a line is necessary to determine the clusters. When $p = 2$, the determination of clusters is also simple, since we can represent the points on a two dimensional chart (using orthogonal or oblique axes) and mark out the subsets of points close together. For $p = 3$, we may use a three dimensional model. But no such visual aid is available for larger values of p.

We may, then raise the question as to whether the points in a p-dimensional space can be reasonably represented in 2 or 3 dimensional spaces without distorting the configuration as specified by the mutual distances. We have seen in Sections 3 and 4, that such a representation is secured through the principal components. Let $\boldsymbol{P}_1, \boldsymbol{P}_2, \boldsymbol{P}_3$ be the first three eigen vectors of $\boldsymbol{\Sigma}$ (or $\boldsymbol{\Sigma}$ with respect to $\boldsymbol{\Lambda}$ in the oblique case) giving the coordinates $\boldsymbol{P}_1'\boldsymbol{X}_i$, $\boldsymbol{P}_2'\boldsymbol{X}_i$, $\boldsymbol{P}_3'\boldsymbol{X}_i$ for the representation of $\boldsymbol{X}_i$ in the best three dimensional space. We may choose, the first, or the first two or all the three coordinates to obtain a visual representation of the points and then determine the clusters of points.

The success of the method obviously depends on how well the configuration of the points is preserved in the reduced space. If $\lambda_1, \ldots, \lambda_p$ are all the eigen values an overall measure of the adequacy of the best q dimensional representation as introduced in Section 3 is $(\lambda_1 + \ldots + \lambda_q)/(\lambda_1 + \ldots + \lambda_p)$. We may examine the values of this measure for $q = 1, 2, 3$ and decide on a suitable value of q. But such a decision may be misleading in individual cases. So it is necessary to obtain a tabulation of all the distances in the original and reduced space for a preliminary comparison. The distance between the i-th and j-th points in the reduced space of q dimensions is computed by the formula

$$d_{ij}^{q} = (\boldsymbol{P}_1'\boldsymbol{X}_i - \boldsymbol{P}_1'\boldsymbol{X}_j)^2 + \ldots + (\boldsymbol{P}_q'\boldsymbol{X}_i - \boldsymbol{P}_q'\boldsymbol{X}_j)^2 \qquad \ldots \quad (16.3)$$

whether the original points are in an orthogonal space or not. The difference $d_{ij} - d_{ij}^{q}$ is always positive and measures the efficiency of reduction in the dimensions of the space with respect to the points i and j. If the differences $(d_{ij} - d_{ij}^{q})$ are uniformly small, i.e., for all i and j, we are on safe grounds in determining the clusters on the basis of the first q principal components. If not we can still use the q principal components for the points where the distances are not distorted. The affinities of the points whose distances with the others are distorted may then be determined by considering their actual distances between themselves and with the others.

17. Tests of significance for principal components

In the previous sections no reference has been made to tests of significance in principal component analysis. The reason is that the problems have not been posed in terms of testing of well-defined hypotheses but in terms of estimating some features of the measurements on the individuals of a population. It is assumed that in any realistic problem, reduction in the number of measurements (by omission of some or by replacing the original measurements by a smaller number of linear functions) entails some loss of information. If so we can only consider null hypotheses which specify the amount of information lost. For instance, a null hypothesis may specify that the ratio of the sum of the last $p-q$ eigen values to the sum of all eigen values of a hypothetical dispersion matrix Σ is π. How can we test such a hypothesis when we have an estimate $\hat{\Sigma}$ of Σ. The relevant statistic appears to be

$$(\hat{\lambda}_{q+1} + \ldots + \hat{\lambda}_p)/(\hat{\lambda}_1 + \ldots + \hat{\lambda}_p) \qquad \ldots \quad (17.1)$$

where $\hat{\lambda}_1, \ldots, \hat{\lambda}_p$ are the eigen values of $\hat{\Sigma}$. It is extremely difficult to derive the distribution of (17.1) even under the assumption that $\hat{\Sigma}$ has a Wishart distribution.

Another null hypothesis of interest is the equality of the last $p-q$ true eigen values. An appropriate statistic to test such a hypothesis is the ratio of the geometric mean to the arithmetic mean of the estimated roots $\hat{\lambda}_{q+1}, \ldots, \hat{\lambda}_p$ (see Bartlett, 1950, 1951a, 1951b; Rao, 1955). The large sample distribution of the logarithm of the ratio suitably standardized is chi-square when $\hat{\Sigma}$ has a Wishart distribution. However, the tests based on the roots $\hat{\lambda}_1, \ldots, \hat{\lambda}_p$ are likely to be sensitive to departure from Wishartness of the distribution of $\hat{\Sigma}$ and no appropriate robust test criteria are known.

18. Use of principal components in testing differences in mean values between groups

It is well known that when multiple measurments are available on samples of individuals from different populations, the total sum of products (S.P.) matrix can be analyzed as due to 'within' and 'between' groups. The analysis of dispersion (which is a generalization of analysis of variance) is indicated as follows (Rao, 1952).

	D.F.	S.P. matrix
Between	$k-1$	B
Within	n	W
	$n+k-1$	T

Differences between the mean values of k populations and the configuration of the mean values when differences exist are examined by the roots of the determinantal equation

$$|B-\theta T| = 0. \qquad \ldots \quad (18.1)$$

For applications the reader is referred to Bartlett (1948); Fisher (1939); Rao (1948, 1952); Williams (1959) etc.

Recently, Roy (1958) proposed stepdown procedures wherby differences are examined in one observed variable first, and then in another observed variable eliminating the regression due to the first, and so on using each time a univariate procedure. There is, however, the problem of fixing the order in which the variables are considered. It is suggested that instead of the original variables, the principal components

$$\boldsymbol{P}_1'\boldsymbol{X}, \ldots, \boldsymbol{P}_q'\boldsymbol{X} \qquad \ldots \quad (18.2)$$

where $\boldsymbol{P}_1, \ldots, \boldsymbol{P}_q$ are the first q eigen vectors of the total S.P. matrix T may be considered in the given order. The performance of such a procedure (Dempster, 1963) is not fully studied except in a very pecial case arising in the comparison of growth curves (Rao, 1958). Although the method suggested is in line with the general philosophy of the principal component analysis it is clear that the first few principal components (18.2) are not designed to summarize as much information as possible on the differences between the mean values of the groups. In actual practice it may turn out that some of the principal components with smaller eigen values are better discriminators between the populations with respect to the differences in means than the first few principal components. This is not so only in special cases such as those considered by Rao (1958). However, the general drawback of the step-down procedures is that they do not enable us to study the configuration of the mean values of the different populations, which in practical problems is more important than merely establishing differences in the mean values.

References

Bartlett, M. S. (1947): Multivariate analysis. *J. Roy. Stat. Soc.* (Supple.), **2**, 176–197.

——— (1950): Tests of significance in factor analysis. *Brit. J. Psych.* (Stat. Sec.), **3**, 77–85.

——— (1951a): The effect of standardization of a χ^2-approximation in factor analysis. *Biometrika*, **38**, 337–344.

——— (1951b): A further note on tests of significance in factor analysis. *Brit. J. Psych*, (Stat. Sec.), **4**, 1–2.

Dempster, A. P. (1963): Multivariate theory for general stepwise methods. *Ann. Math. Stat.*, **34**, 873–883.

Fisher, R. A. (1939): The sampling distribution of some statistics obtained from non-linear equations. *Ann. Eugen.*, **9**, 238–249.

Frisch, R. (1929): Correlation and scatter in statistical variables. *Nordic Stat. J.*, **8**, 36–102.

Hotelling, H. (1933): Analysis of a complex statistical variable into principal components. *J. Educ. Psych.*, **26**, 417–441, 498–520.

——— (1935): The most predictable criterion. *J. Educ. Psych.*, **26**, 139–142.

——— (1936a): Simplified calculation of principal components. *Psychometrika*, **1**, 27–35.

——— (1936b): Relations between two sets of variates. *Biometrika*, **28**, 321–377.

Jolicoeur, P. and Mosimann, J. E. (1960): Size and shape variation in the painted turtle, a principal component analysis. *Growth*, **24**, 339–354.

Kendall, M. G. (1939): The geographical distribution of crop productivity in England. *J. Roy. Stat. Soc.*, **102**, 21–

Mahalanobis, P. C. (1936): On the generalized distance in statistics. *Proc. Nat. Inst. Sci. India*, **12**, 49–55.

Mahalanobis, P. C., Majumdar, D. N. and Rao, C. R. (1949): Anthropometric survey of the United Provinces, 1941. A statistical survey. *Sankhyā*, **9**, 90–327.

Majumdar, D. N. and Rao, C. R. (1958): Bengal anthropometric survey, 1945. A statistical study. *Sankhyā*, **19**, 201–408.

Mosimann, J. E. (1958): An analysis of allometry in Chelonian shell. *Rev. Can. Biol.*, **17**, 137–228.

Pearson, K. (1901): On lines and planes of closest fit to systems of points in space. *Phil. Mag.*, **2** (sixth series), 559–572.

Penrose, L. S. (1947): Some notes on discrimination. *Ann. Eugen.*, **13**, 228–237.

Rao, C. R. and Shaw, D. C. (1948): On a formula for the prediction of Cranial capacity. *Biometrika*, **4**, 247–253.

Rao, C. R. (1948): Tests of significance in multivariate analysis. *Biometrika*, **35**, 58–79.

——— (1952): *Advanced Statistical Methods in Biometric Research.* John Wiley and Sons.

——— (1955): Estimation and tests of significance in factor analysis. *Psychometrika*, **20**, 93–111.

——— (1958): Some statistical methods for comparison of growth curves. *Biometrics*, **14**, 1–17.

——— (1960): Multivariate analysis: An indispensable aid in applied research. *Sankhyā*, **22**, 317–338.

——— (1961): Some observations on multivariate statistical methods in anthropological research. *Bull. Int. Stat. Inst.*, **38**, 99–109.

——— (1962a): Use of discriminant and applied function in multivariate analysis. *Sankhyā*, **22**, 317–338.

——— (1962b): Problems of selection with restrictions. *J. Roy. Stat. Soc.*, (series B), **24**, 401–405.

——— (1964): *Problems of Selection Involving Programming Techniques*, IBM Symposium on Statistics.

Roy, J. (1958): Step-down procedure in multivariate analysis. *Ann. Math. Stat.*, **29**, 1177–1187.

Simonds, J. L. (1963) : Applications of characteristic vector analysis to photographic and optical response data. *J. Optical Society of America*, **53**, 968–974.

Stone, R. (1947) : An interdependence of blocks of transactions. *J. Roy. Stat. Soc.* (Supple.), **9**, 1–3-.

Torgerson, W. S. (1958) : *Theory and Methods of Scaling*, John Wiley and Sons.

Wernimont, G. (1963) : *Use of the Computer to Compare Spectrophotometric Curves*, Management Systems Development Department. Eastman Kodak Company.

Whittle, P. (1953) : On principal components and least square methods of factor analysis. *Skand. Aktuarictidskr.*, **36**, 223–239.

Williams, E. G. (1958) : *Regression Analysis*, John Wiley and Sons.

Paper received : November, 1964.

5

Reprinted from *Growth,* **24**(4), 339–354 (1960)

SIZE AND SHAPE VARIATION IN THE PAINTED TURTLE.[1] A PRINCIPAL COMPONENT ANALYSIS

PIERRE JOLICOEUR AND JAMES E. MOSIMANN[2]

Walker Museum, University of Chicago
and
Institut de Biologie, Université de Montréal

(Received for publication July 11, 1960)

INTRODUCTION

The concepts of size and shape are fundamental to the analysis of variation in living organisms. And yet, as noted by Simpson, Roe and Lewontin (1960), there is at present no general agreement on practical definitions of size and shape. The adoption of a generally applicable method of measuring size and shape variation would be particularly timely as so much effort is being devoted to the study of morphological and physiological variation within and between plant and animal populations (Anderson, 1954; Prosser, 1955).

Parting biometrical variation into size and shape components is often highly desirable as the size of most organisms is more affected than their shape by fluctuations of the external environment; size variation is also more likely to reflect heterogeneity of samples with respect to age-composition. Shape tends thus generally to provide more reliable indications than size on the internal constitution of organisms; this makes the analysis of size and shape a basic step in the study of biometrical variation.

The present study of the Painted Turtle (*Chrysemys picta marginata*) is an attempt to evaluate the applicability of principal component analysis to size and shape variation in groups of living organisms. The beginnings of principal component analysis are probably to be found in the works of Karl Pearson (1901). The statistical properties of principal components were investigated in detail by

[1] Contribution No. 52 from l'Institut de Biologie, Université de Montréal, Montréal, Canada.

[2] Present address: Department of Biostatistics, Johns Hopkins University, Baltimore, Maryland, U.S.A.

Hotelling in 1933. Anderson (1958) has given the most comprehensive recent exposition. Principal component and related forms of multivariate analysis have been used by a number of authors (Blackith, 1960; Teissier, 1955; Wright, 1954) in attempts to describe complex growth patterns in terms of a minimum number of basic trends. Recent biometrical studies of character-complexes (Bailey, 1956; Kraus and Choi, 1958; Olson and Miller, 1958) have made increasingly clear that, whenever the discovery of genetical, developmental and functional relationships is contemplated, joint consideration of all characters becomes indispensable and multivariate statistical techniques are indicated. However, the relationships of principal components with the concepts of size and shape do not appear to have been ever examined in detail. This is the main purpose of the present note. A comparative analysis of sex dimorphism in several species of turtles will be presented later.

Among the vertebrates, turtles are perhaps those that are most readily suitable to studies of variation in overall size and shape. Due to a rigid bony carapace, consistent measurements of body dimensions can be obtained. Further, interesting changes of proportions occur in the shell during growth (Mosimann, 1958). Sexual dimorphism in body size and proportions is common. Finally, for aquatic species like the one studied here, shape is probably of major adaptive significance because of the importance of streamlining for locomotion in a dense fluid medium.

Material and Data

Eighty-four specimens of Midland Painted Turtles (*Chrysemys picta marginata*) were collected in a single day (August 2, 1956) from a small stagnant pond of the St. Lawrence Valley, at Coteau Landing, 35 miles southwest of Montreal, Canada. This material can therefore safely be assumed to represent a single local population. Specimens are preserved in the Department of Biology of the University of Montreal. Only individuals for which the sex was externally discernible are included in the present analysis, one abnormal male being excluded. Carapace dimensions were measured in three mutually perpendicular directions of space: length, maximum width, and height. More detailed definitions of these measurements have been given before (Mosimann, 1958). Thus each specimen is represented in this study by a set of three measurements

(length, width, height)

hereafter designated

$$X = (X_1, X_2, X_3)$$

for convenience. Measurements rounded to the nearest millimeter are listed in Table 1. Several facts are explicit from examination of the data. (1) In general, long individuals are also wide and high, and inversely, short individuals are narrow and low. Clearly this reflects the fact that length, width and height are influenced by one general factor of variation, size. (2) However, some individuals of the same length show different widths or heights and these differences of proportions constitute shape variation. (3) In addition it can be seen that females attain a larger size than males and also (4) that they tend to be higher relative to length.

TABLE 1
CARAPACE DIMENSIONS OF PAINTED TURTLES (*Chrysemys picta marginata*) IN MM.

24 Males			24 Females		
length	width	height	length	width	height
93	74	37	98	81	38
94	78	35	103	84	38
96	80	35	103	86	42
101	84	39	105	86	40
102	85	38	109	88	44
103	81	37	123	92	50
104	83	39	123	95	46
106	83	39	133	99	51
107	82	38	133	102	51
112	89	40	133	102	51
113	88	40	134	100	48
114	86	40	136	102	49
116	90	43	137	98	51
117	90	41	138	99	51
117	91	41	141	105	53
119	93	41	147	108	57
120	89	40	149	107	55
120	93	44	153	107	56
121	95	42	155	115	63
125	93	45	155	117	60
127	96	45	158	115	62
128	95	45	159	118	63
131	95	46	162	124	61
135	106	47	177	132	67

BIVARIATE ANALYSIS

Whereas the preceding points can be detected by inspection of the data, they are brought out more clearly by plotting pairs of measurements of the various individuals as dots in cartesian coordinate systems. In the resulting bivariate scatter diagrams (Figures 1 and 2), the facts noted above are now graphically evident: (1) dots are spread along the diagonal of each diagram according to size differences and (2) scattered about it because of shape variation. (3) In both diagrams females extend further to the right than males since they grow larger and (4), in the first diagram, they tend to be above males of equal length because of greater relative height.

These phenomena have been made considerably more obvious by outlining the clusters of dots with ninety-five per cent equal frequency ellipses. The equation of the latter can be written in matrix notation as:

$$(X - \overline{X})\, W^{-1}\, (X - \overline{X})' = c^2$$

where $X = (X_1, X_2)$, the pair of coordinates of the locus of the ellipse; $X = \begin{bmatrix} X_1 \\ X_2 \end{bmatrix}$, the column-vector obtained by transposing X; $\overline{X} = (\overline{X}_1, \overline{X}_2)$, the pair of coordinates of the center of the ellipse and the row-vector of sample means; $W = \begin{bmatrix} W_{11} & W_{12} \\ W_{21} & W_{22} \end{bmatrix}$, the inverse of the matrix of the equation of the ellipse and the covariance matrix of the sample. The value of c^2 is equal to that of chi-square with two degrees of freedom (Anderson, 1958: 54), which is 5.99 at the 95 per cent level. The elements of the bivariate mean vectors and covariance matrices for Figures 1 and 2 are all included in the trivariate statistics (Table 2).

To determine an equal frequency ellipse, the direction cosines $U = \begin{bmatrix} U_{11} & U_{12} \\ U_{21} & U_{22} \end{bmatrix}$ of its principal axes and the diagonal form $\Lambda = \begin{bmatrix} \lambda_1 & 0 \\ 0 & \lambda_2 \end{bmatrix}$ of its covariance matrix are calculated, the subscripts in U and Λ always being ordered by convention so that $\lambda_1 > \lambda_2$. In this study, the elements of matrix $U = \begin{bmatrix} \cos\theta & \sin\theta \\ -\sin\theta & \cos\theta \end{bmatrix}$ were obtained from 5-place trigonometric tables after evaluating angle θ from $\tan 2\theta = \frac{2\, W_{12}}{W_{11} - W_{22}}$ and Λ was then calculated as $\Lambda = UWU'$, where

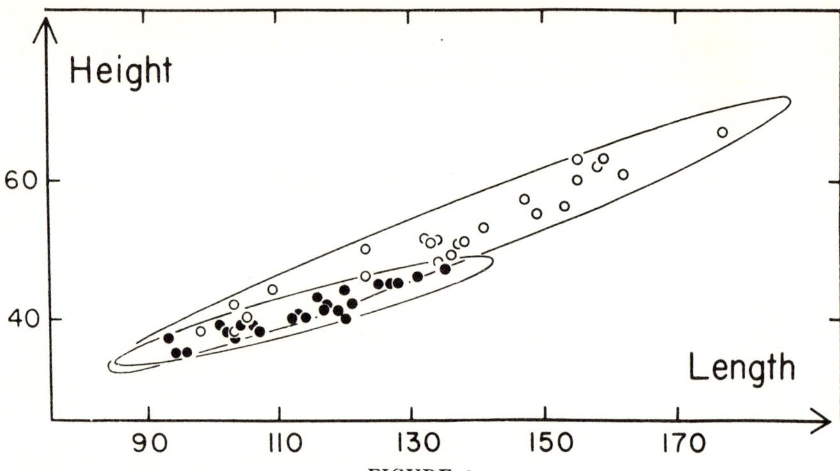

FIGURE 1
Bivariate scatter diagram of carapace height and length. Males, solid circles; females, hollow circles.

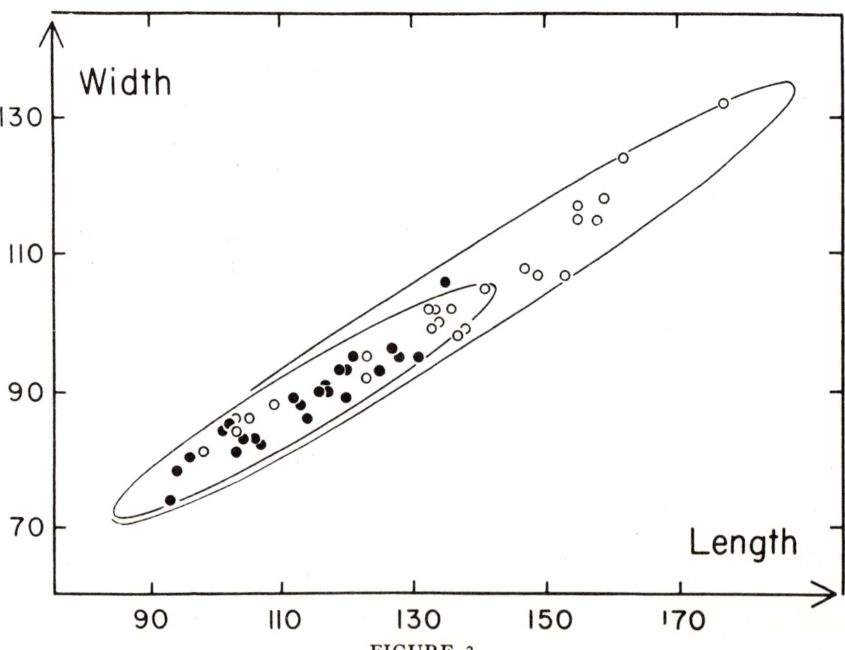

FIGURE 2
Bivariate scatter diagram of carapace width and length. Males, solid circles; females, hollow circles.

U' is the transpose of matrix U. The principal components of the covariance matrix W are the variances λ_1 and λ_2 of the observations along the principal axes of their equal frequency ellipses, the direction cosines of these axes being (U_{11}, U_{12}) for the first and (U_{21}, U_{22}) for the second.

Providing the coordinate axes are graduated in the same scale, the extremities of the principal axes of a 95 per cent equal frequency ellipse are determined by the following vector equations:

$$\text{major axis} \qquad (X_1, X_2) = (\overline{X}_1, \overline{X}_2) \pm \sqrt{5.99\lambda_1}\,(U_{11}, U_{12})$$
$$\text{minor axis} \qquad (X_1, X_2) = (\overline{X}_1, \overline{X}_2) \pm \sqrt{5.99\lambda_2}\,(U_{21}, U_{22}).$$

The above equations lay bare the analogy of bivariate with univariate confidence intervals. After these extremities have been plotted, the ellipse can be graphed either by means of an ellipsograph (such as the one made by Riefler) or following a standard drafting procedure.

Were the parameter mean vector and covariance matrix known, and were the distribution of our observations exactly bivariate normal, then equal frequency ellipses would be true confidence intervals. Although this is not the case, our ellipses are estimates of true confidence intervals. Moreover, their main raison d'être in the present study is their descriptive validity, and the latter is confirmed by visual comparison with the clusters of dots of the data. On the other hand, with small samples, the effect of extreme observations is considerable and equal frequency ellipses should be used with caution. Their biometrical utilization has been thoroughly discussed by Defrise-Gussenhoven (1955).

Trivariate Analysis

While bivariate scatter diagrams with equal frequency ellipses have brought out size and shape differences rather satisfactorily, the present problem is basically trivariate. The well-justified popularity of bivariate techniques stems no doubt largely from the relative simplicity of bivariate computations as well as from the ease of preparation of bivariate scatter diagrams. Graphing is considerably more difficult in three than in two dimensions, and it becomes geometrically impossible when more than three variables are analyzed jointly. Algebraically and numerically, however, multivariate statistical techniques remain identical in pattern beyond the bivariate level.

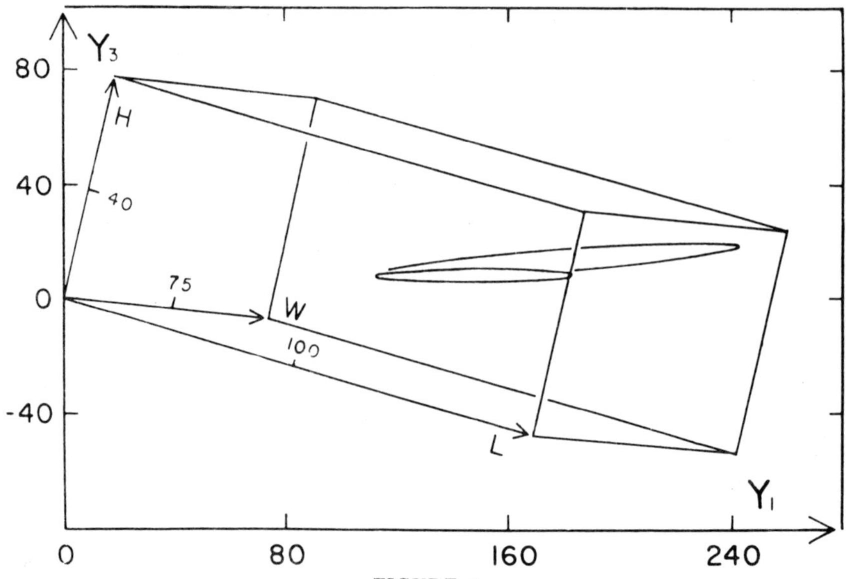

FIGURE 3
The trivariate scatter diagram of carapace length, width and height viewed perpendicularly to the first and third principal axes of males.

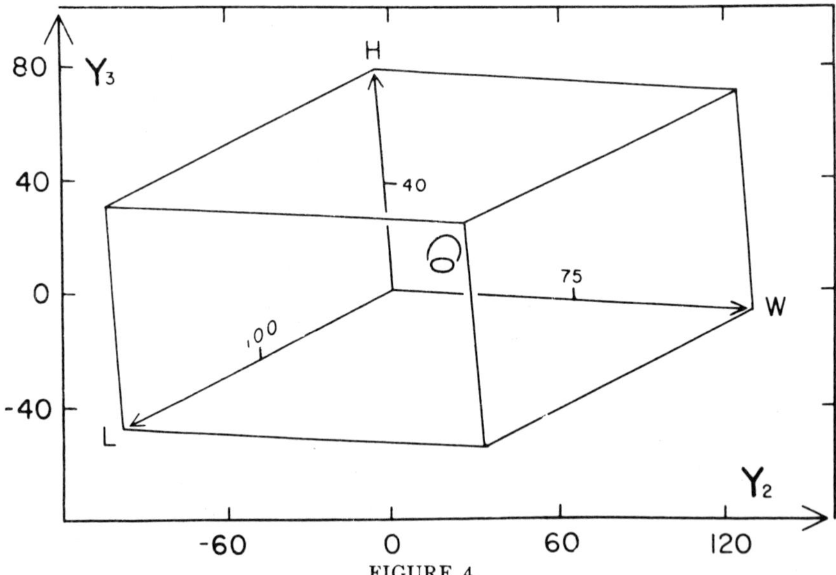

FIGURE 4
The trivariate scatter diagram of carapace length, width and height viewed perpendicularly to the second and third principal axes of males.

Inasmuch as the frequency distribution of observations is approximately multivariate normal, a sample can be summarized by its mean vector $\bar{X}$ and covariance matrix W. In the trivariate case,

$$\bar{X} = (\bar{X}_1, \bar{X}_2, \bar{X}_3) \text{ and } W = \begin{bmatrix} W_{11} & W_{12} & W_{13} \\ W_{21} & W_{22} & W_{23} \\ W_{31} & W_{32} & W_{33} \end{bmatrix}$$

The trivariate statistics relative to (length, width, height) in the present study are listed in Table 2. The elements of the mean vector of females are greater than those of males, reflecting the greater size of the former sex. The elements of covariance matrices are also greater in females, which reflects a greater range of size variation.

Earlier in this paper the analysis was restricted to that of the marginal distributions of the variables two by two. All components of $\bar{X}$ and W were perforce ignored at any one time except those corresponding to two of the variables. Thus the components of $\bar{X}$ other than $(\bar{X}_1, \bar{X}_3)$ and those of W other than $\begin{bmatrix} W_{11} & W_{13} \\ W_{31} & W_{33} \end{bmatrix}$ were automatically discarded in the bivariate analysis of (X_1, X_3). In other words, carapace length and height were analyzed as if width had not been measured, etc. Multivariate techniques alone can take all variables into account jointly and thus provide a unified analytical approach.

TABLE 2

MEAN VECTORS $\bar{X}$ AND COVARIANCE MATRICES W

	24 Males			24 Females		
	length	width	height	length	width	height
$\bar{X}$	(113.38	88.29	40.71)	(136.00	102.58	51.96)
W	138.77	79.15	37.38	451.39	271.17	168.70
	79.15	50.04	21.65	271.17	171.73	103.29
	37.38	21.65	11.26	168.70	103.29	66.65

With the present trivariate data, a joint scatter diagram of all variables is fortunately still possible. Slanted views of the tridimensional scatter diagram have been calculated and illustrated (Figures 3 and 4), the clusters of male and female dots being represented by ellipsoids.

Analogously with bivariate principal components, multivariate principal components are the variances of observations along the principal axes of multidimensional equal frequency ellipsoids. In the present case, with three variables, the latter are tridimensional ellipsoids. The evaluation of the direction cosines $U = \begin{bmatrix} U_{11} & U_{12} & U_{13} \\ U_{21} & U_{22} & U_{23} \\ U_{31} & U_{32} & U_{33} \end{bmatrix}$ of the principal axes and of the corresponding covariance matrix $\Lambda = \begin{bmatrix} \lambda_1 & 0 & 0 \\ 0 & \lambda_2 & 0 \\ 0 & 0 & \lambda_3 \end{bmatrix}$ is an eigenvalue and eigenvector problem. Its best general solution is probably the Jacobi method of matrix diagonalization (White, 1958: 395), a generalization of the procedure already described for bivariate covariance matrices The results of principal component analysis of our data are given in Table 3.

TABLE 3
COVARIANCE MATRICES Λ AND MATRICES OF DIRECTION COSINES U OF THE PRINCIPAL AXES

	24 Males			24 Females		
Λ	195.28	0.00	0.00	680.40	0.00	0.00
	0.00	3.69	0.00	0.00	6.50	0.00
	0.00	0.00	1.10	0.00	0.00	2.86
U	.84012	.49190	.22854	.81263	.49549	.30676
	—.48811	.86938	—.07696	—.54537	.83213	.10062
	—.23654	—.04690	.97049	—.20540	—.24907	.94645

TYPES OF VARIATION CORRESPONDING TO THE PRINCIPAL COMPONENTS

The types of variation occurring in the directions spanned by the principal axes can best be judged from the equations of the latter. Thus the equation of the first principal axis of males is

$$\frac{X_1 - \overline{X}_1}{U_{11}} = \frac{X_2 - \overline{X}_2}{U_{12}} = \frac{X_3 - \overline{X}_3}{U_{13}} = Y_1 - \overline{Y}_1$$

or

$$\frac{X_1 - 113.38}{.84012} = \frac{X_2 - 88.29}{.49190} = \frac{X_3 - 40.71}{.22854} = Y_1 - 147.99$$

upon substitution of the numerical values given in Table 3. The value of Y_1 is obtained by rotation of X as explained later. In fact, for a given set $X = (X_1, X_2, X_3)$, the value of Y_1 is equal to that of the linear compound $U_{11}X_1 + U_{12}X_2 + U_{13}X_3$. It is noteworthy that the direction cosines of the Y_1-axis are all positive. Because of this, the equation of the major axis corresponds to a simultaneous increase (or decrease) of all variables. For instance, a positive deviation of 100 units in Y_1 corresponds approximately to deviations of (+84, +49, +23) millimeters in (X_1, X_2, X_3), that is in carapace (length, width, height). For this reason, and since growth is generally defined as the increase in size of an organism, the first principal component is interpreted as a growth trend. The use of a linear compound like $Y_1 = U_{11}X_1 + U_{12}X_2 + U_{13}X_3$ as a size measure should be very practical in many types of studies, although it calls for careful distinction from the usual volumetric or ponderal definitions of size. Size was for instance defined as a volume estimate from a non-linear compound ($Y_1 = k\ X_1\ X_2\ X_3$) and compared with direct volume measurements by Mosimann (1958).

A feature of special significance is the difference in relative magnitude of the direction cosines of male and female major axes, particularly those corresponding to carapace height. U_{13} equals .22854 and .30676 in males and females respectively. A positive deviation of 100 units in size thus corresponds approximately to a deviation in height of +31 mm. in females, but only to +23 mm. in males. This expresses the lower relative growth rate of carapace height in males, a phenomenon noticeable in Figure 1 and brought out previously by regression analysis (Mosimann, 1958).

The second principal axis of males has the equation

$$\frac{X_1 - \overline{X}_1}{U_{21}} = \frac{X_2 - \overline{X}_2}{U_{22}} = \frac{X_3 - \overline{X}_3}{U_{23}} = Y_2 - \overline{Y}_2$$

or

$$\frac{X_1 - 113.38}{-.48811} = \frac{X_2 - 88.29}{.86938} = \frac{X_3 - 40.71}{-.07696} = Y_2 - 18.28$$

with numerical values. The direction cosines of this second axis differ in sign, and this trend of variation corresponds to a joint increase of one variable and decrease of others. A positive deviation of 100 units in Y_2 would correspond approximately to deviations of (—49, +87, —8) millimeters in carapace (length, width, height). This second principal component is thus a trend of shape variation. The direction cosines of the third principal axis also differ in sign, and the third principal component is similarly interpreted as shape variation.

The results of the analysis of size and shape variation of the present data are summarized in Table 4. It is seen that females are much more variable than males in size and slightly more so in shape.

TABLE 4
SIZE AND SHAPE VARIATION

	24 Males			24 Females		
Principal axes	1st (major)	2nd (intermediate)	3rd (minor)	1st (major)	2nd (intermediate)	3rd (minor)
Magnitude of variance	195.28	3.69	1.10	680.40	6.50	2.86
% of total	97.61	1.84	0.55	98.64	0.94	0.41
Nature of variation	*Size* joint variation of all dimensions	*Shape* contrast of length vs. width mostly	*Shape* contrast of length vs. height mostly	*Size* joint variation of all dimensions	*Shape* contrast of length vs. width mostly	*Shape* contrast of length and width vs. height

SCATTER DIAGRAMS OF SIZE AND SHAPE

The principal axes of an ellipsoid are mutually perpendicular. The principal axes of equal frequency ellipsoids can therefore be used as new orthogonal coordinate axes. Heretofore the sets of measurements (length, width, height) have been interpreted as dots in the orthogonal coordinate system of variables $X = (X_1, X_2, X_3)$ or

$$X' = \begin{bmatrix} X_1 \\ X_2 \\ X_3 \end{bmatrix} \text{ in column-vector notation.}$$

The new coordinates Y of the same dots on principal axes with direction cosines U are given by $Y' = UX'$

$$\text{or} \quad \begin{bmatrix} Y_1 \\ Y_2 \\ Y_3 \end{bmatrix} = \begin{bmatrix} U_{11} & U_{12} & U_{13} \\ U_{21} & U_{22} & U_{23} \\ U_{31} & U_{32} & U_{33} \end{bmatrix} \begin{bmatrix} X_1 \\ X_2 \\ X_3 \end{bmatrix}$$

in extended matrix notation. The operation transforming the X- into the Y-coordinates is commonly referred to as a rotation. The sets of values of the new variables have been represented in two-dimensional scatter diagrams (Figures 5 and 6). The latter constitute views of the tridimensional diagram (Figures 3 and 4) seen in the direction of principal axes.

To determine 95 per cent equal frequency ellipses of groups of dots in bivariate scatter diagrams of the elements of $Y = (Y_1, Y_2, Y_3)$, the covariance matrices of these new variables must be known. These could of course be calculated directly from the values of Y. However, it is much simpler to rotate the X-covariance matrices to the Y-axes. Thus a sample with covariance matrix W on the X-axes has covariance matrix UWU' on the Y-axes since $Y' = UX'$. Upon rotation to the male principal axes, the male covariance matrix assumes its diagonal form, Λ, while the female covariance matrix remains non-diagonalized to the extent to which it differs from that of males. The reverse situation occurs when rotating to female principal axes.

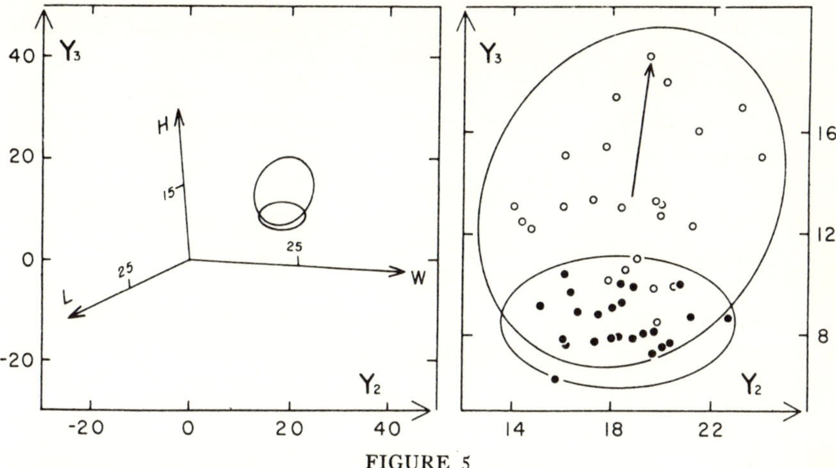

FIGURE 5
View perpendicular to the second and third principal axes of males; blown-up portion of Figure 4.

The cross sections of the male (Figure 5) and female (Figure 6) ellipsoids in the plane of (Y_2, Y_3) are both flattened in the direction of height. This suggests that there is less shape variation in carapace height than in carapace length and width. This may have no biological significance, however, since the order of magnitude of carapace height is smaller than that of length and width and all dimensions are expressed in millimeters. The influence of scales of measurement on principal components is discussed by Anderson (1958: 279). Standardizing variables (Teissier, 1955: 345) is certainly often preferable when only one group of observations is considered; the standard deviation of each variable is taken then as its unit and the covariance matrix becomes the correlation matrix. In the present case, however, this was not done as it would have led to different scales of measurement for males and females.

The cluster of female observations is clearly above that of males in Figures 5 and 6; this expresses the markedly greater carapace height of females. Although principal component analysis brings out well this difference of shape between the sexes, analyzing multivariate differences between groups of observations calls generally for discriminant functions. Discriminatory analysis leads to trends of maximum variation between groups of observations in the same way as principal components lead to trends of maximum variation within groups. The use of discriminant functions has been illustrated and discussed recently by Jolicoeur (1959). However, in the present study, the major interest lies in the nature and the magnitude of variation within local populations, and this calls for principal component rather than discriminatory analysis.

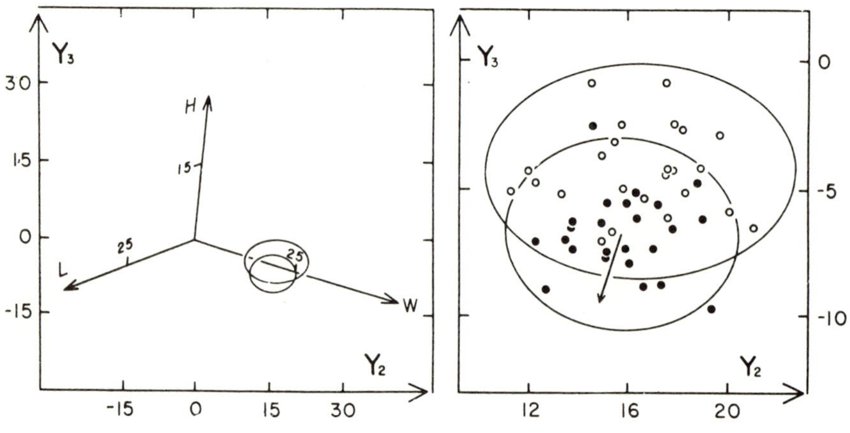

FIGURE 6
View perpendicular to the second and third principal axes of females.

Divergence in Shape During Growth

The major axis of each sex projects merely as a point on the plane of its own second and third principal axes (Figures 5 and 6). Correspondingly, size variation is absent in the scatter diagram of each sex with respect to its second and third principal components. On the other hand, male and female major axes are not parallel and the size variation of each sex is present with respect to the second and third principal components of the other sex. Thus large females tend to congregate upward from their mean in Figure 5 and large males downward from their mean in Figure 6.

The extent and the manner in which males and females diverge in shape as they increase in size can be judged from the projection of the major axis of each sex on the plane of the second and third axes of the other. These projections are indicated by arrows (upwards from the mean of females in Figure 5; downwards from the mean of males in Figure 6). Each arrow corresponds to a length of $\sqrt{5.99\lambda_1}$ units and its extremity can thus be visualized as the tip of an ellipsoid whose projections are the 95 per cent equal frequency ellipses represented. It is clear from this graphical analysis that there is a progressive divergence of relative carapace height between the sexes during growth. Also, carapace width appears to grow at a slightly higher rate in females.

Conclusions

Painted Turtles exhibit considerable variation in carapace dimensions. While the general nature and the magnitude of variation can be perceived by inspection of the data, they are made much more explicit by bivariate scatter diagrams and equal frequency ellipses. When considering multiple characters, however, multivariate statistical techniques are necessary to a unified analytical approach. Applied to carapace length, width and height, principal component analysis discloses three uncorrelated and mutually perpendicular trends of variation. The first principal component corresponds to a direction of size increase and can therefore be interpreted as a growth trend. The second and third principal components correspond to disjoint variation of the various characters and are consequently interpreted as trends of shape variation. Female painted turtles vary much more than males in size and slightly more in shape. While small turtles of both sexes are of comparable size and shape, females reach a greater size and their carapace becomes relatively higher than that of males.

The divergence in shape of males and females during growth is expressed by the difference in direction between the major axes of their covariance matrices. But covariance matrices are necessarily unequal when their principal components are not identical in direction and in length. The very fact that groups of living organisms have unequal covariance matrices can thus be of considerable biological significance since the way in which shape differences relate to size differences is of major importance in all biological disciplines directly or indirectly concerned with development and growth. Assuming covariance matrices of sets of data to be equal, as often done in theoretical statistics, would therefore frequently be unrealistic and misleading from the viewpoint of the biologist. Proper attention can be paid to growth divergences between groups of living organisms if principal component analysis is used. The latter seems the most promising method for the study of size and shape variation in complexes of biometrical characters.

The use of straight or curved lines is very common in relative growth studies and it has been customary to regard departures of data from growth curves as having little biological meaning. Such an attitude is difficult to maintain, however, when deviations of biometrical characters from growth curves are found to be correlated. The description of relative growth data in terms of several axes of variation, one of size and the others of shape, appears therefore more appropriate than their description in terms of a simple growth curve. The latter would generally constitute an over-simplification and would result in a loss of biological information.

At first, principal component analysis is most likely to be regarded by the non-mathematician as a highly arbitrary set of manipulations. Such a reaction should be readily dismissed, however, as soon as the geometrical meaning of the method is considered: principal component analysis merely leads to new angles of viewing data, angles best suited to disclose the nature and the magnitude of size and shape variation.

Summary

The applicability of principal component analysis to size and shape variation in living organisms is illustrated by a study of male and female painted turtles. Numerical and geometrical aspects are discussed in detail and it is concluded that principal component analysis of size and shape variation would be appropriate in relative growth studies whenever multiple biometrical characters are considered.

Acknowledgments

Research reported herein was supported by the National Research Council of Canada. Assistance in collecting and measuring specimens was furnished by J. R. Bider. Pierre Bernier assisted in the calculations. To the above we are grateful.

References

1. Anderson, E. 1954. Efficient and inefficient methods of measuring specific differences. Chapter 6, 93-106 in *Statistics and Mathematics in Biology*. Iowa State College Press, Ames.
2. Anderson, T. W. 1958. An Introduction to Multivariate Statistical Analysis. xii + 374. John Wiley & Sons, Inc. New York.
3. Bailey, D. W. 1956. A comparison of genetic and environmental principal components of morphogenesis in mice. *Growth,* **20**, 63-74.
4. Blackith, R. E. 1960. A synthesis of multivariate techniques to distinguish patterns of growth in grasshoppers. *Biometrics* **16**, 28-40.
5. Defrise-Gussenhoven, E. 1955. Ellipses équiprobables et taux d'éloignement en biométrie. *Bull. Inst. Royal Sci. Nat. Belgique,* **31**. Bruxelles.
6. Hotelling, H. 1933. Analysis of a complex of statistical variates into principal components. *Jour. Educ. Psych.,* **24**, 417-441, 498-520.
7. Jolicoeur, P. 1959. Multivariate geographical variation in the wolf *Canis lupus* L. *Evolution,* **13**, 283-299.
8. Kraus, B. S., and S. C. Choi. 1958. A factorial analysis of the prenatal growth of the human skeleton. *Growth,* **22**, 231-242.
9. Mosimann, J. E. 1958. An analysis of allometry in the chelonian shell. *Rev. Can. Biol.,* **17**, 137-228.
10. Olson, E. C., and R. L. Miller. 1958. Morphological Integration. xv + 317. University of Chicago Press, Chicago.
11. Pearson, Karl. 1901. On lines and planes of closest fit to systems of points in space. *Philos. Mag.,* Ser. 6, **2**, 559-572.
12. Prosser, C. L. 1955. Physiological variation in animals. *Cambridge Philos. Soc., Biol. Reviews,* **30**, 229-262.
13. Simpson, G. G., Roe, A., & Lewontin, R. C. 1960. Quantitative Zoology. (rev. ed.) vii + 440. Harcourt, Brace and Company, Inc. New York.
14. Teissier, G. 1955. Sur la détermination de l'axe d'un nuage rectiligne de points. *Biometrics,* **11**, 344-356.
15. White, P. A. 1958. The computation of eigenvalues and eigenvectors of a matrix. *Jour Soc. Industr. Appl. Math.,* **6**, 393-437.
16. Wright, S. 1954. The interpretation of multivariate systems. Chapter 2, 11-33 in *Statistics and Mathematics in Biology*. Iowa State College Press, Ames.

6

Reprinted from *Ecological Aspects of the Mineral Nutrition of Plants* (British Ecol. Soc. Symp. 9), I. H. Rorison, ed., Blackwell Scientific Publications, Oxford, 1969, pp. 37–66

THE APPLICATION OF ORDINATION TECHNIQUES

R. GITTINS

Institute of Statistics, North Carolina State University, Raleigh

INTRODUCTION

Study of the causal factors determining the distribution of plants and vegetation is a prime objective of ecology. The number of factors affecting plants, however, is very large and an initial problem in many investigations is to identify those factors which are most likely to be of overriding importance in determining the occurrence of particular species and kinds of vegetation. Ordination is a means of analyzing field observations for the purpose of recognizing such factors. The ultimate object of studies of this kind is to explain ecological behaviour in terms of the physiology of the individual plant. The term ordination was introduced and defined by Goodall (1954a) as, 'an arrangement of units in a uni- or multi-dimensional order'. In the context of vegetation, the units involved are generally either samples of vegetation or plant species. Information characterizing these entities is obtained by observation and measurement in the field. The resulting data form the starting point for ordination. These data usually consist of estimates of the composition of a number of vegetation samples in terms of the abundances of the species represented. Ordination involves associating these entities—either the stands *or* species—in some way such that the arrangement arrived at provides insight into the ecological processes which may have generated the observations. The assumption is made at the outset that a latent structure or organization of some kind exists within the observation set.

Ordination is an exploratory technique of greatest value in the intitial stages of a study. Its use is particularly appropriate where the connections between species' distribution and environmental variation are difficult to discern and crucial experiments difficult to conceive. It is the large number of variables involved and the complexity of their inter-relations which are responsible for these difficulties. In situations of this kind, ordination has proved valuable as a means of creating concepts or hypotheses bearing on the nature and effects of the major controls or sources of variation at work. The aim of analysis is to reveal precisely those relationships between plants and environment which are most likely to repay closer investigation.

Once these relationships have been provisionally identified, their effects in particular cases can be examined by controlled experiment. In this way the measurement foundations for a given ecological situation can gradually be established.

Recent developments in this field stem largely from the work of Goodall (1954b), Curtis (e.g. Bray and Curtis 1957; Curtis, 1959) and Dagnelie (1960, 1965a, b). The name ordination embraces a group of methods rather than a single technique. A number of these methods have been in existence for almost 40 years. It was not until the appearance of the studies mentioned, together with those of Greig-Smith (1957, 1964), however, that a general awareness of the possibilities of the approach spread to ecologists. The essential features of ordination will be illustrated here by reference to principal components analysis.

PRINCIPAL COMPONENTS ANALYSIS

Several excellent accounts of principal components analysis in ecological and more general biological contexts are available (see, for example, Kendall 1957; Reyment 1961, 1963; Seal 1964; van Groenewoud 1965; Orloci 1966, 1967). These, however, all assume at least some familiarity with the procedures of multivariate statistical analysis. An attempt is therefore made here to convey the essential ideas of components analysis in a relatively non-technical way. Recourse will be made to a geometric treatment wherever possible.

(1) *Geometric properties of field observations*

The initial observations in studies of vegetation can frequently be expressed in the form:

		Individuals (stands)				
		1	2	3	...	N
Attributes (species)	1	X_{11}	X_{12}	X_{13}	...	X_{1N}
	2	X_{21}	X_{22}	X_{23}	...	X_{2N}
	3	X_{31}	X_{32}	X_{33}	...	X_{3N}
	.	.	.	.	...	.
	n	X_{n1}	X_{n2}	X_{n3}	...	X_{nN}

where the X_{ji} represent the contributions of n species or other variates ($j = 1, 2, \ldots n$) to N samples (stands) of vegetation ($i = 1, 2, \ldots N$). An array of this kind is referred to as a data matrix. Any particular entry X_{ji} constitutes an *observed value*, the subscripts j and i identifying respectively the row and column in which the observation is located. In general, the X_{ji} consist of estimates of the representation of species j in the ith stand, the estimates being in terms of any suitable measure (cover, density, etc.). The *rows* of the matrix, therefore, indicate variation in the representation of the jth species over the N stands examined, while the columns indicate the composition of the ith stand with respect to the n species observed. Attributes† other than species, such, for example, as environmental characteristics of the sites may on occasion also be included in the data matrix.

The principal objectives in the analysis of vegetation are frequently: (i) to describe its composition and structure, indicating perhaps how it might be classified, and (ii) to identify the causes or processes which underlie and determine the behavior of the individual species which collectively form the vegetation. It is with the second of these that we are concerned here. Since the data matrix contains essentially all the information available for analysis, we must endeavor to operate on this in such a way that this objective is realized. We are faced, therefore, with the task of obtaining insight into the nature of the processes underlying differences in species' behavior, by means of suitable treatment or manipulation of the matrix of estimates of their abundance. Before turning to consider this problem, it will be helpful to consider some geometric properties of the data matrix. Although the observed values may be, and frequently are, analysed in their original form, it will be convenient for our purpose to consider them in standard form. This involves replacing each observed value X_{ji} by a new value z_{ji} where $z_{ji} = (X_{ji} - \bar{X}_j)/s_j$, $\bar{X}_j$ being the sample mean and s_j the sample standard deviation for species j. The set of all values z_{ji} ($i = 1, 2, \ldots N$) is called an attribute j in standard form. This transformation results in the origin of measurement being placed at the mean value of each attribute, and in the original unit or units of measurement being replaced by a new unit, the standard deviation. The end result of analysis is not affected by this transformation.

A useful concept where many variates are considered simultaneously is that of the sample space. By means of this, the behavior or variation of an attribute over a number of stands can be represented by a point or by a

† The terms attribute and variate are used interchangeably.

vector in higher dimensional space. This mode of representation is especially valuable where the attributes turn out to be merely different functions of a small number of independent variables. Where a number of attributes have been measured over *N* individuals, we may set up a Euclidean space of *N* dimensions, one dimension for each of the *N* individuals (stands). The entries of each row z_j: $(z_{j1}, z_{j2}, \ldots z_{jN})$ of the data matrix in standard form can then be regarded as the coordinates of a point representing the *j*th

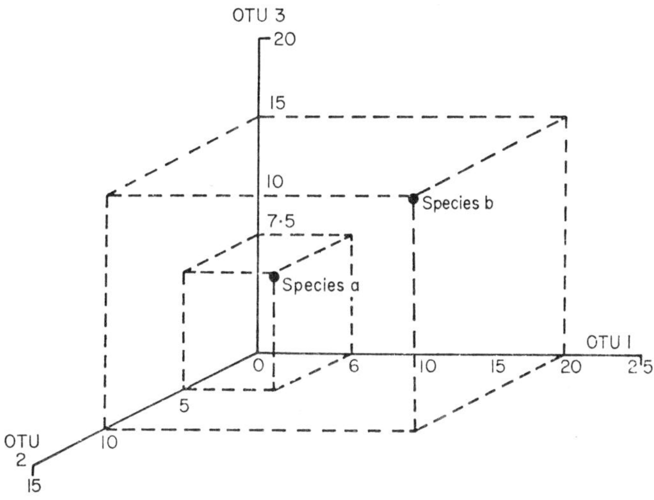

FIG. 1. Point representation of two species in sample space (N=3). The coordinate axes correspond to stands, the scales being in terms of any suitable measure of abundance, e.g. % cover. The position of points is uniquely determined by the abundance of the species in each stand. For OTU read stand. (Modified from van Groenewoud, 1965, by courtesy of *Ber. geobot. Inst. ETH, Stiftg. Rubel, Zürich*).

attribute in the *N* space. The position of each attribute will be uniquely determined by its particular combination of observed values. Species or attributes with similar sets of values will therefore occupy adjacent positions in the sample space. It follows that the positions and spatial relationships of the points are related to the behavior of the attributes corresponding to them over the vegetation samples examined. The closer the points, the greater the ecological similarity of the attributes and *vice versa*. If each point is joined to the origin of the coordinate system by a line, the resulting configuration is called the vector representation of the variables. In this way, the entire observation set can be regarded as determining a set of points or vectors corresponding to variates in a space, the dimension of which equals the number of stands examined. Examples of the point and vector representation of hypothetical sets of observations are shown diagrammatically in Figs. 1 and 2. Although an attempt is made to show only

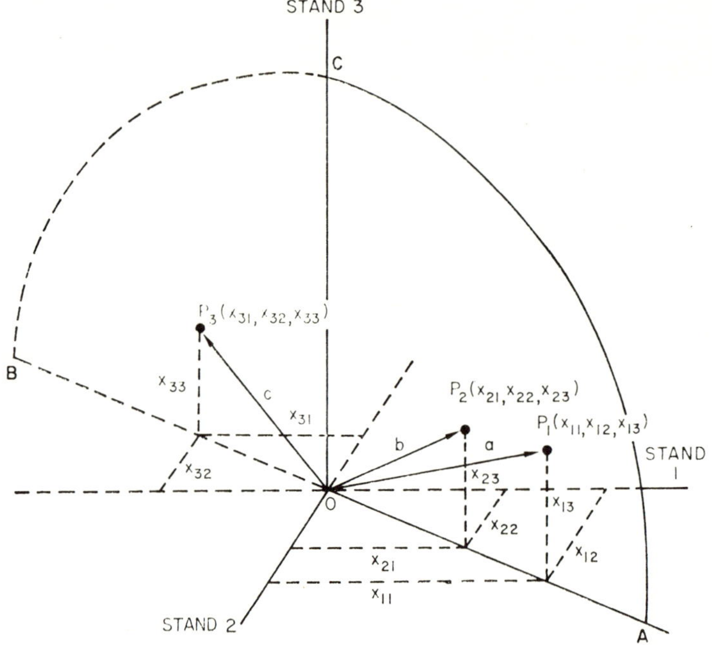

FIG. 2. Vector representation of three species (P_1, P_2, P_3) in sample space ($N=3$). The position of points is determined by species' abundance in the stands, e.g. P_1 by X_{11}, X_{12}, X_{13}. The vectors may then be obtained by drawing a line from the origin to each of the points. The vectors shown all lie in the plane ABC, i.e. within a sub-space of 3-dimensional sample space. (From van Groenewoud, 1965, by courtesy of *Ber. geobot. Inst. ETH, Stiftg. Rübel, Zürich.*)

three dimensions here (which correspond to three stands), no difficulty arises in practice in dealing with many dimensions (stands), at least if electronic computational facilities are available.†

† An alternative geometric representation of the data matrix is possible. A space of n dimensions may be postulated where each dimension corresponds to an attribute (species). The column entries z_i: $(z_{1i}, z_{2i}, \ldots z_{ni})$, $i=1, 2, \ldots N$ can then be used as coordinates locating the ith *stand* in n dimensional attribute-space. (See Goodall, 1963; van Groenewoud, 1965; Williams and Dale, 1965). In the original model the points generally correspond to species, while in the second, they correspond to stands.

(2) *Geometric structure of a correlation matrix*

Interest in the point or vector representation of the field observations centers on the spatial relationships of the points. These relations may be measured and analyzed in terms either of the distance between pairs of points or in terms of the angular separation between pairs of vectors linking points to the origin (see, for example, van Groenewoud 1965). We will consider here only the latter. There is a simple relationship between the

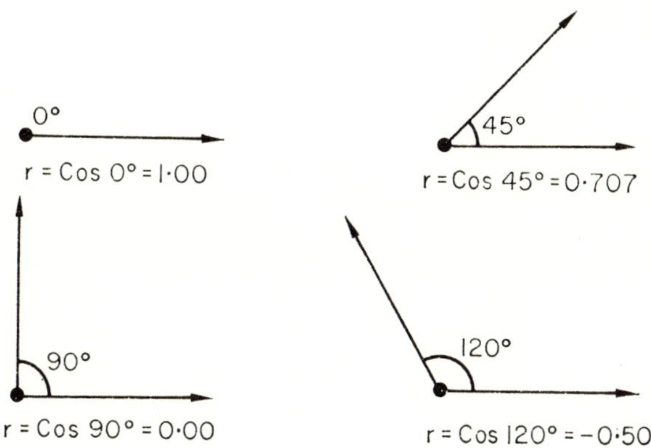

FIG. 3. Vector representation of the correlation coefficient, *r*, for selected values of *r* (see text). (From Fruchter, 1954, by courtesy of D. Van Nostrand Company, Inc.)

vector representation of a pair of species and the correlation coefficient relating the observed values of the species over a number of stands. If the variates are expressed in standard form, the cosine of the angle between the vectors in *N* space is identical to the value of the correlation coefficient between the variates. This may be expressed

$$\cos \phi_{jk} = r_{jk} \qquad (j, k = 1, 2, \ldots n), \tag{1}$$

where ϕ_{jk} is the angle of separation between vectors *j* and *k* and r_{jk} is the correlation coefficient between the representation of variates *j* and *k* over the *N* stands. The correlation coefficient, *r*, can therefore be used for analyzing the relations between attributes conceived of as points in sample space at the tips of vectors extending from the origin. The relationship between the angular separation of vectors and the value of the corresponding coefficient is an inverse one. The smaller the angle between vectors, the larger the correlation and *vice versa*. Consider, for example, two attributes occupying the same position in the sample space so that the angular separation of the vectors linking them to the origin is 0°. This would indicate perfect correlation between their observed values over the samples, since the cosine of 0° is +1, which, by equation (1), corresponds to a correlation coefficient of +1 also. Similarly, two points separated by an angle of 90° represent uncorrelated attributes, the cosine of 90° being zero, corresponding by (1) to a correlation coefficient of the same magnitude. In Fig. 3 the vector representation of these and a number of other values of the correlation coefficient are shown.

ATTRIBUTE	1	2	3	4
1	1·00	0·80	0·96	0·60
2	0·80	1·00	0.60	0·00
3	0·96	0·60	1·00	0·80
4	0·60	0·00	0·80	1·00

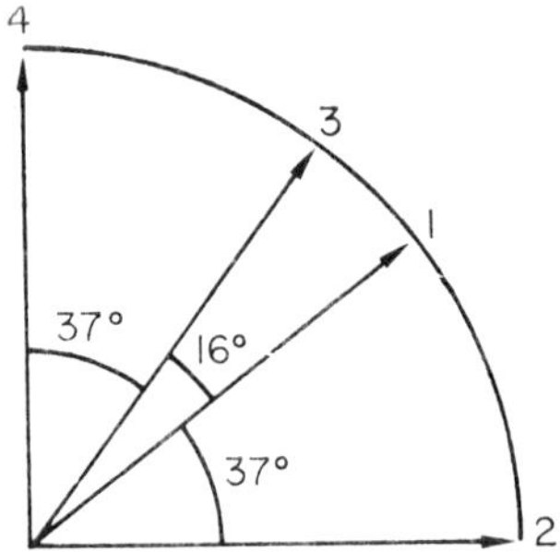

FIG. 4. Vector representation of a correlation matrix. Values of the correlation coefficient between variates considered in all possible pairs are contained in the table (matrix); the corresponding vector representation of the matrix is shown below. (Cos 16°=0·96, cos 37°=0·80, cos 53°=0·60, cos 90°=0·00). (From Fruchter, 1954, by courtesy of D. Van Nostrand Company, Inc.)

The expression of the resemblance between two variates by a pair of vectors or by a single correlation coefficient can easily be extended to cover the inter-relations between many variates. A matrix of correlation coefficients calculated between the observed values of the variates considered in all possible pairs, will, in fact, contain much of the information embodied in the sample space model. Construction of the vector model of the observation set is not itself required. In fact, there is an exact correspondence between the vector representation of a data matrix and the matrix of inter-correlations between the variates. A simple, hypothetical example of this involving four variates is illustrated in Fig. 4.

In most applications involving real data, correlations are frequently found between certain variates. Such correlations reflect the tendency for species to vary together in representation and hence for stands to resemble one another in composition to a greater or lesser extent. In terms of the

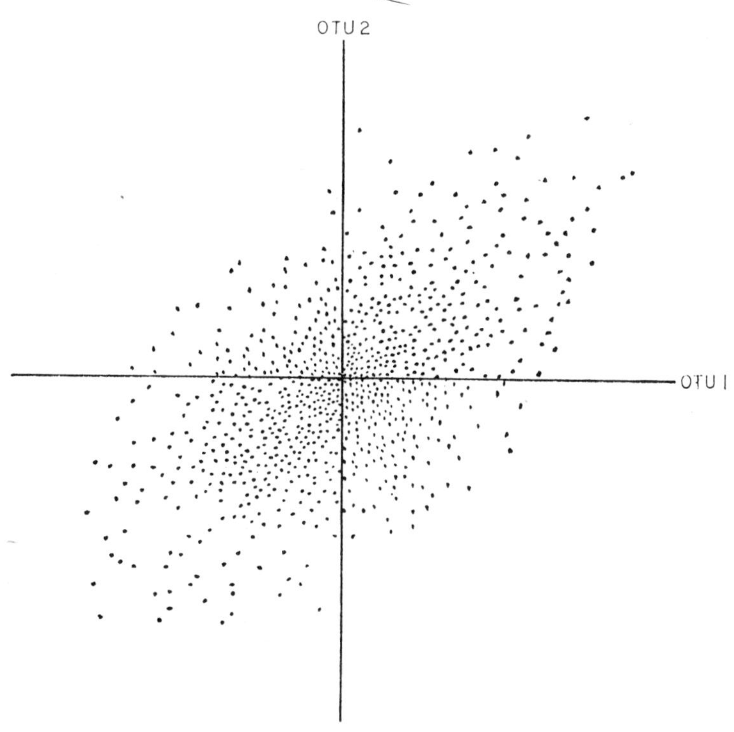

FIG. 5. Point representation of attributes in sample space ($N=2$). The existence of correlations leads to the aggregation of points into an ellipse-shaped scatter. Correlations between species are reflected by the acute angles subtending pairs of points; correlation or similarity between stands by the tendency for points to approximate a straight line at 45°. Diagrammatic. For OTU read stand.

point representation of attributes, this means that the points tend to lie in a more or less well defined ellipsoidal cluster and are not distributed throughout N space. In terms of vectors, it means that the vectors tend to be associated in groups or into 'cones' having rather small generating angles situated at the origin. In Figs. 5 and 6 the existence of inter-correlations between a number of attributes in N space are shown diagrammatically in point and vector form for the special cases where $N=2$ and $N=3$. The existence of inter-correlations of this kind has considerable practical significance. In the following section we will be concerned with a means of exploiting the existence of such correlations.

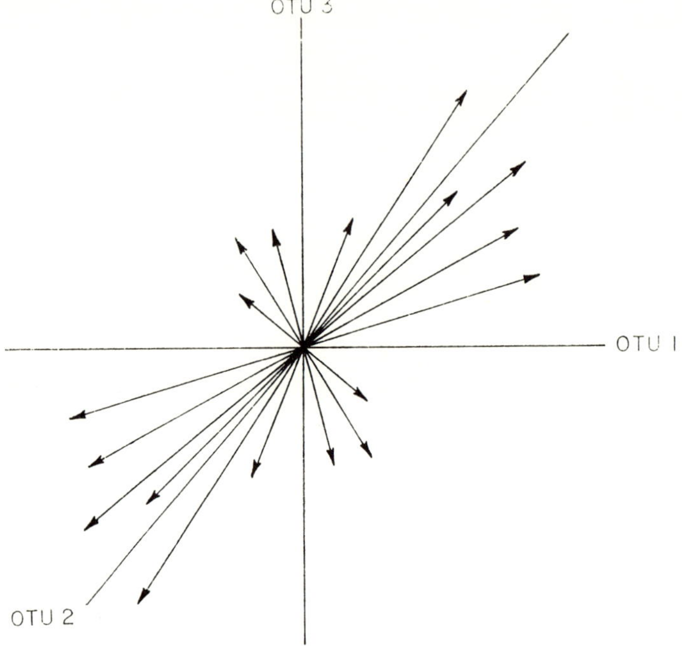

FIG. 6. Vector representation of attributes in sample space ($N=3$). The existence of strong correlation between certain variates is indicated by the acute angles between vectors. Diagrammatic. For OTU read stand.

(3) *Principal axes transformation*

Computation of the correlation matrix between variates normally forms the first step in ordination.† The existence of more or less strong correlations in the matrix implies that the *n* points (or vectors) are contained within

† The covariance matrix calculated between species may sometimes be employed in place of the correlation matrix (van Groenewoud 1965; Yarranton 1967.) For a discussion of the suitability of covariances and correlations in the present context, reference may be made to Lawley and Maxwell (1963), Seal (1964), Pearce (1965), and Morrison (1967).

Ordination may also proceed from a matrix calculated between stands (i.e. between *columns* of the data matrix) rather than between species (rows), provided an appropriate similarity coefficient is used. (See van Groenewoud 1965; Gower 1966, 1967; Orloci 1966, 1967.)

a restricted region of the N space. This enables new coordinate axes to be strategically placed through the major dimensions of the sub-region in which the points actually lie. The points can subsequently be described more efficiently with respect to the new axes than the original ones. Principal components analysis is essentially a technique for introducing a new coordinate system of this kind within the dispersion of points. The new coordinate axes are referred to variously as principal components or

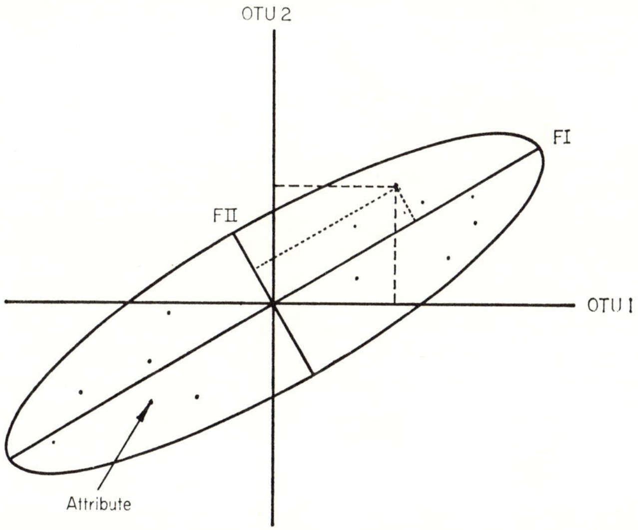

FIG. 7. Principal components. The positions of points are determined by reference to the coordinate axes corresponding to stands (marked OTU 1 and OTU 2). The principal axes or components of the scatter (F I, F II) provide an alternative and more convenient reference frame. The coordinates (loadings) of one species with respect to the original axes and the principal components are indicated.

principal axes. This form of analysis probably represents the most successful ordination technique currently employed in studies of vegetation.

As noted above, the correlations found in practice imply that the points in the sample spaces involved tend to form ellipsoidal clusters. From Fig. 7 the effectiveness of choosing as new reference axes those coinciding with the major and minor axes of an ellipse might be apparent. This is partly because the axes of an ellipse (or, in higher dimensional space, hyperellipsoid) can be ranked such that they become progressively shorter and therefore less important for descriptive purposes. For the particular case illustrated ($N=2$), the difference in lengths of the major and minor axes is obvious. The ordinates of points expressed with respect to the first (major) axis, provide a description of the distribution of points which for many purposes might be an acceptable approximation of their exact relations. Although information would be lost in ignoring the positions of points relative to the shorter axis, the ease of description which would result might be considered to more than outweigh this. Just how much information was lost would depend on the difference in length of the axes. This disparity in turn would itself be dependent on the strength of the correlations between variates. The higher the inter-correlations, the more elongated the point scatter and the greater the difference in the lengths of its axes. For example, if no correlation existed (i.e. $r_{jk}=0$ for all j, k), the scatter of points would be circular (a sphere or hypersphere where $N\geqslant 3$) and all reference axes would be of equal length and importance. In the limiting case where all correlation is supposed perfect (i.e. $r_{jk}=\pm 1$ for all j, k) the points would fall on the line (plane or hyperplane where $N\geqslant 3$) corresponding to the principal axis. The difference in length is then at its maximum, the second (and subsequent) axis vanishing entirely. In practice the degree of correlation found is always between these extremes. This renders certain axes or dimensions of less descriptive importance than others, and, provided a suitable reference frame is chosen, can simplify the tasks of description and interpretation. A suitable reference frame is one coinciding with the principal axes of the hyperellipsoid formed by the points. Another useful property of this set of axes is that those variates which are correlated with respect to the original set of axes are un-correlated in the new system; that is, the co-ordinates of these variates are transformed to independence by the particular change of axes involved. This may render subsequent ecological interpretation of the results more straightforward than would otherwise be the case.

Principal components analysis leads to the establishment of a reference system of exactly the kind described. Reference to Fig. 7 shows that this involves simply a rotation of the original reference axes about the origin. However, it can also be appreciated from the figure that the original axes might conceivably be rotated to any position in the plane shown. Some criterion is therefore required which will lead to a uniquely determined set of axes. The criterion adopted in component analysis relates to the proportion of the total variance attributable to each new axis or component. The requirement is that the first component be positioned in the point scatter such that it accounts for as much of the total variance as possible (i.e. it is located in such a way that the sum of squares of the perpendicular projections of points onto it is the maximum possible); the second component is then located such that it accounts for as much of the remaining variance as possible—which leads to its being at right angles to the first;

and so on for subsequent components until all the variance is accounted for. Geometrically, the resulting axes are those which coincide with the principal axes of the postulated hyperellipsoid.† The lengths of successive components (axes) are proportional to that part of the total variance accounted for by each. It may be that the first two or three components account for almost the whole of the variation, for example for as much as 80 or 90% of the total. The variation can then be represented approximately by the first two or three reference axes and in certain circumstances it may be permissible to neglect the remainder. In such cases this has the effect of reducing the dimensionality of the system from an N dimensional system to one of more manageable number. It is the achievement of precisely this economy which is a major object of component analysis.

The ordination is complete when the coordinates of points are specified with respect to the principal components. It is usual in addition to indicate the variance, i.e. the length, or importance, of successive components. This is related to the 'latent root' or 'eigenvalue' associated with each component. It may be worth emphasizing that, while the coordinates of points differ with respect to the original and rotated axes, the configuration of points itself is unaffected by the transformation. In particular, the distances and angular relations between points are invariant, i.e. preserved, under this transformation. The same information occurs in each case, it is merely more conveniently displayed by the principal components.

(4) *Stand ordination*

In many cases analysis is carried a stage further leading to an ordination of stands. To accomplish this, the standardized observed values of species in a given stand are multiplied by their coordinates on a particular component Summation of the products over the species then yields the coordinate of the stand in relation to that component. This process repeated for all stands and all components results in an ordination of the stands with respect to the components extracted from the between-species correlation matrix. The species' coordinates—or loadings—on the components, function here as *a posteriori* weighting coefficients in describing relations among stands.

† Recently, analyses of ecological data in which the principal components were eventually rotated to other positions in sample or attribute space have appeared (Dagnelie, 1965b; Ivimey-Cook and Proctor, 1967; Webb *et al.*, 1967). Webb *et al.* found the use of *oblique* axes helpful in their analysis. The results of these analyses indicate that rotations of these kinds may well become more important in ordination studies.

Thus, starting from a point or vector representation of species in sample space, we have arrived at an ordination of stands (in component space) comparable in some ways to that referred to in the footnote on p. 41 above. Stands and species can therefore each be arranged or ordered in the component space. Ordinations of both kinds are of ecological interest.

Although the account above has been based largely on illustrations in two- or three-dimensional space, generalization to higher dimensional spaces is straightforward. In practice, of course, ordination is not performed graphically in the way outlined, the equivalent algebraic operations normally being executed by digital computer. Algebraically components analysis may be expressed as follows:

$$F_j = a_{1j}z_1 + a_{2j}z_2 + a_{3j}z_3 + \ldots + a_{nj}z_n \qquad (j = 1, 2, \ldots, n), \tag{2}$$

where F_j is the jth principal component of the observed variates $z_1, z_2, \ldots, z_n$, and the $a_{1j}\, a_{2j}, \ldots, a_{nj}$ are coefficients weighting the relative importance of each species or other variate z_j in the derived component. From equations (2) it is clear that the components F_j are weighted sums of the original variates. Moore (1965) provides a clear account of the nature of the weighting coefficients and of their possible interpretation in biological terms. Equations (2) can be rewritten in the form:

$$z_j = a_{j1}F_1 + a_{j2}F_2 + a_{j3}F_3 + \ldots + a_{jn}F_n \qquad (j = 1, 2, \ldots, n), \tag{3}$$

where the n species z_j are represented as weighted sums of the components, $F_1, F_2, \ldots, F_n$. Using this relation, the observed values of any species j over the i stands are given by the expression:

$$z_{ji} = a_{j1}F_{1i} + a_{j2}F_{2i} + a_{j3}F_{3i} + \ldots + a_{jn}F_{ni} \qquad (j = 1, 2, \ldots, n;\ i = 1, 2, \ldots, N)$$

All n components F_j are required to reproduce the observations z_{ji} exactly. The components, however, are always calculated in order of decreasing magnitude or importance. This enables the observed values z_{ji} to be approximated ($z_{ji}^{\star}$) by the first m, say, components, F_j:

$$z_{ji}^{\star} = a_{j1}F_{1i} + a_{j2}F_{2i} + a_{j3}F_{3i} + \ldots + a_{jm}F_{mi}. \tag{4}$$

Usually m is much smaller than n, the number of observed variates.

It is the numerical values of the F's and a's in equations (2) and (4) respectively which are of practical importance. The calculated values for the F's are the coordinates of stands in relation to the principal axes and which thus provide a stand ordination. Similarly, the a's, when suitably scaled, may be employed as coordinates of species on the principal axes to yield a species ordination. The basic problem is to obtain estimates for the coefficients a_{ji}. The required estimates are the elements of the latent vectors of the correlation matrix between variates (see, for example, Orloci, 1966). The vectors, appropriately scaled, yield a species ordination, and, together with the initial observations z_{ji} lead to the determination of the F's in equations (2).† Thus, while equations (2) are solved explicitly, there is rarely any need in practice to complete the solutions to equations (4). In equations (4), which attempt to describe or represent the structure of the observations z_{ji} mathematically, using a small number of derived 'factors' or components F_j, it is the possible significance of the components F_j ecologically which attracts attention. Further details of the algebraic and computational aspects of components analysis are beyond the scope of this account. Reference may be made to the works cited on p. 38 for additional information bearing on these aspects. Other useful accounts of components analysis and related methods are provided by Cooley and Lohnes (1962), Horst (1965) and Harman (1967).

(5) *The interpretation of results*

Before considering some ecological results of ordination, it is of interest to examine the results of a components analysis. The results usually take the form of a table of the kind shown in Table 1. This analysis relates to an area of grassland vegetation. The field observations consisted of estimates of the abundance of thirty species and the expression of three soil variables in 45 stands (i.e. $n=33$, $N=45$). The main body of the table consists of numbers which, when read by columns, relate to the principal components when read by rows, correspond to the variates. Since the smaller components are generally of little interest, only the first five or ten components are usually involved in such tables. The table is related to equation (4), the row entries comprising the a coefficients of the relevant components (F's) of the equation. These entries are interpretable both as the coordinates (loadings) and as the correlations of the variates with the principal com-

† Orloci (1967) in a valuable paper points out that the required stand and species ordinations are sometimes more conveniently calculated by an alternative route. Here the latent vectors of the between *stand* similarity matrix are the basic requirement. The latent vectors themselves provide the stand ordination while their elements may be used as coefficients weighting the row entries of the data matrix to yield the components scores of species (i.e. the species ordination). This approach is especially valuable where the number of species exceeds the number of stands. Identical ordinations result from either route provided (1) the data matrix is suitably standardized, (2) appropriate similarity functions are used and (3) the latent vectors are suitably scaled.

ponents. The column labeled h^2 contains the sum of the squares of the entries of each row

$$\left(h_j^2 = \sum_{i=1}^{5} a_{ji}\right).$$

This value indicates the proportion of the total original variance of a variate which is preserved by the m components retained. When all n components are taken into account, the value of h^2 is always 1·00. Thus, in the table, the nearer a given h^2 value approaches unity, the more completely the variate in question is accounted for by the components shown. At the foot of the table, the sum of the squares of each column is shown. These values are the variances or latent roots of the principal components. They show clearly the principal of maximum contribution to the total variance of each successive component. Together the five components account for almost 73% of the variance of the observations. Since the latent roots are proportional to the lengths of the components, they enable the shape of the dispersion of points to be visualized. In this case they show that, while the dispersion may be considered to form a hyperellipsoid, this is not as elongated as is sometimes found.

The similarity of the table to a data matrix is worth noticing. This is not really surprising since all that has been done is to relate the points to a new set of coordinate axes. Apart from the smaller number of columns, the chief remaining difference concerns the nature of the columns. In a data matrix, the columns relate to actual stands (showing the abundance of each species present), and, in the sample space model of this, we set up a correspondence between the columns and the coordinate axes, letting each stand be represented by an axis. In the matrix of species' component loadings, the columns define the rotated coordinate axes of the sample space, the column entries indicating the location of species with respect to these. In sample space, therefore, the coordinate axes of the data matrix correspond to real, physical entities—the stands. It remains to be considered whether any similar correspondence can be shown between the rotated axes (principal components) and external reality. This topic is discussed in some detail below.

The results of a stand ordination are similar in appearance to those of a species analysis. In Table 2, the corresponding stand ordination of the grassland data is given. Stands are listed at the left and the entries of the table read by rows are the coordinates of each stand on the principal components. These coordinates are often referred to as the *component scores* of stands. Although much insight into species' behaviour and vegetation structure can be obtained from such tables, interpretation is often aided

TABLE I

Principal components analysis of grassland vegetation, Anglesey, North Wales. Loadings of species and soil variates (attributes) on the first five components

Variate	a_1	a_2	a_3	a_4	a_5	Variance h^2
1. *Helianthemum chamaecistus*	0·775	−0·407	0·208	0·155	−0·123	0·849
2. *Thymus drucei*	0·869	−0·001	−0·310	−0·243	0·091	0·919
3. *Lotus corniculatus*	0·195	−0·654	−0·075	0·106	0·095	0·492
4. *Plantago lanceolata*	−0·374	−0·643	−0·095	0·115	−0·321	0·679
5. *Galium verum*	−0·064	0·559	−0·081	0·409	0·239	0·548
6. *Carex flacca*	0·023	−0·798	0·320	−0·103	−0·004	0·750
7. *C. caryophyllea*	0·499	−0·615	0·202	0·382	0·034	0·815
8. *Poterium sanguisorba*	−0·447	−0·687	−0·052	0·170	−0·131	0·721
9. *Hieracium pilosella*	0·404	−0·401	0·320	−0·114	0·180	0·472
10. *Luzula campestris*	−0·626	−0·568	0·184	0·282	−0·190	0·864
11. *Taraxacum laevigatum*	0·561	−0·061	−0·239	0·377	−0·063	0·522
12. *Viola riviniana*	−0·262	−0·356	0·230	−0·188	−0·320	0·386
13. *Briza media*	0·761	−0·364	0·009	0·266	−0·070	0·787
14. *Koeleria cristata*	0·660	0·037	0·269	0·357	0·003	0·637
15. *Anthoxanthum odoratum*	0·044	−0·365	0·523	−0·234	−0·181	0·496
16. *Sieglingia decumbens*	0·281	−0·368	0·251	−0·449	0·420	0·655
17. *Dactylis glomerata*	−0·774	0·508	0·044	−0·102	0·098	0·879
18. *Agrostis tenuis*	−0·818	−0·391	0·040	−0·085	0·231	0·884
19. *Helictotrichon pubescens*	−0·902	−0·023	0·253	0·132	−0·058	0·899
20. *Phleum bertolonii*	−0·421	0·802	0·168	−0·123	0·127	0·880
21. *Holcus lanatus*	−0·679	−0·296	−0·440	−0·371	−0·385	0·909
22. *Centaurea nigra*	−0·721	−0·378	−0·279	−0·118	−0·059	0·759
23. *Trifolium pratense*	−0·918	−0·069	0·004	0·228	0·019	0·900

24. *T. repens*	−0·718	−0·277	−0·429	−0·101	−0·320	0·889
25. *Anthyllis vulneraria*	−0·041	0·002	−0·438	0·377	0·120	0·350
26. *Cladonia impexa*	0·796	−0·084	−0·291	−0·225	−0·234	0·831
27. *Pseudoscleropodium purum*	0·919	−0·141	−0·227	0·030	−0·067	0·921
28. *Dicranum scoparium*	0·905	−0·057	−0·066	−0·127	−0·184	8·877
29. *Rhytidiadelphus squarrosus*	−0·106	−0·134	−0·141	0·612	0·370	0·561
30. *Hypochaeris radicata*	−0·662	−0·359	−0·269	0·087	−0·123	0·662
31. Soil depth	−0·800	−0·176	0·152	−0·167	0·340	0·838
32. Soil phosphate	−0·495	0·493	0·342	0·328	−0·308	0·807
33. Soil potassium	−0·011	0·377	0·486	0·292	−0·389	0·615
Variance (latent root)	12·257	5·784	2·315	2·143	1·557	24·056
Percent of total variance	37·14	17·53	7·01	6·49	4·72	72·90
Accumulated percent	37·14	54·67	61·68	68·18	72·90	—

Total variance (trace): 33·00

by plotting the positions of species or stands in relation to pairs of components. Figures 8 and 9 are, respectively, ordinations of species and stands plotted with reference to the first two components of Tables 1 and 2. Interest in scatter diagrams of this kind centers on the spatial relations (distances) between points, and on their position in relation to the origin. The closer the points, the greater the ecological similarity of the entities they represent, and *vice versa*. Strong negative correlation is implied by points which are diagonally opposite; such stands or species may well be

TABLE 2

Principal components analysis of grassland vegetation, Anglesey, North Wales. Component scores of stands (individuals) on the first five components

Individuals (Stands)	F_1	F_2	F_3	F_4	F_5
1	4·096	0·989	1·956	−0·678	−0·455
2	−3·224	7·897	−0·499	1·292	0·937
3	3·702	0·201	1·415	0·014	−0·217
4	1·312	−0·292	−2·854	0·457	2·032
5	−5·513	−3·445	1·756	−1·267	−0·453
6	3·803	0·153	1·602	−0·716	−0·851
7	−5·292	−1·275	1·198	0·042	−1·990
8	0·877	−1·709	−1·972	0·183	1·331
9	−3·677	8·197	−0·631	0·991	0·201
10	−5·435	−2·734	1·886	−0·986	0·359
11	2·672	0·933	−0·514	−1·089	−0·659
12	2·361	−0·269	−0·381	−1·145	−0·879
13	−3·909	−1·634	−0·529	−0·448	−2·444
14	2·095	−0·212	−0·629	−1·324	1·376
15	2·775	−2·001	0·319	0·690	−2·350
16	1·961	−0·924	−0·199	−0·075	−1·053
17	0·512	−1·818	−2·091	0·419	1·281
18	−1·371	0·257	−2·435	−3·777	0·268
19	−4·707	2·917	−0·679	−0·754	−1·515
20	−7·412	−3·386	2·354	−0·360	3·491
21	2·839	−0·698	1·717	0·613	−2·198
22	3·789	2·303	1·748	−0·313	0·576
23	−4·148	−0·095	−0·642	−0·488	−0·368
24	1·025	−0·102	−0·099	−2·422	−1·626
25	−5·184	−0·255	2·523	0·112	1·414
26	3·981	1·124	1·860	1·449	0·606
27	2·230	0·149	0·328	−2·572	−1·313
28	−3·193	3·688	−1·555	−0·195	−0·676
29	−2·077	−2·519	−1·156	2·658	−0·947

TABLE 2—*continued*

Individuals (Stands)	F_1	F_2	F_3	F_4	F_5
30	0·366	−2·147	−1·521	2·488	−0·515
31	2·978	−1·870	−0·907	2·089	−0·332
32	1·643	−1·646	−1·426	1·430	1·609
33	1·993	1·036	−0·473	−2·054	0·027
34	3·088	2·929	0·788	−0·829	0·509
35	−6·375	1·143	2·970	1·452	−0·033
36	2·708	0·575	−0·008	1·067	0·916
37	3·436	0·770	1·118	1·444	0·453
38	3·441	−0·168	−0·540	0·454	2·036
39	1·186	−0·982	−0·600	−1·067	1·105
40	−2·585	−2·665	−1·785	2·675	−0·011
41	−2·017	−0·502	−1·758	−2·530	0·571
42	3·739	0·936	2·152	−0·855	0·509
43	4·314	−0·803	1·059	1·716	0·295
44	−0·199	−2·000	−1·290	1·579	−0·231
45	−3·234	2·023	1·573	−0·896	−0·786

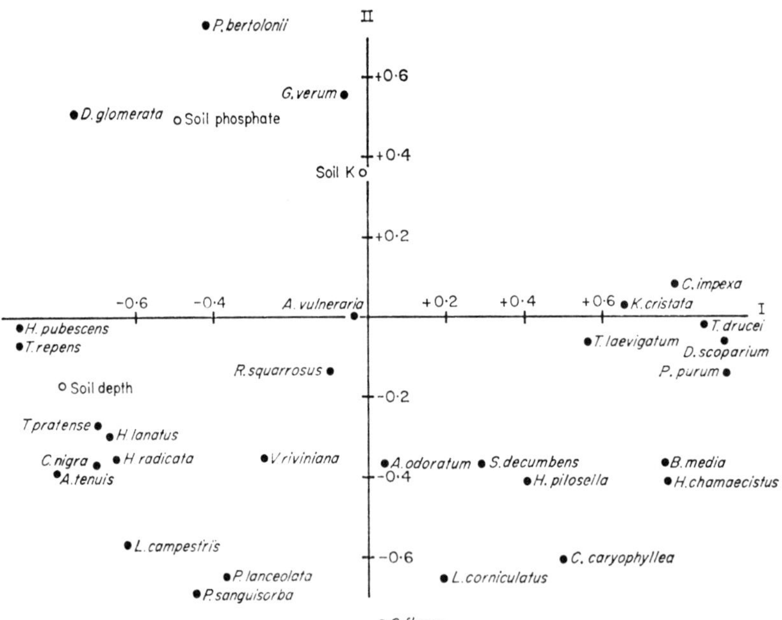

FIG. 8. Grassland vegetation, Anglesey, North Wales. Principal components analysis; projection of species and three soil variables on components I and II. (The variates represented are listed in Table 1.)

ecologically alike to the extent that they are responding to a common influencing factor of some kind (e.g. soil base status, perhaps, or soil moisture content). They differ, however, in reacting to this influence in opposite senses. The distance of a point from the origin is given by h, which is the standard deviation of the variable. It follows from the interpretation of h^2 given above, that, the further a point is from the origin, the more the entity in question is accounted for by, or related to, the components shown.

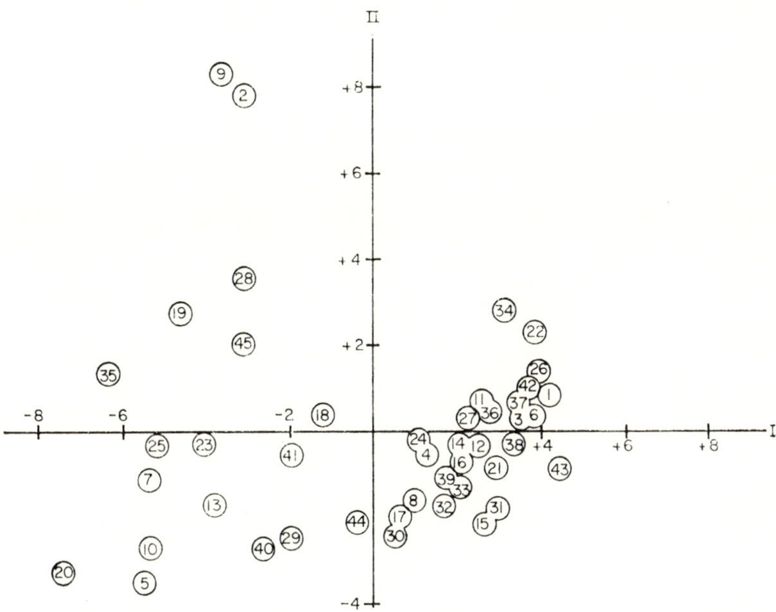

FIG. 9. Grassland vegetation, Anglesey, North Wales. Principal components analysis; projections (component-scores) of stands on components I and II.

The construction of three-dimensional models provides an even more effective aid to the interpretation of results than the preparation of scatter diagrams. In these, species or stands, represented by small balls supported by rods, are located by their coordinates in the 3-space spanned by the first three components.

These various devices are all attempts to portray the sample or attribute spatial conceptions of the field observations. Some distortion of relationships between points arises in these representations from the failure to use all the information available. As a result, however, the salient features of the data stand out more clearly, which in turn may simplify their ecological assessment.

The possible ecological significance of the new reference frame is also of interest. Although the principal components strictly speaking are purely mathematical constructions, it is frequently found that they correspond also to ecological features of the analysis. Indeed, only where the reference axes can be considered to reflect ecological factors or processes of some kind, is the ordination likely to be regarded as completely satisfying. Identification of the ecological factors involved is a matter for interpretation, and, as such, it is dependent in part on the judgement of the investigator. Usually it involves an attempt to evaluate the results of the ordination against information external to the analysis.

In species' ordinations where there are at least some species whose general ecological requirements are known, the interpretation of components is usually straightforward. An indication of the ecological control or influence expressed by different components may then be provided by species which have a high loading on one or other of them. As an example, consider the species ordination shown in Fig. 8. Here *Thymus drucei* and *Trifolium repens* are among the species with high loadings on the first axis. They differ between themselves, however, in occupying positions at opposite ends of the axis. The 'preference' of *T. drucei* for shallow, well-drained soils and of *T. repens* for deeper soils with more balanced water relationships is well known from general field experience. Using this knowledge, component I may, as an initial hypothesis, be regarded as an expression of the possible control exerted by soil depth and the other conditions which go with it, on species' behaviour. Since this axis accounts for more of the total variance of the data than any other (see Table 1) the factor-complex of soil depth may, with some justification, be regarded as the major ecological control operative in the community. The same process of interpretation can be extended to other axes or components although components after the first three or four usually prove difficult to relate to ecological factors. Since successive axes are perpendicular to one another, it is to be expected that they will correspond to habitat factors which operate largely independently of each other, as van Groenewoud (1965) has pointed out. Where environmental variates are included in species' ordinations, interpretation is simplified. In Fig. 8, for example, the positions of soil depth and exchangeable soil potassium in relation to axes I and II respectively, provide clear indications of the habitat influences which may possibly underlie these components. In cases where nothing is known of the ecology of the species involved, insight gained in this way of the nature of the chief ecological factors operative and of species' response to them, although of a provisional nature, is especially valuable. A wise choice of environmental variables for inclusion, however, requires the exercise of experience and care.†

In stand analyses, the possible ecological meaning of the reference frame may be explored by simple or multiple regression analysis. This involves regression of the intensity of an ecological factor over the stands, on the position of stands along a particular axis or axes. Familiarity with the vegetation is helpful in deciding which environmental variables may most profitably be tested in this way with a given axis. In some instances, correlation might be considered to provide a more suitable test than regression. As a preliminary step, it may be worthwhile to plot the levels of the factor against stand positions on a component. An example of this, taken from van Groenewoud (1965), is shown in Fig. 10(a). Here a measure of light conditions (lux hours/day × 1000) in fifteen stands of Swiss forest vegetation, is plotted against stand positions along the first axis of an ordination. A well defined linear relationship is apparent, which is also statistically significant (van Groenewoud, 1965). This suggests that the axis may correspond to the effect of light on vegetation composition. Species' behavior can be examined in the same way. In Fig. 10 (b, c), the representation (% cover) of *Picea abies* and *Abies alba* in relation to stand positions on the same axis are shown. The behavior of the species is evidently linearly related to the component, although in opposite directions. From the nature of the component, it follows that the behavior of the species may be regarded provisionally as controlled largely by variation in light. There is a second approach to the identification of components from a stand ordination. This is to plot the distribution of stand component-scores at the appropriate stand positions on a base-map of the area involved. Particular environmental

† There is a difference of opinion concerning the inclusion of variates measured on different scales in the same analysis. This arises partly from the fact that the components are not invariant under changes of scale in the variates. Thus, for example, it may be held undesirable for, say, species' abundance estimated as percentage cover, soil cations expressed as m.e./100g dry soil and soil acidity expressed in pH units, to be analyzed together, even when the original units of measurement are first standardized. See Anderson (1958, p. 279); Lawley and Maxwell (1963, p. 47); Seal (1964, p. 117); Pearce (1965, p. 195); Williams and Dale (1965, p. 56); Cattell (1966, p. 221); Morrison (1967, p. 223). Nevertheless, certain analyses based on such data have yielded apparently consistent results (Ferrari *et al.*, 1957; Sokal *et al.*, 1961; Rayner, 1966; cf. also Fig. 8 above). Further work on this aspect of ordination is required. A quite different approach to the joint analysis of species and environmental variates of equal or perhaps of greater promise is canonical correlation analysis (see Kendall, 1957; Anderson, 1958; Cooley and Lohnes, 1962; Dagnelie, 1965b).

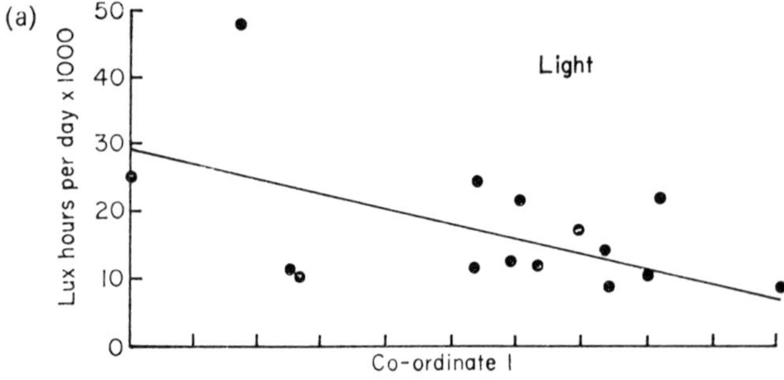

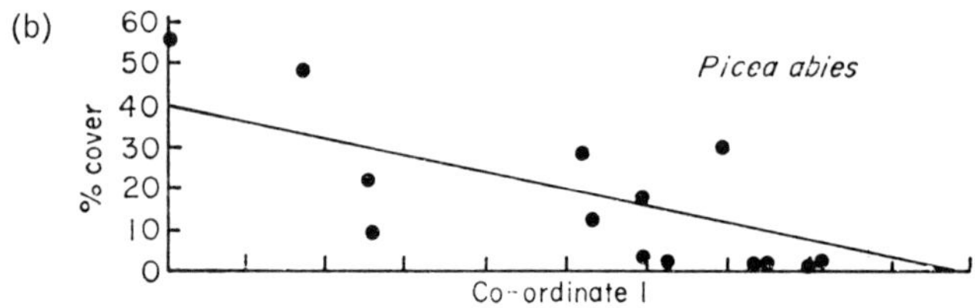

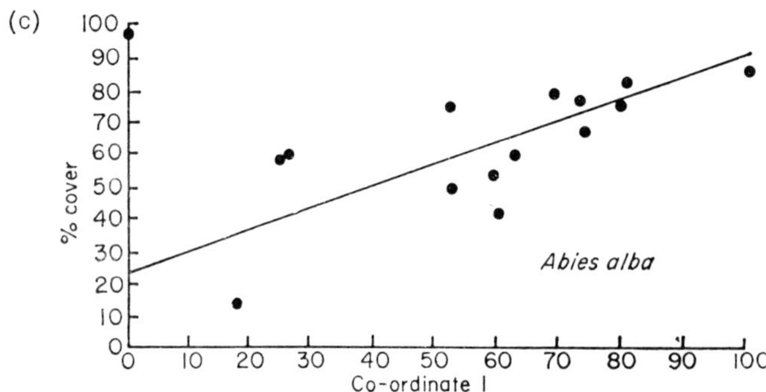

FIG. 10. Forest vegetation, Switzerland. (a) Interpretation of components from stand ordinations; relation of light conditions (lux hours/day) of stands to component-scores of stands on component I. (b, c) species' behavior (% cover) in relation to component I. (From van Groenewoud, 1965, by courtesy of *Ber. geobot. Inst. ETH, Stiftg. Rübel, Zürich*).

variables can be treated in the same way and the resulting distribution maps contoured and compared. The maps may then suggest the existence of connections between habitat factors and components, leading to identification of the latter (see Goodall 1954b). A refinement of this technique is provided by trend-surface analysis. By means of this, the total variance of a mapped variate can be sub-divided into portions, each attributable to separate causes. The method consists of fitting the observed values of mapped variates such as component-scores or habitat factors, to polynomial functions of the map coordinates of the stands examined. Functions of increasing degree, each accounting for a specific portion of the total variation of the variate, may be fitted. The surfaces corresponding to the functions are readily mapped and may then be examined and compared for their ecological implications. This approach may be most useful in the case of components whose nature has proved difficult to establish by the methods mentioned previously. (See Gittins 1968.)

Interpretation of the reference frame enables the nature of the major ecological controls to be appreciated, at least in a provisional way. It therefore turns out that the principal components, which were arrived at initially for the purpose of descriptive economy, are also of value as *explanatory* variables of ecological significance. Thus, it is the interpretation or explanation of relationships embodied in the observations, together with their efficient description, which comprise the basic objectives of ordination. Once these aims are accomplished, the ground is prepared for the use of experimental methods. The components of ordinations of stands and species extracted from the same set of field observations, generally lead to similar ecological conclusions. The chief difference concerning such components is that, in the stand ordination it is a highly loaded stand which exemplifies the component, whereas in the species ordination it is a highly loaded species (or other variate). Despite their similarities, both ordinations have intrinsic interest since they each contain information not present in the other. This, however, draws attention to an inherent weakness of the approach. This is that, within the framework of the spatial models of the observation set (pp. 39–41 above), there is no way of simultaneously manipulating individuals and attributes so as to arrive at a *unified* representation of the essential information of the system. This has led Williams and Dale (1965), to entertain the idea of abandoning spatial models altogether.

(6) *An investigation of factors influencing grassland yield*

As an example of the use of ordination, reference is made to a study by Ferrari, Pijl and Venekamp (1957), of the yield of 50 stands of grassland vegetation in the Netherlands. Appreciable differences in yield between stands were apparent, and the study was conducted in order to identify

TABLE 3

Principal components analysis (centroid approximation) of variates from grassland in the Guelderland Valley, The Netherlands. Loadings of variates on the first four principal components (modified from Ferrari, Pijl and Venekamp, 1957, by courtesy of *Netherlands J. agri. Sci.*)

Variate	a_1	a_2	a_3	a_4	h^2
1. pH (KCl)	0·20	0·24	0·41	−0·09	0·27
2. Content of organic matter	−0·38	0·75	0·02	−0·15	0·73
3. Content of clay particles	−0·10	0·80	0·04	−0·20	0·69
4. Content of fine sand	0·17	0·68	−0·13	0·13	0·53
5. U figure (specific surface)	−0·04	0·88	−0·14	0·13	0·81
6. P-citr (P status)	0·42	0·08	0·45	−0·20	0·43
7. K-HCl (K status)	0·62	−0·04	0·06	−0·04	0·39
8. MgO content (Mg status)	0·27	0·61	0·44	0·02	0·64
9. Thickness of humus layer	0·43	−0·11	−0·02	0·22	0·25
10. Distance of farm (management)	−0·34	0·05	−0·09	−0·34	0·24
11. Ground water level	0·52	−0·39	−0·09	0·01	0·42
12. Fluctuation ground water level	−0·29	0·59	−0·05	0·17	0·46
13. Nitrogenous fertilization	0·59	0·05	0·45	0·11	0·57
14. Phosphatic fertilization	0·37	0·01	0·84	0·00	0·84
15. Potassic fertilization	0·43	0·18	0·71	0·04	0·72
16. Grade of quality pasture	0·57	−0·09	−0·04	0·20	0·37
17. *N* content of grass	0·36	−0·02	0·00	0·63	0·53
18. P_2O_5 content of grass	0·47	−0·13	0·14	0·58	0·59
19. K_2O content of grass	0·68	−0·13	−0·08	0·30	0·58
20. MgO content of grass	−0·03	−0·05	0·22	0·15	0·07
21. Spring yield of % of annual yield	0·90	−0·01	−0·09	−0·31	0·91
22. Spring yield	0·98	0·03	−0·02	−0·11	0·97
23. Annual yield	0·75	0·07	0·03	−0·38	0·71
Variance (latent root)	5·66	3·48	2·12	1·14	12·72

the factors responsible for the differences. Twenty-three soil and vegetational attributes were measured, so that $n=23$ and $N=50$. The variates are shown in Table 3, from which it can be seen that the majority fall roughly into five categories: physical soil characteristics, soil base status, fertilizer application, mineral content of grass and dry weight yield. Yield refers to the overall or total yield of stands, the contributions of individual species to this not being determined. The data were analysed by an approximation of the principal components technique. The account which follows is based on results presented by Ferrari *et al.* (1957), but does not necessarily express their views.

The results of the ordination of attributes are summarized in Table 3 and Fig. 11. Table 3 contains the component loadings of the variates, while

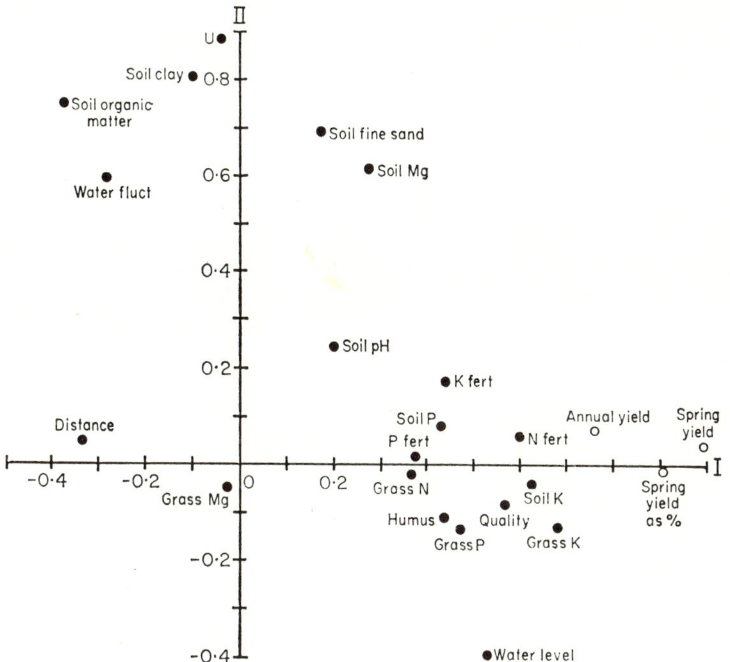

Fig. 11. Grassland vegetation, Guelderland Valley, The Netherlands. Principal components analysis (centroid approximation); projections of attributes on components I and II. Attributes represented are listed in Table 3. Original data from Ferrari *et al.* (1957).

Fig. 11 shows the variates plotted in relation to the first two components. From Table 3, it can be seen that each component is characterized by a few variates with conspicuously high loadings. These enable the components to be interpreted as follows: F_1: yield; F_2: soil physical properties; F_3: fertilizer application; F_4: mineral composition of grass. Variates with loadings of this kind on F_1, (21, 22, 23) and F_2, (2, 3, 4, 5) respectively, are uniquely accounted for by these components. This is apparent from the near-zero loadings of the variates on all other components. In particular, since the variates of each set have near-zero loadings on the component characterized by the other, the variates concerned are evidently independent. On the other hand, variates exemplifying F_3, (6, 13, 14, 15) and F_4, (17, 18, 19) are not uniquely associated with these components; they also have relatively high loadings on F_1. Similarly, the variances of ground water level and fluctuation (11, 12), are distributed over components F_2 and F_1. The ecological implication of these relationships is that, while yield might conceivably be dependent on variates correlated with F_1, it seems likely to be largely independent of variation in the soil physical properties examined. Other variates with high loadings on F_1, and therefore to which yield might be related, are soil potassium status (7) and distance of farm (10). The latter is a convenient index of farm management. Many of these points are brought out by Fig. 11. The most striking feature of this figure is the clear distinction between variates associated either with F_1 or F_2. Attention is also drawn to the relatively large distance of most of the variates from the origin, and to the very high loadings of the indices of yield on F_1 and their near-zero loadings on F_2. Of the variates which are associated with F_1 (see Fig. 11), experience suggests that soil potassium and phosphate status (7, 6), soil moisture relations (11, 12) and distance (10) are those most likely to influence yield. Soil potassium and phosphate, with positive loadings, are evidently related to yield in a direct sense. Distance (equivalent to poor management), on the other hand, with a negative loading, is inversely related to yield. The inverse relationship between distance and soil potassium and phosphate status is also worth noticing. The independence of all three variates from soil physical properties is clearly implied by their extremely small loadings on F_2. The behavior of the measures of soil moisture differs in an important respect from that of the variates just considered. Although ground water level and fluctuations have moderately high loadings (in opposite senses) on F_1, and are therefore correlated with yield, they are also related to F_2. Soil moisture relations are evidently dependent on the soil physical properties measured. Therefore it seems that, while the soil physical properties do not themselves influence yield, they are of significance in their partial control of soil moisture conditions. The contrast in the behavior of the two indices of soil moisture (Fig. 11) is a feature of some interest and can be explained in the following way. The opposite loadings of the variates on F_2 apparently reflects that drier soils are those which experience the greatest fluctuations in water level. The differing effects of these conditions on yield is shown by the contrast in the loadings of the variates on F_1. Dry soils with fluctuating water conditions are shown by the negative loading of ground water fluctuation (12) on F_1 to be inversely related to yield; moist soils with balanced water relations are shown by the positive loading of ground water level (11) on F_1 to be directly related to yield.

The relationships considered by no means exhaust all those provided. They are, however, sufficient to indicate the kind of information to be obtained from ordination. The salient features of the analysis may now be drawn together. The direct (i.e., immediate) controls of yield appear to be soil potassium and phosphate status and soil moisture. These factors can themselves be related to more remote influences. Soil potassium and phosphate are likely to be dependent on fertilizer applications, an aspect of management, and soil moisture on external water supply and soil physical properties. Yield may therefore be regarded ultimately as a function of management, water supply and soil texture, through the control exerted by them on soil potassium and phosphate status and soil moisture conditions. On the basis of these results, Ferrari *et al.* (1957) made a selection of five variates as the starting point for an analysis designed to explore the relationships exposed further. In retrospect, many of the relationships found may appear largely self-evident. This is partly because the ecology of temperate pastures is already reasonably well known, and partly a reflection of the nature of the data used. Overall yield, as a measure of vegetation, provides no information on the effects of differences in species composition on yield. Appreciable differences in yield are likely to exist between species with respect to the controls identified (cf. Fig. 8). In general, such differences between species with respect to the controls operative in a given community, and, indeed, the nature of the controls themselves, are likely to be poorly delineated. There seems little reason to doubt that ordination could do much to resolve them.

ACKNOWLEDGEMENTS

The greater part of this work was carried out at the Department of Entomology of the University of Kansas and this paper is contribution 1404 from that department. Financial aid was provided in part by Public Health Service Research Grant GM-11935 from the National Institute of General Medical Sciences to Robert R. Sokal. I should like to acknowledge my indebtedness to Dr F. James Rohlf, of the Entomology Department, The University of Kansas, for certain of the ideas expressed here. I am also much indebted to Dr R.R.Sokal, for making my visit to Kansas possible and for his continued support and encouragement.

The analyses of grassland data from North Wales were performed at the Computation Center of the University of Kansas (GE 625), using the FORTRAN IV Numerical Taxonomy SYStem of multivariate statistical programs (NT—SYS) by F. James Rohlf, J. Kishpaugh and R. Bartcher of the Entomology Department, The University of Kansas, Lawrence, Kansas.

REFERENCES

ANDERSON T.W. (1958). *An introduction to multivariate statistical analysis.* New York.

BRAY J.R. and CURTIS J.T. (1957). An ordination of the upland forest communities of southern Wisconsin. *Ecol. Monogr.* **27**, 325–349.

CATTELL R.B. (1966). The meaning and strategic use of factor analysis. In *Handbook of multivariate experimental psychology* (ed. R.B. Cattell), pp. 174–243. Chicago.

COOLEY W.W. and LOHNES P.R. (1962). *Multivariate procedures for the behavioral sciences.* New York.

CURTIS J.T. (1959). *The vegetation of Wisconsin: an ordination of plant communities.* Madison, Wisconsin.

DAGNELIE P. (1960). Contribution à l'étude des communautés végétales par l'analyse factorielle. (Engl. summ.). *Bull. Serv. Carte phytogéogr. Sér. B,* **5**, 7–71, 93–195.

DAGNELIE P. (1965a). L'étude des communautés végétales par l'analyse statistique des liaisons entre les espèces et les variables écologiques: principes fondamentaux. (Engl. summ.). *Biometrics,* **21**, 345–361.

DAGNELIE P. (1965b). L'étude des communautés végétales par l'analyse statistique des liaisons entre les espèces et les variables écologiques: un example. (Engl. summ.). *Biometrics,* **21**, 890–907.

FERRARI TH.J., PIJL H. and VENEKAMP J.T.N. (1957). Factor analysis in agricultural research. *Netherlands J. Agric. Sci.* **5**, 211–221.

FRUCHTER B. (1954). *Introduction to factor analysis.* New York.

GITTINS R. (1968). Trend-surface analysis of ecological data. *J. Ecol.* **56**, 845–69.

GOODALL D.W. (1954a). Vegetational classification and vegetational continua. *Angew. PflanzSoz.* **1**, 168–182.

GOODALL D.W. (1954b). Objective methods for the classification of vegetation. III. An essay in the use of factor analysis. *Aust. J. Bot.* **2**, 304–324.

GOODALL D.W. (1963). The continuum and the individualistic association. (French summ.). *Vegetatio,* **11**, 297–316.

GOWER J.C. (1966). Some distance properties of latent root and vector methods used in multivariate analysis. *Biometrika,* **53**, 325–338.

GOWER J.C. (1967). Multivariate analysis and multidimensional geometry. *The Statistician,* **17**, 13–28.

GREIG-SMITH P. (1957). *Quantitative plant ecology.* 1st edn. London.

GREIG-SMITH P. (1964). *Quantitative plant ecology.* 2nd edn. London.

HARMAN H.H. (1967). *Modern factor analysis.* 2nd edn. Chicago.

HORST P. (1965). *Factor analysis of data matrices.* New York.

Ivimey-Cook R.B. and Proctor M.C.F. (1967). Factor analysis of data from an east Devon heath: a comparison of principal component and rotated solutions. *J. Ecol.* **55**, 405–413.

Kendall M.G. (1957). *A course in multivariate analysis.* London.

Lawley D.N. and Maxwell A.E. (1963). *Factor analysis as a statistical method.* London.

Moore C.S. (1965). Inter-relations of growth and cropping in apple trees studied by the method of component analysis. *J. hort. Sci.* **40**, 133–149.

Morrison D.F. (1967). *Multivariate statistical methods.* New York.

Orloci L. (1966). Geometric models in ecology. I. The theory and application of some ordination methods. *J. Ecol.* **54**, 193–215.

Orloci L. (1967). Data centering: a review and evaluation with reference to component analysis. *Syst. Zool.* **16**, 208–212.

Pearce S.C. (1965). *Biological statistics: an introduction.* New York.

Rayner J.H. (1966). Classification of soils by numerical methods. *J. Soil Sci.* **17**, 79–92.

Reyment R.A. (1961). Quadrivariate principal component analysis of *Globigerina yeguaensis. Stockh. Contr. Geol.* **8**, 17–26.

Reyment R.A. (1963). Multivariate analytical treatment of quantitative species associations: an example from palaeoecology. *J. Animal Ecol.* **32**, 535–547.

Seal H. (1964). *Multivariate statistical analysis for biologists.* London.

Sokal R.R., Daly H.V. and Rohlf F.J. (1961). Factor analytical procedures in a biological model. *Univ. Kans. Sci. Bull.* **42**, 1099–1121.

van Groenewoud H. (1965). Ordination and classification of Swiss and Canadian coniferous forests by various biometric and other methods. *Ber. geobot. Inst. ETH, Stiftg. Rubel, Zürich 36*, 28–102.

Webb L.J., Tracey J.G., Williams W.T. and Lance G.N. (1967). Studies in the numerical analysis of complex rain-forest communities. I. A comparison of methods applicable to site/species data. *J. Ecol.* **55**, 171–191.

Williams W.T. and Dale M.B. (1965). Fundamental problems in numerical taxonomy. *Adv. bot. Res.* **2**, 35–68.

Yarranton G.A. (1967). Principal components analysis of data from saxicolous bryophyte vegetation at Steps Bridge, Devon. I. A quantitative assessment of variation in the vegetation. *Canad. J. Bot.* **45**, 93–115.

Part II

FACTOR ANALYSIS

Editors' Comments on Papers 7 Through 18

7 **THURSTONE**
Multiple Factor Analysis

8 **CATTELL**
Factor Analysis: An Introduction to Essentials: I. The Purpose and Underlying Models

9 **RAO**
Estimation and Tests of Significance in Factor Analysis

10 **McDONALD**
The Theoretical Foundations of Principal Factor Analysis, Canonical Factor Analysis and Alpha Factor Analysis

11 **KAISER**
The Varimax Criterion for Analytic Rotation in Factor Analysis

12 **CATTELL and MUERLE**
The "Maxplane" Program for Factor Rotation to Oblique Simple Structure

13 **JENNRICH and SAMPSON**
Rotation for Simple Loadings

14 **McDONALD and BURR**
A Comparison of Four Methods of Constructing Factor Scores

15 **STROUD**
An Application of Factor Analysis to the Systematics of Kalotermes

16 **REYMENT**
Multivariate Analytical Treatment of Quantitative Species Associations: An Example from Palaeoecology

17 **WALLACE and BADER**
Factor Analysis in Morphometric Traits of the House Mouse

18 **ATCHLEY**
Components of Sexual Dimorphism in Chironomus *Larvae (Diptera: Chironomidae)*

The basic assertion of factor analysis is that a variable can be decomposed into a weighted linear sum of hypothetical common factors or underlying causes and a component unique to that variable. It is only the common portion or "communality" of a variable that one seeks to explain by the factors, and hence the method is concerned with reproducing the intercorrelations among variables rather than their respective variances as in principal-components analysis. It is this distinction that separates the two methodologies. In matrix notation, the fundamental equation of factor analysis is $Z = AF + U$, where A is a factor pattern matrix giving the regressions or loadings of each variable onto the independent factors, F contains the hypothetical factors, U is a diagonal matrix of unique variances, and Z contains the original data. The equation bears a striking resemblance to that for multiple regression, but in factor analysis the number of predictor variables as well as the predictor variables themselves are unknown.

Several levels of indeterminancy arise in factor analysis. First, because there are more factors to be estimated than the number of original variables, the fundamental equation is inherently indeterminant. That is, there is no unique mathematical solution to the factor-analysis question and one must rely upon various estimation procedures. Second, once the common factor space has been delineated, the axes may be placed in an infinite number of directions that will equally explain the common variation. To overcome this problem, one usually rotates the axes to new positions, corresponding to a simple structure solution (see below). Even so, different criteria for simple structure may lead to different results, especially as some simple-structure solutions retain orthogonality of factors while others allow for correlated or oblique factors. Finally, the hypothetical factors themselves cannot be uniquely determined from the common factor solution, and the scores of the objects with respect to the underlying patterns of variation must be estimated. Hence, while the goal of factor analysis is clear, the appropriate methodology for achieving the goal is not. As a result, factor analysis is not a single method but rather a complex battery of techniques addressing a common question.

Because of the plethora of methodologies encompassed by the term *factor analysis,* we could not reprint or discuss every important development in the field in these few pages. Therefore, our coverage will be necessarily short in some areas and we encourage the reader to consult the various texts on factor analysis for more comprehensive treatments, such as Thurstone (1947), Horst (1965), Harman (1967), Rummel (1970), and Mulaik (1972).

The use of factor-analytic procedures to explore interrelationships among variables began with Spearman (1904), who was interested in accounting for human mental abilities by a single underlying factor. However, Thurstone (Paper 7), in generalizing Spearman's approach to more than one causal factor, can be credited with founding the modern school of "multiple factor analysis." Thurstone's paper provides a historical background to the factor-analysis philosophy, but because it antedated many later and more analytical treatments of the problem, we have also included here a more recent review paper by Raymond Cattell (Paper 8), which provides an excellent nonmathematical introduction to factor analysis. Although we do not reprint it here, the reader may also wish to consult the sequel paper by Cattell (1965), in which he discusses the uses of factor analysis in modern research.

Many factor-analytic procedures, such as those exemplified by Cattell, are largely nonstatistical in that they require initial estimates of the communalities and provide few tests of significance for either the number of factors or the factor loadings. For example, the widely utilized principal-factor solution successively reestimates the communalities until the solution converges to the weighted eigenvectors of a reduced correlation matrix, containing common variances rather than total variances on the principal diagonal. In addition, the appropriate number of factors to retain during these iterations is determined by various "rules of thumb" such as those discussed by Kaiser (1960) and Sokal (1959), and one has little knowledge of the significance level of the resultant factor loadings. Hence, principal-factor analysis is generally utilized solely as an empirical process to explore intercorrelations among variables whose underlying causes are little understood, rather than as a statistical methodology.

Alternative methods that circumvent many of the problems of principal-factor analysis have been suggested. One such alternative is the "minimum residuals factor analysis" of Harman and Jones (1966). This least-squares solution, originally proposed by Thomson in 1934, is essentially equivalent to that for principal factors, while not requiring initial estimates of the communalities. One still has the problem of the number of factors; however, statistical tests have been suggested for this (e.g., see Harman, 1967). Another approach to resolving indeterminancy is Guttman's "image analysis" (e.g., Guttman, 1956a), which offers a concrete formulation of the communality of a variable. Other statistical approaches include "alpha factor analysis," proposed by Kaiser and Caffrey (1965), and "maximum-likelihood factor analysis," in which Lawley (1940 and later) provided maximum-likelihood estimates for the factor loadings.

Generally speaking, these statistical methods should be preferred over principal-factor analysis; however, as Jones (1964) has pointed out, they require rather restrictive assumptions that are unlikely to hold on biological data. For example, alpha factor analysis assumes that the variables are samples while the objects represent a statistical population; maximum-likelihood factor analysis assumes that the variables are a population and the objects are samples; and image analysis assumes that both the variables and the objects represent statistical populations. No statistical methods are available for the common situation in biology in which both the variables and objects are samples. Nevertheless, the statistical approaches represent important theoretical developments in the field and deserve attention from data analysts. Because maximum-likelihood estimates are generally preferred by statisticians, and the assumptions of this factor technique seem most realistic to biological data, we concentrate on this methodology here.

Lawley (1940, 1941, and later) is largely responsible for developing maximum-likelihood factor analysis (see also Lawley and Maxwell, 1963, for additional discussions on this subject). However, C. R. Rao provided a more direct solution to maximum-likelihood factor analysis in Paper 9 under the name "canonical factor analysis" and he can be credited with publicizing the methodology into more popular light. Unfortunately, as was also the case with Lawley's solution, convergence to the desired canonical factors was slow and uncertain, and few data analysts have adopted the method. Jöreskog (1967) apparently has now provided a more efficient algorithm for this problem, and we may expect greater usage of the method in the future. To this date, however, we are aware of only one application of maximum-likelihood factor analysis to biological data (Sinha and Lee, 1970).

As with principal-factor analysis, we felt that a more recent review paper, in addition to the above paper, would help to summarize the field. We have selected an excellent paper by R. P. McDonald (Paper 10) as a theoretical comparison among several types of factor analyses, including the principal-factor solution. A more empirical assessment of various factor methodologies on biosystematic data can be found in Fisher (1970).

By whatever method one obtains an initial factor solution, the axes are usually rotated to simple-structure positions such that each variable is accounted for by fewer than the total number of factors and each factor is not associated with all variables. Although Thurstone proposed these concepts as early as 1947, analytic solutions to the rotation problem did not appear until several years later, when various procedures were independently proposed by Carroll (1953), Neuhaus and Wrigley (1954), Saunders (1953), and Ferguson (1954). These "quartimax" procedures were virtually equivalent in seeking a joint maximization of the variance of squared factor loading of the variables. Subsequently, Kaiser (Paper 11) reviewed these methods and argued that better simple structure could be obtained by jointly maximizing the variance of squared loading for each factor. His "varimax" technique has been proved to yield good simple structure under diverse conditions and has been widely adopted for orthogonal analytic rotation.

Alternatively, simple-structure solutions that permit more general oblique rotations have been proposed. One approach, due to Thurstone (1947, 1954), uses direct counts of high loading onto specially constructed axes in determining an optimal oblique coordinate system. Basically, each vector, representing either a variable or a factor, can be thought of as a hyperplane in data space containing this variable and other closely associated variables. If one constructs new axes perpendicular to these hyperplanes, the number of variables in the hyperplanes corresponds to the number of low variable loadings onto these new axes. When the reference axes are chosen so as to maximize the number of variables in their hyperplanes, the solution will closely approximate Thurstone's original concept of simple structure. Once these best reference axes have been determined by hyperplane counts, they can be readily transformed back into an oblique factor-pattern matrix. Sokal (1958b) developed one of the first analytic solutions to these ideas under the name "Mass Modification of Thurstone's Analytic Method" (MTAM), but the similar "Maxplane" procedure proposed by Cattell and Muerle (Paper 12) has received wider attention. We reprint the latter paper here as a general discussion of the technique; however, Eber (1966) provided the algorithm generally utilized for this rotation. Other oblique solutions attempt to maximize a continuous function of simple structure, paralleling the functions for orthogonal solutions. Two such methods are Carroll's oblimin solution (Carroll, 1953 and later developments related in Harman, 1967) and Jennrich and Sampson's (Paper 13) "direct oblimin."

We reprint the latter paper here because it represents an important breakthrough in obtaining oblique solutions directly from a primary factor-pattern matrix rather than through factor structure. Two additional approaches commonly encountered in the literature are the Harris–Kaiser method (Harris and Kaiser, 1964), which uses a series of orthogonal transformations to reach oblique simple structure, and the "promax" method of Hendrickson and White (1964), which utilizes a procrustes transformation onto a specially designed target matrix to achieve simple structure. The area of oblique analytic rotation is still an active area of research, and the reader should be alerted that newer solutions will probably continue to appear for some time (e.g., the "functionplane" method of Katz and Rohlf is due to appear in a forthcoming issue of *Psychometrika*).

Finally, after an acceptable factor solution has been obtained, one often wishes to compute the projections or scores of the objects onto these factors. As previously discussed, these scores cannot be determined exactly, and therefore one must resort to various estimation procedures. The most widely utilized method in biological literature is a regression procedure proposed by Thurstone (1935) and called the "complete estimation method" by Harman (1967). Two alternative procedures are the method of Bartlett (1937) and a technique called the "ideal variable method" by Holzinger and Harman (1941) and the "least-squares method" by Horst (1965). Again, rather than reproduce these original presentations here, we have chosen a recent paper by McDonald and Burr (Paper 14) in which they review the various procedures, utilizing a common terminology, and offer a comparative assessment of their respective properties. The reader may also wish to consult Horn (1965) for a more empirical assessment of these factor score procedures.

The field of multivariate analysis has long been in need of increased theoretical unification. Jöreskog (1970, 1973) has recently made an important step toward this unification by developing a general model for covariance structures from which many other multivariate procedures, including factor analysis, can be derived as special cases. In spite of its apparent mathematical complexity, we had intended to include the shorter 1970 paper here, but we could not gain permission to do so from *Biometrika*, and, unfortunately, we did not have remaining space available to reprint the longer 1973 paper. Nevertheless, we encourage interested readers to consult these important developments toward a generalized multivariate theory.

Biologists have largely utilized factor analysis as an exploratory tool to provide *a posteriori* insight into the underlying causes of correlations among variables. Hence, it has been a technique for generating theories rather than for evaluating hypotheses in normal scientific inquiry. Moreover, while factor analysis was designed for sets of homogeneous variables, many biologists have utilized it on heterogeneous data such as in geographic variation analyses, and even on Q-mode analyses of correlations among objects. Nevertheless, the results usually seem to be intuitively meaningful, and thus the apparent empirical success of the method will likely result in its continued usage in these circumstances.

The major biological applications of factor analysis have been with morphometric data, usually in systematics, although occasional applications can be found in other fields, including behavior (e.g., Wenner et al., 1967) and ecological ordination of animals (e.g., Reyment, Paper 16) and plants (e.g., Dagnelie, 1960). Stroud (Paper 15) can be credited with first utilizing factor-analytic procedures in systematics by analyzing a matrix of correlations among morphological characters in termites. Sokal and Hunter (1955), Wallace and Bader (Paper 17), and Atchley (Paper 18) represent additional applications of factor analysis to morphometric data. Sokal (1958a) appears to have been the first to apply factor analysis to Q-mode matrices, followed by Morishima and Oka (1959); and Rohlf and Sokal (1962) have discussed the utility of this approach in revealing taxonomic relationships. The most profuse applications of factor analysis in biology have been in geographic variation studies. Beginning with Sokal's (1962) study on geographic variation in an aphid, numerous papers have appeared in the literature on such diverse organisms as ticks (e.g., Thomas, 1968), blackbirds (e.g., Power, 1970), and biting midges (e.g., Atchley, 1971). Atchley's paper offers an example of such an application, and demonstrates multiple-regression procedures in relating geographic trends in morphometrics to environmental components.

Paralleling the profusion of theoretical developments in factor analysis, a wide variety of analytic procedures are encountered in the applied literature. We have therefore attempted to chose a few papers that not only represent clear applications of factor analysis to a variety of types of data, but also reflect this methodological diversity. The paper by Stroud, which introduced factor analysis to systematics, utilized Thurstone's centroid approximation to principal-factor analysis followed by a graphical rotation to simple structure. This paper also offers a rather detailed introduction to the technique which is not generally found in more recent papers. A second paper concerning ecological ordination in ostracods by Reyment (Paper 16) utilized the image factor analysis of Jöreskog (1962), a procedure similar to Guttman's image analysis, with no subsequent rotation. A third example is the paper by Wallace and Bader (Paper 17) on morphometrics in the house mouse, in which they utilized principal-factor analysis with varimax rotation to orthogonal simple structure. Finally, we include a study by Atchley (Paper 18) which utilized the maxplane oblique simple-structure solution of Cattell and Muerle (Paper 12) to rotate an initial principal-factor solution.

BIBLIOGRAPHY

Atchley, W. R. 1971. A comparative study of the causes and significance of morphological variation in adults and pupae of *Culicoides:* a factor analysis and multiple regression study. *Evolution 25:*563–583.

Bartlett, M. S. 1937. The statistical concept of mental factors. *British J. Psychol. 28:*97–104.

Browne, M. W. 1969. Fitting the factor analysis model. *Psychometrika 34:*375–394.

Carroll, J. B. 1953. An analytic solution for approximating simple structure in factor analysis. *Psychometrika 18:*23–38.

Cattell, R. B. 1965. Factor analysis: an introduction to essentials. II. The role of factor analysis in research. *Biometrics 21:*405–435.

Dagnelie, P. 1960. Contribution à l'étude des communautés végétates par l'analyse factorielle. *Bull. Serv. Carté. Phytogéog. B 5:*7–71, 93–195.

Eber, H. W. 1966. Toward oblique simple structure: Maxplane. *Multivariate Behav. Res. 1:*211–218.

Ferguson, G. A. 1954. The concept of parsimony in factor analysis. *Psychometrika 19:*218–290.

Fisher, D. 1970. A comparison of the various techniques of multiple factor analysis applied to biosystematic data. Ph.D. Thesis, University of Kansas.

Guttman, L. 1956a. Image theory for the structure of quantitative variates. *Psychometrika 18:*277–296.

———. 1956b. Best possible systematic estimates of communalities. *Psychometrika 21:*273–285.

Harman, H. H. 1967. *Modern Factor Analysis.* University of Chicago Press, Chicago. 474 pp.

———, and W. H. Jones. 1966. Factor analysis by minimizing residuals (MINRES). *Psychometrika 31:*331–368.

Harris, E. W., and H. F. Kaiser. 1964. Oblique factor solutions by orthogonal transformation. *Psychometrika 29:*347–362.

Hendrickson, A. E., and P. O. White. 1964. Promax: a quick method for rotation to oblique simple structure. *British J. Stat. Psychol. 17:*65–70.
Holzinger, K. J., and H. H. Harman. 1941. *Factor Analysis.* University of Chicago Press, Chicago. 417 pp.
Horn, J. L. 1965. An empirical comparison of methods for estimating factor scores. *Educ. Psychol. Measurement 25:*313–322.
Horst, P. 1965. *Factor Analysis of Data Matrices.* Holt, Rinehart and Winston, Inc., New York. 730 pp.
Jones, L. E. 1964. Multivariate statistical analyzer. Unpublished write-up for factor analysis program.
Jöreskog, K. G. 1962. On the statistical treatment of residuals in factor analysis. *Psychometrika 27:*335–354.
———. 1967. Some contributions to maximum likelihood factor analysis. *Psychometrika 32:*443–482.
———. 1970. A general method for analysis of covariance structures. *Biometrika 57:*239–251.
———. 1973. In P. R. Krishnaiah (ed.), *Multivariate Analysis,* Vol. 3, pp. 263–285. Academic Press, New York.
Kaiser, H. F. 1960. The application of electronic computers to factor analysis. *Educ. Psychol. Measurement 20:*141–151.
———, and J. Caffrey. 1965. Alpha factor analysis. *Psychometrika 30:*1–14.
Katz, J. O., and F. J. Rohlf. 1974. Functionplane—a new approach to simple structure rotation. *Psychometrika,* in press.
Lawley, D. N. 1940. The estimation of factor loadings by the method of maximum likelihood. *Proc. Roy. Soc. Edinburgh 60:*64–82.
———. 1941. Further investigations in factor estimation. *Proc. Roy. Soc. Edinburgh 61:*176–185.
———. 1943. The application of the maximum likelihood method to factor analysis. *British J. Psychol. 33:*172–175.
———. 1958. Estimation in factor analysis under various initial assumptions. *British J. Stat. Psychol. 11:*1–12.
———, and A. E. Maxwell. 1963. *Factor Analysis as a Statistical Method.* Butterworth, London. 117 pp.
Morishima, H., and H. Oka. 1959. The patterns of interspecific variation in the genus *Oryza:* its quantitative representation by statistical methods. *Evolution 14:*153–165.
Mulaik, S. A. 1972. *The Foundations of Factor Analysis.* McGraw-Hill Book Company, New York. 453 pp.
Neuhaus, J. O., and C. Wrigley. 1954. The quartimax method: an analytic approach to orthogonal simple structure. *British J. Stat. Psychol. 7:*81–91.
Power, D. M. 1970. Geographic variation of the red-winged blackbird in central North America. *Univ. Kansas Mus. Publ. 19:*1083.
Rohlf, F. J., and R. R. Sokal. 1962. The description of taxonomic relationships by factor analysis. *Syst. Zool. 11:*1–16.
Rummel, R. J. 1970. *Applied Factor Analysis.* Northwestern University Press, Evanston, Ill. 617 pp.
Saunders, D. R. 1953. An analytic method for rotation to orthogonal simple structure. *Educ. Testing Res. Bull.* 53–10.

Sinha, R. H., and P. J. Lee. 1970. Maximum likelihood factor analysis of natural arthropod infestations in stored grain bulks. *Res. Population Ecol. 12:*51–60.

Sokal, R. R. 1958a. Quantification of systematic relationships and of phylogenetic trends. *Proc. Xth Intern. Congr. Entomol. 1:*409–415.

———. 1958b. Thurstone's analytic method for simple structure and a mass modification thereof. *Psychometrika 23:*237–257.

———. 1959. A comparison of five tests for completeness of factor extraction. *Trans. Kansas Acad. Sci. 62:*141–152.

———. 1962. Variation and covariation of characters of alate *Phemphigus populi-transversus* in eastern North America. *Evolution 16:*227–245.

———, and P. E. Hunter. 1955. A morphometric analysis of DDT-resistent and non-resistent house fly strains. *Ann. Entomol. Soc. Amer. 48:*499–507.

Spearman, C. 1904. General intelligence objectively determined and measured. *Amer. J. Psychol. 15:*201–293.

Thomas, P. A. 1968. Variation and covariation in characters of the rabbit tick *Haemaphysalis leporispalustris. Univ. Kansas Sci. Bull. 47:*829–862.

Thomson, G. H. 1934. Hotelling's method modified to give Spearman's *g. J. Educ. Psychol. 25:*366–374.

Thurstone, L. L. 1935. *The Vectors of the Mind.* University of Chicago Press, Chicago. 266 pp.

———. 1947. *Multiple Factor Analysis.* University of Chicago Press, Chicago. 535 pp.

———. 1954. An analytic method for simple structure. *Psychometrika 19:*173–182.

Wenner, A. M., P. H. Wells, and F. J. Rohlf. 1967. An analysis of the waggle dance and recruitment in honey bees. *Physiol. Zool. 40:*317–344.

7

Reprinted from *Psychol. Rev.*, **38**, 406–427 (1931)

MULTIPLE FACTOR ANALYSIS

BY L. L. THURSTONE

The University of Chicago

The two-factor problem of Spearman consists in the analysis of a table of intercorrelations for the discovery of some general factor that is common to all of the variables in the table. Spearman differentiates three types of factors, namely, a general factor which is common to all of the variables, group factors which are common to some of the variables but not to all of them, and specific factors that are peculiar to single variables alone. In practice, the Spearman two-factor methods meet with the difficulty that group factors are frequently encountered. The two-factor methods are not applicable to situations that involve group factors except in indirect ways. This is a serious limitation on Spearman's technique since many important psychological problems involve a complex of variables that are known from the nature of the problem to contain group factors. The present multiple factor methods in no way contradict the Spearman two-factor methods which are very ingenious and powerful in the situations to which they apply. The present multiple factor method may be thought of as supplementary to the Spearman two-factor method in that we do not have any restrictions as to the number of general factors or the number of group factors.

It is the purpose of this paper to describe a more generally applicable method of factor analysis which has no restrictions as regards group factors and which does not restrict the number of general factors that are operative in producing the intercorrelations. Our first question concerns the number of general, independent, and uncorrelated factors that are operative in producing a given table of intercorrelations for any number of variables. In our terminology general factors will include what Spearman calls general and group factors.

This question can be answered by methods that will here be described and which are applicable to any table of intercorrelations. We may consider three examples to illustrate the nature of this first problem. If we have a table of intercorrelations for a battery of motor tests it is of considerable psychological interest to know how many independent motor abilities it is necessary to postulate in order to account for the whole table of intercorrelations. If this turns out to be three, then our next task would be to hunt about for the nature of these three motor abilities. If we have a table of intercorrelations of the interests of eighteen professions it would be of considerable importance to know how many independent interest factors it is necessary to postulate in order to account for the whole table of intercorrelations. This refers to the tables published by E. K. Strong. We have applied our methods to his data and we have found that his table of intercorrelations can be accounted for by postulating four general interest factors which turn out to be (1) interest in science, (2) interest in language, (3) interest in people, and (4) interest in business. Again, Professor Moore[1] has prepared a table of intercorrelations of 48 psychotic symptoms on the basis of his work with about four hundred patients with various psychoses. A general factor analysis of the type here discussed would enable us to know how many general factors or mental disease entities it is necessary to postulate in order to account for the whole table of correlations of psychotic symptoms. If this should turn out to be five, for example, then we should be justified to look for five fundamentally different psychoses.

Our next problem is to assign a weight or loading of each of the general factors to each of the variables. For example, in the table of interest-correlations above referred to we should assign four loadings, one for each of the four general factors, to each of the eighteen professions. It then turns out that Engineering, for example, has a high loading of interest in science, a rather low loading of interest in language. The profession of law has just the reverse loadings, namely low

[1] T. V. Moore, The empirical determination of certain syndromes and underlying praecox and manic depressive psychoses, *Amer. J. Psychiat.*, 1930, **9**, 719–738.

for science and high for language. The ministry is loaded high for interest in people and in language but low for science. Finally, we should want to be able to assign to each individual subject a quantitative rating in the form of a standard score for each of the general factors or abilities that have been isolated.

Let there be n factors. In this explanation n will be assumed to be three. It can be any number.

Let there be N individuals in a group, all of whom have taken w tests.

Let a, b, c, d, etc., represent the tests.

Let the three factors be represented by numerals 1, 2, 3.

S_a = standard score of one individual in test a.

S_b = standard score of one individual in test b.

S_c = standard score of one individual in test c.

$$S_a = \frac{X_a - m_a}{\sigma_a} = \text{the usual definition of a standard score.}$$

The standard score S_a of an individual in test a depends on (1) his rating in each of the three abilities or factors 1, 2, 3, and (2) the weight or loading of each of these abilities in test a. For example, if test a calls for much of ability No. 1 and very little of abilities 2 and 3, and if one subject has a low rating in ability No. 1 and average or high ratings on the other two abilities, then this subject may be expected to do poorly on test a. The loadings of the three general factors in each test and for each subject may be represented by the following notation.

Let

x_1 = standard score of an individual in ability No. 1.

x_2 = standard score of an individual in ability No. 2.

x_3 = standard score of an individual in ability No. 3.

and let

a_1 = loading of ability No. 1 in test a,

a_2 = loading of ability No. 2 in test a,

a_3 = loading of ability No. 3 in test a.

Then we shall assume that the standard score of each individual subject is a sum of the products of his standard score in each ability and the loading of the ability in each test. This assumption leads to the following fundamental equations.

$$S_a = a_1x_1 + a_2x_2 + a_3x_3 \tag{1}$$

and

$$S_b = b_1x_1 + b_2x_2 + b_3x_3.$$

Strictly speaking, we should add to each of these two expressions a term to account for those additional general factors, beyond three, which are here ignored, and also a term to account for the specific factor, peculiar to the particular test. However, our object is to ascertain how many general and independent factors it is necessary to postulate in order to account for a whole table of intercorrelations and we shall therefore intentionally ignore these additional specific factors as well as those minor group factors which may not appreciably affect the correlations.

We want to express the correlation between tests *a* and *b* in terms of the standard scores in the three abilities and the loadings of the three abilities in each of the two tests. For this purpose we shall need the product S_aS_b. Then

$$S_aS_b = a_1b_1x_1^2 + a_2b_2x_2^2 + a_3b_3x_3^2 + \text{cross products.}$$

The correlation r_{ab} can be expressed simply as

$$r_{ab} = \frac{\Sigma S_a \cdot S_b}{N},$$

because S_a and S_b are both standard scores so that the standard deviations of the given scores are all unity. Then

$$\frac{\Sigma S_aS_b}{N} = a_1b_1\frac{\Sigma x_1^2}{N} + a_2b_2\frac{\Sigma x_2^2}{N} + a_3b_3\frac{\Sigma x_3^2}{N},$$

in which the cross products vanish because x_1, x_2, x_3 are uncorrelated. But

$$\frac{\Sigma x_1^2}{N} = \frac{\Sigma x_2^2}{N} = \frac{\Sigma x_3^2}{N} = 1,$$

since x_1, x_2, x_3 are all standard scores in the three abilities. Therefore

$$r_{ab} = a_1b_1 + a_2b_2 + a_3b_3. \tag{2}$$

This is one of the fundamental equations. Here the correlation between two tests is expressed in terms of the loadings of the three abilities in the two tests. By analogy we may write equation (2) for any pair of tests, as

$$r_{ac} = a_1c_1 + a_2c_2 + a_3c_3$$

and so on.

Another fundamental equation can be derived as follows:

$$S_a = a_1x_1 + a_2x_2 + a_3x_3.$$

Squaring,

$$(S_a)^2 = a_1^2x_1^2 + a_2^2x_2^2 + a_3^2x_3^2 + \text{cross products.}$$

Summing and dividing by N

$$\frac{\Sigma S_a^2}{N} = a_1^2\frac{\Sigma x_1^2}{N} + a_2^2\frac{\Sigma x_2^2}{N} + a_3^2\frac{\Sigma x_3^2}{N}.$$

The cross products vanish because x_1, x_2, x_3 are uncorrelated by definition. But

$$\frac{\Sigma x_1^2}{N} \doteq \sigma_1^2 = 1$$

and

$$\frac{\Sigma S_a^2}{N} = 1 \text{ by definition.}$$

Hence

$$a_1^2 + a_2^2 + a_3^2 = 1, \qquad (3)$$

and similarly by analogy for every other test, as

$$b_1^2 + b_2^2 + b_3^2 = 1.$$

Equations (3) come about because x_1, x_2, x_3 and S_a, S_b are standard scores.

Still another fundamental equation that we shall need can be derived as follows:

$$\begin{array}{rcl} r_{aa} &=& a_1^2 + a_2^2 + a_3^2 = 1. \\ r_{ab} &=& a_1b_1 + a_2b_2 + a_3b_3 \\ r_{ac} &=& a_1c_1 + a_2c_2 + a_3c_3 \\ r_{ad} &=& a_1d_1 + a_2d_2 + a_3d_3 \\ \cdot & \cdot & \cdot \quad \cdot \quad \cdot \quad \cdot \quad \cdot \\ \hline \Sigma r_{ak} &=& a_1\Sigma k_1 + a_2\Sigma k_2 + a_3\Sigma k_3 \end{array} \qquad (4)$$

In the summation equation Σk_1 = sum of all loadings of the first factor in all of the tests of the set. The notation k refers to each of the tests in succession, *i.e.*, k takes values from 1 to w when there are w tests in the series. But this summation equation may be written for each one of the tests as follows

$$\begin{aligned}
\Sigma r_{ak} &= a_1\Sigma k_1 + a_2\Sigma k_2 + a_3\Sigma k_3 \\
\Sigma r_{bk} &= b_1\Sigma k_1 + b_2\Sigma k_2 + b_3\Sigma k_3 \\
\Sigma r_{ck} &= c_1\Sigma k_1 + c_2\Sigma k_2 + c_3\Sigma k_3 \\
&\cdot \quad \cdot \quad \cdot \quad \cdot \quad \cdot \quad \cdot \\
\hline
\Sigma r_{kk} &= (\Sigma k_1)^2 + (\Sigma k_2)^2 + (\Sigma k_3)^2 \qquad (5)
\end{aligned}$$

in which

Σr_{kk} = sum of all intertest correlations in the whole table. Since the full table is symmetrical it follows that each intertest correlation enters twice in this sum. Also, the full table of intercorrelations includes the correlation of each test with itself and this is here regarded as unity.

Σk_1 = sum of all loadings of factor No. 1 in all tests.

Σk_2 = sum of all loadings of factor No. 2 in all tests.

Σk_3 = sum of all loadings of factor No. 3 in all tests.

Note that equation (3) is the equation of a sphere and that a_1, a_2, a_3 are the three coördinates of a point on the surface of the sphere. The space order of the sphere is equal to the number of postulated general factors. If five factors were postulated we should have a five-dimensional sphere, and so on.

Also note that each test is here really a point on the surface of a sphere. Hence, if there are three factors essentially operative in producing the intercorrelations, then it should be possible to locate each test as a point on the surface of a ball.

Note also that each correlation is the cosine of the central angle subtended by the two points a and b. Hence, if three factors are operative, then each test would be represented by a point on the surface of the ball, and the points would be so allocated that the cosine of each central angle is equal to the intercorrelation between the respective pair of tests. This will be shown in more detail later.

Our problem is to ascertain the coördinates a_1, a_2, a_3 for test a, the coördinates b_1, b_2, b_3 for test b, and so on for each of the w tests in the whole series. Then, if our determinations are correct, it should be possible to calculate the correlation coefficients by equations (2). These calculated coefficients should agree with the observed correlations within reasonable experimental error.

If we had all of the tests allocated to as many points on the surface of a ball we could not determine the coördinates for any of these points without first deciding where our coördinate axes are to be drawn. The location of these axes is arbitrary and not at all given by the intercorrelations because the latter are merely the cosines of the angular separations between all pairs of points on the surface of the ball.

One simple plan would be to draw the axis OX through one of the tests which might be more or less arbitrarily chosen, such as a. Then if this x-axis represents the first factor, it follows of course that the xyz coördinates of point a are $(+1, 0, 0)$ and hence that

$$a_1 = +1, \quad a_2 = 0, \quad a_3 = 0.$$

The y-axis must of course be at right angles to the x-axis but it could be drawn through the origin in any direction in the plane at right angles to the x-axis.

We might now revolve the sphere around the x-axis until any arbitrarily selected second test b lies in the xy-plane. Then it is clear that the z-coördinate of point b must be zero so that b_3 in equation (3) vanishes. Therefore equation (3) becomes

$$b_1^2 + b_2^2 = 1$$

and

$$a_1 = 1.$$

The correlation between a and b is

$$r_{ab} = a_1b_1 + a_2b_2 + a_3b_3,$$

but since $a_1 = 1$, $a_2 = 0$, $a_3 = 0$, $b_3 = 0$, this reduces to

$$r_{ab} = b_1.$$

But

$$b_1 = \cos\phi$$

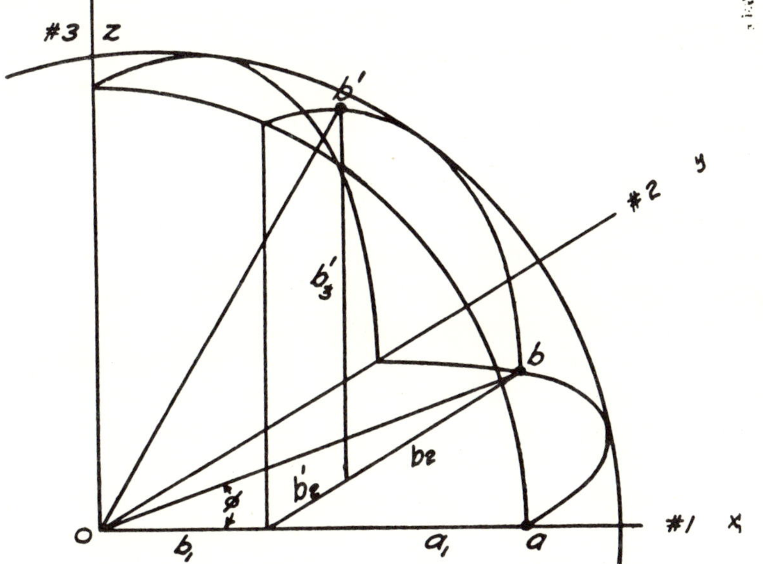

Fig. 1. This diagram shows the relation between the correlation coefficient and the central angle between the two tests or points. Let two points which represent any two tests be designated a and b'. Let the x-axis, representing the first general factor, pass through the point a. Then its coördinates are (1, 0, 0). Therefore $a_1 = 1$ while a_2 and a_3 are both zero. Revolve the sphere about the x-axis so that the point b' is in the horizontal xy-plane at b. Then the z-coördinate of the point b is evidently zero while $b_1^2 + b_2^2 = 1$. Let the angle aob be designated ϕ. Then, clearly, $\cos\phi = b_1$ since the radius of the sphere is unity. The correlation between tests a and b can be written as follows.

$$r_{ab} = a_1b_1 + a_2b_2 + a_3b_3.$$

Since a_2 and a_3 are both zero, the second and third terms vanish and since $a_1 = 1$, it follows that $r_{ab} = b_1$. The angle aob is equal to the angle aob'. Hence the correlation coefficient is equal to the cosine of the central angle between the points that represent the two tests.

and hence the correlation between tests a and b is the cosine of the central angle between them. Since the sphere can be revolved in a similar manner for any pair of tests we see that the correlation coefficient for any pair of tests is the cosine of the central angle between them. This is true for any number of factors that may be postulated and hence for any space order.

Our problem can now be restated as follows. We have a table of intercorrelations which are cosines of central angles between pairs of points. Every test is represented by a point on the surface of a sphere. Since all of the correlation coefficients are subject to experimental errors, we cannot trust the accuracy of any one of them. Our problem is to find the best fitting allocations of the points on the surface of the sphere. The least square methods are here altogether too unwieldy. They are out of the question, undoubtedly. We shall therefore apply an adaptation of the principles of curve fitting by the method of averages. Just as in the method of averages we deal with summations of partial sets of observations, so we shall here deal with summations for partial sets of tests or points. In fact, we shall define the x-axis so that it passes through the origin and through the center of gravity of a partial set of points. The y-axis will be so determined that the xy-plane contains the x-axis and the center of gravity of a second set of points. This procedure can be continued for any number of dimensions or factors. The number of dimensions of the sphere is the same as the number of general factors that are postulated. This procedure will now be described in more detail.

In locating the coördinate axes there are several criteria and several quantitative tests that may be applied. These criteria will be considered in a separate paper, and we shall here describe only one such criterion by which the axes may be located. Assume that you have before you the sphere with a point designated on its surface for each test or variable in the correlation table. Our first problem is to locate the x-axis through this sphere. The projection of a point on this axis will be the loading of the first factor in that test. Suppose that we arbitrarily pass the x-axis through test *a*. Then this test will have loading of unity for the first factor and zero for all of the other factors. The projection of test *b* on this axis will then be the loading of the first factor in test *b*. In dealing with actual data we shall always have a residual composed of minor general factors that are ignored and specific factors peculiar to each test or variable. For this reason we want to account for as much as possible of the correlations in terms of the smallest possible number of general independent factors. We therefore want to pass the *x*-axis through this sphere so as to maximize the projections of all the points on it.

We first find that one test which has the highest average correlation with all the other tests, disregarding sign. We disregard sign because it does not matter here whether the projection is positive or negative. The sign is arbitrary and either projection serves to determine the correlation. The sum of each column in the correlation table, disregarding sign, is recorded. Let the test with the highest sum be designated test *a* and let the sum be designated $\Sigma|r_{ak}|$. This test is evidently more nearly like the rest of the tests in the series than any other single test. The average distance from it to all the other tests is, in general, smaller than the corresponding average separation of each other test from all the rest. We have found now the most representative single test in the battery. Of course we could now simply pass the *x*-axis through this test and thereby make sure that we have nearly maximized the projections of all the tests on this axis. But we do not want to define any of the axes by any single test or variable. We therefore make a list of all the tests that correlate positively with test *a*. These tests will all lie in a hemisphere with test *a* at its pole. This is clear because all positive correlations correspond to angles less than 90 degrees, and all correlations that are negative correspond to angles greater than 90 degrees. We can now be fairly sure, but not absolutely so, that this partial group of tests have something in common that is a conspicuous and important general factor in the whole set of tests. We shall define the *x*-axis so that it passes through the center of the sphere, of course, and also through the center of gravity of all the tests in the hemisphere with *a* at its pole.

Let us designate all the tests which correlate positively

with test a by the general notation s. The notation s refers then to each of the tests in this partial group, taken in succession. The correlation of test a with each of the tests in this sub-group s may then be written as

$$r_{as} = a_1 s_1 + a_2 s_2 + a_3 s_3. \tag{6}$$

Summing, we have

$$\Sigma r_{as} = a_1 \Sigma s_1 + a_2 \Sigma s_2 + a_3 \Sigma s_3, \tag{7}$$

in which Σs_1 is the sum of the projections of the points in the set s on the first axis, Σs_2 is the sum of the projections of the points in this set on the second or y-axis, Σs_3 is the sum of the projections on the third or z-axis.

The center of gravity of the points in the set s is a point inside the sphere and since we have taken a set of tests in a single hemisphere we are certain that this center of gravity will not lie at or near the center of the sphere. If that should happen our subsequent determinations would be very inaccurate. The coördinates of the center of gravity of the set of points s will evidently be

$$\frac{\Sigma s_1}{N_s}, \quad \frac{\Sigma s_2}{N_s}, \quad \frac{\Sigma s_3}{N_s}$$

with reference to the three axes that are still to be chosen. Now let us pass the first axis, the x-axis, through the center of gravity of the set of points s. But if this center of gravity lies on the x-axis, then it is clear that the y- and z-coördinates must be zero so that

$$\Sigma s_2 = \Sigma s_3 = 0.$$

The correlation between the test a and any one of the other tests in the set s can then be simplified because the last two terms in equation (7) vanish. The expression for this correlation becomes then

$$\Sigma r_{as} = a_1 \Sigma s_1 \tag{8}$$

or, rewriting this explicitly for the loading a_1, we have

$$a_1 = \frac{\Sigma r_{as}}{\Sigma s_1} \tag{9}$$

By analogy with equation (5) we can write the following corresponding equation for the set s,

$$(\Sigma s_1)^2 + (\Sigma s_2)^2 + (\Sigma s_3)^2 = \Sigma r_{ss} \tag{10}$$

in which Σr_{ss} is the sum of all the correlation coefficients in the full table of the s tests, including self correlations taken as unity, and recording each coefficient twice in the table since it is symmetrical. But since Σs_2 and Σs_3 are zero by the location of our x-axis, we have the simpler relation

$$(\Sigma s_1)^2 = \Sigma r_{ss} \tag{11}$$

or

$$\Sigma s_1 = \sqrt{\Sigma r_{ss}}$$

and from this we know the loading of the first general factor in each test as follows:

$$a_1 = \frac{\Sigma r_{as}}{\sqrt{\Sigma r_{ss}}},$$

$$b_1 = \frac{\Sigma r_{bs}}{\sqrt{\Sigma r_{ss}}},$$

$$\cdot \quad \cdot \quad \cdot$$

$$w_1 = \frac{\Sigma r_{ws}}{\sqrt{\Sigma r_{ss}}}. \tag{12}$$

It should be noted that these relations are valid even though the test b, for example, is not a member of the particular set of points s. Hence the above simple relation enables us to determine the loading of the first general factor in each of the w tests.

We now want to determine the loading of the second general factor in each of the tests. Imagine again the sphere with the w points allocated to its surface. We have now passed the x-axis through this sphere. The y-axis must of course be at right angles to the x-axis because we assume that the general factors are uncorrelated. The angle between the first and second axes must therefore be 90 degrees. If we revolve the sphere about the x-axis with test a somewhere in the vicinity of the pole, it is clear that the y-axis which shall represent the second general factor will pierce the surface of the sphere somewhere in the equator. We shall now select a pivot test for the second factor which shall serve the same function in locating the second factor that test *a* served in locating the first factor. This second test will clearly lie in an equatorial band and it will have a low correlation with test *a*. We therefore tabulate separately all the tests that have a low correlation with test *a*. In doing so we must decide how wide an equatorial band is to be included. The band should be wide enough to include all tests that are essentially different from test *a*, but the band should not be so wide as to include those tests which are heavily loaded with the first factor which is already represented by the *x*-axis. The width of the equatorial band selected will depend also in part on the number of tests in the whole series of intercorrelations. Let us decide to use an equatorial band not wider than 30 degrees on either side of the equator. An angle of this size has a cosine of .50 so we make a list of all tests that correlate less than .50, either positive or negative, with test *a*.[2]

In order to find the most representative test in the equatorial band, we inspect again the summations $\Sigma|r_{ak}|$, $\Sigma|r_{bk}|$, $\Sigma|r_{ck}|$, $\cdots$ $\Sigma|r_{wk}|$, disregarding sign. We find the test *b* in this list which has the highest sum of its correlations with the other tests. In the equatorial band this test is the most representative of all the tests in the whole series *w*. By selecting this one as our second pivot test we will, in general, but not necessarily, maximize the projections of all the tests on the second axis. Let this second pivot test, so selected, be test *b*.

We now tabulate, as before, all those tests in the whole series *w* which correlate positively with test *b*. Let this second sub-group of tests be designated by the notation *t*. We have now a set of tests *t* which lie in a hemisphere with test *b* as its pole. Consider the center of gravity of this set of points *t*. It will be a point inside the sphere but it will

[2] In case all the intercorrelations in the initial table are positive it is clear that all the points lie in one quadrant or octant of the surface and hence the selection of the subgroups *s*, *t*, *u*, must be adapted to the distribution of the initial coefficients as to positive and negative values.

certainly not coincide with the center of the sphere. That is to be avoided in the interest of accuracy of subsequent determinations. Now we shall locate the y-axis so that the xy-plane contains the x-axis and also the center of gravity of the set of points t. By so doing we make the z-coördinate of the center of gravity of the set t vanish. By this procedure in locating the successive axes we reduce the number of unknowns so that the loadings may be readily determined.

The correlation between test a and any one of the tests in the set t may be written

$$r_{at} = a_1t_1 + a_2t_2 + a_3t_3. \tag{13}$$

Summing for all the tests in the set t, as before, we have

$$\Sigma r_{at} = a_1\Sigma t_1 + a_2\Sigma t_2 + a_3\Sigma t_3. \tag{14}$$

Here the values of Σt_1, Σt_2, Σt_3, are the sums of the projections of the points in the set t on each of the three coördinate axes.

The coördinates for the center of gravity of the set of points t are evidently

$$\frac{\Sigma t_1}{N_t}, \ \frac{\Sigma t_2}{N_t}, \ \frac{\Sigma t_3}{N_t},$$

but the z-coördinate vanishes by so locating the y-axis that the xy-plane contains the center of gravity for the set of points t. Therefore

$$\Sigma t_3 = 0$$

and hence

$$\Sigma r_{at} = a_1\Sigma t_1 + a_2\Sigma t_2 \tag{15}$$

Since all the x-coördinates are now known, so is also Σt_1 for the set t. Therefore we can solve for a_2 in the above equation in the form

$$a_2 = \frac{\Sigma r_{at} - a_1\Sigma t_1}{\Sigma t_2}. \tag{16}$$

The summation Σt_2 is known from the relation

$$\Sigma r_{tt} = (\Sigma t_1)^2 + (\Sigma t_2)^2, \tag{17}$$

which can be written by analogy with equation (5). This equation enables us to determine the loadings of the second factor in all of the tests.

The loadings of the second factor in the remaining tests are determined by equations analogous to (16), namely

$$b_2 = \frac{\Sigma r_{bt} - b_1 \Sigma t_1}{\Sigma t_2}, \tag{16}$$

$$\cdot \quad \cdot \quad \cdot \quad \cdot$$

$$w_2 = \frac{\Sigma r_{wt} - w_1 \Sigma t_1}{\Sigma t_2}.$$

When the loadings of the first two factors in each of the tests have been determined it would be possible to determine the loading of the third factor from the relations

$$a_1^2 + a_2^2 + a_3^2 = 1,$$
$$b_1^2 + b_2^2 + b_3^2 = 1,$$
$$c_1^2 + c_2^2 + c_3^2 = 1,$$

and so on provided that we were certain that three factors were sufficient to determine the correlation coefficients. However, we must assume that in any ordinary situation there are residuals composed of additional general factors of minor importance, perhaps, and specific factors peculiar to each of the tests. On this account we shall not make use of the above equation for determining the loadings of the third factor. These loadings will be determined in a manner similar to that used for the previous loadings.

We make a list of all the tests that correlate between $+.50$ and $-.50$ with *both* tests *a* and *b*. In this manner we shall be sure that we have a list of tests represented by points which are in the general vicinity of a right angle from both *a* and *b*. We now leave the restricted three dimensional sphere and proceed by analogy into higher dimensions. We select that one test in this list which has the highest sum Σr_{kk}, disregarding sign. Let this test be test *c*. This will be the pivot test for the sub-group of tests which are to be used for determining the third general factor.

After having found the pivot test for the third general factor we list all of the tests that have positive correlations with test *c*. Let this sub-group be designated *u*. These will all lie in a hemisphere with test *c* at its pole. We shall so locate the third axis or factor that the center of gravity of the set of points u has finite values for the first three coördinates, and so that the higher numbered coördinates of this center of gravity will vanish. This is merely carrying out the same procedure as before. The correlation between test a and any one of the tests in set u will then be

$$r_{au} = a_1 u_1 + a_2 u_2 + a_3 u_3. \tag{17a}$$

Summing we have

$$\Sigma r_{au} = a_1 \Sigma u_1 + a_2 \Sigma u_2 + a_3 \Sigma u_3. \tag{18}$$

This equation can be solved for a_3 since all the other values are known. The values of Σu_1 and Σu_2 are known because we have the loadings of the first and second factors in each of the tests and we have listed the tests that belong in the set u. The summation Σr_{au} is known directly from the given correlation coefficients. The value of Σu_3 can be obtained from the equation

$$\Sigma r_{uu} = (\Sigma u_1)^2 + (\Sigma u_2)^2 + (\Sigma u_3)^2, \tag{19}$$

which is written by analogy with equation (5). Hence

$$a_3 = \frac{\Sigma r_{au} - a_1 \Sigma u_1 - a_2 \Sigma u_2}{\Sigma u_3}, \tag{20}$$

$$b_3 = \frac{\Sigma r_{bu} - b_1 \Sigma u_1 - b_2 \Sigma u_3}{\Sigma u_3},$$

$$\cdot \quad \cdot \quad \cdot \quad \cdot \quad \cdot \quad \cdot$$

$$w_3 = \frac{\Sigma r_{wu} - w_1 \Sigma u_1 - w_2 \Sigma u_3}{\Sigma u_3}.$$

In the same manner we may extend the procedure to any number of factors that may be necessary to account for a given table of intercorrelations. However, it is not advisable to carry this procedure so far as to determine the correlations within the errors of measurement because in that case the last factors are likely to be merely the loadings that are necessary to adjust for chance errors. Our purpose is to discover only those principal factors that are truly operative in producing the correlation coefficients. Hence this procedure should be carried out far enough to lock the coefficients with discrepancies somewhat greater than the chance errors in the given coefficients.

Finally, when the loadings have been determined for each test, one should calculate all of the coefficients by equations (2). These can then be compared with the given experimental coefficients. The discrepancies $d = r_e - r_c$ in which r_c = calculated r, and r_e = experimental or observed r, should then be calculated. A frequency distribution of these discrepancies will indicate fairly well how closely the factors account for the given coefficients. In no case should one expect the standard deviation of this distribution to be the same as the average standard error of the given coefficients. The standard deviation of the discrepancies will be the larger because it may be generally assumed that the limited number of general factors that are postulated do not include the minor group factors and specific factors that together will make the residuals greater than those expected by chance errors alone.

In order to facilitate the application of the procedure we shall list here the successive steps together with the fundamental equations that are necessary in solving for the successive values.

The following outline has been extended to include four factors in order to illustrate the method.

(1) Prepare a full table of intercorrelations for the w tests. Assume that all self correlations are unity and record them so. The full table will be symmetrical in that every coefficient will occur in two cells. Thus the coefficient r_{ab} will occur in column a opposite b and also in column b opposite a. Record the sums $\Sigma|r_{ak}|$, $\Sigma|r_{bk}|$, $\Sigma|r_{ck}|$, etc., for all columns, disregarding sign. The subscript k refers to each of the w tests in the whole series taken in succession.

(2) Find that test a which has the highest sum for its column, namely $\Sigma|r_{ak}|$.

(3) Make a list of all the tests that correlate positively with test a. This is done by listing all tests with positive correlations in column a. Let this sub-group of tests

be designated s. The sum Σr_{ss} is the *algebraic* sum of all the coefficients in a full table of the tests in group s. Then

(4) $$(\Sigma s_1) = \sqrt{\Sigma r_{ss}}$$

(5) $$a_1 = \frac{\Sigma r_{as}}{\sqrt{\Sigma r_{ss}}}$$

$$b_1 = \frac{\Sigma r_{bs}}{\sqrt{\Sigma r_{ss}}}$$

$$\cdot \quad \cdot \quad \cdot$$

$$w_1 = \frac{\Sigma r_{ws}}{\sqrt{\Sigma r_{ss}}}$$

The loading of the first factor in every test is now determined.

(5*a*) Determine the algebraic sum Σs_1 from the list of first factor loadings and see that it agrees with the value found in step 4.

(6) Make a list of all tests that correlate with test a within the range $\pm .50$. These tests are in general different from test a. Select that test b in this list which has the highest sum in its column, namely $\Sigma|r_{bk}|$, disregarding sign.

(7) Now make a list of all the tests that correlate positively with test b and let this sub-group be designated t.

(8) Determine (Σt_1). This can be done since we know the first factor loading in each test and we have a list of the tests in group t. Also determine Σr_{tt} for the sub-group t.

(9) Determine

$$\Sigma t_2 = \sqrt{\Sigma r_{tt} - (\Sigma t_1)^2}.$$

(10) Determine

$$a_2 = \frac{\Sigma r_{at} - a_1 \Sigma t_1}{\Sigma t_2},$$

$$b_2 = \frac{\Sigma r_{bt} - b_1 \Sigma t_1}{\Sigma t_2},$$

$$\cdot \quad \cdot \quad \cdot \quad \cdot$$

$$w_2 = \frac{\Sigma r_{wt} - w_1 \Sigma t_1}{\Sigma t_2}.$$

(10*a*) Determine the sum Σt_2 from the list of second factor loadings and make sure that it agrees with value found in step 9.

(11) Make a list of all tests that correlate low, say within the range $\pm .50$, with *both* tests *a* and *b*. If there are no such tests, then you need no additional general factors and this procedure need not be carried any further. Select that one test *c* in this restricted list that has the highest sum $\Sigma|r_{ck}|$, disregarding sign.

(12) Make a list of all tests that correlate positively with test *c* and let this sub-group be designated *u*.

(13) Determine Σu_1 and Σu_2. This can be done since we have the first and second factor loadings for each test and we have a list of all tests in the sub-group *u*. Also determine Σr_{uu} which is the algebraic sum of all coefficients in the full table for group *u*.

(14) Determine

$$\Sigma u_3 = \sqrt{\Sigma r_{uu} - (\Sigma u_1)^2 - (\Sigma u_2)^2}.$$

(15) Determine

$$a_3 = \frac{\Sigma r_{au} - a_1\Sigma u_1 - a_2\Sigma u_2}{\Sigma u_3},$$

$$b_3 = \frac{\Sigma r_{bu} - b_1\Sigma u_1 - b_2\Sigma u_2}{\Sigma u_3},$$

$$\cdot \quad \cdot \quad \cdot \quad \cdot \quad \cdot$$

$$w_3 = \frac{\Sigma r_{wu} - w_1\Sigma u_1 - w_2\Sigma u_2}{\Sigma u_3}.$$

(15*a*) Determine the sum Σu_3 from the list of third factor loadings and make sure that it agrees with the value found in step 14.

(16) Make a list of all tests that correlate low, say within the range $\pm .50$, with the tests *a*, *b*, and *c*. If there are no such tests, then you need no additional factors. If there are several such tests, continue as follows. Select that one test *d* in this restricted list that has the highest sum $\Sigma|r_{dk}|$, disregarding sign.

(17) Make a list of all tests that correlate positively with test *d* and let this sub-group be designated *v*.

(18) Determine Σv_1, Σv_2, and Σv_3. This can be done since we have the first, second, and third factor loadings in each of the w tests and we have a list of the tests in sub-group v. Also determine the algebraic sum Σr_{vv}.

(19) Determine

$$\Sigma v_4 = \sqrt{\Sigma r_{vv} - (\Sigma v_1)^2 - (\Sigma v_2)^2 - (\Sigma v_3)^2}.$$

(20) Determine

$$a_4 = \frac{\Sigma r_{av} - a_1\Sigma v_1 - a_2\Sigma v_2 - a_3\Sigma v_3}{\Sigma v_4},$$

$$b_4 = \frac{\Sigma r_{bv} - b_1\Sigma v_1 - b_2\Sigma v_2 - b_3\Sigma v_3}{\Sigma v_4}$$

$$\cdot \quad \cdot \quad \cdot \quad \cdot \quad \cdot \quad \cdot \quad \cdot$$

$$w_4 = \frac{\Sigma r_{wv} - w_1\Sigma v_1 - w_2\Sigma v_2 - w_3\Sigma v_3}{\Sigma v_4}.$$

It is evident that this procedure can be extended in cycles of five steps for each new factor. Thus a new factor is started in steps 1, 6, 11, 16, 21, and new factors would appear in steps 26, 31, 36, and so on, if the process were continued. The process should be continued until no new tests appear in the lists prepared in each of the steps just enumerated.

In this manner one can make some rational estimate of the number of factors that will be needed to account for any given table of coefficients. Note that this estimate can be made by preparing the lists of tests called for in steps 6, 11, 16, 21, 26, 31, and so on until the list vanishes, before any calculations are started. The number of factors can thus be estimated merely by inspection of the coefficients and without any calculating whatever. This enables one to lay out in advance the data sheets for a postulated number of general and independent factors.

As stated at the outset there are several other criteria that may be applied in selecting the general factors or coördinate axes which are of importance in special cases where unusual clustering of the tests or variables may be suspected. Some special methods of locating the coördinate axes will be considered in a separate paper.

We have described a method of multiple factor analysis

by which it is possible to ascertain how many general, independent, and uncorrelated factors it is necessary to postulate in order to account for a whole table of intercorrelations. The method is free from any limitations about group factors. Objective methods have been described for locating the general factors and it is probable that, except in unusual distributions of the tests, these methods will prove adequate.

After the factor loadings have been determined for all of the tests or variables, it is of considerable interest to describe or name the general factors that have been isolated. This can be done best by noting which tests or variables have a relatively high positive loading with the first factor and which have a low or negative loading with this factor. By this inspection it is possible to name the first factor although the statistical procedures do not of course concern these matters of describing or naming the factors. In a similar manner one might inspect the second factor loadings to ascertain which tests have positive loadings and which have negative loadings with this factor. A name might then be found for this factor and so on.

Finally, it may be of interest to assign the standard score in each of the factors to each of the individual subjects. If each factor represents an ability of some kind, then it would be of interest to be able to assign to each person a standard score in each of the factors or abilities that have been isolated. In the following list of equations the only unknown factors are x_1, x_2, x_3, and x_4. These standard scores in the four abilities for any one person may be determined by solving the following observation equations for the four unknown standard scores by the method of least squares or by the method of averages. This procedure is laborious but it is at least possible to solve the problem of individual standard scores in the several independent abilities.

$$\begin{aligned}
S_a &= a_1x_1 + a_2x_2 + a_3x_3 + a_4x_4, \\
S_b &= b_1x_1 + b_2x_2 + b_3x_3 + b_4x_4, \\
S_c &= c_1x_1 + c_2x_2 + c_3x_3 + c_4x_4, \\
&\cdot \quad \cdot \quad \cdot \quad \cdot \quad \cdot \quad \cdot \quad \cdot \\
S_w &= w_1x_1 + w_2x_2 + w_3x_3 + w_4x_4.
\end{aligned}$$

It is probable that these methods of multiple factor analysis will be useful in discovering how many factors underlie a given table of correlation coefficients and in discovering their general nature. Several applications of the method to different types of correlation problems are under way and these will be reported in subsequent papers.

[MS. received March 9, 1931]

8

Reprinted from *Biometrics,* **21**(1), 190–215 (1965)

FACTOR ANALYSIS: AN INTRODUCTION TO ESSENTIALS

I. THE PURPOSE AND UNDERLYING MODELS

RAYMOND B. CATTELL

University of Illinois, Urbana, Illinois, U. S. A.

1. INTRODUCTION

Factor analysis aims to explain observed relations among numerous variables in terms of simpler relations. The simplification may consist of producing a set of classificatory categories, or creating a smaller number of hypothetical variables. Such resolution is so central to all scientific work that we must pause in a moment to ask about the relation of factor analysis to investigatory methods in general. However, it may be said forthwith that its most valuable functions lie in the biological and behavioral sciences, where a great array of phenomena are multiply determined and where the conceptual independent variables are not easily located and agreed upon.

For example, in the behavioral sciences, a student of delinquency or mental disorder may have doubts whether there is such a single thing as delinquency or mental disorder and must have an open mind as to where the important causal influences lie in the bewildering array of possibilities. In such a situation he may take a set of variables, each semantically entitled to be called a form of delinquency, and measure each on members of a population and then seek for the functional unities by computing correlation coefficients for each pair of variables over the list of people concerned. Into this correlation matrix—as we designate the square matrix of coefficients shown in Table 1 below—he may also introduce measures of putative causal influences, seeking in the subsequent factor analysis to find those underlying 'hypothetical variables' or factors among them which are of theoretical importance. Or again a biologist concerned with nutrition might correlate—over many animals, each fed in a different way—various signs of health and disease and various ingredients in the food. The vitamins, for example, might historically have been located—perhaps more economically of research time than they were—as factors each with a certain pattern of deficiency signs and of relative presence in various foods.

Of the two aims of factor analysis commonly mentioned, namely, the seeking of basic underlying influences, and the development of classificatory schemes, some confusion repeatedly arises about its role in the latter. Actually, factor analysis is a way of classifying manifestations or variables, but not, immediately, the producer of a taxonomy of organisms. Even for the former purpose it may be preferable simply to group variables by the observed clusters of correlations among them (each cluster of intercorrelating variables sometimes being called a syndrome or surface trait.)

After the essential nature of factor analysis has been described, we shall return to the question of how best to proceed to a taxonomy of organisms, i.e., to classification problems *per se*. It will then be seen that cluster search, based on measures derived from prior factor analyses, but using a coefficient different from the correlation coefficient, is probably our clearest path to types and categories. Right here, therefore, let us put categorising and taxonomy into perspective as only a secondary and non-essential part of factor analysis and concentrate upon understanding its primary aim and logical nature, namely, the isolation and development of hypothetical constructs out of observed phenomena.

As a research instrument, factor analysis happens to have been a child of psychology, though as a mathematical model it is part of the general development of mathematics. Spearman, in 1904, being interested to prove his psychological theory that all forms of intellectual performances spring from a single general mental capacity, developed a proof that if a matrix of correlations among performances takes a certain form—a hierarchy of values—then the inter-relationships of all the variables involved could be accounted for by a single underlying general ability factor, plus a factor specific to each performance. This two-factor theory (Spearman [1904; 1927]) was generalized in the next twenty years, principally by Garnett [1919] and Thurstone [1931], into principles for *multiple* factor analysis, capable of analysing correlation matrices into as many common factors behind the variables as may be necessary to account for all the observed correlations. Factor analysis thus became a generalized method for making invisible influences visible, at least to a first approximation of their form. Textbooks developing these methods in various directions have appeared in the last thirty years by Burt [1941], Cattell [1952], Fruchter [1954], Henrysson [1957], Harman [1960], Holzinger and Harman [1941], Kelley [1935], Thomson [1951], Thurstone [1947] and other psychologists. Systematical contributions have also been made by mathematical statisticians such as Ahmavaara [1954], Anderson and Rubin [1956],

Bartlett [1948], Guttman [1953; 1954], Hotelling [1933], Lawley [1940; 1956; 1958], Rao [1952; 1955] and others. From psychology the method has been adopted in sociology (Gouldner [1957]), economics (Quenouille [1957]), entomology (Sokal and Hunter [1955]), semantics (Osgood [1957]), anthropology (Driver [1956]), physical education (Cureton [1947]), physiology (Mefferd [1965]) and in other fields also. In some, notably medicine, its use has lagged far behind its potential capacity to contribute. There, as to some extent in biology also, the obstacle to cross fertilization seems to lie in the remoteness of mathematical training from the main stream of the curriculum in that area.

To see factor analysis in perspective let us note that experimental designs, in scientific method generally, may be classified according to several parameters, of which the following are generally the most important: (1) The number of variables chosen for study. This gives a dimension from bivariate to multivariate. (2) The amount of manipulation employed. This ranges from observation without interference to complete manipulative control. Parenthetically, it is an elementary error to confuse this with the concept of 'controlled experiment'; 'manipulative' is a better term. (3) Simultaneity as contrasted with known time sequence of measurements. This is the parameter which decides whether we can infer causal connection.

The 'classical' experiment of the physical sciences is—as with Galileo at the Tower of Pisa—bivariate, manipulative and with known time sequence. Because factor analysis differs from the traditional design in using *many* variables simultaneously, i.e., it is a multivariate design, a mistaken stereotype has arisen in some quarters that it must also be non-manipulative and have no dealings with time sequence (i.e., with causal concepts). Research designs do not fall into this simple dichotomy of 'classical' and 'multivariate,' but into at least 2^3 types, even if one considers only the above three parameters. It happens, as history, that most factor analyses to date *have* been non-manipulative and *have* measured only simultaneously, e.g., in psychology by considering only a set of response variables, instead of response and stimulus variables together, but there is nothing intrinsic to the model of factor analysis which prevents its use, as a multivariate method, along with manipulative control of variables, or with sequential timing of observations so as to permit inference about causality.

However, the preponderant association of non-manipulative treatment with factor analytic methodology is not entirely an accident. There happen to be many research situations where it is physically or ethically impossible to manipulate, or where the scientific objectives would themselves be lost by manipulative interference, but if one can resort to factor analysis, this veto on the 'classical' *manipulative* type of experiment need not bring research to a halt. Even without it, let us note, astronomy grew to a mature science without any manipulation at all. In physiology the manipulative approach, as in extirpating the adrenals to find out what adrenalin does, produces secondary effects which make it inferior to a factor analytic approach. For in the latter the natural variation in some hormone concentration permits the emergence of that influence as a factor having correlations with a wide range of manifestations. In psychology, since it is not ethically acceptable to experiment manipulatively with profound human emotions, e.g., to provoke insanity, those who continued to think of 'experiment' only in classical manipulative terms gradually found themselves forced either out of human psychology into animal psychology or into emotionally trivial areas, e.g., into finding out more and more about less and less in cognitive perception. Indeed, much of this manipulative experiment moved into the special senses, or even out of psychology into neurology.

The special appeal of the factor analytic design at this point is that it presents one of the few methods capable of teasing out what *would* happen through manipulation, when for these various reasons manipulation is impossible. It seeks conclusions by statistical finesse rather than manipulative control. However, manipulative control and the arrangement of observations in particular time sequences in order to establish causal connections are by no means barred. This and certain broader questions concerning the role of factor analysis among other experimental and analytic designs are best discussed afresh below, after the method itself has been more fully specified. With this brief setting in history and method we shall therefore proceed to the method itself.

2. THE BASIC THEOREMS IN REPRESENTING VARIABLES BY FACTORS

Typically the experimenter will have gathered measurements on n variables, consisting of traits, performances, etc., over N persons or organisms, N being appreciably larger than n for reasons apparent later. Wishing to know 'what goes with what' and even 'what causes what' among these variables, carefully chosen from his special field of interests, he will now intercorrelate the n variables, as they vary over the N organisms. He must do this for all possible $\frac{1}{2}n(n - 1)$ pairings of variables, thus producing a square, symmetrical correlation matrix, which we may symbolize as **R**. (See Table 1). The process of factor analysis is designed to resolve this *correlation matrix* into an $n \times k$

TABLE 1

RELATION OF CORRELATION MATRIX AND FACTOR MATRIX

	v_1	v_2	v_3	$\cdots$	v_n
v_1	h_1^2				
v_2	r_{12}	h_2^2			
v_3	r_{13}	r_{23}	h_3^2		
v_4	r_{14}	r_{24}	r_{34}	etc.	
		R	A matrix of correlation coefficients		
v_n	r_{1n}	r_{2n}			

=

	f_1	f_2	$\cdots$	f_k
v_1	v_{11}	v_{12}		v_{1k}
v_2	v_{21}	v_{22}		
v_3	v_{31}	v_{32}	etc.	
		$\mathbf{V}_0$	Factor Matrix	
v_n	v_{n1}	v_{n2}		

×

	v_1	v_2	$\cdots$	v_n
f_1				
f_2				
		$\mathbf{V}_0'$	Transposed factor matrix	
f_k				

In **R** the off-diagonal elements are the correlation coefficients, the diagonal elements are the communalities.

factor matrix, where the number of factors, k, is usually much smaller than n, the number of variables. This resolution can for the moment be regarded as a purely mathematical procedure, in which we take the mathematician's word for it that the relations among the n variables can be treated equally well as the relations of the n variables to these k new factors. Scientificially, these factors may be regarded as underlying influences or dimensions which, in further measurement, can be substituted for the more numerous variables and which largely account for the correlations among the latter. For example, in a concrete illustration (i.e., a 'plasmode'[1]) worked out by Cattell and Dickman [1962] the n variables were 32 different physical behavior measures (e.g., height of bounce, distance of rolling, collision effect, etc.) on a variety of balls ($N = 100$) and the factors which emerged to 'explain' the behavior were weight, volume, elasticity and length of suspending string when the balls were used as pendula. The discovered factors were thus precisely those which a physicist would require in his equations for representing the n diverse performances.

It will be assumed that the reader is familiar with the correlation coefficient and the **R** matrix containing all correlations of variates, and that all we need to explain at this point is the mathematical relation of the factor matrix to this correlation matrix. The relation is most succinctly represented by matrices, as in Table 1, which shows that the factor matrix has to be one which, when post multiplied by its transpose, restores[2] the correlation matrix, as expressed in equation 1.

The values v_{11}, v_{21}, etc., in the first column of the factor matrix are called the *factor loadings* for variables v_1, v_2, etc., in factor 1, and v_{12}, v_{22}, etc., are similarly the loadings of factor 2 on these variables. Initially, they are to be considered *correlations* of the newly discovered factors with the variables. The factors f_1, f_2, etc., are *mutually orthogonal* (essentially, uncorrelated) and in this connection the loadings are also the *weights* to be given to the factors in regression coefficients to estimate the variables.

The symbol **V** will be used to represent all matrices relating *variables to factors*, but will bear a different subscript to specifiy particular kinds of relation. $\mathbf{V}_0$ stands for the orthogonal factors, before rotation.

[1] The term plasmode (from 'form' and 'measurement') is reserved for a concrete, real example which is known to operate according to the form of the mathematical model. By measuring and analysing data from such a plasmode one may check properties of the mathematical model. It is, however, confusing to call such a plasmode a model, in and of itself, since it contains, for example, random error not present in the pure mathematical model.

[2] To avoid digression in the text a technicality is omitted here, namely, that **R** is strictly what is called the *reduced* **R** matrix. That is to say, it does not have unities in the diagonal, as would any experimental matrix without error of measurement, but communalities instead, as explained later. This reduced matrix $\mathbf{R} = \mathbf{R}_0 - \mathbf{U}^\delta$, where $\mathbf{U}^\delta$ is the diagonal matrix of squared specific factor loadings and $\mathbf{R}_0$ is the original matrix with unities in the diagonals.

In matrix notation the relation of Table 1 can be expressed:

$$\mathbf{R} = \mathbf{V}_0\mathbf{V}_0' . \tag{1}$$

However, for the reader unused to terse matrix notation, the reason why the above matrix relation should hold may require a more step by step argument. This goes back to one of the early formulae used by Spearman: If variables, v_1 and v_2 , have only one factor in common, namely, f, then

$$r_{12} = r_{1f}r_{2f} , \tag{2}$$

where r_{1f} and r_{2f} are correlations of the variables with f, and r_{12} is the direct correlation of the two variables. The essential point is that two variables will correlate more highly with each other the more they possess the factor in common. Obviously this relation will continue if they have more than one factor in common, provided that the factors are all orthogonal to (*i.e.* independent of) one another, so we can write:

$$r_{12} = r_{1f_1}r_{2f_1} + r_{1f_2}r_{2f_2} + \cdots + r_{1f_k}r_{2f_k} .$$

If the reader will follow out how the matrix multiplication in equation (1) worked out in Table 1 (a row of $\mathbf{V}_0$ by a column of $\mathbf{V}_0'$) he will see that the matrix arrangement achieves precisely this sum of products for every r cell in **R**.

3. THE COMPUTATIONS IN FACTOR EXTRACTION

What has not yet been explained here is how one determines this particular $\mathbf{V}_0$ having quantities such that the back multiplication, $\mathbf{V}_0\mathbf{V}_0'$, will yield the experimentally given **R**. In other words, reversing the above direction, how does one computationally obtain $\mathbf{V}_0$ from **R**?

The essential rationale is best perceived by taking the simplest case—that of a single factor—as in equation (2). If one added the correlations of v with all other variables, i.e., v_2 to v_n , and took the mean, one would obtain a value that depended upon the extent to which v_1 was correlated with f, the sole common factor, i.e., the mean would be proportional to r_{1f} , the factor loading of v_1 . The loadings of the variables, in other words, would be proportional to the column sums of the **R** matrix.

In more complex cases there could be several common factors. The procedure just given can be used to find approximate loadings for the most important factor, but further computation is needed before exact values will emerge. The method is to take weighted totals of the columns of **R**, the weights being the approximate values of the loadings. From there, better approximations are found and so on until the values converge. The loadings of the first factor having been found, its effect on **R** is removed using equation (2) and the process is repeated on the 'residual matrix' in order to find further factors. The number of factors required is shown by the number of times this has to be done until an 'exhausted' residual matrix appears which is essentially only a set of zeros.

This brief sketch of what is done computationally may help the scientist who wishes to keep his feet on the ground operationally. But the mathematician will prefer to think of the whole factor extraction procedure as 'analysing the matrix **R** for its latent roots and vectors.' (Again one must be reminded this is a 'reduced **R**,' the full meaning of which can only become clear after a further point is made below). Then, if **M** is a matrix consisting of the *latent vectors* of **R**, and $\mathbf{L}^{\delta}$ is a diagonal matrix of what are called the *latent roots* of **R**, the relation to the correlation matrix is:

$$\mathbf{R} = \mathbf{M}\mathbf{L}^{\delta}\mathbf{M}'$$

whence it follows that

$$\mathbf{V}_0\mathbf{V}_0' = \mathbf{M}\mathbf{L}^{\delta/2}\mathbf{L}^{\delta/2}\mathbf{M}',$$

where $\mathbf{L}^{\delta/2}$ is a similar diagonal matrix of the square roots of the latent roots. Consequently,

$$\mathbf{V}_0 = \mathbf{M}\mathbf{L}^{\delta/2}.$$

The square roots of the successive latent roots ($\mathbf{L}^{\delta/2}$) thus tell us about the relative 'sizes' of the successive factors taken out. The squares of the loadings of a given factor (i.e., the summed squares down the factor column) totalled over the given set of variables, in fact tells us whether it is a big or a small factor in terms of its contribution to these variables as a whole.

Although the general nature of computations on the correlation matrix has been described, actually they divide into two basic methods of calculation which are in common use in extracting factors, i.e., in reducing a correlation matrix to a factor matrix. They are (1) the principal axes method, and (2) the centroid method. The former has properties which recomment it to mathematicians; the latter was introduced by Thurstone as a substantial labour-saving approximation to the principal axes method. They are aptly referred to by Burt [1941] respectively as the *weighted* summation and the *simple* summation methods. For in the actual computing by the former one adds any v_i column in **R** (as described for the unweighted centroid method above) with a special weighting in the principle axis method. Before electronic computers became widely available the vast additional labour of the principal-axes method far outweighed its slight advantages. Today it is more practicable and more widely used, though its advantages are trivial for most purposes.

4. FACTOR MODELS AND THE COMMUNALITY PROBLEM

It has been said above that the number of factors necessary to explain correlations is fixed by the data itself, in the sense that when the factor extraction process described leaves a residual R matrix of essentially zero we have accounted for all covariance originally present. At this point, however, it is necessary to lime-light the difference between the unreduced R_0 and the reduced R matrix (so far used) and this indeed exposes what unfortunately still remains the Achilles' Heel of factor analysis.

Any experiment actually gives only the off-diagonal correlations in R, i.e., the triangle of $\frac{1}{2}n(n - 1)$ experimentally obtainable correlations. Since the correlation of each test with itself would not be meaningful here, we are left with an empty diagonal, but for factor extraction we need a complete matrix. Two main alternatives exist. One is to put unities down the diagonal on the ground that, except for unreliability of measurement, any variable would correlate perfectly with itself. We shall call this the 'closed model,' though some prefer to call it the method of 'component analysis' and to reserve 'factor analysis' for the other or 'open model.'

The other alternative is to insert values into the diagonal, known as *communalities*, by which is meant the amount of the variance of the variable accounted for by the common factors together. Mostly this will be less than the whole variance, so it will leave a residue to be uniquely accounted for by a specific factor. We shall call this the factor model.

Using the closed model, we need not miss by very much our goal of accounting for many variables by a few factors. Although the number of factors will in general equal the number of variables, some will prove to make a negligible contribution. Using the open model there will usually be a limited number, k, of common factors. The reason that scientists prefer this model is that the other is intrinsically unlikely to fit nature. Admittedly the common factors which could enter as influences into any one variable are really likely to be extremely numerous rather than few. The 'flower in the crannied wall' could well prove to be influenced by most things in the universe, but it is absurd to expect all of these influences to be substantially represented in any small sample set of variables. Consequently, in an R of, say, 50 or even 100 variables, the correlation of any variable with itself due to the common factors represented among those variables is likely to be decidedly short of unity. Part of its variance must be set aside to be accounted for by common factors which will only appear when an extremely large number of variables has been coupled with it in a correlation matrix. As will become evident, the setting aside of part of its variance in a mathematical 'specific factor' means only that we set aside in this receptacle, f_u, the variance that rightly belongs to common factors yet unrepresented. The term 'open model' means simply that variance is left in the f_u's which is open to later resolutions. The closed model, on the other hand, assumes a perfection within the small set of variables sampled at the time and, with mathematical completeness, accounts for all the variance of each variable in common factors peculiar to that restricted sample of variables.

In symbolic representation by the open model, therefore, the full variance of any actually measured variable, v_i, is going to be partly an expression, usually written h^2, for its communality, and partly a u_i^2 written for a specific factor.[3] To be realistic our symbolism must ultimately allow for several specific factors because even in a small sample of variables there is unlikely to be only one. However, mathematically for most purposes the several specifics required by a truly adequate model can be treated together as one, f_u. Further, however, we must have a term for the error of measurement which goes into any real observation, thus:

$$\sigma_i^2 = h_i^2 + \sigma_{ui}^2 + \sigma_{ei}^2 \tag{5}$$

This can alternatively be written out in more detail expanding h^2 into the sum of the variance on the several common factors:

$$\sigma_i^2 = \sigma_{i1}^2 + \sigma_{i2}^2 + \cdots + \sigma_{ik}^2 + \sigma_{ui}^2 + \sigma_{ei}^2 \tag{6a}$$

which corresponds to the resolution of a variable into factors thus:—

$$v_i = b_{i1}f_1 + b_{i2}f_2 \cdots + b_{ik}f_k + b_{iu}f_u + b_{ie}f_e \tag{6b}$$

This is set out, along with similar rows of loadings for other variables, as a complete factor matrix in Table 2, and one should remember that strictly the matrix in Table 1 should have been written in this way, though we simplified it to show the common factors only. The specifics will, in any case, do nothing in the matrix multiplication to affect the

[3] Although, to fit traditional symbolism we retain u to represent each specific or unique factor, we have dropped the term unique in writing. For it is confusing to the psychologist or biologist who reserves the term for what is unique to a particular organism, as a 'unique trait' (Allport [1938]; Cattell [1946]).

TABLE 2

Variables	Common Factors	Specific Factors	Error Factors
1	1 $\cdots$ k	1 $\cdots$ n	1 $\cdots$ n
1	r_{11} r_{12} $\cdots$ r_{1k}	u_1	e_1
2	r_{21} r_{22} $\cdots$ r_{2k}	u_2	e_2
3		u_3	
$\vdots$		$\ddots$	$\ddots$
n	r_{n1} r_{n2} $\cdots$ r_{nk}	u_n	e_n

R matrix, except in the diagonal, where the reduced matrix will have h^2 instead of one as in the unreduced.

Accepting this open model, with communalities, we must face the problem of how to choose the communality values to be put into the experimentally obtained **R** matrix. For this purpose the correlation of a variable with itself is no good, since, as just seen in (5), it represents self correlation due to the specific factors as well as the common factors. It thus represents only an upper bound. The communality of a variable being the most that it has in common with other tests, one can see that the square of its multiple correlation with all other variables would constitute a lower limit. This was formally demonstrated by Merrill Roff [1936] and has been generalized by Guttman [1956], but the true communality lies between these upper and lower bounds and there is no royal road to reach this true value.

The problem is aggravated, or, at least, made more complex by the fact that the communalities and the number of factors taken out are mutually and inextricably inter-dependent. For, the values put in will obviously determine when the residual correlation matrices (*p.* 197 above) drop to zero. Since the estimated h^2 and the estimated k are mutually inter-dependent and cannot be solved simultaneously one has to start with one and allow the other to fit it. The Lawley [1940] and Rao [1952] 'maximum likelihood solution' goes to and fro between communalities and number of factors until it hits on the combination which yields a smallest residual. However, if one is to fix one and then allow the other to adjust to it, the better solution is to start by fixing the number of factors—in ways to be discussed below. Then it can be shown (Albert [1944]) that provided the k to be tried is less than $\frac{1}{2}n$ we can find, by iteration,[4] communalities that exactly fit the off-diagonals to give that number of factors. At this point of convergence the factor matrix will very closely reinstate the original correlation matrix, with the given communalities in the diagonal, and the given agreed number of factors by equation (1) above.

With this first sketch of the whole process of factor analysis before us, let us defer to the next section, since it is rather complex, the means by which the above decision on the number of factors is made. Thus we can here scrutinise the fuller implications of our choice between the two models. Placing ones in the diagonal and taking out as many common factors as variables, which we have called the closed model, is generally preferred, as stated by mathematicians, for its simplicity and elegance. It also has an attractive completeness in that it will not only give those factor loadings which restore the unreduced correlation matrix, as seen in Table 2, but it will also give factor *scores* which will exactly reproduce the test performance scores (i.e., the values of variables for individuals).

The open model, using a reduced matrix, with communalities in the diagonal, can produce fewer common factors than variables, as we have seen (Table 2). This model will restore the reduced correlation matrix, but it will not, even if one extracts as many common factors as variables, enable one to reinstate the test scores from the factor scores by anything closer than an estimate because the specific factor contribution cannot be known. Indeed, except when the correlation matrix is large relative to the number of factors, there even remains a further blemish, namely that there is some free play regarding the weights of the variables by which the factors are to be identified and the estimate attempted.

Yet the elegance of the closed model from the unreduced matrix is a meretricious neatness. For, as we have seen, it is wildly unlikely that the small sample of variables employed will actually represent within themselves all the real common influences required to account for all the variance of all the performances. The trick of putting ones in the diagonals, though comforting in accounting fully for the variance of variables, perpetrates a hoax, for actually it really drags in all the specific factor and error variance (Table 2) to inflate specious, incorrect common factors. The trick has thus included in the common factor space much variance about which we know nothing more than is

[4] Iteration in this case requires that one start with a best guess at the communalities usually by Burt's method [1941] or by Roff's lower bound [1936] by the multiple correlation. After factoring the **R** matrix with these inserts one obtains a new resultant set of communalities which are a more convergent approximation. With these one starts factoring again, and so proceeds until the change in communality between two successive factorings is trivial.

known about specific factors. This intrusion quickly becomes evident from practice when we try to match factors from one research with those of another so treated. Commonly, any two such compared researches (when one is testing a hypothesis that a factor in one is the same as a factor in the other) will be designed to have many but by no means all variables in common. In such an experiment the common factors in one research cannot possibly be made accurately to match those in a second, even theoretically, because they derive from (or are 'contaminated with') a different set of specific variable factors in each.[5]

By the second or open model one does safely set aside this specific factor variance and so avoids the possibility of distorting the common factors. However, as stated above, one must beware of assuming that there is really just one specific to each variable or that one knows much about these specifics. The value $(1 - h^2)$ is not, as is frequently said, "the specific factor variance" but is also error, which may be represented by several error factors, and probably several specific factors are involved in the former. One can separate specific factor and error variance by knowing the reliability coefficient of the test.

$$\sigma_e^2 = 1 - r_r \tag{7a}$$

$$\sigma_u^2 = 1 - h^2 - \sigma_e^2 = r_r - h^2 \tag{7b}$$

where r_r is the reliability coefficient, i.e., the correlation coefficient of a variable with itself on repeated measurement.

However, we cannot trace these u_{j1}, u_{j2}'s, etc., because each of these contributions comes from a different common factor which cannot be identified so long as we have no variable in the matrix to share it and make it a common factor. The 'specific factor' is thus really a question mark or confession of ignorance only to be filed away under that symbol as a reminder of variance needing further investigation. Even the convenience of writing it in the model as orthogonal to the common factors and the other specifics is a simplifying fiction. For the common factors, if they eventually appear, will be correlated to some degree (see notation below) and the present specifics are therefore only an orthogonal projection of that variance.

More refined or extensive discussion of models is possible only after two further conceptions have been introduced in the two following sections.

5. DECIDING UPON THE NUMBER OF FACTORS

Obviously it is highly important for the successful use of factor

[5] Any factor estimate is a linear function of the variables (or, if we wish not to speak of estimates, a factor is a linear function of the h's or common factor components in the variables). Consequently the factor from one set of variables cannot be identical, except by some quite special relation among the variables, with that from another overlapping but not identical set.

analysis in developing scientific theory that it be capable of reliably showing the exact number of factors at work in a given area. This capacity becomes of crucial importance also for additional reasons, notably that it is necessary for fixing communalities, as just stated, and also because an error on this point may lead to the 'degenerative factor fission' explained below.

To the problem of deciding the number of factors there have long been three main approaches.

(1) *By mathematical concepts*: aiming to give communalities which will exactly fix the rank of the matrix. This approach accepts the correlations as dependable, exact numbers, without any regard to their statistical stabilities or properties, and has usually aimed at discovering communalities which will give the minimum possible rank. Roff's suggestion of filling each cell in the diagonal with the square root of the multiple correlation of that variable with every other variable in the **R** matrix, has been advocated by Guttman. It is the best known proposition here and we have seen that it gives a value equal to the lowest possible bound for the communality. To this type of approach it may be cogently objected that our real aim is not in fact to minimize the number of factors. The latter is certainly a definite goal but not necessarily the right one. What the scientist is seeking is the best possible estimate of the true number of factors actually operating, as far as can possibly be gauged from the experimentally given evidence. The latter amounts at first sight only to the obtained 'off-diagonal' correlations, i.e., the original correlation matrix with nothing in the diagonal. However, the definiteness of the lowest rank proposal has appealed excessively to many perplexed by the ambiguities of the real situation.

(2) *By statistical evaluation*: These look for some sort of 'best fit,' in the light of probability, having regard to the size of sample of organisms and variables. Such methods as those of Bartlett [1951], Burt [1952], Lawley [1940; 1956], Rao [1952] and others represent this criterion, and, unlike the purely mathematical approach, they concentrate on the statistical implications. The maximum likelihood method may be regarded as simultaneously exploring changes in factor number and communalities to produce a fit to the original correlation matrix which is satisfactory at a given significance level. Incidentally, one must beware of assuming that the usual factor extraction processes turn out the earliest factors free of error and that at some later point in extraction they begin to extract pure error factors. The error variance is extracted along with true variance from the beginning, though in the rotation process later one may be able to rotate (see below) the error

variance into strictly error factors. Normally the proportion of error variance will stay constant, and one cannot therefore eliminate error variance without losing true variance by any supposedly cautious and conservative policy of stopping factor extraction at an early stage.

(3) *By the factor structure criterion*: The rationale here requires a brief digression to sketch in the state of affairs described more fully by the present writer elsewhere [1957]. Shifting from the mathematical to the scientific setting one must first recognize that in any experimentally gathered data the number of influences producing covariance among the variables is likely to be larger than the number of variables. Just because mathematically the rank of a matrix cannot possibly exceed its order there is no excuse for ignoring the real state of affairs, which the model must try to fit. For example, a mathematician might agree that it is reasonable to deal with, say, thirty factors influencing the performance of a hundred men on eighty variables. But if we next take only two of those same variables (and even only two men) it will usually follow that in fact they are still influenced by the same thirty distinct influencing factors, not merely by the one common factor to which we are now mathematically limited. Consequently, the aim of the experimenter should be to take out as many factors as mathematics permits, and then sort them out (in ways to be described below) into (a) true factors, and (b) error factors, i.e., factors created by correlations from chance-distributed experimental errors, which correlations are not zero in the sample but which should be in the population as a whole, and (a) (i) True factors large enough to be worth dealing with, and (a) (ii) trivial true factors. The greatest that this total number of factors can be considered to be, in practice, is n, the order of the matrix. We can never tie down the influences beyond this number. Moreover, we may not even be able to deal with n, for if we wish to obtain convergent, unique communalities, as will usually be required, we must stop at $\frac{1}{2}n$. Fortunately, in most researches this is usually large enough to have included all factors, error or real, except those of quite trivial variance.

The notion of separation of error from true factors and of arranging factors in order of increasing triviality to lead to a cutting point, cannot be explained until rotation is understood.

6. ROTATION AND THE UNIQUE RESOLUTION OF A FACTOR ANALYSIS.

Having performed an analysis according to the above principles and extracted the number of common factors constituting a matrix accounting for, perhaps, 99% of the covariance to be explained, we find ourselves with a V_0 matrix, as in Table 1 and Table 2, which multiplied as V_0V_0', very closely restores the original correlations. However, these factors may be, indeed, almost certainly are, quite remote from correspondence with the patterns of any of the real influences behind the data. Indeed, the arrangement of factors as they come fresh from the computer is affected by such accidental matters as the raw score scaling (in the case of the principal axis solution) and the order in which variables have been reflected for computational convenience (in the case of the centroid extraction).

Mathematically, in fact, any such matrix can be converted into any one of an infinite series of equally 'accidental' mathematically equivalent factor matrices (with the same number of factors). Though infinite these will nevertheless be quite different from the infinite series from some other matrix. By 'equivalent' we mean that one set can be linearly transformed into the other; or, if we think geometrically, rotated; or, since we are concerned about the original experimental correlations, can be equally exact in restoring the correlations and test scores. Before deciding how a choice is to be made from among these, let us look at the rotational transformation in itself.

The linear transformation from the original V_0 to a particular rotation V_m which is taken to represent some unique resolution defined by conditions yet to be set can be written:

$$V_m = V_0L_m \tag{8}$$

where L_m is a $k \times k$ transformation matrix and V_m is the $n \times k$ 'rotated' matrix. Most readers will probably find this 'rotational' change more intelligible in its geometric form. Figure 1 represents an example with two factors only for easy illustration in a single paper plane. The coordinates represent the centroids or principal axes as they come from computation and are initially placed, just as one conventionally places them vertically and horizontally on the page. The variables have been plotted as points (the end points of vectors), using their factor loadings given by the V_0 to fix their projection and positions here, in the usual graph-plotting way. Each variable is thus a vector in itself, in the plane or space defined by the factors which should strictly be called 'Reference Vectors.' The reader will notice that these vector end points do not lie on the circumference of a circle, fixed by the unit radius of the unit length reference vectors, because the length of each is only h (and h^2, the communality, is less than one). This is because the common factors do not account for all the variance of most variables. The drawing deals only with the test projections in the common factor space. The factors (axes) are of unit length by definition, but the variables would only be a unit length if their specific and error factors could also be included.

The rotation decided upon in Figure 1 is a simple structure rotation (explained below) to the position shown by the heavy solid lines. Naturally the projections of the performance variable points on the new Reference Vectors (or 'Factors') *A* and *B* will be different from those on 1 and 2. (Consider, for example, *j*.) They will be different in ways which could be found non-graphically, by using the transformation matrix, $\mathbf{L}_m$, in (8), the entries in which, incidentally consist simply of the sines and cosines of the angles which the new reference vectors bear to the old.

How then does one decide on the particular rotation positions and get to this unique transformation, $\mathbf{L}_m$, which converts $\mathbf{V}_0$ to a meaningful $\mathbf{V}_m$? It depends upon what principle of unique resolution one regards as most meaningful. Obviously, one could reach it by visual rotation to some preferred position as in Figure 1, writing down the sines and cosines after shifting or in other ways but this leaves un-

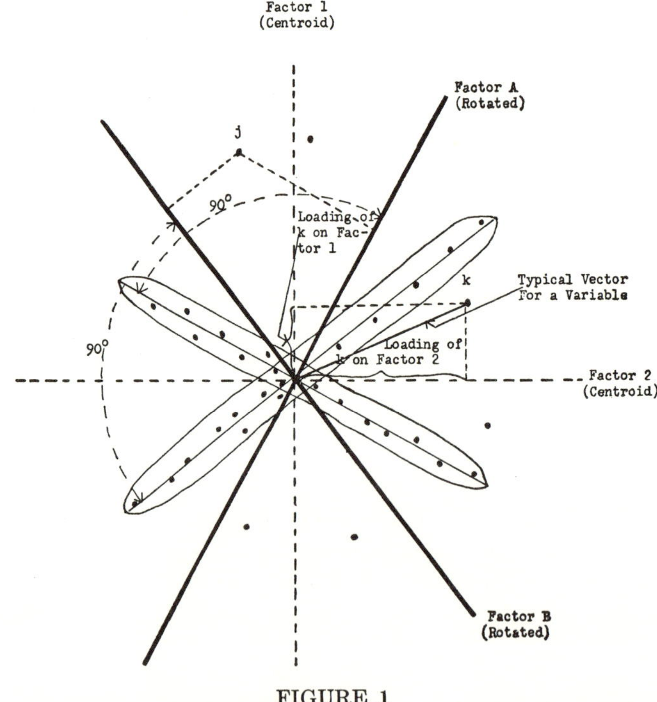

FIGURE 1

ROTATION TO SIMPLE STRUCTURE

explained why the rotation in Figure 1 is a preferred position, i.e., why it was stopped where it was. Naturally, as far as the fixing of the positions of the points (reproducing the correlations of variables) is concerned either coordinate axis pair is as good as the other, or as any of the infinite number of other possible positions. All restore the correlations, i.e., the cosines of the angles between the tests, equally well, and the 'test configuration', as an interlocking set of vectors, remains unchanged under the transformation. But the meaning of the factors themselves—as deduced from what variables they correlate with and affect—will change with every rotation.

What the experimenter is looking for is a rotation that will cause factors to correspond to certain scientific entities, which can be regarded as explanatory of the observed covariation of variables. Although the rotational location of such factors can, in special cases, be aided by information or conditions external to the factor analysis the most general principles available for such resolution, which depend only on intrinsic properties of the factor analysis, are two, as follows.

(1) *Simple Structure.* This is a specific application of the general principle of hypothesis-choice stated by Newton, as *natura est simplex,* or by Occam, in a wider context as '*Entia non sunt multiplicanda sine necessitate.*' One view of this ideal of 'simplest explanation' as it affects the present aim of explaining a configuration of variables, is that it is reached when each factor affects only a few variables, i.e., that it loads or correlates with the smallest possible number. In common sense terms we may say that given a wide sample of variables we should expect any true natural influence to affect only a few of them rather than all indiscriminately. In terms of points like that in Figure 1 this means that there should be a possible rotation such that each factor has zero projection on most variables. In a three-dimensional space, for example, we should be able to see planes of variables, which could be perceived as 'crystalized out' from the irregular mass of points. Similarly, hyper-planes could appear with more than three dimensions. Experience verifies that this undoubtedly occurs in real data and, in two-dimensional space, there appear the blurred lines, blurred from a true line by intrusion of measurement error, as shown in Figure 1. On finding such nebulae of points one rotates until the new factors (technically 'Reference Vectors') can be placed as perpendiculars to these planes, as shown at A and B in Figure 1. At that point they affect, i.e., correlate with, fewest variables.

(2) *Confactor Rotation.* In the definition of simple structure above the phrase: 'A wide sample of variables' will be noted. If the experimenter's choice of variables is not fortunate in being wide and repre-

sentative of all possible variables one may have a situation where a real factor does happen to operate on all or most of the variables in the sample. In that case, obviously, no hyperplane of unaffected variables can be discovered whereby its reference vector position can be fixed. The alternative and more general concept of the 'confactor principle' was therefore proposed by the present writer [1944; 1955; 1965] to avoid this and other difficulties in simple structure. Although confactor rotation may be theoretically a broader and more satisfying principle, it is limited in practice by yet unsolved technical difficulties. In any case, it requires that we do two distinct experiments, on two samples, each from a population in which the magnitude of influence of the operating factors would be expected to be markedly different from the other, though the same variables would be measured in each. The principle invoked is that the loadings of any factor X on variables, $a, b \cdots n$, would be expected to change in a lawful, proportional way from Experiment A to Experiment B—since the factor as a whole changed in the magnitude of its influence from Experiment 1 to Experiment 2. Thus the profile of loadings of each factor in A should be matchable with a profile in B, if the correct position has been found jointly in the two studies. It has been shown (Cattell [1944]) that if it is possible to find positions in A and B such that a proportional, parallel change of loadings on all variables holds between them, then these positions are unique. No other randomly 'spun' position can again have any such simple proportional relation. The situation is analogous to that in a combination lock where both cylinders are appropriately rotated, and, like such a position, it takes a lot of finding. Briefly, the present state of development is that the solution works for orthogonal factors, but has not yet been made to work for oblique factors (to be explained below). Since most scientific data produce the latter, the simple structure resolution method has to remain the daily work horse for the great majority of factor analyses.

The alternative to rotating in terms of the quite general hypothesis that the data contains simple structure positions is to rotate directly to test some specific hypothesis. Such hypothesis testing rotation is a comparatively simple procedure requiring little discussion, and is described below, but let us note that the two general principles of simple structure and confactor rotation are primarily hypothesis generating. They utilize to the full the special advantage which resides in factor analysis of obtaining 'answers' more structured than most statistical analyses can provide to the general question 'What structure is there?' This type of question cannot be directly asked in analysis of variance of the more simple methods of experiment and analysis.

Rotation has been discussed so far nominally in terms of orthogonal rotations, in which the rotated factors are kept in a framework of mutually uncorrelated axes, preserving the form in which they stood in the original centroid. However, this model is altogether too restricted to permit a fit to naturally occurring causes, which are, more often than not, correlated or 'oblique'. The development of rotational techniques in factor analysis has indeed gone a long way, for in the early days the centroid axes were either left as they stood or arbitrarily rotated to some alleged meaningful position of axes in which they were still orthogonal. Experience in locating naturally-existing hyperplanes, as soon as sufficient scientific data became available, showed, however, what would have been theoretically expected, namely, that in most scientific fields factors are oblique. For in our interacting universe most influences should surely show some degree of systematic mutual correlation. Thus temperature and pressure are independent[6] factor concepts, but if they are obtained by factoring weather data over certain months, say in the sub-arctic, they will be correlated, so if we enforce a structure in which axes must be kept uncorrelated, then our entities are no longer pure. This is illustrated by Dickman's ball plasmode[7] in Figure 2, where, in spite of beach balls and golf balls being included in the population of balls, there is a substantial correlation of the weight and volume factors. The confounding of loading patterns which arises from enforced orthogonality is shown here by the fact that a vector made to be uncorrelated with weight would thus be a very arbitrary mixture of negative weight and volume. As pointed out in footnote 6, in factor analysis we must distinguish between the independence of a pair of concepts in our minds and their statistical independence in nature. The inappropriateness of rigid orthogonality is brought out again when one reflects that the correlations among factors will vary with circumstances and sample. If we want to deal in factor analysis not merely with artificial, mathematical factors of arbitrary constitution and peculiar to one matrix but factors recognizably identical as concepts across experiments and factor analyses, we must allow them to follow simple structure to whatever obliqueness is indicated. (This is quite compatible, however, with orthogonality appearing as a special case.)

---

[6] As many as three meanings of independent need to be kept clear here: (1) Independent as possible conceptual entities in the mind of man; (2) Statistically independent or orthogonal (as factors) in the sense of uncorrelated in incidence in a natural population of people or occasions; and (3) Orthogonal (as factors) in the sense that the loading patterns are uncorrelated. Except in the principal axis solution before rotations all factors have some correlation in this third sense.

[7] As pointed out in footnote 1, a plasmode will enable us to test for modifications in the model, evaluate allowance for error of measurement, etc., required in practical use of the model, and throw light on methodology and the model in general.

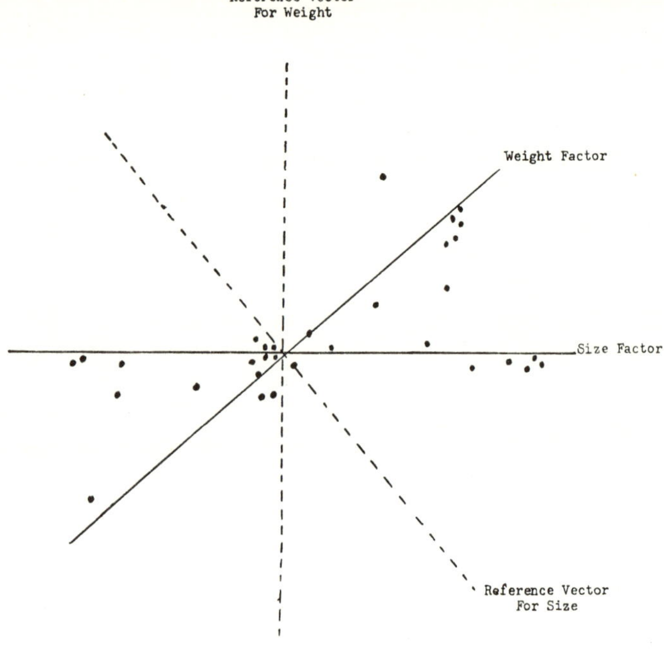

FIGURE 2

Primary Factor Pattern: Size And Weight

From Table 4.4, *Factorial Validity of a Rating Instrument*,
By Kern W. Dickman, p. 96.

Because rotation is the last of a long series of steps in factor analysis there has been a tendency, in actual published studies, for it to be incompletely pursued, either from failure to perceive its vital importance or through fatigue. Even with computers, obtaining a good, definitive, unimprovable rotation generally takes longer than all the previous steps put together. Determined attempts have been made to produce analytical methods, with or without computer programs, to find the best simple structure. Those that have survived to become embodied in computer programs work on one of two or three principles. The most common, as in quartimax (Wrigley, Saunders and Neuhaus, [1958]), varimax (Kaiser [1958]), oblimax (Pinzka and Saunders [1954]), oblimin (Carroll [1958]), etc., aims to maximize the fourth power of the loadings, which amounts essentially to maximizing the scatter among loadings. Since a few highs mean several lows ($\sum a^2$ staying constant) this leads to finding a position in which there are many low loadings. One of two objections can be raised to most analytic programs of this type: (1)

That they aim to prevent a variable being simultaneously well loaded on two factors; despite the fact that 'cooperative' factors (Cattell [1952]), i.e., factors with a significantly similar loading pattern, tend to occur systematically in nature and would be suppressed by this condition, and, (2) The percentage of low loadings, say those below 0.25, is only moderately correlated with the number of variables actually in the hyperplane, say within ± 0.10, yet the former function is made a substitute for the latter.

The second principle used in analytical programs is to count and maximize the total of variables in the actual hyperplane, as in the maxplane program (Cattell and Muerle [1960]). This simulates the 'trial and error' search of a human observer in visual rotation, adopting as hyperplane width the ± 0.10 band, or whatever is the best estimate of the standard error of a zero loading. Whatever analytical program is used, many researches are likely to be denied their scientific function in these current years through the fashionable superstition that a machine rotation must be more objective and unique in its scientific meaning. A simple structure rotation should certainly be 'blind,' in the sense that the experimenter does not know which points are which variables, but the objectivity of a rotation has little to do with mere machine programming. The solution is a true test of a hypothesis when performed, either by machine or visual rotation (a) 'blindly,' so that the experimenter's prejudices[8] (or ideas of 'meaningfulness') cannot perpetuate themselves by distorting the outcome, and (b) with a demonstration, by a plot of total hyperplane counts from successive improvements, that this count has genuinely reached a plateau, which no one can improve upon by any further trial and error search. The acceptability of a rotation, even after this thorough trial and error procedure to a maximum possible count, still depends further on the verdict of a test of statistical significance being applied to the obtained simple structure. Tests of simple structure significance, such as that of Bargmann [1953], or of Cattell and Pawlik [In Press] can show whether the obtained concentration of variables in hyperplanes betokens a structure that would very rarely occur randomly. By photographing plots made by a computer on an oscilloscope, as in the rotoplot program (Cattell and Foster [1963]), most present analytical program results can in fact be improved appreciably in hyperplane fit, and it is the evidence presented by obtaining the maximum hyperplane count in this way which can alone justify the conclusion that the unique simple structure has been attained. Occasionally, two almost equally good

[8] 'Rotation for meaning' is self-contradictory if the aim of basic research is to find meaning. In applied science, rotation to a position already known from basic research is of course a different proposition.

resolutions, decidedly better than others, can be found; which is an illustration of the fact that one can find more hyperplanes than factors. Such a discovery generally shows that insufficient factors have been taken out, unless these extras are what have been called geometers' hyperplanes.[9] In the very rare instances where a genuine alternative simple structure exists it usually makes sense also as an equally acceptable conceptual explanation.

With this grasp of the rotation problem as such the reader can now return to complete his understanding of the explanation (Section 5 above) of how the number of factors can be decided by what was there called the third or structural criterion. If the true dimensionality of an experimental space is, say, k, the projection of the tests in the true hyperplanes in such a space into a space less than k will result in each hyperplane having a broader bandwidth than in the true space. From this the principle emerges that a plot of the mean hyperplane width, from the best achieveable rotation position at each successive extraction, should produce a curve which falls and then flattens out when the true dimensionality is reached. In practice one would expect this to be complicated by the fact that if more factors are taken out than exist, i.e., if one uses dimensions for quite trivial variance there sets in a 'degenerative fission' whereby a factor on hyperplane x splits into two factors on roughly the same hyperplane, which then tends to include more zeros. This point can be noticed, however, by a comparatively sudden increase in the mean angle among factors (Cattell [1957]). Although the structural criterion of number of factors as here defined has some practical aspects of technique still to be worked out, it may theoretically be the most defensible procedure, since the purely mathematical must be rejected and the statistical offers no solution to the separation of small true and large error factors. In the structural criterion, on the other hand, there is some hope that sufficiently controlled rotation may at least locate the hyperplanes of all but the most trivial factors, without contaminating the form of the important factors by throwing this trivial variance into them.

BIBLIOGRAPHY

Ahmavaara, Y. [1954]. The mathematical theory of factorial invariance under selection. *Psychometrika 19*, 27–38.

Albert, A. A. [1944]. The minimum rank of a correlation matrix. *Proc. Nat. Acad. Sci.* 144–46.

---

[9] With n variables in k space one can always find ${}^{n}C_{k-1}$ hyperplanes with $(k - 1)$ variables exactly lying in each. If one has gone as far as taking out ½ n factors, this gives a hyperplane percentage near 50. Since, even in some good rotations, with suitably widely sampled variables in a good research design, the count runs typically in the 60 to 80 percent region, some real danger of being confused by these bogus hyperplanes exits. Cattell and Gorsuch [1963] have shown empirically that true simple structure, however, cannot be reached in random material.

Allport, G. W. [1938]. *Personality; a psychological interpretation.* New York: Henry Holt and Co.

Anderson, T. W., and Rubin, H. [1956]. Statistical inference in factor analysis. *Proc. of the Third Berkeley Symp. on Math. Statistics and Probability 5,* 11–50.

Bargmann, R. [1953]. The statistical significance of simple structure in factor analysis. (Mimeographed). Frankfurt-Main (Germany): *Hochschule fuer Internationale Paedagogische Forschung.*

Bartlett, M. S. [1948]. Internal and external factor analysis. *Brit. J. Psychol. Stat. Sec. 1,* 73–81.

Bartlett, M. S. [1951]. A further note on tests of significance in factor analysis. *Brit. J. Psychol. Stat. Sec. 4,* 1–2.

Burt, C. [1941]. *The factors of the mind: an introduction to factor analysis in psychology.* New York: Macmillan.

Burt, C. [1952]. Tests of significance in factor analysis. *Brit. J. psychol. Stat. Sec. 5,* 109–33.

Carroll, J. B. [1958]. Oblimin rotation solution in factor analysis. Computing program for the IBM 704, Harvard University.

Cattell, R. B. [1944]. 'Parallel proportional profiles' and other principles for determining the choice of factors by rotation. *Psychometrika 9,* 267–83.

Cattell, R. B. [1946]. *Description and measurement of personality.* New York: World Book Co.

Cattell, R. B. [1952]. *Factor analysis.* New York: Harper.

Cattell, R. B. [1957]. *Personality and motivation structure and measurement.* New York: World Book Co.

Cattell, R. B. [1962]. The basis of recognition and interpretation of factors. *Educ. and psychol. Measmt. 22,* 667–97.

Cattell, R. B. [1965]. The configurative method for surer identification of personality dimensions, notably in child study. *Psych. Rep. 16,* 269–70.

Cattell, R. B., and Cattell, A. K. S. [1955]. Factor rotation for proportional profiles: analytical solution and an example. *Brit. J. Stat. Psychol. 8,* 83–92.

Cattell, R. B., and Dickman, K. [1962]. A dynamic model of physical influences demonstrating the necessity of oblique simple structure. *Psychol. Bull. 59,* 389–400.

Cattell, R. B., and Foster, M. J. [1963]. The rotoplot program for multiple, single-plane, visually-guided rotation. *Behav. Sci. 8,* 156–65.

Cattell, R. B., and Gorsuch, R. L. [1963]. The uniqueness and significance of simple structure demonstrated by contrasting organic 'natural structure' and 'random structure' data. *Psychometrika 28,* 55–67.

Cattell, R. B., and Muerle, J. L. [1960]. The 'maxplane' program for factor rotation to oblique simple structure. *Educ. Psychol. Measmt. 20,* 569–90.

Cattell, R. B., and Pawlik, K. [In press]. The use of algebraically mutually dependent variables in factor analytic research.

Cureton, T. K. [1947]. *Physical fitness, appraisal and guidance.* St. Louis: C. V. Mosby.

Dickman, K. W. [1960]. *Factorial validity of a rating instrument.* Doctoral thesis, Univ. of Illinois.

Driver, H. E. [1956]. An integration of functional, evolutionary and historical theory by means of correlation. *Suppl. to Internat. J. Amer. Linguistics 22,* 1–35.

Fruchter, B. [1954]. *Introduction to factor analysis.* New York: D. Van Nostrand Co.

Garnett, J. C. M. [1919]. On certain independent factors in mental measurement. *Proc. Roy. Soc. Lon. 46*, 91–111.

Gouldner, A. W. [1957]. Cosmopolitans and locals: Toward an analysis of latent social roles. *Admin. Sci. Quart. 2*, 281–306.

Guttman, L. [1953]. Image theory for the structure of quantitative variates. *Psychometrika 18*, 277–96.

Guttman, L. [1954]. Some necessary conditions for common-factor analysis. *Psychometrika 19*, 149–61.

Guttman, L. [1956]. 'Best possible' systematic estimates of communalities. *Psychometrika 21*, 273–85.

Harman, H. H. [1960]. *Modern factor analysis*. Chicago: Univ. of Chicago Press.

Henrysson, S. [1957]. *Applicability of factor analysis in the behavioral sciences: a methodological study*. Stockholm: Almqvist and Wiksell.

Holzinger, K. J., and Harman, H. H. [1941]. *Factor analysis*. Chicago: Univ. of Chicago Press.

Hotelling, H. [1933]. Analysis of a complex of statistical variables into principal components. *JEP 24*, 417–41, 498–520.

Kaiser, H. F. [1958]. The varimax criterion for analytic rotation in factor analysis. *Psychometrika 23*, 187–200.

Kelley, T. L. [1935]. Essential traits of mental life. *Harvard Studies in Education, 26*. Cambridge, Mass.: Harvard U. Press.

Lawley, D. N. [1940]. The estimation of factor loadings by the method of maximum likelihood. *Proc. Roy. Soc. Edin. A 60*, 64–82.

Lawley, D. N. [1956]. Tests of significance for the latent roots of covariance and correlation matrices. *Biometrika 43*, 128–36.

Lawley, D. N. [1958]. Estimation in factor analysis under various initial assumptions. *Brit. J. Stat. Psychol. 11*, 1–12.

Mefferd, R. B. [1965]. Multivariate methods in physiology. Chap. 16 in Cattell (Ed.) *Handbook of Multivariate Experimental Psychology*. Chicago: Rand McNally.

Osgood, C. E., Suci, G. J., and Tannenbaum, P. H. [1957]. *The measurement of meaning*. Urbana, Ill.: Univ. of Illinois Press.

Pinzka, C., and Saunders, D. R. [1954]. Analytic rotation to simple structure, II: Extension to an oblique solution. *Research Bull. RB-54-31*. Princeton: Educational Testing Service.

Quenouille, M. H. [1957]. *The analysis of multiple time series*. London: Griffin; New York: Hafner.

Rao, C. R. [1952]. *Advanced statistical methods in biometric research*. New York: Wiley.

Rao, C. R. [1955]. Estimation and tests of significance in factor analysis. *Psychometrika 20*, 93–111.

Roff, M. [1936]. Some properties of the communality in multiple factor theory. *Psychometrika 1*, 1–6.

Sokal, R. R., and Hunter, P. E. [1955]. A morphometric analysis on D.D.T. resistant vs. non-resistant house fly strains. Annals of the *Ent. Soc. Amer. 48*, 499–507.

Spearman, C. [1904]. General intelligence objectively determined and measured. *Amer. J. Psychol. 15*, 201–93.

Spearman, C. [1927]. *The abilities of man*. London: Macmillan.

Thomson, G. H. [1951]. *The factorial analysis of human ability*. New York: Houghton Mifflin Co.

Thurstone, L. L. [1931]. Multiple factor analysis. *Psychol. Rev. 38*, 406–27.

Thurstone, L. L. [1947]. *Multiple factor analysis*. Chicago: Univ. of Chicago Press.

Wrigley, C., Saunders, D. R., and Neuhaus, J. O. [1958]. Application of the quartimax method of rotation to Thurstone's primary mental abilities study. *Psychometrika 23*, 151–70.

9

Reprinted from *Psychometrika,* **20**(2), 93–111 (1955)

ESTIMATION AND TESTS OF SIGNIFICANCE IN FACTOR ANALYSIS

C. RADHAKRISHNA RAO

VISITING RESEARCH PROFESSOR, UNIVERSITY OF ILLINOIS

A distinction is drawn between the method of principal components developed by Hotelling and the common factor analysis discussed in psychological literature both from the point of view of stochastic models involved and problems of statistical inference. The appropriate statistical techniques are briefly reviewed in the first case and detailed in the second. A new method of analysis called the canonical factor analysis, explaining the correlations between rather than the variances of the measurements, is developed. This analysis furnishes one out of a number of possible solutions to the maximum likelihood equations of Lawley. It admits an iterative procedure for estimating the factor loadings and also for constructing the likelihood criterion useful in testing a specified hypothesis on the number of factors and in determining a lower confidence limit to the number of factors.

1. *Introduction*

Whatever may be the arguments for or against factor analysis as a tool in psychological research, the statistical problems it involves have been of considerable interest to the statistician mainly because of their complexity. Two important contributions on the statistical side are by Hotelling (**8**), who introduced the principal component analysis, and Lawley (**11**, **12**), who provided a test criterion for judging the significance of factors in addition to working out the maximum-likelihood equations of estimation. These two authors were, however, considering two different problems, both of which seem to have important application. They are sometimes considered as two possible formulations of the same problem providing the same answer. In theory it helps to make a distinction between the two. The term principal component analysis (PCA) should be used for Hotelling's formulation of the problem and its solution; the term factor analysis should be used for the specialized formulation considered in psychological literature and for the various solutions offered (see also **10**). Lawley was considering the latter problem under the assumption that the variables (test scores) are normally distributed.

Illustrations have appeared from time to time to show that PCA gives nearly the same relative magnitudes of factor loadings as any effective method of factor analysis. This is true only when what have been termed as communalities are very nearly equal for all the tests as shown in section 3.1 of this paper.

The PCA is sometimes modified (**3**, p. 114; **7**) by the insertion of communalities in the diagonal of the correlation matrix. This method, called the principal factor analysis (PFA), seems to provide a valid approach to the problem of factor analysis; however, it carries with it the flavor of principal component analysis intended to explain the variations in the standardized scores. It will be seen that an alternative approach developed in section 3.2 explains most effectively the correlations between the test scores in a battery. This method may be called a canonical factor analysis (CFA). Formulas for estimation are detailed in section 4.

The tests of significance associated with component analysis and factor analysis also differ to some extent. In the former case interest chiefly lies in the magnitude of, or the relationship between, the latent roots of the hypothetical matrix of raw correlations or those corrected for attenuation. In factor analysis it is the decomposition of the correlation matrix as the sum of a diagonal matrix and a positive semi-definite matrix. The differences in nature of these tests are sometimes fundamental (section 2.2). The tests of component analysis are contained in Hotelling's paper (**8**), and an appropriate test for factor analysis is given by Lawley (**11**). An alternative form of Lawley's test yielding slightly more precise results is given in section 4.2. It is also shown (section 4.3) that the test criterion can be calculated during the process of estimation and used in obtaining a lower confidence limit to the number of factors.

Recently Bartlett (**1**, **2**) proposed a test involving the latent roots of the correlation matrix intended to study "the correlation structure in relation to the variance of the measurements." The exact nature of the hypothesis for which Bartlett's test is applicable and the conditions under which it is valid are examined in section 2.2. It appears that this test does not provide a complete answer to either form of analysis under consideration.

For a full account of tests of significance in factor analysis developed up to 1952, the reader is referred to Burt (**3**).

The author of this article is not concerned here in examining which of the methods, component or factor analysis, is relevant in problems of psychological research or whether both methods provide rather similar numerical results (not identical in general) leading to the same psychological interpretation. The main emphasis is on the differences in statistical techniques needed in these two cases and a detailed examination of the methods for factor analysis.

2. *Problems of Factor and Component Analyses*

2.1 *Factor Analysis*

Factor analysis postulates an underlying structure of a set of measurements in terms of hypothetical variables (non-observable) depending on what are called common and specific or individual factors. If $x_1, \cdots, x_p$ denote p different measurements on an individual, then x_i is written

$$x_i = z_i + s_i \qquad (i = 1, \cdots, p), \tag{2.1.1}$$

where z_i, the variables depending on common factors, and s_i, the variables depending on specific factors, satisfy the following conditions of zero covariance:

$$\text{cov}(z_i, s_i) = 0, \quad \text{cov}(z_i, s_j) = 0, \quad \text{cov}(s_i, s_j) = 0 \quad (i \neq j). \tag{2.1.2}$$

Sometimes, another independent variable representing unreliability in the measurement x_i is added to $(z_i + s_i)$, but for purposes of factor analysis based on unrepeated test scores of individuals, this variable can be combined with s_i without loss of generality. If such repeated test scores are available, then a more comprehensive analysis of the common and specific factors is possible. This latter analysis is not considered here.

From the structural setup (2.1.1), (2.1.2) it follows

$$\begin{aligned} V(x_i) &= V(z_i) + V(s_i) \\ \sigma_{ii} &= \gamma_{ii} + \delta_{ii}. \end{aligned}$$

$$\begin{aligned} \text{cov}(x_i, x_j) &= \text{cov}(z_i, z_j) + \text{cov}(z_i, s_j) + \text{cov}(s_i, z_j) + \text{cov}(s_i, s_j) \\ &= \text{cov}(z_i, z_j) \\ \sigma_{ij} &= \gamma_{ij} \qquad (i \neq j). \end{aligned}$$

If Σ, Γ, Δ denote the dispersion matrices of the vector variables $\underset{\sim}{x}$, $\underset{\sim}{z}$, $\underset{\sim}{s}$, then

$$\Sigma = \Gamma + \Delta,$$

where Δ is a diagonal matrix.

It is seen from the above analysis that any correlation between x_i, x_j is solely due to the correlation between z_i, z_j. What we can actually observe are the values of the variables $\underset{\sim}{x}$ on a group of individuals but not $\underset{\sim}{z}$, $\underset{\sim}{s}$ which are not operationally defined, but whose hypothetical existence is postulated. We thus obtain an estimate of the matrix Σ. The subject of factor analysis is mainly concerned with the estimation of the matrix Γ starting with an estimate of Σ. The object is not to find any matrix Γ satisfying the condition $\Sigma = \Gamma + \Delta$ but the one which has the least complexity leading to a parsimonious description of the relationships between the observable variables $\underset{\sim}{x}$. The complexity, when defined as the rank of the matrix Γ, has a special significance for the problems on which this technique is applied, as shown in the subsequent sections of this paper.

Some of the statistical problems of factor analysis are:

(a) to estimate the minimum rank of the dispersion matrix (variances and covariances) Γ of the variables $z_1, \cdots, z_p$ occurring in the structural equations (2.1.1, 2.1.2),

(b) to test any hypothesis specifying the minimum rank of Γ,

(c) to estimate *a basis of the common factor space* (defined below),

(d) to predict the value of any common factor from the observed set $x_1, \cdots, x_p$ for any individual.

The statement that the rank of Γ is $k < p$ implies that the variables $z_1, \cdots, z_p$ can be expressed as linear combinations of k independent variables only. To bring out the precise meaning of such a dependence let us consider the entire *vector space* of elements consisting of all linear combinations of the set $z_1, \cdots, z_p$ of variables introduced into the different tests of a battery with the restriction that any two variables differing by a constant represent the same element. We may call any element of this space a common factor variable or simply a factor variable unless otherwise specified. The vector product of any two elements f and g of this space is defined by $\text{cov}(f, g)$ and the square of the norm of f by variance of f, $V(f)$.

Vectors $f_1, f_2, \cdots, f_r$ of this space are said to be independent if no linear combination $a_1 f_1 + a_2 f_2 + \cdots + a_r f_r$ (all $a_i \neq 0$ simultaneously) has zero norm. A vector space is said to be finite dimensional if all its elements can be expressed as linear combinations of a finite number of elements. In such a case there is a minimal number of such elements called the rank of the space. A set of such elements necessarily independent (to be minimal) is called a basis. A basis of a vector space is not, however, unique but its rank is.

We can always choose a basis such that its elements are orthogonal (zero vector product), implying that the chosen factor variables are uncorrelated. A convenience provided by such a choice is that a basis can be simply represented by a set of correlations between the measurements and factors. If $Z_1, \cdots, Z_k$ is an orthogonal basis then each z_i can be expressed in terms of Z_j.

$$z_i = a_{i1} Z_1 + \cdots + a_{ik} Z_k . \tag{2.1.3}$$

The covariance of x_i with Z_j is a_{ij}, the coefficient of Z_j in the representation (2.1.3). This may be regarded as a correlation coefficient once x_i and Z_j are properly standardized. A basic set of factors or equations (2.1.3) can be represented by the matrix of correlations

$$\begin{bmatrix} a_{11} & \cdots & a_{1k} \\ \vdots & & \vdots \\ a_{p1} & \cdots & a_{pk} \end{bmatrix}, \tag{2.1.4}$$

which is also called the factor loading matrix. Such a basis is not unique because the choice of Z_i, or the representation (2.1.3), is not unique. A basis is just meant to generate the entire space of factor variables which are linear combinations of some hypothetical variables introduced into the various tests. From this point of view one basis is as good as any other. Of course, given one basis, orthogonal or oblique, the other can be derived by a linear transformation.

The choice of a suitable basis which is "psychologically meaningful" has largely rested with the psychologist, perhaps rightly so. But it is quite conceivable, once the psychological meaning is translated to mean some precisely stated restrictions on the basis, that its choice will turn out to be a problem of statistical estimation. In this sense the graphical methods of rotation of factor loadings advocated by Thurstone (**17**) and the quadrimax method of Neuhaus and Wrigley (**13**) are statistical methods of factor analysis, where the number of zero or small loadings is maximized. Such a restriction, or even the orthogonality of a basis, may not be the most helpful in leading to a suitable psychological interpretation or the discovery of "real entities." The choice of the restrictions to be imposed on the basis is perhaps a problem for psychological research. An objective method developed in this connection by Cattell (**4**, **5**) seems to have interesting possibilities from a statistical point of view.

2.2. *Principal Components*

It was pointed out by Sir Cyril Burt that this method was originally put forward by Karl Pearson in 1901. But the statistical problems of estimation and testing connected with the principal components were first considered by Hotelling.

Hotelling (**8**) considers two types of problems:

First, without assuming a decomposition of the measurements as in (2.1.1), hypotheses are framed in terms of the latent roots of the correlation matrix with a view to studying the shape of the scatter of the standardized scores in the p-dimensional space or alternatively the relative importance of the different principal components in explaining the total variance. For instance, if some of the calculated roots are not significantly different then the components corresponding to them may be considered equally important.

Let us consider a specific hypothesis that the $(i + 1)$th to the $(i + r)$th roots of the population correlation matrix (denoted by ρ) are equal. The value of i may be 0, 1, $\cdots$ to $(p - r)$.

This hypothesis imposes a restriction on ρ, viz., that it admits the decomposition

$$\rho = \theta + \lambda I, \tag{2.2.1}$$

where θ is a matrix of rank $(p - r)$, λ is the common value of the roots from $(i + 1)$th to $(i + r)$th, and I is the unit matrix.

Any such hypothesis or a similar one based on the dispersion matrix Σ instead of ρ can be tested by a likelihood-ratio criterion Λ, provided sample size is moderately large. Exactly how large the sample size should be is a matter for further investigation. The statistic ($-2 \log_e \Lambda$) is distributed in large samples as χ^2 with degrees of freedom equal to the number of restrictions on the free parameters imposed by the hypothesis.

The number of restrictions in a hypothesis of the form (2.2.1) is equal to the number of restrictions on the symmetric matrix θ with rank $(p - r)$ minus one for the unknown λ. Only $(p - r)$ rows and columns in θ are independent; the rest of the elements depend on them. Therefore, the number of restrictions on the elements of θ is

$$(p - r)(p - r + 1)/2,$$

and with one less we have

$$(p - r - 1)(p - r + 2)/2 \tag{2.2.2}$$

degrees of freedom for the χ^2 approximation.

If A is the estimated dispersion matrix from observations on n individuals, then the test criterion for the hypothesis (2.2.1) with Σ instead of ρ is

$$-(n - 1) \log \frac{| A |}{| \hat{\Sigma} |}, \tag{2.2.3}$$

where $| \hat{\Sigma} |$ is the estimated dispersion matrix under the conditions of the hypothesis. The latent roots of $\hat{\Sigma}$

$$\mu_1, \mu_2, \cdots, \mu_i, \mu_{i+1}, \cdots, \mu_{i+r}, \mu_{i+r+1}, \cdots, \mu_p$$

are connected with the latent roots of A in the following way:

$$\lambda_j = \mu_j \qquad (j = 1, \cdots, i, i + r + 1, \cdots, p),$$

$$\mu_{i+1} = \frac{\lambda_{i+1} + \cdots + \lambda_{i+r}}{r}.$$

Since

$$| A | = \lambda_1 \cdots \lambda_p, \qquad | \hat{\Sigma} | = \mu_1 \cdots \mu_p,$$

the ratio $| A |/| \hat{\Sigma} |$ is

$$\frac{\lambda_{i+1} \cdots \lambda_{i+r}}{\left(\dfrac{\lambda_{i+1} + \cdots + \lambda_{i+r}}{r}\right)^r}, \tag{2.2.4}$$

which is a suitable power of the ratio of the geometric to the arithmetic mean of the $(i + 1)$th to $(i + r)$th roots of A. From this point of view it would appear, by choosing $i = (p - r)$ in (2.2.4), that Bartlett's (1) test using the dispersion matrix instead of the correlations is valid for judging the significance of equality of the least r roots.

Unfortunately the test does not seem to reduce to the form (2.2.4) in terms of the roots of the observed correlation matrix R when the hypothesis is as stated in (2.2.1) in terms of the population correlation matrix ρ. The effect of standardizing the variables by the *sample standard deviations* is not properly allowed for by a criterion of the form (2.2.4). This is also partly revealed by Bartlett's own evaluation of the degrees of freedom by the expectation method in a simple case. They depend on the unknown correlations and reach the value (2.2.2) only in a limiting case, while for a genuine likelihood ratio this is not expected. The exact evaluation of the test criterion depends on complicated equations which require further investigation.

Secondly, Hotelling considers the problem of "testing the variances of components against the variance to be expected on account of the inaccuracy of the tests as revealed by their self-correlations or reliability coefficients." For this purpose a test score is thought of as made up of two parts, a true score with variance unity and a random error. Thus

$$x_i = X_i + \epsilon_i \qquad (i = 1, \cdots, p), \tag{2.2.5}$$

with the conditions $\text{cov}(\epsilon_i, \epsilon_j) = 0$, $(i \neq j)$. The hypothesis stated above is interpreted to imply that the true scores X_i are linearly dependent, i.e., "the scatter diagram of the true scores will lie in a flat space of smaller dimensionality immersed in the p-dimensional space." If independent estimates of the variances of ϵ_i are available, either from an external source or by re-tests on individuals, there is no need to consider the true scores as random variables in order to test the above hypothesis. The general multivariate tests of dimensionality developed in more complicated situations are directly applicable for this problem. The non-stochastic model on the scores corrected for unreliabilities used in testing the second hypothesis provides a strong contrast to tests in factor analysis where, of necessity, all the variables (the common and specific factors) involved are considered to be stochastic, which makes the problem more complex.

3. *Special Characterizations of a Basis in Factor Analysis*

Using the vector notation

$$\underset{\sim}{x} = (x_1, \cdots, x_p), \qquad \underset{\sim}{z} = (z_1, \cdots, z_p), \qquad \underset{\sim}{s} = (s_1, \cdots, s_p).$$

The equation (2.1.1) can be written

$$\underset{\sim}{x} = \underset{\sim}{z} + \underset{\sim}{s}. \tag{3.1}$$

The dispersion matrix of $\underset{\sim}{x}$ (using D for dispersion) is

$$D(\underset{\sim}{x}) = D(\underset{\sim}{z}) + D(\underset{\sim}{s}),$$

or

$$\Sigma = \Gamma + \Delta, \tag{3.2}$$

where Σ, Γ and Δ are defined by equation (3.2). The covariance of $\underline{z}$ and $\underline{s}$ is zero because of conditions (2.1.2). The matrix Γ is positive semi-definite with rank $k < p$ and Δ is a positive-definite diagonal matrix. The equation (3.2) supplies the fundamental decomposition of the dispersion matrix Σ in terms of those of the hypothetical variables postulated by a factorial structure. If the rank of Γ is $k < p$, the space of common factors has a basis of k independent factors as shown in section 2.1. For a proper identification of the space and "an orderly selection of independent factors" there is a need to characterize a basis in a convenient way. A basis so characterized need not admit a psychological interpretation, for only mathematical and statistical convenience is being sought at this stage. A basis once obtained can always be transformed to meet other requirements. Two special characterizations are discussed here.

3.1 *First Characterization*

Let $\underline{l} = (l_1 , \cdots , l_p)$ be a vector of arbitrary coefficients giving rise to a new factor variable

$$\underline{l}\underline{z}' = l_1 z_1 + \cdots + l_p z_p .$$

The variation in the variable x_i explained by the factor variable $\underline{l}\underline{z}'$, is

$$\frac{\text{cov}^2 (x_i , \underline{l}\underline{z}')}{V(\underline{l}\underline{z}')} = \frac{(l_1\gamma_{1i} + \cdots + l_p\gamma_{pi})^2}{\underline{l}\Gamma\underline{l}'}, \tag{3.1.1}$$

assuming that $\underline{l}\ \Gamma\ \underline{l}'$, the variance of $\underline{l}\underline{z}'$, is not zero. The total variation explained in all the variables is

$$\frac{\left\{\sum_1^p (l_1\gamma_{1i} + \cdots + l_p\gamma_{pi})^2\right\}}{\underline{l}\Gamma\underline{l}'} = \frac{\underline{l}\Gamma\Gamma\underline{l}'}{\underline{l}\Gamma\underline{l}'}. \tag{3.1.2}$$

Let us choose $\underline{l}$ such that (3.1.2) is a maximum. Differentiating with respect to the vector $\underline{l}$ (see **14**, p. 21), the equation leading to the optimum value λ of the ratio (3.1.2) and the vector $\underline{l}\ \Gamma$ is

$$\underline{l}\Gamma\Gamma - \lambda\underline{l}\Gamma = 0$$

or eliminating $\underline{l}\Gamma$

$$| \Gamma - \lambda I | = 0, \tag{3.1.3}$$

where I is the identity matrix. This shows that λ is the maximum latent root of Γ and $\underline{m} = \underline{l}\Gamma$ is the latent vector corresponding to it. Since the vector $\underline{m}$ satisfies the equation

$$\underline{m}\Gamma = \lambda\underline{m},$$

and

$$\underline{m} = \underline{l}\Gamma,$$

so that

$$\underline{m}\Gamma = \lambda \underline{l}\Gamma, \tag{3.1.4}$$

the vector $\underline{m}$ itself can be taken to be a solution of $\underline{l}$. We thus obtain the first factor variable as a linear combination of $z_1, \cdots, z_p$. From the theory of canonical roots and vectors (**14**, p. 24), it would then follow that the second factor variable, which explains the highest proportion of the residual variation independently of the first, is the linear combination corresponding to the second canonical vector. There are as many linear combinations as there are non-zero roots λ, which is equal to the rank of the matrix Γ. The linear combinations of $z_1, \cdots, z_p$ supplied by the canonical vectors of zero roots of λ vanish identically, indicating the dependence of the factor variables associated with the measurements $x_1, \cdots, x_p$.

The factor loading of the variable x_i on the first factor chosen above is the correlation between the two. The covariance is

$$\text{cov}\,(x_i, \underline{l}\underline{z}') = l_1\gamma_{1i} + \cdots + l_p\gamma_{pi} = \lambda_1 l_i,$$

and if the variables x_i are initially chosen to have unit variance the correlation is

$$\frac{\lambda_1 l_i}{\sqrt{\underline{l}\Gamma\underline{l}'}} = \frac{\lambda_1 l_i}{\sqrt{\lambda_1 \underline{l}\underline{l}'}} = \frac{\sqrt{\lambda_1}\, l_i}{\sqrt{l_1^2 + \cdots + l_p^2}}. \tag{3.1.5}$$

The factor loadings are then the elements of the first canonical vector suitably standardized. Similarly the factor loadings of any other factor are derived from the canonical vector defining the factor.

Even after exhausting all the independent factor variables, there still remains some variation left in $\underline{x}$ to be explained by the specific factors unless the number of independent common factors is equal to p. In the problem originally considered by Hotelling, the successive components explaining variation in $\underline{x}$ were not confined to the common factor portion $\underline{z}$ but were also functions of the specific factors $\underline{s}$, which then are equivalent to linear functions of $\underline{x}$. Hotelling's principal components are, therefore, important in problems where the total variation of a measurement vector $\underline{x}$ is sought to be accounted for, to the maximum amount possible, by a smaller number of linear functions of $\underline{x}$. The principal components of Hotelling are derived from the latent vectors of the matrix $\Sigma = \Gamma + \Delta$ instead of Γ alone as used above. It may be observed that when

$$\Delta = \delta^2 I,$$

i.e., when all the specific variables have the same variance δ^2, a latent vector $\underline{l}$ of $\Gamma + \Delta$ satisfying the equation

$$\underline{l}(\Gamma + \delta^2 I) = \mu \underline{l}$$

also satisfies the equation

$$\underline{l}\Gamma = (\mu - \delta^2)\underline{l} = \lambda \underline{l}$$

and is therefore a latent vector of Γ. The principal component analysis of Hotelling is thus a method of factor analysis with the factor loadings inflated keeping the same relative magnitudes, when all the specific variances are the same.

There is some arbitrariness in the above characterization of the basic set of factors because instead of maximizing the sum of the variations explained in $x_1, \cdots, x_p$ we could maximize a weighted sum and arrive at a different basis and consequently a different set of factor loadings. When the variables $x_1, \cdots, x_p$ are chosen to have unit variances the method adopted is equivalent to using reciprocals of total variances as weights.

The quantity δ_i^2, the residual variance of x_i unexplained by the factor variables, satisfies the equation

$$\sigma_{ii} = \frac{\lambda_1 l_i^2}{\underline{l}\underline{l}'} + \frac{\lambda_2 m_i^2}{\underline{m}\underline{m}'} + \cdots + \delta_i^2, \tag{3.1.6}$$

where $\underline{l}, \underline{m}, \cdots$ are the latent vectors defining the factors $\underline{l}\underline{z}', \underline{m}\underline{z}', \cdots$ and the subscript i relates to the ith element in the vectors.

The best formula for predicting a factor variable such as $\underline{l}\underline{z}'$ from the observed measurements $\underline{x}$ is obtained by the method of regression (16). If $\underline{k}\underline{x}'$ is the predicted value, then $\underline{k}$ satisfies the equation

$$\underline{k}\Sigma = \underline{l}\Gamma, \qquad \underline{k} = \underline{l}\Gamma\Sigma^{-1} = \lambda_1 \underline{l}\Sigma^{-1}, \tag{3.1.7}$$

and similarly for other factors. The characterization of the basis considered here together with methods of estimation is known as the principal factor analysis (PFA) (7).

3.2 *Second Characterization*

Instead of asking for a factor variable which explains as much of variation as possible of $\underline{x}$, we may pose the problem in a different way. What is that factor variable which is predictable from $\underline{x}$ with the maximum possible precision? Or in other words, what is that factor variable which is maximally related to $\underline{x}$? The solution to this problem depends on a canonical correlation analysis of the hypothetical factor variables $\underline{z}$ with the measurable variables $\underline{x}$ of which $\underline{z}$ constitute a part.

If $\underline{l}\underline{z}'$ and $\underline{q}\underline{x}'$ represent two linear combinations of factor variables and test scores, then according to the theory of canonical correlations (8) the correlation (or its square) between the two linear functions

$$\frac{(\underline{l}\Gamma\underline{q}')^2}{(\underline{l}\Gamma\underline{l}')(\underline{q}\Sigma\underline{q}')} \tag{3.2.1}$$

has to be maximized. Using the algebra developed in a similar genetic problem (**14**), the optimum value of the correlation ν is found to be a root of the equation

$$| \Gamma - \nu^2\Sigma | = 0, \tag{3.2.2}$$

or

$$| \Sigma - \lambda\Delta | = 0, \qquad \lambda = 1/(1 - \nu^2). \tag{3.2.3}$$

The vectors $\underline{l}$ and $\underline{q}$ are proportional and satisfy the same equation

$$\underline{l}(\Sigma - \lambda\Delta) = 0, \qquad \underline{q}(\Sigma - \lambda\Delta) = 0. \tag{3.2.4}$$

That factor variable which is highly correlated with $\underline{x}$ is $\underline{l}\underline{z}'$, where $\underline{l}$ is the latent vector corresponding to the largest root of the determinental equation (3.2.2). The second factor variable, uncorrelated with the first and possessing the highest correlation with $\underline{x}$ is $\underline{m}\underline{z}'$, where $\underline{m}$ is the latent vector corresponding to the second root of (3.2.2), and so on. We get as many factors as the number of non-zero values of ν^2 or values of λ greater than unity which is the same as the rank of Γ.

For any factor $\underline{l}\underline{z}'$ as determined above

$$\underline{l}\Gamma\underline{l}' = \nu^2\underline{l}\Sigma\underline{l}' = \frac{\lambda - 1}{\lambda}\underline{l}\Sigma\underline{l}' = (\lambda - 1)\underline{l}\Delta\underline{l}'$$

$$\text{cov}\,(x_i\,,\underline{l}\underline{z}') = l_1\gamma_{1i} + \cdots + l_p\gamma_{pi} = (\lambda - 1)l_i\delta_i^2\,.$$

The correlation between x_i and $\underline{l}\underline{z}'$,

$$\frac{(\lambda - 1)l_i\delta_i^2}{\sqrt{(\lambda - 1)\underline{l}\Delta\underline{l}'\sigma_{ii}}}\,, \tag{3.2.5}$$

is the factor loading of x_i on the factor $\underline{l}\underline{z}'$. This is again an element of $\underline{l}$ multiplied by a constant. It can be shown that the same factor loadings are obtained if instead of $(x_1\,,\cdots,x_p)$ we consider $(c_1x_1\,,\cdots,c_px_p)$ with the variables arbitrarily scaled. In the previous case it is necessary to reduce the variables $(x_1\,,\cdots,x_p)$ to unit standard deviation before proceeding to derive factors in order to achieve uniqueness of factor loadings.

To predict the factor measurements we use the regression equation as in (3.1.7). In this case it turns out that $\underline{l}\underline{z}'$, $\underline{m}\underline{z}'$, $\cdots$, defined by the latent vectors of (3.2.3), can be best predicted by $\underline{l}\underline{x}'$, $\underline{m}\underline{x}'$, $\cdots$, avoiding the complication of multiplication by Σ^{-1} necessary in the case of factors defined in the earlier characterization of the basic set (3.1.7).

The residual variance δ_i^2 in x_i unexplained by the factor variables satisfies the equation

$$\sigma_{ii} = \frac{(\lambda_1 - 1)}{\underline{l}\Delta\underline{l}'}l_i^2\delta_i^4 + \frac{(\lambda_2 - 1)}{\underline{m}\Delta\underline{m}'}m_i^2\delta_i^4 + \cdots + \delta_i^2\,, \tag{3.2.6}$$

similar to the formula (3.1.6) in the earlier case. This second characterization of a basis together with methods of estimation may be called canonical factor analysis (CFA) to bring out its connection with the theory of canonical correlations.

Factor analysis thus fits in a general theory of canonical correlations involving two sets of variables: one set being observable and the other set, observable as in multiple regression; dummy as in multiple discrimination; or hypothetical as in problems of genetic selection.

3.3 *Which Is a Better Characterization?*

This question is meaningless if we are dealing with the true values of the dispersion elements satisfying the conditions of a given rank of the factor-variable space because one can be transformed into the other, and in fact they may be replaced by any other basic set and they all serve the same purpose.

But this is no longer true when we have only *estimates* of the dispersion elements and factors are estimated by formally substituting for Σ the estimated quantities and choosing Δ to satisfy the equation (3.1.6) in the first case (PFA) and (3.2.6) in the second case (CFA). Which then is a better estimate of a basis?

From the point of view of statistical estimation, PFA gives a least-squares estimate (**16**, p. 119) and CFA, a maximum-likelihood estimate, when normality of the distribution of the observations is assumed. At present there is not much to choose between the two methods except for the following reasons. The maximum-likelihood estimation leads in general to better results when the distribution of the variables is specified. No suitable test based on the least-squares estimates is available while there exists an easily computable test criterion on the basis of maximum-likelihood estimates. No further computations are needed to obtain the factor measurements if the factors are estimated by CFA; in fact, in this method, factors are deduced from a description of their measurement.

There is another logical argument which may have to be borne in mind in deciding the issue. A rigorous hypothesis concerning the number of independent factor variables is perhaps never true, and a test of this null hypothesis can detect its falsehood only when there is a serious departure. If then by following a rule of behavior (as determined by a test criterion) we decide to extract a certain number of factors, any method of estimation may be looked upon as providing only a summary of all the factors in terms of a few dominant ones having a definite existence with magnitudes bigger than standard errors calculable from the observations. It is then of interest to examine whether one method of estimation leads to a better summary than the other and at the same time has low errors of estimation.

From this viewpoint, PFA may be thought of as providing the best k

(given number not necessarily exhaustive) factors explaining the maximum possible variance in the measurements while CFA, the best k factors which have in some sense highest possible correlations with the measurements. This may mean that while the first set attempts to explain as much as possible of the variations in the individual measurements, the latter set focuses on the correlations. Perhaps the psychological interest chiefly lies in the latter set, which offers a better explanation of the correlations between the measurements.

4. *Estimation and Tests of Significance for Factors*

4.1 *Estimation of Factor Loadings*

Let $A = (a_{ij})$ denote the observed dispersion matrix of the vector variable $\underline{x}$. This is sufficient for the estimation of Γ and Δ, the two components of the population dispersion matrix Σ. Following the equations (3.2.3, 3.2.4) of the second characterization, we have on substituting A for Σ

$$| A - \lambda\Delta | = 0,$$

$$\underline{l}(A - \lambda\Delta) = 0, \tag{4.1.1}$$

where $\underline{l}$ is a latent vector corresponding to the latent root λ. From the point of view of mechanical computations it is convenient to solve for

$$\underline{b} = \underline{l}\Delta^{1/2}, \qquad \underline{b}\underline{b}' = 1, \tag{4.1.2}$$

in which case $\underline{b}$ is the latent vector of

$$| \Delta^{-1/2}A\Delta^{-1/2} - \lambda I | = 0. \tag{4.1.3}$$

Let us suppose, for the sake of illustration, that we are extracting two factors. If $\underline{b} = (b_1, \cdots, b_p)$ and $\underline{c} = (c_1, \cdots, c_p)$ are the first two latent vectors of (4.1.3) corresponding to the roots λ_1 and λ_2, then the equation (3.2.6) gives

$$a_{ii} = [(\lambda_1 - 1)b_i^2 + (\lambda_2 - 1)c_i^2 + 1]\delta_i^2, \tag{4.1.4}$$

or

$$\delta_i^2 = \frac{a_{ii}}{(\lambda_1 - 1)b_i^2 + (\lambda_2 - 1)c_i^2 + 1} = \frac{a_{ii}}{g_i^2}, \tag{4.1.5}$$

where g_i is defined by the last part of equation (4.1.5). The equation (4.1.3) can now be written in terms of the observed correlation matrix R instead of the dispersion matrix A

$$| GRG - \lambda I | = 0, \tag{4.1.6}$$

where the elements g_i of the diagonal matrix G satisfy

$$g_i = \sqrt{(\lambda_1 - 1)b_i^2 + (\lambda_2 - 1)c_i^2 + 1}. \tag{4.1.7}$$

The computational problem is then to solve for g_i's satisfying the equations (4.1.6) and (4.1.7), where λ_1, λ_2, are the latent roots of (4.1.6) and $\underline{b}$, $\underline{c}$, the latent vectors. A tentative method is to start with a trial matrix G and obtain successive approximations by solving (4.1.6) for λ_1, λ_2, and $\underline{b}$, $\underline{c}$, and substituting in (4.1.7). The process is repeated until the g_i converge.

A better approximation to g_i is obtained by using the formula

$$g_i = \sqrt{\left(\frac{\lambda_1}{\lambda_e} - 1\right)b_i^2 + \left(\frac{\lambda_2}{\lambda_e} - 1\right)c_i^2 + 1}, \tag{4.1.8}$$

where

$$\lambda_e = \frac{[\Sigma g_i^2]_0 - \lambda_1 - \lambda_2}{p - 2}, \tag{4.1.9}$$

the summation $[\Sigma g_i^2]_0$ refers to the g_i^2 at the previous stage used in equation (4.1.7) to obtain λ_1, λ_2.

The two formulas (4.1.7) and (4.1.8) should agree towards the final stages when convergence is expected to be slow. But in the initial stages (4.1.8) may accelerate convergence.

The estimated factor loadings on the first and second factors at any stage of approximation are

$$\sqrt{\lambda_1 - 1}\,\underline{b}G^{-1}, \qquad \sqrt{\lambda_2 - 1}\,\underline{c}G^{-1}.$$

The same method holds good for any number of factors. The estimates of factor loadings obtained from equations (4.1.6, 4.1.7) can be shown to satisfy the maximum-likelihood equations of Lawley (**11**, **12**) and thus constitute one out of a number of possible solutions. The equations (4.1.6, 4.1.7) are in a proper shape to admit an iterative procedure for solution. The use of equation (4.1.7) seems to avoid a difficulty which may occur in the iterative procedures. The iterative method given by Lawley (**16**, p. 130) may suffer a breakdown on the initial iteration due to an improperly chosen trial set of factor loadings leading to imaginary values of the quantities commonly designated by "h_1, h_2, $\cdots$." This may also occur in PFA with the guessed communalities at the first stage.

4.2 *Tests of Significance and Estimation of Number of Factors*

It is also necessary to lay down some rules for determining the number of factors to be estimated. This is partially answered by any reasonable test for a specified number of factors. We determine that number of factors for which the chosen test does not show significance, while for any smaller number the hypothesis is contradicted. If the level of significance is based on the 5 per cent level, then this method leads us to a lower confidence limit to the number of factors. That is, we can assert, with a risk of only 5 per cent, that the number of factors is at least as large as that discovered by the above procedure. This is no doubt an objective rule for determining the lower limit to the number of factors, but in practice it may be better to extract one or two more factors, depending on the magnitude of the residual roots. If one or two such roots are sufficiently bigger than unity (though not significantly so) it may be worth while to extract the factor corresponding to them also.

The hypothesis we propose to test is that the population dispersion matrix admits the decomposition

$$\Sigma = \Gamma + \Delta, \tag{4.2.1}$$

where Δ is a diagonal matrix with positive terms and Γ is a positive semi-definite matrix of rank $k < p$.

The test criterion we use is derived by the principle of likelihood ratio, assuming that the observations are normally distributed.

The exact distribution of the test criterion is not known but in large samples (− 2 log) of the likelihood ratio is distributed as χ^2 with degrees of freedom equal to the number of independent restrictions on the elements of Σ imposed by the hypothesis (4.2.1). This hypothesis specifies the rank of the matrix $\Sigma - \Delta$ for suitably chosen Δ. If its rank is k, then by fixing the first k rows and columns the rest of the elements can be computed, which implies $(p - k)\ (p - k + 1)/2$ restrictions. Allowing for p unknown values in Δ, the number of restrictions is equal to

$$\frac{(p-k)(p-k+1)}{2} - p = \frac{(p-k)^2 - p - k}{2}. \tag{4.2.2}$$

The test based on the likelihood-ratio criterion is

$$-(n-1)\log\frac{|A|}{|\hat{\Sigma}|}, \tag{4.2.3}$$

where $\hat{\Sigma}$ is the estimated dispersion matrix using the maximum-likelihood equations of section 4.1. The multiplying coefficient $(n - 1)$, where n is the sample size, may be replaced by the more appropriate value for the χ^2 approximation to hold when n is not large,

$$\left(n - 1 - \frac{2p+5}{6} - \frac{2k}{3}\right),$$

where p is the number of variables and k is the number of factors (1). Since the roots of the equation $|\Sigma - \lambda\Delta| = 0$ corresponding to k factors are estimated by $|A - \lambda\Delta| = 0$ or the equivalent forms (4.1.3), (4.1.6), it follows that the roots of $|\hat{\Sigma} - \lambda\Delta| = 0$ are

$$\lambda_1, \cdots, \lambda_k, 1, 1, \cdots, 1, \tag{4.2.4}$$

while the roots of $|A - \lambda\Delta|$ are, in descending order of magnitude,

$$\lambda_1, \cdots, \lambda_k, \lambda_{k+1}, \cdots, \lambda_p, \tag{4.2.5}$$

and, therefore,

$$\frac{| A |}{| \hat{\Sigma} |} = \lambda_{k+1} \cdots \lambda_p , \tag{4.2.6}$$

which is the product of the least $(p - k)$ roots, at the last stage of iteration of the equation (4.1.6), $| GRG - \lambda I | = 0$.

The χ^2 test is

$$-(n - 1) \log (\lambda_{k+1} \cdots \lambda_p) \tag{4.2.7}$$

with $[(p - k)^2 - p - k]/2$ degrees of freedom apart from the slight refinement in the multiplying coefficient.

4.3 *A Modified Criterion and Its Practical Use*

It may be recalled that the likelihood-ratio criterion is the ratio of the maximum likelihood under the restrictions of the hypothesis (4.2.1) to that without any restrictions on Σ. It is of interest to examine how the ratio (or its logarithm) of the likelihoods is converging to the maximum value (or the negative of log ratio to its minimum value) during the iterative process. Fortunately this can be expressed in terms of the roots $\lambda_{k+1} , \cdots , \lambda_p$ at any stage of the iterative process

$$-(n - 1)[\log (\lambda_{k+1} \cdots \lambda_p) - (p - k) \log \lambda_e], \tag{4.3.1}$$

where λ_e of (4.1.9) is the arithmetic mean of $\lambda_{k+1} , \cdots , \lambda_p$. [Strangely the sequence (4.3.1) of statistics (likelihood ratios), which converge ultimately to the test criterion (maximum-liklihood ratio), resembles Bartlett's (1) ratio test but, of course, the roots λ_i are obtained differently and the ratios are used with different degrees of freedom. From this analysis it would appear that Bartlett's ratio is an initial approximation to the actual test criterion.] (4.3.1) converges to

$$-(n - 1) \log (\lambda_{k+1} \cdots \lambda_p)$$

at the final stage when

$$(p - k) = \lambda_{k+1} + \cdots + \lambda_p . \tag{4.3.2}$$

Suppose that (4.3.1) is not significant as χ^2 with $[(p - k)^2 - p - k]/2$ degrees of freedom, at any stage, then the same conclusion is reached even after completing the iterative process. If a test of significance is the only aim of analysis, then, sometimes, iteration can be stopped at some stage. Even if the result is significant, it is possible to terminate the computations provided the change in (4.3.1) at that stage is small from one cycle of operations to another.

The modified criterion is extremely useful in practice when the object of the analysis is to estimate the number of factors (lower confidence value) as well as the factor loadings.

Before proceeding with the cycle of operations for estimation, let us fix some high value of k as the number of factors and calculate the roots after one or two iterations. At this stage, find that value of r for which Λ_r (with d.f. $[(p - r)^2 - p - r]/2$) is not significant, but Λ_{r-1} is. This shows that the number of factors is not greater than r. We may set the number of factors provisionally at r and continue the process of estimation. Each time we may calculate Λ_{r-1} and Λ_r to see whether Λ_{r-1} becomes not significant at any stage. If it is not significant, there is a case for switching over to $(r - 1)$ factors instead of r.

5. *Summary*

The experimental situation and the nature of the data on which the technique of factor analysis can be successfully employed may be stated as follows. Each of the p measurements on an individual has a linear regression on a common set of a few hypothetical variables or factors. The deviations from regression for any two measurements are uncorrelated. The factor analysis seeks the smallest number of independent hypothetical variables necessary to explain the intercorrelations between the measurements.

If R is the observed correlation matrix, the computational problem of factor analysis depends on the solution of the diagonal matrix G satisfying the equations

$$| GRG - \lambda I | = 0, \tag{5.1}$$

$$g_i = [(\lambda_1 - 1)a_{1i}^2 + \cdots + (\lambda_k - 1)a_{ki}^2 + 1]^{1/2}, \tag{5.2}$$

where k is the number of factors assumed, $\lambda_1, \cdots, \lambda_k$, are the first k largest roots of (5.1) and $\underset{\sim}{a}_j = (a_{j1}, \cdots, a_{jp})$ is the latent vector corresponding to the root λ_j. Once G is found to satisfy the equations (5.1, 5.2), then the factor loadings are given by

$$(\lambda_j - 1)\underset{\sim}{a}_j G^{-1} \qquad (j = 1, \cdots, k),$$

and the test of the hypothesis that k factors are adequate to explain the intercorrelations is

$$\chi^2 = -(n - 1) \log_e (\lambda_{k+1} \cdots \lambda_p)$$

with $[(p - k)^2 - p - k]/2$ degrees of freedom. The lower confidence limit to the number of factors is the smallest value of k for which χ^2 is not significant.

Some research remains to be done to find an elegant computational technique for solving the equations (5.1, 5.2). The method available at present is to guess suitable values of g_i, substitute in (5.1) and obtain better approximations to g_i by using (5.2). This process is continued until convergence is secured. Unfortunately this appears to be a slow process unless the initial values of g_i are very near the true values. Even with a good set of trial values the problem can be best tackled only on an electronic computer when large numbers of variables are involved. A suitable program for Illiac is being written by Mr. Golub of the Digital Computer Laboratory at the University of Illinois. A numerical example solved on a tentative program is reported below. Full details will be presented soon.

First it may be noted that the relation between g_i and the communality h_i^2 for the ith variate is

$$g_i = 1/\sqrt{1 - h_i^2} ,$$

so that good trial values of g_i are available once the communalities are approximately determined by an initial factorization of the correlation matrix by a simpler method, such as the centroid. Another method suggested in the literature is to choose the squared multiple correlation as an estimate of the communality. In many cases it is sufficient to start with the initial approximation $g_i = 1/2$.

Second, although the test involves the product of the roots at the final stages of convergence, it is useful to compute at intermediate stages the statistic

$$\chi^2 = -(n - 1)[\log_e (\lambda_{k+1} \cdots \lambda_p) - (p - k) \log_e (\lambda_{k+1} + \cdots + \lambda_p)],$$

which, when not significant, implies the nonsignificance of the ultimate χ^2. We could stop at any stage after this, provided further iterations do not considerably alter the factor loadings.

The following correlation matrix was presented by Davis (6) in an attempt to study factors of comprehension in reading.

1.00								
.72	1.00							
.41	.34	1.00						
.28	.36	.16	1.00					
.52	.53	.34	.30	1.00				
.71	.71	.43	.36	.64	1.00			
.68	.68	.42	.35	.55	.76	1.00		
.51	.52	.28	.29	.45	.57	.59	1.00	
.68	.68	.41	.36	.55	.76	.68	.58	1.00

Assuming a single factor, the χ^2 was calculated and found to be significant. This indicated more than one factor. Under the hypothesis of two factors the value of χ^2

$$-(n - 1)[\log (\lambda_3 \cdots \lambda_9) - (9 - 2) \log (\lambda_3 + \cdots + \lambda_9)]$$

came down to 29.73 at an early stage of iteration. This being less than 30.1, the 5 per cent significance value of χ^2 with 19 degrees of freedom, the hypothesis of two factors stands unrejected. So the data admit an interpretation in terms of two significant factors only. A fairly stablilized set of factor loadings are

Factor 1	.845	.817	.477	.401	.669	.891	.834	.651	.833
Factor 2	−.309	−.084	.012	.153	.161	.145	.081	.122	.080

I wish to thank Dr. C. F. Wrigley, who read the manuscript and offered some helpful comments.

REFERENCES

1. Bartlett, M. S. Tests of significance in factor analysis. *Brit. J. Psychol., Statist. Sect.*, 1950, **3**, 77–85.
2. Bartlett, M. S. A further note on tests of significance in factor analysis. *Brit. J. Psychol., Statist. Sect.*, 1951, **4**, 1–2.
3. Burt, C. Tests of significance in factor analysis. *Brit. J. Psychol., Statist. Sect.*, 1952, **5**, 109–133.
4. Cattell, R. B. Parallel proportional profiles. *Psychometrika*, 1944, **9**, 267–283.
5. Cattell, R. B. The description and measurement of personality. Yonkers, New York: World Book Co., 1946.
6. Davis, F. B. Fundamental factors of comprehension in reading. *Psychometrika*, 1944, **9**, 185.
7. Holzinger, K. J., and Harman, H. H. Factor analysis. Chicago: Univ. Chicago Press, 1941.
8. Hotelling, H. Analysis of a complex of variables into principal components. *J. educ. Psychol.*, 1933, **24**, 417–441, 498–520.
9. Hotelling, H. Relations between two sets of variates. *Biometrika*, 1936, **28**, 321–377.
10. Kendall, M. G. Factor analysis. *J. roy. stat. Soc., Series B*, 1950, **12**, 60.
11. Lawley, D. N. The estimation of factor loadings by the method of maximum likelihood. *Proc. roy. Soc. Edin.*, 1940, **60**, 64–82.
12. Lawley, D. N. Further investigations in factor estimation. *Proc. roy. Soc. Edin.*, 1941, **61**, 176–185.
13. Neuhaus, J. O., and Wrigley, C. F. The quadrimax method: an analytic approach to orthogonal simple structure. Manuscript on file in the Univ. Illinois Library, 1953.
14. Rao, C. R. Advanced statistical methods in biometric research. New York: Wiley, 1952.
15. Rao, C. R. Discriminant functions for genetic differentiation and selection. Sankhyā, 1953, **12**, 229.
16. Thomson, G. H. The factorial analysis of human ability. London: Univ. London Press 5th ed., 1951.
17. Thurstone, L. L. A new rotational method in factor analysis. *Psychometrika*, 1938, **3**, 199–218.

Manuscript received 2/19/54

Revised manuscript received 5/11/54

10

Reprinted from *British J. Math. Stat. Psychol.*, **23**, Pt. 1, 1–21 (May 1970)

THE THEORETICAL FOUNDATIONS OF PRINCIPAL FACTOR ANALYSIS, CANONICAL FACTOR ANALYSIS, AND ALPHA FACTOR ANALYSIS

By Roderick P. McDonald[1]

University of New England, N.S.W., Australia

It is shown that PFA, CFA and AFA are particular cases of a scale-invariant factoring procedure based on variance ratios of certain weighted combinations of variables. Standard derivations in the literature are shown, in contrast, to have unsatisfactory features. It is suggested that the choice between PFA, CFA and AFA involves relatively independent choices of features of each, and that in most cases CFA is to be preferred.

1. Introduction

Over a period of approximately 60 years, a tremendous amount of attention has been given to the fundamental properties of the common factor analysis model, to basic procedures for fitting the model to real data, and to derived problems of estimation and hypothesis testing on the basis of finite samples. Many factoring methods have been described in the literature, most being well summarized in such texts as Burt (1940), Thurstone (1947), Harman (1960) and Horst (1965). The majority of these methods can now be regarded as of historical interest only.

On a cursory examination of the present state of theory, it seems reasonable to suggest that of currently fashionable procedures (fashionable in theoretical writings, if not yet in practice), there are three especially notable methods among the proper iterative techniques. These are: principal factor analysis (PFA) (see, for example, Rao, 1955), canonical factor analysis (CFA) (due to Rao, 1955; see also Harris, 1962, for useful comments), and alpha factor analysis (AFA) (due to Kaiser & Caffrey, 1965). This paper is concerned principally with these three methods.

There are four main reasons for examining the three systems here, in what is partly an expository paper. The first is that the question of what, precisely, characterizes each system itself needs clarification.

The second reason is that, in terms of a recent treatment of weighted combinations of variables (McDonald, 1968), it is possible to give a unified account of these systems, which has already proved useful both in teaching and in theoretical work.

[1] Present address: Ontario Institute for Studies in Education, Toronto 5, Ontario, Canada.

The third reason is that it is possible to suggest a somewhat more straightforward derivation of each of the systems than is given in the standard accounts of them. In particular, Rao's account of PFA and of CFA is subject to formal difficulties.

The fourth reason relates to recent work by Harris. Harris (1967) has argued that in the present state of theoretical knowledge about factor analysis, it is desirable in practice to employ several different factoring methods on any given data matrix, and to accept only those factors that are obtained independently of method. In contrast, Harris (1967), and also McDonald & Burr (1967), show that a purely theoretical analysis of the factor score problem leads to a reasonably definitive choice of one method (Bartlett, 1937) of factor score construction, over competitive methods. The points of comparison made below, on the factoring problem itself, while not yet pointing to a definitive choice of method, at least indicate the further research needed if one is to avoid costly and wasteful proliferation of computing procedures.

While the central objective here is to examine PFA, CFA and AFA, it is convenient to add remarks about principal component analysis (PCA). Reference will also be made to two procedures that are closely parallel, respectively, to CFA and AFA, namely weighting variables to maximize composite reliability, and obtaining the principal components of variables corrected for attentuation. (The temptation to label these, respectively, MCR and PCC, will be resisted!)

In this paper, problems relating to finite samples will not be considered directly. Except where otherwise stated, it is assumed that the population covariance matrix is known, and that the covariances are ' explained ' precisely by a known minimum number of common factors.

2. The Foundations of the Models

Let $\mathbf{y}' = [y_1, \ldots, y_n]$ be a vector of real valued random variables, each component of which has a mean equal to zero, and finite, non-zero variance. Write

$$\mathbf{y} = \mathbf{c} + \mathbf{s} + \mathbf{e}, \tag{1}$$

where

$$\mathbf{c} = \mathbf{F}\mathbf{x}, \tag{2}$$

and

$$\mathscr{E}\{\mathbf{y}\} = \mathscr{E}\{\mathbf{c}\} = \mathscr{E}\{\mathbf{s}\} = \mathscr{E}\{\mathbf{e}\} = \mathbf{0}. \tag{3}$$

The matrix $\mathbf{F}$ is a $n \times r$ (non-random) matrix of *common factor loadings*, of rank $r < n$, i.e. a *factor pattern*. The *orthogonal* common factor model is assumed, in which the vector of (random) *common factor scores* $\mathbf{x}' = [x_1, \ldots, x_r]$, has covariance matrix $\mathscr{E}\{\mathbf{x}\mathbf{x}'\} = \mathbf{I}$. The common factor scores $\mathbf{x}$, the *specific factor scores* $\mathbf{s}' = [s_1, \ldots, s_n]$, and the errors $\mathbf{e}' = [e_1, \ldots, e_n]$, are assumed mutually orthogonal, that is

$$\mathscr{E}\{\mathbf{xs}'\} = \mathbf{0}, \tag{4}$$

$$\mathscr{E}\{\mathbf{xe}'\} = \mathbf{0} \tag{5}$$

and

$$\mathscr{E}\{\mathbf{se}'\} = \mathbf{0}, \tag{6}$$

so that by eqn. (2),

$$\mathscr{E}\{\mathbf{cs}'\} = \mathscr{E}\{\mathbf{ce}'\} = \mathbf{0}, \tag{7}$$

that is, every component of the *common part* $\mathbf{c}$ is orthogonal to every component of the *specific factor* $\mathbf{s}$, and to every component of the *error* $\mathbf{e}$.

Let

$$\mathbf{t} = \mathbf{c} + \mathbf{s}, \tag{8}$$

and

$$\mathbf{u} = \mathbf{s} + \mathbf{e}, \tag{9}$$

that is to say, $\mathbf{t}$ is what is usually called the (vector) *true score*, and $\mathbf{u}$ the *unique factor*, and

$$\mathbf{y} = \mathbf{t} + \mathbf{e} \tag{10}$$

$$\mathbf{y} = \mathbf{c} + \mathbf{u}. \tag{11}$$

Define the covariance matrices

$$\mathbf{C} = \mathscr{E}\{\mathbf{yy}'\}, \tag{12}$$

$$\mathbf{C}_c = \mathscr{E}\{\mathbf{cc}'\} = \mathbf{FF}', \tag{13}$$

$$\mathbf{C}_t = \mathscr{E}\{\mathbf{tt}'\}, \tag{14}$$

$$\mathbf{S}^2 = \mathscr{E}\{\mathbf{ss}'\}, \tag{15}$$

$$\mathbf{E}^2 = \mathscr{E}\{\mathbf{ee}'\}, \tag{16}$$

$$\mathbf{U}^2 = \mathscr{E}\{\mathbf{uu}'\} = \mathbf{S}^2 + \mathbf{E}^2. \tag{17}$$

In the classical common factor model, it is assumed that $\mathbf{S}^2$, $\mathbf{E}^2$ and $\mathbf{U}^2$ are diagonal matrices, at least non-negative definite, and in general positive definite. (This is in contrast to certain generalizations on common factor analysis, suggested by McDonald, 1968.)

Define, further, the matrices

$$\mathbf{V}^2 = \text{Diag}\ \{\mathbf{C}\}, \tag{18}$$

$$\mathbf{H}^2 = \text{Diag}\ \{\mathbf{C}_c\}, \tag{19}$$

$$\mathbf{T}^2 = \text{Diag}\ \{\mathbf{C}_t\} = \mathbf{H}^2 + \mathbf{S}^2, \tag{20}$$

and

$$\mathbf{C}_0 = \mathbf{C} - \text{Diag}\ \{\mathbf{C}\}. \tag{21}$$

Then $\mathbf{C}$ is the covariance matrix of the variables; $\mathbf{C}_c$ is the covariance matrix of their common parts; $\mathbf{C}_t$, of their true parts; $\mathbf{S}^2$, of their specifics; $\mathbf{E}^2$, of their errors; $\mathbf{U}^2$, of their unique parts. Further, $\mathbf{V}^2$ is a diagonal matrix of observed variances; $\mathbf{H}^2$ of *common variances*, or *communalities*; $\mathbf{T}^2$, of *true variances*. Lastly, the matrix $\mathbf{C}_0$ contains the observed covariances, with zeros replacing variances in the diagonal.

From the above, the following relations are easily obtained:

$$\mathbf{C}=\mathbf{C}_0+\mathbf{H}^2+\mathbf{S}^2+\mathbf{E}^2, \tag{22}$$

$$\mathbf{C}=\mathbf{C}_c+\mathbf{U}^2=\mathbf{F}\mathbf{F}'+\mathbf{U}^2, \tag{23}$$

$$\mathbf{C}=\mathbf{C}_0+\mathbf{T}^2+\mathbf{E}^2=\mathbf{C}_t+\mathbf{E}^2. \tag{24}$$

Unless otherwise stated, it is assumed in the sequel that $\mathbf{C}$, $\mathbf{C}_t$ are positive definite, that $\mathbf{C}_c$ is non-negative definite of rank $r<n$, and that $\mathbf{S}^2$, $\mathbf{E}^2$, $\mathbf{U}^2$, $\mathbf{V}^2$, $\mathbf{H}^2$, $\mathbf{T}^2$ are positive definite (diagonal) matrices.

Consider the weighted combination of variables

$$\eta=\mathbf{w}'\mathbf{y}, \tag{25}$$

where $\mathbf{w}'=[w_1, \ldots, w_n]$ is a non-null vector of real weights. By eqns. (1), (10) and (11),

$$\eta=\eta_c+\eta_s+\eta_e, \tag{26}$$

$$\eta=\eta_c+\eta_u, \tag{27}$$

$$\eta=\eta_t+\eta_e, \tag{28}$$

where

$$\eta_c=\mathbf{w}'\mathbf{c}, \tag{29}$$

$$\eta_s=\mathbf{w}'\mathbf{s}, \tag{30}$$

$$\eta_e=\mathbf{w}'\mathbf{e}, \tag{31}$$

$$\eta_u=\mathbf{w}'\mathbf{u}, \tag{32}$$

$$\eta_t=\mathbf{w}'\mathbf{t}. \tag{33}$$

It follows that

$$\text{var}\{\eta\}=\mathbf{w}'\mathbf{C}\mathbf{w}, \tag{34}$$

$$\text{var}\{\eta_c\}=\mathbf{w}'\mathbf{C}_c\mathbf{w}, \tag{35}$$

$$\text{var}\{\eta_u\}=\mathbf{w}'\mathbf{U}^2\mathbf{w}, \tag{36}$$

$$\text{var}\{\eta_t\}=\mathbf{w}'\mathbf{C}_t\mathbf{w}. \tag{37}$$

Now let $\mathbf{D}^2$ be a diagonal, positive definite $n\times n$ matrix, which may be identified as required with one of $\mathbf{V}^2$, $\mathbf{U}^2$, $\mathbf{H}^2$, $\mathbf{S}^2$, $\mathbf{T}^2$ or $\mathbf{E}^2$. Without making a specific identification of $\mathbf{D}^2$, it is easy to develop a 'generalized' factoring procedure, which has certain desirable properties.

Consider the ratio of quadratic forms

$$\lambda=(\mathbf{w}'\mathbf{C}_c\mathbf{w})/(\mathbf{w}'\mathbf{D}^2\mathbf{w}). \tag{38}$$

Since $\mathbf{C}_c$ is non-negative definite, of rank $r<n$, and $\mathbf{D}^2$ is positive definite, λ has in general r non-zero stationary values, with respect to non-null $\mathbf{w}$. One procedure for obtaining these is to write

$$\mathbf{q}=\mathbf{D}\mathbf{w}, \tag{39}$$

so that eqn. (38) becomes

$$\lambda=(\mathbf{q}'\mathbf{D}^{-1}\mathbf{C}_c\mathbf{D}^{-1}\mathbf{q})/\mathbf{q}'\mathbf{q}. \tag{40}$$

It is well known that the r non-zero stationary values of λ, say $\lambda_1, \ldots, \lambda_r$, are the r non-zero eigenvalues of the matrix $\mathbf{D}^{-1}\mathbf{C}_c\mathbf{D}^{-1}$, and the stationary points are given by the corresponding r eigenvectors, say $\mathbf{q}_1, \ldots, \mathbf{q}_r$. Let

$$\mathbf{\Lambda} = \text{Diag}\{\lambda_1, \ldots, \lambda_r\},$$

where the eigenvalues, in general distinct, are arranged in descending order of magnitude, and $\mathbf{Q} \equiv [\mathbf{q}_1, \ldots, \mathbf{q}_r]$, so that

$$\mathbf{Q}'\mathbf{Q} = \mathbf{I}. \tag{41}$$

Then

$$\mathbf{D}^{-1}\mathbf{C}_c\mathbf{D}^{-1} = \mathbf{Q\Lambda Q}', \tag{42}$$

so that

$$\mathbf{C}_c = \mathbf{DQ\Lambda Q'D}. \tag{43}$$

Of the possible matrices $\mathbf{F}$ that satisfy eqn. (23), choose

$$\mathbf{F} = \mathbf{DQ\Lambda}^{1/2}, \tag{44}$$

so that

$$\mathbf{C}_c = \mathbf{FF}', \tag{45}$$

consistent with eqn. (23).

Note that, by eqns. (41) and (44),

$$\mathbf{F}'\mathbf{D}^{-2}\mathbf{F} = \mathbf{\Lambda} \tag{46}$$

is a diagonal matrix. For each appropriate choice of $\mathbf{D}$, this property fixes $\mathbf{F}$ with respect to rotation, i.e. with respect to $r \times r$ orthonormal transformations, provided that the eigenvalues are all distinct.

Note, further, that

$$\begin{aligned}\mathscr{E}\{\mathbf{yx}'\} &= \mathscr{E}\{(\mathbf{Fx}+\mathbf{s}+\mathbf{e})\mathbf{x}'\}\\ &= \mathbf{F},\end{aligned} \tag{47}$$

so that by eqn. (18) the correlation of each of the n variables y_j $(j = 1, \ldots, n)$, with each of the common factor scores x_p $(p = 1, \ldots, r)$ is given by the (j, p)th element of the matrix

$$\mathbf{G} \equiv \mathbf{V}^{-1}\mathbf{F} = \mathbf{V}^{-1}\mathbf{DQ\Lambda}^{1/2}. \tag{48}$$

In Thurstone's classical terminology, $\mathbf{G}$ is the *factor structure*.

Further, the common factor scores $\mathbf{x}$ are given, in terms of the common parts $\mathbf{c}$ of the variables, by

$$\mathbf{x} = \mathbf{D}^{-1}\mathbf{Q}'\mathbf{\Lambda}^{-1/2}\mathbf{c}. \tag{49}$$

By a specialization of theory given by McDonald (1968), it may be shown that the stationary values of λ, and, more importantly, the factor structure $\mathbf{G}$, and the factor scores $\mathbf{x}$, are invariant under any rescaling of any or all variables by multiplication by constants (other than zero), i.e. under any arbitrary transformation of the observed variables by a non-singular diagonal matrix. These desirable properties, taken together, will be regarded as defining *scale invariance*, aspects of which have been noted by Kaiser & Caffrey (1965) as properties of

CFA and of AFA. More generally, scale invariance is a property of any results based on eqns. (38) to (49).

At this point, by making the appropriate identification of $\mathbf{D}^2$, it is easy to set down the special cases represented by PFA, CFA and AFA. These may be summarized, in an obvious notation, as follows:

$$\lambda_p = (\mathbf{w}'\mathbf{C}_c\mathbf{w})/\mathbf{w}'\mathbf{V}^2\mathbf{w}), \tag{38p}$$

$$\lambda_c = (\mathbf{w}'\mathbf{C}_c\mathbf{w})/(\mathbf{w}'\mathbf{U}^2\mathbf{w}), \tag{38c}$$

$$\lambda_a = (\mathbf{w}'\mathbf{C}_c\mathbf{w})/(\mathbf{w}'\mathbf{H}^2\mathbf{w}), \tag{38a}$$

yielding

$$\lambda_p = (\mathbf{q}'\mathbf{V}^{-1}\mathbf{C}_c\mathbf{V}^{-1}\mathbf{q})/\mathbf{q}'\mathbf{q}, \tag{40p}$$

$$\lambda_c = (\mathbf{q}'\mathbf{U}^{-1}\mathbf{C}_c\mathbf{U}^{-1}\mathbf{q})/\mathbf{q}'\mathbf{q}, \tag{40c}$$

$$\lambda_a = (\mathbf{q}'\mathbf{H}^{-1}\mathbf{C}_c\mathbf{H}^{-1}\mathbf{q})/\mathbf{q}'\mathbf{q}, \tag{40a}$$

where

$$\mathbf{F}_p = \mathbf{V}\mathbf{Q}_p\mathbf{\Lambda}_p^{1/2}, \tag{44p}$$

$$\mathbf{F}_c = \mathbf{U}\mathbf{Q}_c\mathbf{\Lambda}_p^{1/2}, \tag{44c}$$

$$\mathbf{F}_a = \mathbf{H}\mathbf{Q}_a\mathbf{\Lambda}_a^{1/2}, \tag{44a}$$

with

$$\mathbf{F}'_p\mathbf{V}^{-2}\mathbf{F}_p = \mathbf{\Lambda}_p, \tag{46p}$$

$$\mathbf{F}'_c\mathbf{U}^{-2}\mathbf{F}_c = \mathbf{\Lambda}_c, \tag{46c}$$

$$\mathbf{F}'_a\mathbf{H}^{-2}\mathbf{F}_a = \mathbf{\Lambda}_a, \tag{46a}$$

also

$$\mathbf{G}_p = \mathbf{V}^{-1}\mathbf{F}_p = \mathbf{Q}_p\mathbf{\Lambda}_p^{1/2}, \tag{48p}$$

$$\mathbf{G}_c = \mathbf{V}^{-1}\mathbf{F}_c = (\mathbf{V}^{-1}\mathbf{U})\mathbf{Q}_c\mathbf{\Lambda}_c^{1/2}, \tag{48c}$$

$$\mathbf{G}_a = \mathbf{V}^{-1}\mathbf{F}_a = (\mathbf{V}^{-1}\mathbf{H})\mathbf{Q}_a\mathbf{\Lambda}_a^{1/2},. \tag{48a}$$

and, finally,

$$\mathbf{x} = \mathbf{V}^{-1}\mathbf{Q}'_p\mathbf{\Lambda}_p^{-1/2}\mathbf{c}, \tag{49p}$$

$$\mathbf{x} = \mathbf{U}^{-1}\mathbf{Q}'_c\mathbf{\Lambda}_c^{-1/2}\mathbf{c}, \tag{49c}$$

$$\mathbf{x} = \mathbf{H}^{-1}\mathbf{Q}'_a\mathbf{\Lambda}_a^{-1/2}\mathbf{c}. \tag{49a}$$

The alternative interpretations of $\mathbf{D}^2$, above, were introduced arbitrarily, with no indication of motive. Indeed, a variety of motives can be offered for each of these. In the following sections, each of the methods is considered in detail. First, however, it is desirable for reference in the sequel, to set down the basic results of principal component analysis.

3. Principal Component Analysis

Consider the ratio of quadratic forms

$$\lambda_{pc} = (\mathbf{w}'\mathbf{C}\mathbf{w})/(\mathbf{w}'\mathbf{V}^2\mathbf{w}). \tag{38pc}$$

Using the notation and procedure of the previous section, this may be written

$$\lambda_{pc} = (\mathbf{q}'\mathbf{V}^{-1}\mathbf{C}\mathbf{V}^{-1}\mathbf{q})/\mathbf{q}'\mathbf{q}, \tag{40pc}$$

where $\mathbf{V}^{-1}\mathbf{C}\mathbf{V}^{-1}$ is the correlation matrix of the observed variables, which in general will be positive definite, with n non-zero eigenvalues, yielding, corresponding to eqn. (43),

$$\mathbf{C}=\mathbf{V}\mathbf{Q}_{pc}\mathbf{\Lambda}_{pc}\mathbf{Q}'_{pc}\mathbf{V}, \tag{43pc}$$

where, in this case, $\mathbf{\Lambda}_{pc}$, $\mathbf{Q}_{pc}$ are $n \times n$ matrices. Writing

$$\mathbf{F}_{pc}=\mathbf{V}\mathbf{Q}_{pc}\mathbf{\Lambda}_{pc}^{1/2}, \tag{44pc}$$

for the principal component 'loadings' yields in this case

$$\mathbf{G}_{pc}=\mathbf{V}^{-1}\mathbf{F}_{pc}=\mathbf{Q}_{pc}\mathbf{\Lambda}_{pc}^{1/2}, \tag{48pc}$$

for the correlations between observed variables and principal component scores, and

$$\mathbf{x}=\mathbf{V}^{-1}\mathbf{Q}'_{pc}\mathbf{\Lambda}_{pc}^{-1/2}\mathbf{y}, \tag{49pc}$$

for the principal component scores themselves.

The derivations of eqns. (38*pc*) to (49*pc*) are analogous to eqns. (38) to (49). However, we are not concerned here with a common factor model; hence, for example, $\mathbf{y}$ replaces $\mathbf{c}$ in (49*pc*).

The conventional treatments of principal component analysis either originate from (40*pc*), or else replace (38*pc*) by

$$\lambda_{pc}=(\mathbf{w}'\mathbf{C}\mathbf{w})/\mathbf{w}'\mathbf{w}. \tag{50}$$

In the former case, it is usually stated that the principal components of the correlation matrix, rather than those of the covariance matrix, are sought because, at least in psychological applications, the 'proper' scale of the observed variables is unknown. In the latter case, it is usually pointed out that the principal components of the covariance matrix are not independent of the scaling of the variables. However, the stationary values of λ_{pc}, the correlations $\mathbf{G}_{pc}$, given by eqn. (48*pc*), and the principal component scores given by eqn. (49*pc*), are certainly invariant under changes of scale.

4. Principal Factor Analysis

There are several alternative derivations of PFA. Certainly the most obvious is to begin with the correlation matrix, $\mathbf{R}=\mathbf{V}^{-1}\mathbf{C}\mathbf{V}^{-1}$, of the observed variables, rather than the covariance matrix, then seek a matrix $\mathbf{U}^2$ such that $\mathbf{R}-\mathbf{U}^2$ is of rank $r<n$, and obtain the principal components of $\mathbf{R}-\mathbf{U}^2$. It is generally held, as in the corresponding case of principal component analysis, that this procedure is *faute de mieux*, and is necessitated by ignorance of the 'proper' scale of the variables. Alternatively, one may seek $\mathbf{U}^2$ such that $\mathbf{C}-\mathbf{U}^2$ is of rank r, and obtain its principal components. In this case, the results are held to lack scale invariance, as discussed earlier. On the other hand, the treatment given in Section 2 makes it possible to say that PFA shares the property of scale invariance, as defined, with CFA and AFA.

Rao (1955) introduces PFA from a somewhat different starting point. The variance in any observed variable y_j, explained by a weighted combination of the common parts of all the variables, $\mathbf{w}'\mathbf{c}$, is equal to

$$[\text{cov}\{y_j, \mathbf{w}'\mathbf{c}\}]^2/\text{var}\{\mathbf{w}'\mathbf{c}\} \quad (j=1, \ldots, n).$$

Let

$$\boldsymbol{\gamma}=\mathbf{C}_c\mathbf{w} \tag{51}$$

be the vector of covariances, $\text{cov}\{y_j, \mathbf{w}'\mathbf{c}\}$ $(j=1, \ldots, n)$. Then the total variance in $y_1, \ldots, y_n$, explained by $\mathbf{w}'\mathbf{c}$, is

$$\mu=\boldsymbol{\gamma}'\boldsymbol{\gamma}/\text{var}\{\mathbf{w}'\mathbf{c}\}=(\mathbf{w}'\mathbf{C}^2{}_c\mathbf{w})/(\mathbf{w}'\mathbf{C}_c\mathbf{w}). \tag{52}$$

Rao (1955) suggests choosing $\mathbf{w}$ such that μ is a maximum. Differentiation of μ with respect to $\mathbf{w}$ yields the eigen equation

$$(\mathbf{C}_c{}^2-\mu\mathbf{C}_c)\mathbf{w}=\mathbf{0}, \tag{53}$$

in which $\mathbf{C}_c$ is, by hypothesis, singular, of rank $r<n$. The singularity of $\mathbf{C}_c$ places a formal difficulty in the way of further rigorous development on these lines. This difficulty, plus the matter of scale invariance, suggests that PFA is better introduced, for purposes of exposition, as in Section 2.

5. Canonical Factor Analysis

Rao (1955) obtains CFA by an application of canonical variate analysis. Let $\mathbf{w}'\mathbf{y}, \mathbf{w}_c{}'\mathbf{c}$ be weighted combinations respectively, of the observed variables $\mathbf{y}$, and of their common parts $\mathbf{c}$. Choose $\mathbf{w}$, and $\mathbf{w}_c$, non-null, to yield the stationary points of the squared correlation between $\mathbf{w}'\mathbf{y}$ and $\mathbf{w}'{}_c\mathbf{c}$, that is, of

$$\rho_1{}^2=\frac{(\mathbf{w}'\mathbf{C}_c\mathbf{w}_c)^2}{(\mathbf{w}'\mathbf{C}\mathbf{w})\,(\mathbf{w}'{}_c\mathbf{C}_c\mathbf{w}_c)}, \tag{54}$$

for $\mathbf{w}'{}_c\mathbf{C}_c\mathbf{w}_c\neq 0$. As in the case of Rao's treatment of PFA, the development of CFA from eqn. (54) is subject to a formal difficulty. Rao (1955) refers to his earlier treatment of a genetic problem (Rao, 1953), for the details of the algebra. In order to show, as Rao (1955, p. 103) asserts, that the vectors $\mathbf{w}$ and $\mathbf{w}_c$ are proportional, it is necessary to assume, contrary to hypothesis, that $\mathbf{C}_c$ is non-singular.

To avoid these difficulties, it seems natural to assume from the outset a special case of canonical variate analysis, in which the stationary points of the squared correlation between *identically* weighted combinations, $\mathbf{w}'\mathbf{y}$ and $\mathbf{w}'\mathbf{c}$, are sought, i.e. the stationary points of

$$\rho_2{}^2=\frac{(\mathbf{w}'\mathbf{C}_c\mathbf{w})^2}{(\mathbf{w}'\mathbf{C}\mathbf{w})\,(\mathbf{w}'\mathbf{C}_c\mathbf{w})}=\frac{\mathbf{w}'\mathbf{C}_c\mathbf{w}}{\mathbf{w}'\mathbf{C}\mathbf{w}}, \tag{55}$$

for $\mathbf{w}'\mathbf{C}_c\mathbf{w} \neq 0$. Since, by eqn. (23),

$$\frac{\mathbf{w}'\mathbf{C}_c\mathbf{w}}{\mathbf{w}'\mathbf{C}\mathbf{w}} = 1\Big/\left(1 + \frac{\mathbf{w}'\mathbf{C}_c\mathbf{w}}{\mathbf{w}'\mathbf{U}^2\mathbf{w}}\right), \tag{56}$$

that is

$$\rho_2{}^2 = 1/(1+\lambda_c), \tag{57}$$

this problem is an equivalent to the starting point for CFA used in Section 2.

In the treatment in Section 2, the 'natural' interpretation of CFA is to say that we seek weights yielding stationary points of the ratio of the variance due to the common parts of the variables, to the variance due to their unique parts, or, what is the same thing, the ratio of total variance to unique variance, since

$$\frac{\mathbf{w}'\mathbf{C}_c\mathbf{w}}{\mathbf{w}'\mathbf{U}^2\mathbf{w}} = \frac{\mathbf{w}'\mathbf{C}\mathbf{w}}{\mathbf{w}'\mathbf{U}^2\mathbf{w}} - 1. \tag{58}$$

Both these interpretations seem as 'natural' as the canonical correlation interpretation, and have the advantage over the latter of being readily generalized to situations of the ***interbattery*** factor analysis type (see McDonald, 1968). It is therefore suggested that the better conceptual basis of CFA is that of Section 2.

It is also appropriate at this point to indicate, as is well known, that Lawley (1940), in his development of maximum-likelihood estimators for the common factor model, found it convenient to impose a basis in the common factor space in which the sample estimates obey a restriction which is the sample analogue of eqn. (46*c*). While Lawley was not concerned initially to build this restriction into the model for the population, if we regard CFA as defined in terms of the stationary points of eqn. (58) or its equivalents, and hence necessarily having the property (46*c*), it seems appropriate, as some investigators do, to attribute CFA to both Lawley and Rao, regarding the former as responsible for the statistical treatment, and the latter as responsible for clarification of the properties of the model itself.

6. Alpha Factor Analysis

In Section 2, the equations of AFA were based on the stationary points of the variance ratio given in eqn. (38*a*), whence of eqn. (40*a*), viz.

$$\lambda_a = \mathbf{q}'\mathbf{H}^{-1}\mathbf{C}_c\mathbf{H}^{-1}\mathbf{q}/\mathbf{q}'\mathbf{q}.$$

In this form, the matrix $\mathbf{H}^{-1}\mathbf{C}_c\mathbf{H}^{-1}$ can be recognized as the correlation matrix of the common parts of the observed variables, so that AFA can be described as yielding the principal components of the correlation matrix of the common parts.

Kaiser & Caffrey (1965) introduce AFA by considering the stationary points of coefficient alpha, for the common parts, i.e.

$$\alpha=\frac{n}{n-1}\left(1-\frac{\mathbf{w}'\mathbf{H}^2\mathbf{w}}{\mathbf{w}'\mathbf{C}_c\mathbf{w}}\right). \tag{59}$$

By eqns. (38*a*) and (59),

$$\alpha=\frac{n}{n-1}\left(\frac{\lambda_a-1}{\lambda_a}\right), \tag{60}$$

whence it may be seen that, since λ_a has r non-zero stationary values, and $m=n-r$ zeros, correspondingly, α has r finite stationary values, and m (negative) infinite stationary values. Of the r finite extrema of α, some number $s \leqslant r$ will be positive, where s is the positive index of $\mathbf{C}_0$, since by eqns. (23) and (59),

$$\alpha=\frac{n}{n-1}\left(\frac{\mathbf{w}'\mathbf{C}_0\mathbf{w}}{\mathbf{w}'\mathbf{C}_c\mathbf{w}}\right). \tag{61}$$

Further, the positive index of $\mathbf{C}_0$ is the same as the number of eigenvalues of the correlation matrix of the observed variables that are greater than unity, since

$$\mathbf{C}_0=\mathbf{V}(\mathbf{R}-\mathbf{I})\mathbf{V}, \tag{62}$$

and $\mathbf{V}$ is positive definite.

Kaiser & Caffrey interpret coefficient alpha as a coefficient of generalizability, which by definition must be positive. (The derivation of alpha as a coefficient of generalizability, and as a lower bound to reliability, is considered in an appendix to this paper.) On this ground, they recommend that in AFA, the common factors retained in the analysis be just those s factors that correspond to values of alpha greater than zero. It is suggested in the appendix that the interpretation of alpha as a coefficient of generalizability has paradoxical consequences. A further consequence is that if we are given a covariance matrix $\mathbf{C}$ and have chosen a matrix $\mathbf{U}^2$ such that $\mathbf{C}_c=\mathbf{C}-\mathbf{U}^2$ is non-negative definite, of minimum rank r, then $s \leqslant r$ of the r common factors ‘ explaining ’ the covariance have positive alpha. Equality is attained if and only if the matrix $\mathbf{H}^{-1}\mathbf{C}_0\mathbf{H}^{-1}$ has no eigenvalue in the interval bounded by zero and minus unity, since

$$\mathbf{C}_c=\mathbf{C}_0+\mathbf{H}^2=\mathbf{H}(\mathbf{H}^{-1}\mathbf{C}_0\mathbf{H}^{-1}+\mathbf{I})\mathbf{H}, \tag{63}$$

and $\mathbf{H}$ is positive definite.

Kaiser & Caffrey state that if there exists $\mathbf{U}^2$ such that $\mathbf{C}_c$ is of rank r, the results of AFA are related to those of CFA, or other factoring methods, by an $r \times r$ transformation. But this can be true, in terms of their decision rule for the number of factors, only if it be the case that no eigenvalue of $\mathbf{H}^{-1}\mathbf{C}_0\mathbf{H}^{-1}$ lies in the interval bounded by zero and minus unity. It is readily verified that this is the same as the condition that would be required in relation to Guttman's well-known lower bound (Guttman, 1954), to the number of common factors, for the bound to be attained.

7. Some Parallel Problems

There are two well-known weighting methods for the case where the error variances of the observed variables, or, what is the same thing, their reliabilities, are known. These are closely related to CFA and AFA respectively. Eqns. (24), (34) and (37) yield, for the reliability of a weighted combination of $y_1, \ldots, y_n$, the expression

$$\rho_{\eta\eta} \leqslant \rho^2_{\eta, \eta_t} = \frac{\mathbf{w}'\mathbf{C}_t\mathbf{w}}{\mathbf{w}'\mathbf{C}\mathbf{w}} = 1 - \frac{\mathbf{w}'\mathbf{E}^2\mathbf{w}}{\mathbf{w}'\mathbf{C}\mathbf{w}}, \tag{64}$$

where ρ_{η, η_t} is the correlation between $\eta = \mathbf{w}'\mathbf{y}$ and its true part $\eta_t = \mathbf{w}'\mathbf{t}$. Compare eqn. (64) with (55), and note that $\mathbf{C}_t - \mathbf{C}_c = \mathbf{S}^2$ is non-negative definite, so that

$$\rho_{\eta\eta} \geqslant \rho_2^2 \tag{65}$$

for every non-null $\mathbf{w}$, with equality if and only if $\mathbf{S}^2 = 0$, i.e. if the variables have no specific components other than their errors. The choice of weights to maximize the reliability of a weighted combination of variables, as given by eqn. (64), has been considered by Thomson (1940), Mosier (1943), Peel (1948) and Green (1950). Here, it may be noted that the CFA weights yield stationary points for what is a lower bound to reliability, suitable for this purpose in the absence of information about the error variances of the variables.

In contrast, from eqn. (24) it follows that

$$\mathbf{R}_c = \mathbf{T}^{-1}\mathbf{C}_t\mathbf{T}^{-1}, \tag{66}$$

is the correlation matrix of the observed variables after correction for attenuation. The eigenvalues of $\mathbf{R}_c$ are the same as the stationary values of

$$\lambda_t = \frac{\mathbf{w}'\mathbf{C}_t\mathbf{w}}{\mathbf{w}'\mathbf{T}^2\mathbf{w}} = 1 + \frac{\mathbf{w}'\mathbf{C}_0\mathbf{w}}{\mathbf{w}'\mathbf{T}^2\mathbf{w}}. \tag{67}$$

Compare eqn. (67) with (38*a*), and note thence that

$$\lambda_a \geqslant \lambda_t, \tag{68}$$

for every non-null $\mathbf{w}$, with equality if and only if $\mathbf{T}^2 - \mathbf{H}^2 = \mathbf{0}$, i.e. $\mathbf{S}^2 = 0$, as in eqn. (65). Thus the AFA weights yield stationary points for what is an upper bound to the magnitude of the components of a correlation matrix, corrected for attenuation.

A well-known defect of the correction for attenuation procedure is that it tends to give maximum weights to variables with the largest error variance. In the limit, it gives infinite weight to a variable whose variance is entirely due to error, and whose true variance is zero. This is true, *a fortiori*, of AFA, since the common variance of a variable is part of its true variance. In AFA, it would seem that common factors are maximally represented by those variables which have the smallest communalities. Kaiser & Caffrey (1965) note, indeed, that variables with zero communalities must be deleted before the analysis is commenced, so that $\mathbf{H}^2$ will be positive definite.

In contrast, the weighting procedure for maximizing reliability, and the CFA procedure, give minimum weight to variables with the largest error variance, or largest unique variance (error plus specific). This would seem desirable for most purposes.

The ' general ' factoring procedure of eqns. (38) to (49) suggests further possibilities, in the case where the error variances, $\mathbf{E}^2$, hence true variances, $\mathbf{T}^2$, are known. Thus, one might consider factoring procedures based on either

$$\lambda_e = (\mathbf{w}'\mathbf{C}_c\mathbf{w})/(\mathbf{w}'\mathbf{E}^2\mathbf{w}), \tag{69}$$

or

$$\lambda_t = (\mathbf{w}'\mathbf{C}_c\mathbf{w})/(\mathbf{w}'\mathbf{T}^2\mathbf{w}), \tag{70}$$

thence proceeding as in eqns. (38) to (49).

While neither of these variants need be taken too seriously, the first at least is not unattractive. That is, for some purposes it may seem desirable to choose a basis in the common factor space that maximizes the ratio of common variance to (known) error variance.

8. Further Discussion

The analysis up to this point has been predicated on the assumptions that the population covariance matrix of the observed variables is known, that it can be precisely ' explained ' by a known minimum number of factors, with a precisely given matrix of uniquenesses. In practice, none of these assumptions would hold. They are, however, essential to a proper examination of the theoretical foundations of the procedures. At this purely theoretical level, a number of considerations have emerged in terms of which PFA, CFA and AFA can be critically compared.

In effect, each of the three procedures can be described as a scale invariant system, based on ratios of quadratic forms (variance ratios) which differ only in the diagonal matrix chosen for the denominator. Of the three, CFA, based on the ratio of common variance to unique variance, seems the most ' natural ' choice. It systematically gives most weight to variables having the smallest uniqueness (largest communality). The reverse is true of AFA, while PFA can be considered intermediate, neither systematically weighting according to communality, nor according to uniqueness. Alternative derivations of PFA, CFA and AFA, in the literature, have been seen to be subject to conceptual difficulties.

Conceptual difficulties, mainly set out in the appendix, have also been noted with regard to the interpretation of coefficient alpha as a coefficient of generalizability. This matter requires further theoretical analysis. Meanwhile, it is at least clear that Kaiser & Caffrey's decision rule for the number of common factors, which is the same as Guttman's weakest lower bound to the number of factors, in general will not yield a non-negative definite reduced covariance matrix *in the case where it is known* that a certain number of factors will ' explain ' the covariances precisely.

Further, at this purely theoretical level, other advantages of CFA over both PFA and AFA can be indicated. McDonald & Burr (1967) have shown that the CFA basis is unique in the properties conferred on constructed ' estimates ' of factor scores (see also Harris, 1967). If and only if the basis given by CFA is chosen, constructed factor scores given by the method of Bartlett (1937) are mutually orthogonal, orthogonal to non-corresponding true factor scores in the model, and have (relative) maximum correlations with the true factors, for each factor in turn. If one adds to this much the well-known fact, mentioned earlier, that the equations of CFA, in sample analogue form, satisfy the maximum likelihood equations of Lawley (1940), it seems possible to say unequivocally that CFA is the best procedure for almost all applications.

However, as soon as the assumptions made for the purposes of a theoretical analysis are abandoned, the issue is no longer clear. Thus, still supposing for the present that the population covariance matrix is known, and that $\mathbf{U}^2$ exists such that $\mathbf{C}_c = \mathbf{C} - \mathbf{U}^2$ is of known minimum rank r, let us now suppose that $\mathbf{U}^2$ is unknown, or what is the same thing, $\mathbf{H}^2$ is unknown. Immediately, questions would need to be raised as to whether $\mathbf{U}^2$ is unique, and determinable. It will be sufficient for the present purposes to make some remarks concerning iterative algorithms for the determination of $\mathbf{U}^2$, given $\mathbf{C}$ and r.

The point seems to deserve emphasis that there is no necessary connection between factor models such as PFA, CFA, and AFA, and the iterative algorithms for determining $\mathbf{U}^2$ (or $\mathbf{H}^2$) commonly associated with them. Rather, PFA, CFA and AFA should each be identified by one sufficient characteristic, say, their distinct (in general) and unique bases in common factor space, or any mathematical equivalent.

Note first that the procedure which Harman (1960, p. 89) has termed *iteration by refactoring* seems equally applicable, in principle, to any of the models considered here. In terms of the ' general ' factoring procedure indicated by eqns. (38) to (49), it amounts to obtaining, at the kth iteration,

$$\mathbf{C} = \mathbf{F}^{(k)}\mathbf{F}^{(k)\prime} + \mathbf{U}^{(k)^2} = D^{(k)}Q^{(k)}\Lambda^{(k)}Q^{(k)\prime}D^{(k)} + U^{(k)^2}, \tag{71}$$

where the quantities $F^{(k)}$, $D^{(k)}$, $Q^{(k)}$, $\Lambda^{(k)}$ are analogues of those in eqns. (43) and (44), and writing

$$\mathbf{H}^{(k+1)^2} = \mathrm{Diag}\{F^{(k)}F^{(k)\prime}\} = \mathrm{Diag}\{D^{(k)}Q^{(k)}\Lambda^{(k)}Q^{(k)\prime}D^{(k)\prime}\}, \tag{72}$$

or, what is the same thing,

$$\mathbf{U}^{(k+1)^2} = V^2 - \mathrm{Diag}\{F^{(k)}F^{(k)\prime}\}. \tag{73}$$

In PFA, $\mathbf{D}^{(k)}=\mathbf{V}$ is a constant, so that

$$\mathbf{H}^{(k+1)^2}=\mathbf{V}^2\,\mathrm{Diag}\{Q_p^{(k)}\Lambda_p^{(k)}Q_p^{(k)\prime}\}, \tag{72p}$$

and

$$\mathbf{U}^{(k+1)^2}=\mathbf{V}^2[\mathbf{I}-\mathrm{Diag}\{Q_p^{(k)}\Lambda_p^{(k)}Q_p^{(k)\prime}\}]. \tag{73p}$$

In CFA, $\mathbf{D}^{(k)}=\mathbf{U}^{(k)}$, so that

$$\mathbf{H}^{(k+1)^2}=\mathbf{U}^{(k)^2}\,\mathrm{Diag}\{Q_c^{(k)}\Lambda_c^{(k)}Q_c^{(k)\prime}\}, \tag{72c}$$

and

$$\mathbf{U}^{(k+1)^2}=\mathbf{V}^2-\mathbf{U}^{(k)^2}\,\mathrm{Diag}\{Q_c^{(k)}\Lambda_c^{(k)}Q_c^{(k)\prime}\}. \tag{73c}$$

In AFA, $\mathbf{D}^{(k)}=\mathbf{H}^{(k)}$, so that

$$\mathbf{H}^{(k+1)^2}=\mathbf{H}^{(k)^2}\,\mathrm{Diag}\{Q_a^{(k)}\Lambda_a^{(k)}Q_a^{(k)\prime}\}. \tag{72a}$$

It is not known under what conditions the sequence $\mathbf{H}^{(1)^2}$, $\mathbf{H}^{(2)^2}$, ..., or $\mathbf{U}^{(1)^2}$, $\mathbf{U}^{(2)^2}$, ..., will converge on the desired matrices.

In the case of PFA, Wrigley (cf. Harman, 1960, p. 89) has extensively tested the above procedure with apparently satisfactory results. Hemmerle (1965, p. 302) reports that it ' has converged in every instance in which it was employed '. This is also my own experience.

In the case of CFA, Rao (1955) suggested two iterative algorithms which are quite distinct from the above. The first of these amounts to obtaining, at the kth iteration,

$$\mathbf{U}^{(k+1)^{-2}}=\mathbf{V}^{-2}[\mathbf{I}+\mathrm{Diag}\{Q_c^{(k)}\Lambda_c^{(k)}Q_c^{(k)\prime}\}]. \tag{74}$$

The second is a modification of this, which is recommended as speeding up the process in the early stages.

There seems to be a growing belief, difficult to document from the published literature, that CFA tends to be badly behaved with respect to iterations. It would be unfortunate if this belief led to an unthinking rejection of CFA itself, in favour of alternatives that lack its ' nice ' conceptual properties. The point here is that there is no necessary connection between CFA and the iterative algorithms suggested by Rao in the context of CFA. Indeed, Hemmerle (1965) obtained better results (faster convergence) with his Method C, which can be identified with the CFA application above (eqn. (73c)) of the general scheme (eqn. (73)), than with Rao's two methods, or that of Lawley (1942). At present it seems likely that the best treatment of sample data for CFA is that recommended by Jöreskog (1967).

In the case of AFA, Kaiser & Caffrey (1965) recommend an iterative algorithm that can be identified with eqn. (72a), i.e. the special case for AFA of the general scheme (72). Harris (1967), at least, has reported one case where this procedure has failed to converge. It would seem, with this and the other two models, that an extensive programme of testing of available algorithms (preferably with constructed data of precisely known structure) is desirable, at least in default of a definitive theoretical analysis.

The next step away from the basic conceptual analysis above is when we suppose that the population covariance matrix **C** is known, and that $\mathbf{U}^2$ exists such that $\mathbf{C}-\mathbf{U}^2$ is non-negative definite of *unknown* minimum rank r. From the remarks in Section 6, it should at least be evident that Kaiser & Caffrey's decision rule for the number of common factors will in general yield $s<r$ common factors, and the $\mathbf{C}-\mathbf{U}^2$ so obtained will not be non-negative definite. In principle, given a well-behaved iterative algorithm, any of the three systems should yield $\mathbf{U}^2$, and the smallest r for $\mathbf{C}-\mathbf{U}^2$ non-negative definite, if we try a sequence of values of r, say increasing from s, regarded, as in Guttman's work, as a lower bound to r.

From one point of view, it might be urged that (except for the question of a decision-rule for the number of factors), the whole question of choosing between PFA, CFA and AFA is trivial at the conceptual level, since the results from any of these should be related to those from any other by an $r \times r$ orthonormal transformation (and since most investigators will rotate the factors anyway, before interpreting). In practice, however, the situation is that we do not have a covariance matrix that is precisely explained by $r \ll n$ common factors. In practice, then, we do not expect that for r non-trivially less than the number of variables, the $n-r$ smallest eigenvalues of $\mathbf{D}^{-1}\mathbf{C}_c\mathbf{D}^{-1}$ as in eqn. (41), will be zero. Rather, as Kaiser & Caffrey (1965, p. 8) point out: ' while CFA obtains a least-squares solution for the residuals from the original intercorrelations, each weighted inversely by the product of their associated unique part standard deviations, AFA obtains a least-squares solution for the residuals from the original intercorrelations, each weighted inversely by the product of their associated common part standard deviations '. (And it may be added, PFA uses equally weighted residuals.) The further implication of this is that CFA treats unique variation as quasi-error variation, and thence follows classical statistical weighted least squares procedures, whereas AFA does not. This is a further aspect of the properties indicated by the variance ratios in eqns. (38*c*), (38*a*), and the parallel cases dealing with errors, in Section 7. In principle, on these grounds, the CFA weighting has a better rational basis than the AFA weighting.

In choosing a factor method, there are, in fact, at least three separate choices to be made, and these are relatively independent. The first is the choice of basis in the common factor space. This is the clearest defining characteristic of the three systems discussed. The second is the choice of an iterative algorithm for the determination of communalities/uniquenesses. The third is the decision rule for the number of common factors. It seems natural to suggest a double criterion, most effectively applicable to CFA. That is, the r factors should be statistically significant, and the eigenvalue corresponding to the rth factor should be sufficiently larger than unity for the correlation between the ' true ' factor score and its best constructed ' estimate ' to be not ' trivially ' small.

APPENDIX

Some comments on coefficient alpha

Kaiser & Caffrey (1965) base their development of AFA on the stationary points of coefficient alpha for a weighted combination of the common parts of the variables, in the common factor model. In their treatment of the question, alpha is regarded as a coefficient of generalizability from the common factors ' underlying ' the observed variables, to corresponding common factors in an infinite domain of content, from which the observed variables are deemed to have been drawn. For a derivation of coefficient alpha as a coefficient of generalizability, Kaiser & Caffrey refer to Cronbach, Rajaratnam & Gleser (1963). However, the latter in turn, refer to a private communication from Kaiser. Cronbach *et al.* remark, with reference to Kaiser's basic assumption, ' just how this constrains the factorial structure of the universe of items is not known ' (*op. cit.*, p. 140). A derivation, not necessarily the same as Kaiser's unpublished treatment, will be given, and the assumptions on which it is based, examined. While this investigation may serve to cast doubt on Kaiser & Caffrey's interpretation of alpha, unfortunately the conclusions reached are not definitive.

Consider two sets of random variables,

$$\mathbf{y}_1' = [y_1^{(1)}, \ldots, y_n^{(1)}]$$

and

$$\mathbf{y}_2' = [y_1^{(2)}, \ldots, y_m^{(2)}] \quad (m \geqslant n)$$

such that

$$\mathscr{E}\{\mathbf{y}_1\} = \mathbf{0}, \quad \mathscr{E}\{\mathbf{y}_2\} = \mathbf{0}.$$

Let

$$\eta_1 = \mathbf{w}_1'\mathbf{y}_1, \tag{A.1}$$

$$\eta_2 = \mathbf{w}_2'\mathbf{y}_2 \tag{A.2}$$

be weighted combinations of the components of $\mathbf{y}_1$, and the components of $\mathbf{y}_2$, where $\mathbf{w}_1$, $\mathbf{w}_2$ are non-null vectors of real weights. Let

$$\mathbf{C}_{11} = \mathscr{E}\{\mathbf{y}_1\mathbf{y}_1'\}, \tag{A.3}$$

$$\mathbf{C}_{22} = \mathscr{E}\{\mathbf{y}_2\mathbf{y}_2'\}, \tag{A.4}$$

and

$$\mathbf{C}_{12} = \mathscr{E}\{\mathbf{y}_1\mathbf{y}_2'\}, \tag{A.5}$$

and write

$$\mathbf{V}_1^2 = \mathrm{Diag}\{\mathbf{C}_{11}\}, \tag{A.6}$$

$$\mathbf{V}_2^2 = \mathrm{Diag}\{\mathbf{C}_{22}\}. \tag{A.7}$$

Then the squared correlation between η_1 and η_2 is given by

$$\rho^2 = \frac{(\mathbf{w}_1'\mathbf{C}_{12}\mathbf{w}_2)^2}{(\mathbf{w}_1'\mathbf{C}_{11}\mathbf{w}_1)(\mathbf{w}_2'\mathbf{C}_{22}\mathbf{w}_2)}. \tag{A.8}$$

Suppose that the vector $\mathbf{y}_2$ is augmented with further variables, so that $m \to \infty$. Kaiser & Caffrey assert that, under certain assumptions, $\rho^2 \to \alpha$ as $m \to \infty$, where

$$\alpha = \frac{n}{n-1}\left\{1 - \frac{\mathbf{w}_1'\mathbf{V}_1^2\mathbf{w}_1}{\mathbf{w}_1'\mathbf{C}_{11}\mathbf{w}_1}\right\}, \tag{A.9}$$

is the well-known coefficient in classical reliability theory.

Assume, in words, that 'the hypothetical average weighted covariance in $\mathbf{C}_{12}$ is equal to the geometric mean of the observable average weighted covariance in $\mathbf{C}_{11}$, and the hypothetical average weighted covariance in $\mathbf{C}_{22}$' (Kaiser & Caffrey, 1965, p. 6, slightly adapted to conform with the present notation). That is, assume that

$$\frac{1}{nm}\mathbf{w}_1'\mathbf{C}_{12}\mathbf{w}_2 = \left\{\frac{1}{n(n-1)}[\mathbf{w}_1'(\mathbf{C}_{11}-\mathbf{V}_1^2)\mathbf{w}_1] \times \left[\frac{1}{m(m-1)}\mathbf{w}_2'(\mathbf{C}_{22}-\mathbf{V}_2^2)\mathbf{w}_2\right]\right\}^{1/2}. \tag{A.10}$$

This assumption can be lent some plausibility by noting that in the special case of equal weights, i.e. where $\mathbf{w}_1' = [1, \ldots, 1]$, $\mathbf{w}_2' = [1, \ldots, 1]$, the stronger joint assumptions (*a*) that the average of covariances in $\mathbf{C}_{12}$ equals the average in $\mathbf{C}_{11}$, (*b*) that the average in $\mathbf{C}_{12}$ equals the average in $\mathbf{C}_{22}$, may seem quite plausible. On the other hand, one notes immediately that the assumption (A.10) cannot possibly hold for every choice of weights $\mathbf{w}_1$ and $\mathbf{w}_2$. (Consider for example, $\mathbf{w}_1 = [1, 0, 0, \ldots, 0]$.) It is not clear that it can hold for at least one, or for more than one choice of weights.

A further assumption, not explicitly noted by Kaiser & Caffrey, or by Cronbach *et al.*, seems necessary. Suppose that, as $m \to \infty$,

$$\frac{1}{m(m-1)}\mathbf{w}_2'(\mathbf{C}_{22}-\mathbf{V}_2^2)\mathbf{w}_2 \to \frac{1}{m^2}\mathbf{w}_2'\mathbf{C}_{22}\mathbf{w}_2. \tag{A.11}$$

In the special case of equal weights, this assumption seems plausible, since it is difficult to imagine augmenting $\mathbf{C}_{22}$ with further rows and columns in such a way that it will not be true that the sum of the m^2 elements of $\mathbf{C}_{22}$ is of the order of m^2, while the sum of the m diagonal elements of $\mathbf{V}_2^2$ is of the order of m. Again, however, it is clear that the assumption (A.11) cannot hold for every choice of weights $\mathbf{w}_2$. (Consider, for example, the case where all the covariances in $\mathbf{C}_{22}$ are equal, and choose weights of fixed magnitude, with alternating signs.) Rearranging terms in eqn. (A.11), and writing $\mathbf{R}_{22}$ for the correlation matrix of $\mathbf{y}_2$, and $\mathbf{q}_2 = \mathbf{V}_2\mathbf{w}_2$, yields

$$\frac{m}{m-1}\left(1 - \frac{\mathbf{q}_2'\mathbf{q}_2}{\mathbf{q}_2'\mathbf{R}_{22}\mathbf{q}_2}\right) \to 1, \tag{A.12}$$

as $m \to \infty$. A sufficient condition for (A.12) to hold, for any non-null $\mathbf{w}_2$, is that

$$\max\left\{\frac{\mathbf{q}_2'\mathbf{q}_2}{\mathbf{q}_2\mathbf{R}_{22}\mathbf{q}_2}\right\} \to 0, \tag{A.13}$$

where the maximum, with respect to $\mathbf{q}_2$, is the largest eigenvalue of $\mathbf{R}_{22}^{-1}$. The problem requires a more refined analysis, however.

Given the assumptions (A.10) and (A.11), the required result, i.e. that $\rho^2 \to \alpha$, easily follows. The squared correlation, in the limit as $m \to \infty$, is regarded as a coefficient of generalizability to an infinite domain of content.

Lord (1958) examined the stationary points of alpha, as given by (A.9), with respect to $\mathbf{w}_1$. Writing $\mathbf{R}_{11}$ for the correlation matrix of $y_1^{(1)}, \ldots, y_n^{(1)}$, yields from eqn. (A.9)

$$\alpha = \frac{n}{n-1}\left(\frac{\lambda-1}{\lambda}\right), \tag{A.14}$$

where

$$\lambda = \frac{\mathbf{q}_1'\mathbf{R}_{11}\mathbf{q}_1}{\mathbf{q}_1'\mathbf{q}_1},$$

with

$$\mathbf{q}_1 = \mathbf{V}_1\mathbf{w}_1. \tag{A.15}$$

Thus the stationary values of α are in one–one correspondence with the eigenvalues of $\mathbf{R}_{11}$. An immediate, and paradoxical, implication is that positive stationary values of alpha correspond to eigenvalues of the correlation matrix that are greater than unity. Hence, if the above argument and assumptions be accepted, there are at most $s < n$ positive (extreme) coefficients of generalizability, corresponding to the s eigenvalues of $\mathbf{R}_{11}$ that are greater than unity and hence $n-s$ negative coefficients. Since it is impossible for ρ^2 to approach a negative limit as $m \to \infty$, one is forced to conclude that, at least for the vectors $\mathbf{w}_1$ corresponding to these negative values, ρ^2 *does not approach* α.

Alpha as a lower bound. Whatever doubts might pertain as to the interpretation of alpha as a coefficient of generalizability of a weighted combination of variables, its applicability as a lower bound to the reliability of a (weighted or unweighted) combination of variables is well understood (Guttman, 1945; Novick & Lewis, 1967; Lord & Novick, 1968). It is, nevertheless, appropriate to point out that all such derivations of alpha as a bound, either assume equal weights, or, what is the same thing, that the weights are fixed constants, prescribed or at least known. In the latter case, as Lord & Novick (1968) point out, one may rescale the variables, so as to absorb the weights. The situation is rather more complex, however, if the weights are to be considered variable, as in the present context of stationary value problems.

Reverting now to the notation of Section 2, the reliability of a weighted combination of variables is given by

$$\rho_{\eta\eta} = \frac{\text{var}\{\eta_t\}}{\text{var}\{\eta\}} = \frac{\mathbf{w}'\mathbf{C}_t\mathbf{w}}{\mathbf{w}'\mathbf{C}\mathbf{w}}. \tag{A.16}$$

Since $\mathbf{C}_t = \mathbf{C}_c + \mathbf{S}^2$, by (22) and (24), the inequality

$$\theta \equiv \frac{\mathbf{w}'\mathbf{C}_c\mathbf{w}}{\mathbf{w}'\mathbf{C}\mathbf{w}} \leqslant \rho_{\eta\eta}, \tag{A.17}$$

clearly holds. If $\mathbf{S}^2$ is positive definite, this is a strict inequality for every non-null $\mathbf{w}$.

It is shown below that, further,

$$\theta \geqslant \alpha, \tag{A.18}$$

with equality if and only if $\mathbf{C}_c$ is of rank one (the Spearman case) and a set of weights is chosen in which the weights are proportional to the inverse of the Spearman factor loadings for each variable. *A fortiori*, $\rho_{\eta\eta} = \alpha$ if and only if, in addition to the conditions just stated, it is also true that $\mathbf{S}^2$ is null, i.e. the unique variance of every variable is entirely error variance. This result is, in one sense, a little more general than that given by Lord & Novick (1968, Theorem 4.4.3), and in another sense a little more specific, since Lord & Novick do not relate their results explicitly to the factor structure of the observed variables, and they consider only the case where the weights are all equal, or, what is the same thing, fixed and known, hence requiring only a rescaling of the variables.

Theorem. For any non-null $\mathbf{w}$,

$$\alpha \leqslant \theta \tag{A.19}$$

where α is given by (A.9), θ by (A.17), with equality if and only if $\mathbf{C}_c = \mathbf{ff}'$, where $\mathbf{f}' = [f_1, \ldots, f_n]$, and $\mathbf{w}' = \gamma[1/f_1, \ldots, 1/f_n]$, where γ is a constant.

Proof. Since $\mathbf{C}_c = \mathbf{FF}'$, where $\mathbf{F}$ is an $n \times r$ matrix of rank $r < n$, (A.17) and (A.9) yield, respectively, in scalar notation,

$$\theta = \sum_{p=1}^{r} \left\{ \sum_j w_j^2 f_{jp}^2 + \sum_{j \neq k}\sum w_j w_k f_{jp} f_{kp} \right\} \Big/ \sigma^2, \tag{A.20}$$

and

$$\alpha = \frac{n}{n-1} \sum_{p=1}^{r} \left\{ \sum_{j \neq k}\sum w_j w_k f_{jp} f_{kp} \right\} \Big/ \sigma^2, \tag{A.21}$$

where

$$\sigma^2 = \mathbf{w}'\mathbf{C}\mathbf{w},$$

and the ranges of j, k are $1, \ldots, n$. Thence may be obtained

$$\theta - \alpha = \sum_{p=1}^{r} \left\{ \sum_j w_j^2 f_{jp}^2 - \frac{1}{n-1} \sum_{j \neq k}\sum f_{jp} f_{kp} \right\} \Big/ \sigma^2. \tag{A.22}$$

The identity, for any numbers $b_1, \ldots, b_n$,

$$\sum_{j \neq k}\sum (b_j^2 + b_k^2) = 2(n-1) \sum_j b_j^2,$$

yields, further,

$$\theta - \alpha = \sum_{p=1}^{r} \left\{ \frac{1}{2(n-1)} \sum_{j \neq k}\sum (w_j^2 f_{jp}^2 + w_k^2 f_{kp}^2) - \frac{1}{n-1} \sum_{j \neq k}\sum (w_j w_j f_{jp} f_{kp}) \right\} \Big/ \sigma^2,$$

$$= \sum_{p=1}^{r} \sum_{j \neq k}\sum \{w_j f_{jp} - w_k f_{kp}\}^2 / (n-1)\sigma^2, \tag{A.23}$$

hence $\theta \geqslant \alpha$. The expression (A.23) will equal zero if and only if

$$w_j f_{jp} = w_k f_{kp} \tag{A.24}$$

for all j, k, p. But these conditions cannot be satisfied for $r > 1$, with $\mathbf{F}$ of rank r, and $\mathbf{w}$ non-null. It then must be that $\mathbf{F}$ is the $n \times 1$ vector, $\mathbf{f}$, say, and that $\mathbf{w}' = \gamma[1/f_1, \ldots, 1/f_n]$, where γ is a constant.

Acknowledgement

The author is indebted to Dr E. J. Burr, of the Department of Mathematics, University of New England, for his advice and criticism.

References

Bartlett, M. S. (1937). The statistical conception of mental factors. *Br. J. Psychol.* **28,** 97–104.

Burt, C. (1940). *The Factors of the Mind.* London: University of London Press.

Cronbach, L. J., Rajaratnam, N. & Gleser, G. C. (1963). Theory of generalizability: a liberalization of reliability theory. *Br. J. statist. Psychol.* **16,** 137–163.

Green, B. F. (1950). A note on the calculation of weights for maximum battery reliability. *Psychometrika* **15,** 57–61.

Guttman, L. (1945). A basis for analyzing test-retest reliability. *Psychometrika* **10,** 255–282.

Guttman, L. (1954). Some necessary conditions for common factor analysis. *Psychometrika* **19,** 149–162.

Harman, H. H. (1960). *Modern Factor Analysis.* Chicago: University of Chicago Press.

Harris, C. W. (1962). Some Rao–Guttman relationships. *Psychometrika* **27,** 247–264.

Harris, C. W. (1967). On factors and factor scores. *Psychometrika* **32,** 363–379.

Hemmerle, W. J. (1965). Obtaining maximum-likelihood estimates of factor loadings and communalities using an easily implemented iterative computer procedure. *Psychometrika* **30,** 291–302.

Horst, P. (1965). *Factor Analysis of Data Matrices.* New York: Holt, Rinehart & Winston.

Jöreskog, K. G. (1967). Some contributions to maximum likelihood factor analysis. *Psychometrika* **32,** 443–482.

Kaiser, H. J. & Caffrey, J. (1965). Alpha factor analysis. *Psychometrika* **30,** 1–14.

Lawley, D. N. (1940). The estimation of factor loadings by the method of maximum likelihood. *Proc. R. Soc. Edin.* **60,** 64–82.

Lawley, D. N. (1942). Further investigations in factor estimation. *Proc. R. Soc. Edin.* **61,** 176–185.

Lord, F. M. (1958). Some relations between Guttman's principal components of scale analysis and other psychometric theory. *Psychometrika* **23,** 291–296.

Lord, F. M. & Novick, M. R. (1968). *Statistical Theories of Mental Test Scores.* Reading, Mass.: Addison-Wesley.

McDonald, R. P. (1968). A unified treatment of the weighting problem. *Psychometrika* **33,** 351–381.

McDonald, R. P. & Burr, E. J. (1967). A comparison of four methods of constructing factor scores. *Psychometrika* **32,** 381–401.

Mosier, C. J. (1943). On the reliability of a weighted composite. *Psychometrika* **8,** 161–168.

Novick, M. R. & Lewis, C. (1967). Coefficient alpha and the reliability of composite measurements. *Psychometrika* **32,** 1–13.

Peel, E. A. (1948). Prediction of a complex criterion and battery reliability. *Br. J. Psychol., Statist. Sect.* **1,** 84–94.

Rao, C. R. (1953). Discriminant functions for genetic differentiation and selection. *Sankhyā* **12,** 229–246.

Rao, C. R. (1955). Estimation and tests of significance in factor analysis. *Psychometrika* **20,** 93–112.

Thomson, G. H. (1940). Weighting for battery reliability and prediction. *Br. J. Psychol.* **30,** 357–366.

Thurstone, L. L. (1947). *Multiple-factor Analysis.* Chicago: University of Chicago Press.

11

Reprinted from *Psychometrika*, **23**(3), 187–200 (1958)

THE VARIMAX CRITERION FOR ANALYTIC ROTATION IN FACTOR ANALYSIS*

HENRY F. KAISER
UNIVERSITY OF ILLINOIS

An analytic criterion for rotation is defined. The scientific advantage of analytic criteria over subjective (graphical) rotational procedures is discussed. Carroll's criterion and the quartimax criterion are briefly reviewed; the varimax criterion is outlined in detail and contrasted both logically and numerically with the quartimax criterion. It is shown that the *normal* varimax solution probably coincides closely to the application of the principle of simple structure. However, it is proposed that the ultimate criterion of a rotational procedure is factorial invariance, not simple structure—although the two notions appear to be highly related. The normal varimax criterion is shown to be a two-dimensional generalization of the classic Spearman case, i.e., it shows perfect factorial invariance for two pure clusters. An example is given of the invariance of a normal varimax solution for more than two factors. The oblique normal varimax criterion is stated. A computational outline for the orthogonal normal varimax is appended.

In factor analysis, an analytic criterion for rotation is defined as one that imposes mathematical conditions beyond the fundamental factor theorem, such that a factor matrix is uniquely determined. Historically, the first such criterion was Thurstone's treatment of the principal axes problem [10]: from any arbitrary factor matrix he suggested rotating under the criterion that each factor successively accounts for the maximum variance. But principal axes have seldom been accepted as psychologically very interesting ([9], p. 139). The rotation problem for psychologically meaningful factors is usually handled judgmentally. Scientifically, however, this procedure is not very satisfactory: the ad hoc quality of subjective rotation makes uniquely determined factors impossible; only factors that are subject to the uncertainties and controversies besetting any a posteriori reasoning can be defined. In contrast, an analytic criterion for rotation would allow factor analysis to become a straightforward methodology stripped of its subjectivity and a proper tool for scientific inquiry.

The Quartimax Criterion

The first analytic criterion for determining psychologically interpretable factors was presented in 1953 by Carroll [1]. In an attempt to provide a

*Part of the material in this paper is from the writer's Ph.D. thesis. I am indebted to my committee, Professors F. T. Tyler, R. C. Tryon, and H. D. Carter, chairman, for many helpful suggestions and criticisms. Dr. John Caffrey suggested the name *varimax*, and wrote the original IBM 602A computer program for this criterion.

I am also indebted to the staff of the University of California Computer Center for help in programming the procedures described in the paper for their IBM 701 electronic computer. Since their installation is partially supported by a grant from the National Science Foundation, the assistance of this agency is acknowledged.

mathematical explication of Thurstone's simple structure, he suggested that for a given factor matrix,

$$f = \sum_{s<t} \sum_{j} a_{js}^2 a_{jt}^2 \tag{1}$$

should be a minimum, where $j = 1, 2, \cdots, n$ are tests, $s, t = 1, 2, \cdots, r$ are factors, and a_{js} is the factor loading of the jth test on the sth factor. It appears that Carroll was motivated in writing (1) primarily by a close inspection of Thurstone's five formal rules for simple structure ([12], p. 335), particularly the requirement that a large loading for one factor be opposite a small loading for another factor.

In his original paper, Carroll provided two numerical examples of the application of his method. Without the restriction of orthogonality, these illustrations gave somewhat equivocal results—while the application of (1) appears to bring one close to the desired simple structure, the criterion has an obvious bias in being too strongly influenced by factorially complex tests.

In the light of later developments, Carroll's criterion should probably be relegated to the limbo of "near misses"; however, this does not detract from the fact that it was the first attempt to break away from an inflexible devotion to Thurstone's ambiguous, arbitrary, and mathematically unmanageable qualitative rules for his intuitively compelling notion of simple structure.

Almost simultaneously with Carroll's development, Neuhaus and Wrigley [7], Saunders [8], and Ferguson [2] proposed what is usually called the quartimax method for orthogonal simple structure. Neuhaus and Wrigley suggest that a most easily interpretable factor matrix, in the simple structure sense, may be found when the variance of all nr squared loadings of the factor matrix is a maximum, i.e.,

$$q_1 = [nr \sum_{j} \sum_{s} (a_{js}^2)^2 - (\sum_{j} \sum_{s} a_{js}^2)^2]/n^2r^2 = \text{maximum}. \tag{2}$$

Saunders' approach requires that the kurtosis (fourth moment over second moment squared) of all loadings and their reflections be a maximum,

$$q_2 = nr \sum_{j} \sum_{s} a_{js}^4/(\sum_{j} \sum_{s} a_{js}^2)^2 = \text{maximum}. \tag{3}$$

While Ferguson, basing his rationale on certain parallels with information theory, calls simply for

$$q_3 = \sum_{j} \sum_{s} a_{js}^4 = \text{maximum}. \tag{4}$$

All these investigators are concerned with attaining a factor matrix with a maximum tendency to have both small and large loadings. While less obviously related to Thurstone's rules than Carroll's criterion, the emphasis on small loadings coincides with Thurstone's requirements of zero loadings. For orthogonal factors, criteria (2), (3), and (4) are equivalent because of the invariance of the sum of the communalities, $\sum_j \sum_s a_{js}^2$. (This term, as well as other constants, disappear when differentiated in the calculus problem involved in finding the required critical point.)

Indeed, it turns out that they are also equivalent to Carroll's criterion in the orthogonal case. Minimizing (1) is equivalent to maximizing (4) since the squared communality of a test is

$$\text{constant} = (\sum_s a_{js}^2)^2 = \sum_s a_{js}^4 + 2 \sum_{s<t} a_{js}^2 a_{jt}^2,$$

and the sum of squared communalities over all tests is

$$\text{constant} = \sum_j \sum_s a_{js}^4 + 2 \sum_j \sum_{s<t} a_{js}^2 a_{jt}^2$$

$$= q_3 + 2f.$$

Thus, since the quartimax criterion plus twice Carroll's criterion is a constant, maximizing q_3 is equivalent to minimizing f.

Neuhaus and Wrigley realized that none of these criteria can be realistically applied without the aid of an electronic computer—the calculations involved are too extensive for a desk calculator or punched card mechanical computers. Consequently, they programmed the quartimax method for the Illiac* and provided a rather extensive numerical investigation of the empirical properties of the quartimax method.

Their results were perhaps more encouraging than Carroll's. Under the restriction of orthogonality Carroll's criterion (or the equivalent quartimax method) does not show nearly so obvious a bias as does Carroll's criterion when the restriction of orthogonality is removed. However, as an explication of orthogonal simple structure, the quartimax method does have a systematic bias which will be more fully examined in the next section.

The Varimax Criterion

From the outset, the above methods consider all nr loadings simultaneously. In every case, however, these criteria may be applied separately to each *row* of the factor matrix and summed over rows for the final criterion because of the invariance of the communalities. For example, Neuhaus and Wrigley could have defined the *simplicity*, say, of the factorial composition of the jth test as the variance of the squared loadings for this test,

$$q_j^* = [r \sum_s (a_{js}^2)^2 - (\sum_s a_{js}^2)^2]/r^2. \tag{5}$$

*The Illiac is the University of Illinois electronic computer. Subsequently, the quartimax criterion has been programmed for the CRC–102A (Neuhaus), and the IBM 701 (Kaiser). The varimax criterion described in the next two sections has been programmed for SWAC at UCLA (Comrey), the IBM 701 (Kaiser), Illiac (Dickman), and the IBM 650 (Vandenberg).

To obtain the total criterion for the entire factor matrix, (5) could then be summed over all tests to give

(6) $$q^* = \sum_i \{[r \sum_s (a_{is}^2)^2 - (\sum_s a_{is}^2)^2]/r^2\}.$$

Maximizing q^* is equivalent to maximizing q_3, again because constant terms vanish when differentiated.

Equation (6) perhaps provides some insight into the quartimax criterion—its aim is to simplify the description of each row, or test, of the factor matrix. It is unconcerned with simplifying the columns, or factors, of the factor matrix (probably the most fundamental of all requirements for simple structure). The implication of this is that the quartimax criterion will often give a general factor. Under requirement (5) there is no reason why a large loading for each test may not occur on the same factor. In practice, this tendency for the quartimax criterion to yield a general factor is most pronounced when the unrotated factor matrix has a pronounced general factor.

From the simple structure viewpoint, an immediate modification of the quartimax criterion is apparent. Let us define the simplicity of a factor as the variance of its squared loadings,

(7) $$v_s^* = [n \sum_i (a_{is}^2)^2 - (\sum_i a_{is}^2)^2]/n^2.$$

And for the criterion for all factors, define the maximum simplicity of a factor matrix as the maximization of

(8) $$v^* = \sum_s v_s^* = \sum_s \{[n \sum_i (a_{is}^2)^2 - (\sum_i a_{is}^2)^2]/n^2\},$$

the variance of squared loadings by columns rather than by rows.

Since a factor is a vector of correlation coefficients, the most interpretable factor is one based upon correlation coefficients which are maximally interpretable. Those correlations which satisfy this condition are patently obvious: correlations of $\pm$ 1, which indicate a functional relationship, and correlations of zero, which indicate no linear relationship. On the other hand, middle-sized correlations are the most difficult to understand. Thus, it is seen why v_s^* in (7) could be maximized for the maximum interpretability or simplicity of a factor, and more generally, why the interpretability of an entire factor matrix could be considered best when (8) is a maximum.

Criterion (8) is the original *raw* varimax criterion [4]. In the original proposal of this criterion, it was shown to be mathematically equivalent, in the orthogonal case, to minimizing

(9) $$c^* = \sum_{s<t} \{[n \sum_i a_{is}^2 a_{it}^2 - (\sum_i a_{is}^2)(\sum_i a_{it}^2)]/n^2\},$$

i.e., minimizing the covariance of pairs of columns of squared loadings and summing over all possible pairs of columns for the criterion. Criterion (9) then bears the analogous relationship to Carroll's criterion (1) that the varimax criterion (8) does to the quartimax criterion (6).

Some distinctions between quartimax and varimax orthogonal solutions can be illustrated numerically. In Table 1 solutions for Thurstone's eleven-variable box problem ([12], pp. 373–375) are given. It will be noted that the quartimax solution [7] could hardly be called a simple structure. There is a large general factor, and the second factor seems only vaguely concerned

Table 1

Thurstone's 11-Variable Box Problem[a]

	Subjective (oblique)			Quartimax			Raw Varimax		
Test	X	Y	Z	X	Y	Z	X	Y	Z
x	90	05	00	68	65	05	91	19	16
y	04	88	01	83	-47	00	05	93	25
z	03	05	79	42	-08	79	11	17	88
xy	62	63	-06	99	11	-04	64	74	20
yz	-05	54	57	71	-40	56	02	65	75
x^2y	82	37	-01	92	41	03	84	51	22
xy^2	35	76	02	96	-18	03	37	86	28
2x + 2y	53	71	-09	100	00	-07	54	82	18
$(x^2 + y^2)^{\frac{1}{2}}$	52	71	-08	99	-01	-07	53	81	18
$(x^2 + z^2)^{\frac{1}{2}}$	52	-07	65	59	38	68	60	09	77
xyz	42	43	43	88	04	45	48	58	65

[a]Decimal points omitted.

with dimensions of boxes. On the other hand, the raw varimax solution closely parallels Thurstone's original subjective solution, given the restriction of orthogonality.

In Table 2 are solutions for Holzinger and Harman's 24 psychological tests ([3], pp. 229–233). Both the quartimax [7] and the raw varimax methods seem to duplicate the subjectively rotated simple structure patterns. But the respective variance contributions of the factors are perhaps more interesting. It is seen that the dispersion of the $\sum_i a_{is}^2$ for the subjective solution is less than the corresponding figures for the two analytic methods. In other words, Holzinger and Harman have made the factors a little more level or

Table 2

Holzinger and Harman's Twenty-four Psychological Tests[a]

	Subjective				Quartimax				Raw Varimax				Normal Varimax			
Test	A	B	C	D	A	B	C	D	A	B	C	D	A	B	C	D
1	10	32	62	20	37	19	60	07	24	20	65	13	14	19	67	17
2	07	15	41	13	24	07	38	04	17	07	42	08	10	07	43	10
3	10	12	53	13	31	01	48	01	22	02	52	06	15	02	54	08
4	15	18	53	12	36	07	46	-01	27	08	52	04	20	09	54	07
5	75	15	26	15	81	14	-02	-04	78	21	12	06	75	21	22	13
6	72	05	28	25	81	03	00	06	78	10	13	14	75	10	23	21
7	81	08	27	11	85	07	-04	-10	84	15	10	00	82	16	21	08
8	54	26	38	14	66	20	20	-04	60	25	31	05	54	26	38	12
9	76	-04	29	30	86	-06	-02	10	84	01	12	19	80	01	22	25
10	28	66	-19	14	23	70	-12	11	17	71	-08	17	15	70	-06	24
11	27	61	-04	29	31	62	01	23	22	63	06	29	17	60	08	36
12	13	72	09	03	16	69	19	-01	06	70	23	04	02	69	23	11
13	24	63	31	02	35	57	32	-08	24	59	39	-01	18	59	41	06
14	23	19	-02	48	32	19	-03	42	26	20	01	46	22	15	04	50
15	11	14	08	50	25	11	10	45	17	11	13	48	12	07	14	50
16	05	22	34	45	29	13	37	37	17	13	41	41	08	10	41	43
17	15	24	-03	62	28	24	02	57	20	23	05	61	14	13	06	64
18	01	39	20	52	22	32	30	47	08	31	32	51	00	26	32	54
19	12	22	18	39	28	18	19	32	19	18	22	36	13	15	24	39
20	31	18	46	29	52	09	35	14	42	12	43	21	35	11	47	25
21	17	46	33	24	35	38	35	14	23	40	40	20	15	38	42	26
22	31	12	40	40	53	04	30	26	44	06	37	32	36	04	41	36
23	31	29	54	25	55	19	44	09	43	21	52	16	35	21	57	22
24	39	46	14	31	49	43	10	20	40	46	18	27	34	44	22	34
$\sum_j a_{js}^2$	343	292	268	236	559	242	196	142	431	260	264	186	350	244	308	236

[a]Decimal points omitted.

even in their contribution to variance than the analytic criteria. Of the two analytic criteria, the raw varimax solution has given a solution which is closer in this respect to Holzinger and Harman's. It is also noteworthy that as a result of these differences the large loadings of the factors with the larger variance contributions for the analytic methods are larger than the large loadings for the smaller factors, and similarly, the small loadings for the larger factors are larger than the small loadings for the smaller factors. Holzinger and Harman's subjective solution does not show this systematic bias; their solution gives a more equitable patterning of factor loadings.

How this bias may be removed is indicated in the next section. This leads to a revision of the varimax criterion, which appears to have more important characteristics than merely satisfying the rules of simple structure.

Factorial Invariance: Normal Varimax

It seems reasonable to attribute the systematic bias seen in both the quartimax and varimax solutions of the Holzinger-Harman data and other examples [4] to the divergent weights which implicitly are attached to the tests by their communalities. When one deals with fourth-power functions of factor loadings, a test with communality 0.6, for example, would tend to influence the rotations four times as much as a test whose communality was 0.3. Thus, while the most obvious weights have been applied to the tests, namely the square roots of their communalities, after the fact it seems that there is probably a better set of weights—weights which would tend to equalize to a greater extent the relative influence of each test during rotation.

There seems no rational basis for choosing among different weighting schemes. Let us then make the agnostic confession of ignorance which pervades any form of correlational analysis. For the purposes of rotation, weight the tests equally, in the sense that the lengths of the common parts of the test vectors have equal length. (The author is indebted to Dr. D. R. Saunders for this suggestion.) The varimax criterion could then be rewritten as

$$v = \sum_{s} \{[n \sum_{j} (a_{js}^2/h_j^2)^2 - [\sum_{j} (a_{js}^2/h_j^2)]^2]/n^2\}, \tag{10}$$

where h_j^2 is the communality of the jth test. In contrast to (7) and (8), where the variance of the squared correlations of the tests with a factor is maximized, the variance of the squared correlations of the common parts of the tests (the reflections of the tests onto the common-factor space) with a factor is now being maximized. [Note from (10) that we are not advocating a permanent weighting of the tests by a weight inversely as the square root of their communalities. During rotation this weighting extends the common part of each test vector to unit length, but after rotation each of these vectors is shortened to its proper length by reweighting directly as the square root of the test's communality.]

As will be seen in Table 2, under this modification the varimax criterion (the *normal* varimax, since rotation is with respect to normalized common parts of tests) has effectively removed the small but disturbing bias in the raw varimax solution of Holzinger and Harman's example. It also has been shown in a number of other examples [6] that the normal varimax does not seem to deviate systematically from what may be considered the best orthogonal simple structure.*

Thus far, however, merely a numerical-intuitive basis for a weighting procedure which leads to "prettier" results has been provided. Such a basis is quite unsatisfactory theoretically. Indeed, this sort of ad hoc thinking could conceivably lead to a different set of judgmentally determined weights for any particular example—a situation as scientifically reprehensible as the subjective graphical methods.

There is a more fundamental rationale for attempting to establish the normal varimax criterion (10) as a mathematical definition for the rotation

*Professor Andrew Comrey has apparently reached the same conclusion in an extensive application of the normal varimax criterion to interitem correlation matrices of the MMPI (personal communication). A further example, available from the writer, is the normal varimax solution of Thurstone's classic PMA study[11] (dittoed).

problem. Consider the situation illustrated in Fig. 1. There are two clusters of tests, each of which is pure in the sense that the reflections of the test vectors of the cluster onto the two-dimensional, common-factor space are collinear. (While these clusters are drawn less than 90° apart, the following argument is perfectly general.)

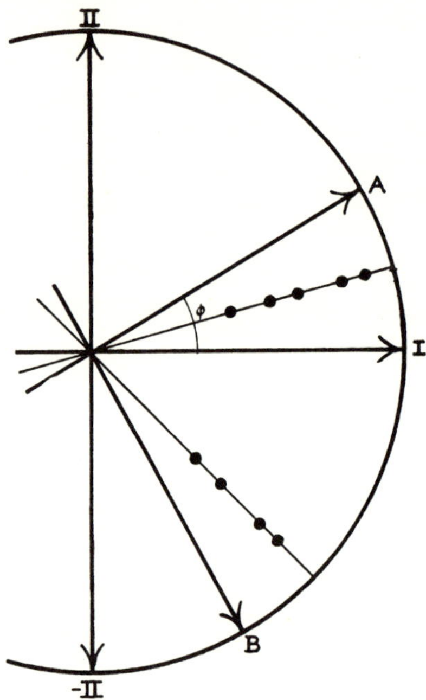

FIGURE 1
Case for which a normal varimax solution is invariant under changes in the composition of the test battery.

It is shown below that the angle of rotation in a plane which maximizes (10) is

$$\phi = \tfrac{1}{4} \arctan \frac{2[n \sum_i u_i v_i - \sum_i u_i \sum_i v_i]}{n \sum_i (u_i^2 - v_i^2) - [(\sum_i u_i)^2 - (\sum_i v_i)^2]}, \tag{11}$$

where

$$u_i = (a_{i1}/h_i)^2 - (a_{i2}/h_i)^2,$$

and

$$v_i = 2(a_{i1}/h_i)(a_{i2}/h_i).$$

Let n_A ($n_A \geq 1$) be the number of tests in the first cluster and n_B ($n_B \geq 1$) be the number of tests in the second cluster ($n = n_A + n_B$). It is readily apparent that all tests of the first cluster have the same values for u_i and v_i.

Let these values be u_A and v_A. Similarly let the values for the second cluster be u_B and v_B. In this case (11) reduces to

$$(12) \qquad \phi = \tfrac{1}{4} \arctan \frac{2n_A n_B(u_A v_A + u_B v_B - u_A v_B - u_B v_A)}{n_A n_B(u_A^2 + u_B^2 - v_A^2 - v_B^2 - 2u_A u_B + 2v_A v_B)} .$$

A most important result is shown in (12). The $n_A n_B$ term may be cancelled, indicating that the angle of rotation does not depend on the number of tests in each cluster, i.e., *for the case illustrated in Fig.* 1, *the normal varimax solution is invariant under changes in the composition of the test battery.*

This invariance property would seem to be of greater significance than the numerical tendencies of the normal varimax solution to define mathematically the doctrine of simple structure. Although factor analysis seems to have many purposes, fundamentally it is addressed to the following problem. Given an (infinite) domain of psychological content, infer the internal structure of this domain on the basis of a sample of n tests drawn from the domain. The possibility of success in such inferences is obviously dependent upon the extent which a factor derived from a particular battery or sample of tests approximates the corresponding unobservable factor in the infinite domain. If a factor is invariant under changing samples of tests, i.e., shows factorial invariance ([12], pp. 360–361), there is evidence that inferences regarding domain factors are correct.

The normal varimax solution, according to the above result, allows such inferences; regardless of the sampling of tests, for the problem shown in Fig. 1 it is possible to infer precisely the domain normal varimax factors. This is not true for either the quartimax or raw varimax solutions since the angle of rotation is a function of n_A and n_B.

Note that domain normal varimax factors are not said to be more *meaningful* than domain factors according to some different criterion; it is suggested that observed normal varimax factors will have a greater likelihood of portraying the corresponding domain factors.

Although one often gets the impression that simple structure is the ultimate criterion of a rotational procedure, it is suggested here that the ultimate criterion is factorial invariance. The normal varimax solution was originally devised solely for the purpose of satisfying the simple structure criteria. But the fact that it shows mathematically this sort of invariance suggests that Thurstone's reasoning was basically directed toward factorial invariance. The principle of simple structure may probably be considered incidental to the more fundamental concept of factorial invariance. This viewpoint renders meaningless the arguments concerning "psychological reality" of general factors, bipolar factors, simple structure factors, etc.

Admittedly, the result (12) is for a special case. The correlations among the variables within each of the two pure clusters must form a perfect Spearman matrix, and the reduced correlation matrix as a whole must be of rank two. Normal varimax does however give invariant results for two such Spearman clusters simultaneously, and consequently the normal varimax criterion is a two-dimensional generalization of the classic Spearman case. Obviously, the next step would be a generalization along the same lines to the r-dimensional case; thus far, however, work on this problem suggests virtually insuperable mathematical difficulties.

To investigate numerically the tendency of the criterion to give factorially invariant solutions for r larger than two, again consider Holzinger and Harman's empirical data. Taking their centroid loadings, the first five tests were rotated then the first six, etc., systematically until the analysis of all 24 tests was reached, as in Table 2. The results of this application of the normal varimax criterion are given in Table 3. There, the normal varimax loadings for the four factors as a function of the changing number of tests are given.

Note that factors A and C are essentially invariant from the outset; the loading changes, while somewhat systematic, are negligible—24 appears to be essentially infinite. On the other hand, factors B and D show more movement before they become stable. The reason for this is readily apparent. For both B and D, this movement occurs only while they are underdetermined, i.e., only while they contain no appreciable loadings. However, once they pick up a test or two with high loadings, their ambiguous definition abruptly stops, and they settle down to exhibit a degree of invariance similar to the

Table 3A

Normal Varimax Loading Changes for Holzinger and Harman's Factor A (n = 5, 6, ..., 24)[a]

Test	n: 5	6	7	8	9	10	11	12	13	14	15	16	17	18	19	20	21	22	23	24
1	19	19	18	19	18	16	16	16	16	16	15	16	15	16	16	15	15	15	14	14
2	12	12	12	13	12	12	12	12	12	11	11	11	11	12	11	11	11	11	10	10
3	15	16	15	16	16	17	17	17	17	16	16	16	16	17	17	16	16	16	15	15
4	21	21	21	22	21	21	21	21	21	21	21	21	21	21	21	20	20	20	20	20
5	79	78	78	79	78	76	75	76	76	76	75	75	75	76	76	75	75	75	75	75
6		78	77	77	78	77	76	77	77	76	75	75	75	75	75	75	75	75	75	75
7			83	84	83	82	82	82	82	82	82	82	82	83	82	82	82	82	82	82
8				59	58	55	55	55	55	55	55	55	55	55	55	55	55	55	54	54
9					82	83	82	83	83	81	80	80	80	80	80	80	80	80	80	80
10						18	15	17	17	16	16	16	15	16	16	16	16	15	16	15
11							19	21	21	19	18	18	18	18	18	18	18	18	18	17
12								03	02	03	03	03	03	03	03	03	03	03	02	02
13									18	19	19	20	20	20	20	19	19	19	18	18
14										23	21	22	21	22	22	22	22	22	22	22
15											11	12	11	12	12	12	12	12	12	12
16												09	09	10	09	09	09	09	09	08
17													13	14	14	14	14	14	14	14
18														01	01	00	00	00	00	00
19															13	13	13	13	13	13
20																35	35	35	35	35
21																	16	16	16	15
22																		36	36	36
23																			35	35
24																				34

Table 3B

Normal Varimax Loading Changes for Holzinger and Harman's Factor B (n = 5, 6, ..., 24)[a]

Test	n 5	6	7	8	9	10	11	12	13	14	15	16	17	18	19	20	21	22	23	24
1	10	10	10	10	07	12	16	19	21	20	20	19	19	19	18	19	19	19	19	19
2	01	01	01	01	02	03	05	07	08	08	07	07	07	07	06	07	07	07	07	07
3	-07	-07	-07	-06	-04	-04	-01	01	02	03	02	02	02	02	02	02	02	02	02	02
4	-02	-03	-02	-01	-04	02	05	07	09	09	09	09	09	08	08	08	09	09	08	09
5	-00	03	05	04	-00	19	22	21	22	22	21	21	20	20	20	20	20	21	20	21
6		-03	-02	-05	03	10	14	12	13	11	10	10	09	08	08	08	09	09	09	10
7			-03	-03	-08	12	16	15	16	16	16	16	15	15	15	15	15	16	16	16
8				10	02	21	24	25	26	26	26	26	25	25	25	25	25	25	25	26
9					03	03	08	05	05	03	02	01	00	-00	-01	-01	-00	00	-00	01
10						73	74	74	74	72	71	71	70	69	69	69	69	70	70	70
11							68	67	67	64	62	62	61	60	59	59	60	60	60	60
12								67	68	69	69	69	69	69	68	69	69	69	69	69
13									57	59	59	59	59	59	59	59	59	59	59	59
14										22	19	19	16	15	15	14	15	15	15	16
15											10	10	07	06	06	06	06	06	06	07
16												12	10	09	08	08	09	09	09	10
17													19	17	17	17	17	17	17	18
18														26	25	25	26	26	26	26
19															14	14	14	15	15	15
20																09	10	10	10	11
21																	37	38	37	38
22																		04	04	04
23																			20	21
24																				44

[a]Decimal points omitted.

Table 3C

Normal Varimax Loading Changes for Holzinger and Harman's Factor C (n = 5, 6, ..., 24)[a]

Test	n 5	6	7	8	9	10	11	12	13	14	15	16	17	18	19	20	21	22	23	24
1	70	70	70	70	70	70	70	69	68	68	68	67	67	67	66	67	67	67	67	67
2	45	44	45	44	44	45	44	44	44	44	44	43	43	43	43	43	43	43	43	43
3	54	54	54	54	54	54	54	54	54	54	54	54	54	54	54	54	54	54	54	54
4	55	55	55	55	55	55	55	55	54	54	54	54	54	54	54	54	54	54	55	54
5	23	23	24	22	23	23	23	21	21	21	21	20	20	20	19	20	20	20	22	22
6		24	25	24	24	24	24	23	23	23	22	21	21	21	20	21	21	21	23	23
7			21	20	21	21	21	20	19	19	19	19	19	18	18	19	19	19	21	21
8				39	40	40	39	38	37	37	37	37	37	36	36	37	37	37	38	38
9					23	23	23	22	22	22	21	20	21	20	19	20	20	20	22	22
10						04	01	-03	-04	-05	-06	-06	-06	-07	-07	-07	-07	-07	-06	-06
11							16	13	11	11	09	09	09	08	07	07	07	07	08	08
12								25	23	23	23	23	23	23	23	23	23	22	23	23
13									41	41	41	41	41	41	41	41	41	41	41	41
14										08	05	04	05	03	02	03	03	03	04	04
15											15	14	15	13	12	13	13	13	14	14
16												42	42	41	40	41	41	41	41	41
17													08	06	04	05	06	06	06	06
18														32	31	31	31	31	32	32
19															22	23	23	23	24	24
20																47	47	47	48	47
21																	42	41	42	42
22																		41	41	41
23																			57	57
24																				22

Table 3D

Normal Varimax Loading Changes for Holzinger and Harman's Factor D (n = 5, 6, ..., 24)[a]

	n																			
Test	5	6	7	8	9	10	11	12	13	14	15	16	17	18	19	20	21	22	23	24
1	01	-00	00	01	-07	01	01	01	02	08	13	14	15	17	19	18	[illegible]	17	17	17
2	01	01	02	01	00	02	02	02	01	05	08	09	09	10	11	11	[illegible]	10	10	10
3	-00	01	01	-00	05	-00	01	-00	-01	02	06	07	07	08	09	09	[illegible]	08	08	08
4	-03	-03	-03	-04	-01	-04	-03	-04	-04	01	05	06	06	07	09	08	[illegible]	08	07	07
5	-00	-06	-03	-02	-09	-06	-06	-05	-05	07	12	12	14	14	15	14	[illegible]	[illegible]	14	13
6		06	09	09	05	05	06	06	05	17	21	21	22	21	22	22	22	[illegible]	22	21
7			-06	-06	-06	-10	-10	-09	-10	03	07	07	09	08	09	09	[illegible]	08	08	08
8				-04	-14	-07	-08	-07	-07	04	10	10	12	12	14	13	[illegible]	[illegible]	12	12
9					15	11	12	12	11	22	26	26	27	25	26	26	[illegible]	26	25	25
10						-00	-07	-00	02	15	19	19	23	24	25	25	[illegible]	[illegible]	24	24
11							07	12	15	27	31	32	35	36	37	37	[illegible]	37	37	36
12								-14	-11	-00	05	05	09	11	13	12	[illegible]	11	11	11
13									-17	-05	00	01	04	07	08	08	07	07	07	06
14										47	49	49	50	50	50	50	[illegible]	[illegible]	50	50
15											49	49	50	50	50	50	[illegible]	50	50	50
16												42	42	44	45	44	[illegible]	44	44	43
17													64	64	64	64	[illegible]	[illegible]	64	64
18														54	55	55	[illegible]	[illegible]	54	54
19															40	40	[illegible]	[illegible]	39	39
20																26	[illegible]	[illegible]	25	25
21																	[illegible]	[illegible]	26	26
22																		35	36	36
23																			22	22
24																				34

[a]Decimal points omitted.

other two factors, which had high loadings from the beginning. For $n = 24$, there appear to be good approximations to the domain normal varimax factors.

The Oblique Case

If the restriction of orthogonality is relaxed, it is impossible to apply directly the quartimax criterion (4) or the normal varimax criterion (10). This is because interfactor relationships are not considered when the criteria are in this form, and when applied all factors will collapse into the same factor—that one factor which best meets the criterion. However, Carroll's version of the quartimax criterion explicitly considers interfactor relationships and an oblique solution is attainable. As suggested by (9), if

$$(13) \qquad c = \sum_{s<t} \{[n \sum_j (a_{js}^2/h_j^2)(a_{jt}^2/h_j^2) - (\sum_j a_{js}^2/h_j^2)(\sum_j a_{jt}^2/h_j^2)]/n^2\},$$

it may be shown that in the orthogonal case $v = -2c$. This alternative form of the normal varimax may then be used to obtain oblique factors. The mathematical problem of minimizing (13) is exactly analogous to Kaiser's [5] treatment for Carroll's criterion. Computationally, the (iterative) solution involves finding the latent vector associated with the smallest latent root of a constantly changing symmetric matrix of order r.

Computational Appendix

To compute an orthogonal normal varimax solution, the following procedure is suggested. The first step is to normalize the rows of the arbitrary reference factor matrix (e.g., principal axes or centroids) by dividing each element by h_j. Rotation to the direction of the normal varimax factors may then be carried out with respect to these normalized loadings.

The criterion (10) will be applied to two factors at a time. For this purpose, the following notation for an orthogonal rotation is convenient.

$$\begin{bmatrix} x_1 & y_1 \\ x_2 & y_2 \\ \vdots & \vdots \\ x_n & y_n \end{bmatrix} \cdot \begin{bmatrix} \cos\phi & -\sin\phi \\ \sin\phi & \cos\phi \end{bmatrix} = \begin{bmatrix} X_1 & Y_1 \\ X_2 & Y_2 \\ \vdots & \vdots \\ X_n & Y_n \end{bmatrix},$$

where x_j and y_j, the present normalized loadings, are constants, and X_j and Y_j, the desired normalized loadings, are functions of ϕ, the angle of rotation.

It is immediately seen that

$$X_j = x_j \cos\phi + y_j \sin\phi, \tag{14}$$

$$Y_j = -x_j \sin\phi + y_j \cos\phi. \tag{15}$$

Thus,

$$dX_j/d\phi = Y_j, \tag{16}$$

$$dY_j/d\phi = -X_j. \tag{17}$$

According to (10), in this plane,

$$n^2 v_{xy} = n \sum (X^2)^2 - (\sum X^2)^2 + n \sum (Y^2)^2 - (\sum Y^2)^2 \tag{18}$$

should be a maximum. Differentiating (18) with respect to ϕ, using (16) and (17), and setting the derivative equal to zero,

$$n \sum XY(X^2 - Y^2) - \sum XY \sum (X^2 - Y^2) = 0. \tag{19}$$

To solve (19) for ϕ in terms of x_j and y_j, substitute the values of X_j and Y_j from (14) and (15), consult a table of trigonometric identities, and, after a good deal of algebraic manipulation,

$$\phi = \tfrac{1}{4} \arctan \frac{2[n \sum (x^2 - y^2)(2xy) - \sum (x^2 - y^2) \sum (2xy)]}{n\{\sum [(x^2 - y^2)^2 - (2xy)^2]\} - \{[\sum (x^2 - y^2)]^2 - [\sum (2xy)]^2\}}. \tag{20}$$

If $u_j = x_j^2 - y_j^2$ and $v_j = 2x_j y_j$, (20) reduces to the form (11) above.

Of course, (11) or (20) is only a necessary condition for a maximum. By taking the second derivative of (18) sufficient conditions for a maximum may be found. These are summarized below.

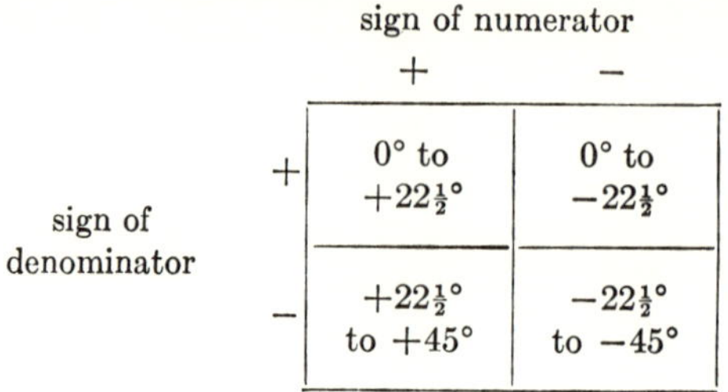

		sign of numerator +	sign of numerator −
sign of denominator	+	0° to $+22\frac{1}{2}$°	0° to $-22\frac{1}{2}$°
	−	$+22\frac{1}{2}$° to +45°	$-22\frac{1}{2}$° to −45°

The sign of numerator and denominator refer to the right-hand member of (20); the values in the cells refer to ϕ.

These single-plane rotations are made on factors 1 with 2, 1 with 3, $\cdots$, 1 with r, 2 with 3, $\cdots$, 2 with r, $\cdots$, $(r - 1)$ with r, 1 with 2, $\cdots$ iteratively until $r(r - 1)/2$ successive rotations of $\phi = 0$ are obtained, i.e., until the process converges. (It was shown [6] that v in (10) cannot be greater than $(r - 1)/r$, and since each successive application of (20) can result only in a non-decrease of v, this iterative procedure must converge.) After convergence, each normalized test vector is restored to its proper length by multiplying by h_j .

Since this article was accepted for publication, the author has prepared a detailed outline for coding an electronic computer program for the varimax criterion. This (dittoed) paper is available from the writer.

REFERENCES

[1] Carroll, J. B. An analytical solution for approximating simple structure in factor analysis. *Psychometrika*, 1953, **18**, 23–38.

[2] Ferguson, G. A. The concept of parsimony in factor analysis. *Psychometrika*, 1954, **19**, 281–290.

[3] Holzinger, K. J. and Harman, H. H. *Factor analysis*. Chicago: Univ. Chicago Press, 1941.

[4] Kaiser, H. F. An analytic rotational criterion for factor analysis. *Amer. Psychologist*, 1955, **10**, 438. (Abstract)

[5] Kaiser, H. F. Note on Carroll's analytic simple structure. *Psychometrika*, 1956, **21**, 89–92.

[6] Kaiser, H. F. The varimax method of factor analysis. Unpublished doctoral dissertation, Univ. California, 1956.

[7] Neuhaus, J. O. and Wrigley, C. The quartimax method: an analytical approach to orthogonal simple structure. *Brit. J. statist. Psychol.*, 1954, **7**, 81–91.

[8] Saunders, D. R. An analytic method for rotation to orthogonal simple structure. Princeton: Educational Testing Service Research Bulletin 53–10, 1953.

[9] Thomson, G. H. *The factorial analysis of human ability*. (5th ed.) New York: Houghton Mifflin, 1951.

[10] Thurstone, L. L. *Theory of multiple factors*. Ann Arbor: Edwards Bros., 1932.

[11] Thurstone, L. L. Primary mental abilities. *Psychometric Monogr. No. 1*, 1938.

[12] Thurstone, L. L. *Multiple-factor analysis*. Chicago: Univ. Chicago Press, 1947.

Manuscript received 11/15/57
Revised manuscript received 1/9/58

12

Reprinted from *Educ. Psychol. Measurement,* **20**(3), 569–590 (1960)

THE "MAXPLANE" PROGRAM FOR FACTOR ROTATION TO OBLIQUE SIMPLE STRUCTURE

RAYMOND B. CATTELL AND JOHN L. MUERLE[1]
Laboratory of Personality Assessment and Group Behavior
and Digital Computer Laboratory, University of Illinois

1. *Characteristics of Existing Analytical Programs*

OBTAINING a unique rotational resolution of factor analyses, either in exploring an entirely new domain or in seeking to confirm hypotheses, depends at present on attaining *simple structure,* since the *proportional profiles* method (Cattell & Cattell, 1955) still lacks completion.

Both theoretical arguments and extensive empirical results show that simple structure and orthogonality are, except for rare accident, incompatible. Available computer programs using analytical methods (whether aiming, incidentally, at oblique *or* orthogonal resolutions of rotation) use either the principle of maximizing fourth powers, as in Wrigley's quartimax (Neuhaus & Wrigley, 1954) and Saunders', Pinzka's, and Dickman's oblimax (Pinzka & Saunders, 1954) or on the distinct principle involved in Kaiser's varimax (1959) and in Carroll's oblimin, biquartimin and his other programs (1953; 1957; 1958). It is generally conceded—and it is certainly the writers' experience—that in the orthogonal programs, Carroll's designs and Kaiser's varimax are best. However, any program which is restricted to orthogonality must be rejected as a basis for general scientific research.

Unfortunately, also, Kaiser's oblique varimax and some of Carroll's programs have the inherent vice that they attempt to avoid

[1] The writers wish to express their indebtedness for advice and help on special subproblems to Kern Dickman, James N. Snyder, Owen White and Neil Wiseman.

solutions in which two factors have very similar loading patterns. Since there seems to be a systematic tendency for a minority of such "cooperative" (Cattell, 1952) simple structure factors actually to occur in nature (Cattell, 1957), the correct solutions are avoided or distorted. In the latter event, there is a tendency for the two reference vectors marking the factors to be forced apart, i.e., to end with a more obtuse angle than the true simple structure requires. As it happens, the other main approach to oblique resolution, oblimax, has an inherent vice with precisely *opposite* results. Here two reference vectors try simultaneously to "cash in" on the fourth powers of any variables which happen substantially (positively) to load both (in the true position), so that the vectors tend to approach. The result is a more acute angle between the reference vectors than the true simple structure requires.

These symmetrical deviations, to obtuseness or acuteness of angle, can readily be seen as necessary corrections on the first visual plots after the analytical methods have done their work. But a *single* visual shift on all factors to these roughly visible hyperplanes is rarely sufficient to correct the inaccuracy, and we have found about ten over-all visually directed shifts generally necessary in reaching good simple structure. Unless a spate of misleading factor analyses is to flood the journals the warning should be heeded that no existing analytical program gives good simple structure and no editor should accept research results simply on the basis that since the rotation is "untouched by hand" it must be machine perfect. Different machine programs give different solutions. The criteria of simple structure remain as before that (a) first, the study has reached *significant* structure, as tested statistically by Bargmann's (1954) or some other test based on the total number of variables brought into the hyperplane; and (b) second, that it not only reaches significance but has the highest *possible* significance as shown by plotting the hyperplane count for successive trials until this "history of the hyperplane" reaches an unimprovable plateau. The analytical program is only an instrument, like visual plotting, for reaching a goal, the success of which endeavor must be *independently* tested.

Incidentally, it seems to have been overlooked by enthusiasts for quick analytical solutions that computer matrix multiplication and photographic reproduction of plots have also greatly shortened and simplified *the procedure of visual rotation,* so that it is no longer unfit to be linked with a good analytical program as the normal final clarifying procedure.

2. *Aims in an Improved Solution*

The weakness of most of the above analytical methods is that they maximize or minimize the wrong function. Despairing of maximizing what is really required—the hyperplane count—they maximize something of which the latter is a far from perfect function. Since the sum of the second powers of projections (the communalities) remains constant, maximizing the sum of the fourth (or some higher) powers amounts to increasing the deviation among loadings. Because some are made high, it is inferable that others are made low, *but there is no guarantee* (either in orthogonal or oblique methods) *that all those which are made low are being forced into the hyperplane.* Indeed, there could be an appreciable difference between the position which maximizes the number of low loadings and that which maximizes the number with zero loadings (or, say, within $\pm.05$, if this be accepted in a given case as twice the standard error of a zero loading).

The five important characteristics which we aimed to build into the new maxplane method, now to be described, were that (1) it would permit obliqueness, comprehending orthogonality as a special case; (2) it should work *directly,* maximizing the hyperplane count, *not* some value of which the hyperplane count is an imperfect function; (3) it should, unlike varimax, etc., put no restriction on the factor patterns and their relationships. For example, we depart from some earlier writers who have added the condition that factorial simplicity rather than complexity of a variable should be a desideratum; (4) it should permit parameters conveniently to be inserted into the program in response to statistical and other properties of the given research, e.g., the adjustment of the width of the bound of a "zero" loading in accordance with the sample size and test reliabilities; (5) where as is typical, there are more true hyperplanes than extracted factors (See Cattell, 1958), it should tend first to select the hyperplanes bearing the factors of largest variance.

In these aims only one condition, Number (2) above, probably requires further discussion. While Thurstone's well-known specification of simple structure in terms of zeros simultaneously in rows and columns is generally a very good one, it appears to us, in the light of psychological experience, a shade too restrictive notably, as indicated above, in desiring factorial simplicity of *a variable*. The definition that simple structure is that which maximizes the *total number* of zero loadings for k hyperplanes in k space is our preferred and simpler definition, providing condition (5) is met (compare Tucker, 1955). In going for the greatest *total* hyperplane count we recognize that in the usual k factor problem the hyperplane counts per factor vary, so that they can normally be ranked from that with fewest variables to that with most, the latter having possibly 20 to 50 per cent more variables than the former (at least in typical psychological data, where factors vary somewhat in pervasiveness of influence). But provided the program is written so that two or more factors cannot "pile in" upon the same largest or larger hyperplane, and provided the rotation will not accept a factor of lesser variance when one of larger variance is available, no further restriction is necessary. There is then only one way to maximize the total hyperplane count, namely, by each factor in succession finding the best possible hyperplane in the subspace in which it is free to maneuver.

It is the employment of this criterion which distinguishes maxplane from all preceding objective programs, e.g., oblimax, quartimax, varimax, minimax.[2] The distinction lies in (a) the *direct* measurement of the hyperplane, and (b) the use of a nonparametric instead of an analytic continuous function, in that we are interested simply in classifying a variable as belonging in, or not belonging in, the hyperplane.

The use of this approach exposes us to two dangers which, however, exist as much in visual rotation based on human judgment as in the machine process. They are:

(1) That when the hyperplane of a factor B is distinctly poorer than for factor A, and the variables in it have little projection on A, factor B may actually get a better hyperplane count by going *very* oblique to factor A. Thereby it cashes in on many of the variables toward the center of the Factor A hyperplane and does better than if it slightly misses its *own*, rather poorly defined, hyperplane. This is a lesser degree of the outright "piling in" of all factors on

[2] Its title, differing in form from those of the previous family of programs, is meant to indicate this generic difference.

the single best hyperplane discussed above, and it may here be stated in advance that we overcome both by prohibiting one reference vector from approaching closer than 35° (cos 35° $= r_{AB} =$ 0.82) to another.

(2) That in the space obtained when, as customarily, fewer than $n/2$ factors are extracted from n variables, there are nevertheless as many true hyperplanes to be found (apart from the merely spurious "mathematical" ones (Cattell, 1958)) as there are common influences at work among the variables. That is to say, there are usually more real factors than n, and certainly more than k, the number of factors extracted. However, most of these are factors of very small variance, and one's aim in rotation is commonly to find the hyperplanes for the first k factors in order of declining variance (see argument in Cattell, 1958).

In the older use of visual rotation, *starting from the centroid* rather than from some spin, e.g., a Landahl transformation (1938), aimed roughly to equalize the variances of the k factors, it is noticeable empirically that one occasionally misses one or two of the factors of middling rank (Cattell, 1957) and proceeds to find hyperplanes of two or three factors *beyond* the first k in variance rank. This is likely to happen when one or two of the last centroids are relatively near a simple structure position for such small variance factors, and in this case one falls into them readily and the last columns of the transformation matrix—λ_n, in $V_n = V_o\lambda_n$—then bring little change.

The chance of omitting an intermediate hyperplane, and landing on one for a smaller factor, always exists, for visual as for machine rotation. The only real escape from it is to find more hyperplanes than one has factor space, and throw away those bearing smallest variance, or to make the goodness of the hyperplane *relative to the factor variance size* a criterion. Our final program takes care of this.

3. *General Description of Steps in the Maxplane Program*

It seems to us generally desirable that the descriptions of computer programs should be written at two levels, which we have exemplified in the present article, as follows:

(1) In *general* terms, which any psychologist with normal statistical training can follow. Such a presentation permits the de-

scription of the essential sequence of steps to be accompanied by "digressions" explaining the psychological and computational reasons for certain procedural steps—for the readers' evaluation of the suitability of a program depends upon these broader considerations not appropriately included in the exact consecutive account of the instructions of the program itself. (2) In *technical detail* as needed by the computer programmer, staying very close to the exact flow chart. A third possibility, of course, is to give the machine tape itself, which, of course, can always be done when the machines are identical. The present section covers the first of these.

The basic maxplane has been worked out to produce a rotation from the orthogonal centroid or principal components[3] position given by the extraction, because this is where most psychologists will find their starting data. But the program can be and has been made also to proceed from an oblimax, varimax, quartimax, etc., position, and if the above argument is correct that it is better to start from factors of more equal variance than the centroids, experience may well prove this to be an advantage. Maxplane is then coupled to, say, varimax, as a means of "polishing" the hyperplanes and discovering the correct obliquities necessary both for the best simple structure and for proceeding to second-order factorizations.

The general design of the maxplane program is such that it mimics very closely indeed the procedures followed by an experienced psychologist in making single plane rotations visually. Incidentally, an alternative design which was considered was that of taking the k-dimensional sphere (unit length vector, in hyperspace) and planning a complete search, at 10° intervals, on axes which

[3] Parenthetically, a recent dictum by Kaiser and Dickman (1959) has, perhaps inadvertently, reroused among the uninitiated the mathematician's idolatry for principal components solution. Possibly advantages of principal components are two, viz., (1) that if unities are placed in the diagonals one can restore the original scores and (2) that greater fractions of the total variance to be taken out come out in the earlier factors. The first is utterly irrelevant, since no psychologist ever has so many variables in a matrix, namely, an infinite number, that every variable will find another with which its communality is unity. To put ones in when this is not the case is to muddy the common factors with specific factors. As to the second advantage, by the time one is approaching $n/2$ factors extracted, as advocated in modern practice (Cattell, 1958), the difference in variance extracted is commonly negligible. The far greater machine cost of the principal components may in special cases be justified by this fine gain. A far more likely consequence of suggesting principal components to busy researchers is that they will keep the same machine time and restrict their matrices—with the numerous defects, especially in rotation, which smaller matrices bring.

would themselves shift 10° at a time. But with about 20 factors the number of positions is so enormous that this method is too long, even for a computer. The presently adopted method has greater economy.

The essential procedures are as follows:

(1) The factor plots are considered systematically in all possible paired combinations, i.e., $F_1F_2 \ldots F_1F_n$; $F_2F_1, F_2F_3, \ldots F_2F_n; \ldots;$ $F_nF_1, \ldots F_nF_{n\text{-}1}$ through $k(k - 1)$ plots. (In the oblique case these coordinates are actually the reference vectors not the final factors).

(2) On each pair it shifts one factor at a time, in five successive 9° shifts. It is not necessary to move beyond 45° in one direction because the shifts on the other factor will later explore downwards to the 45° position, and there is a similar approach on the negative direction of shifts. Thus all 360° on the plot of these two factors are explored at 9° intervals through 10 shifts, 5 forward, 5 backward, on each of the two factors.

(3) At each position the number of variables in the hyperplane (of prescribed width, e.g., ±.10; see below) is counted, and the information stored. On the machine this requires that what would normally, to the human rotator, be only "envisaged" shifts, evaluated by eye, shall *actually be computed,* i.e., for each position the rotated, V_n, matrix is calculated (for that one factor) and a count made of the variables showing correlations with the reference vector falling within the prescribed band.

(4) A count is also made of the loadings beyond ±.50. For, it will be remembered, we wish the evaluation to rest both on the goodness of the hyperplane count and the variance magnitude of the factor. In most personality studies it can be taken that the true salient variables (Cattell & Baggaley, 1954) for a factor will be above 0.50 and that one is not much interested in a factor with lesser definition. The relative importance of these considerations can be adjusted by altering the parameter for the above lower bound for "high loading salients." Obviously, the prime consideration must be that the hyperplane count for a factor position is only to be rejected if the factor proves of really trivial variance. For most studies with which psychologists are concerned a bound at ±.50 would perhaps weight the high loading count about 1/2 to a 1/10 of the weight given the hyperplane count.

(5) On the stored results, a scanning is carried out to select the

best shifts (combined hyperplane and high variance count) of Factor 1 on each of the other factors. Then the tangents defining the angular shifts productive of the best hyperplane count for Factor 1 with respect to each of the other factors are put together in a column of the "shift matrix." (The S matrix which multiplies and changes the transformation matrix.) The same is done for the remaining factors, except that no reference vector is allowed to take up a position, even if it would produce some slight improvement of its hyperplane, if that would bring it closer than 35° to the best position already assumed by an earlier factor.

(6) The multiplication by the shift matrix producing the new transformation matrix is now carried out simultaneously for all factors giving the new rotated matrix—the first, V_1, following the unrotated orthogonal matrix V_o. At this point it may be desirable to remind the reader of the relations of the shift matrix, S, the transformation matrix, λ_n, the orthogonal unrotated matrix V_o, the oblique rotated matrix V_n, and the reference vector cosine matrix C_{rn}, defining the angles reached among the reference vectors.

$$V_n = V_0\lambda_n \qquad (1)$$

$$\lambda_{n+1} = \lambda_n S_{n \to (n+1)} D_{(n+1)} \qquad (2)$$

$$C_{rn} = \lambda_n'\lambda_n \qquad (3)$$

The chief difference of the procedure in the above first cycle from unrotated to rotated matrix from that followed in the usual visual judgment rotation (besides the actually computing of envisaged positions, just described) is that the changes are worked out for the first factor before those for the second factor are studied, in order that the computer may know what positions must not be approached too closely by the second factor, and so on. Incidentally, it will be noted that D_{n+1} in equation (2) above has to be introduced to renormalize each column of the new, λ_{n+1}, transformation matrix, just as in the visual procedure.

(7) The cycle constituted by (1) through (5) is repeated to proceed from V_1 to V_2, except that the adjustments are now carried out *with increased refinement.* The refinement consists in (a) narrowing the width of the hyperplane in which the count is made, and (b) making counts after smaller trial shifts—4° instead of 9°—in the reference vector and accompanying hyperplane. Both of these

changes are based on the assumption that convergence is taking place.

As regards (a), the convergence of hyperplane width, it is evident that the hyperplane as seen in any one plot will at first not be seen "edgewise," but will consist of an ellipse. To recognize the position of this ellipse against a background of randomly scattered points it is necessary that the width of the band within which the first count is made should be approximately as large as the short axis of the ellipse. On the second cycle, however, the ellipse will have narrowed, by reason of the adjustments in other planes, and it becomes appropriate to count now for a narrower band. Finally, in some later cycle, the program shifts to counting in that final narrowest possible width which is fixed as four times the standard error ($\pm\ 2\sigma$) of a zero loading for the given study (to include 95 per cent of the true count). In regard to (b), the reduction of the angle of search, it is common sense to use one's restricted total of possible trial positions to a narrower range on each successive occasion, since one is presumably "on" the target generally and the problem is to be increasingly precise about the position of the bull's eye.

(8) The program must have a built in stopping point, analogous to the personal decision that the "history of the hyperplane count" has reached a plateau. At present the program is set to stop when the current cycle fails to produce a 1 per cent in total count. This is the value we have found acceptable for "plateau attained" in visual shifting, and it seems similarly to be reached in about 12 to 36 shifts, which is a reasonable amount of machine time. (A 36-variable, 12-factor problem takes about 20 minutes per cycle.) It may be noted here that, as with oblimax, etc., convergence is not direct and continuous but proceeds in what may be roughly described as dampened oscillations in a curve rising to a plateau.

At present the program is so constructed that successive and final widths of hyperplane count can be easily changed, each time to suit the sample size, etc., but the angular shifts, the definition of a high loading, the perfection accepted for stopping, and so on, are permanent features of the program.

4. *Technical Description of Instructions In The Maxplane Program*

Although the actual program has been written specifically for Illiac, at the University of Illinois Digital Computer Laboratory, where it has the number KSL 1.97 in the Statistical Library, it will now be described to transcend the idiosyncrasies while indicating necessary precision for other computers.

The characteristics required of a computer on which a maxplane type program is to be used are speed and a large storage capacity. Speed is required since large amounts of data must be iteratively processed using, for one thing, many multiplications (in general, the most time consuming computer operation next to division). A large storage capacity is required either as directly accessible storage or as back-up storage since large matrices must be processed and reprocessed. Reasonably high speed input-output equipment is desirable again because of the large amount of data to be read in and written out. Some sort of high speed plotting equipment is desirable so that the computer can perform the tedious plotting of the output data as a part of the program. Floating point arithmetic is not, in general, required or even desirable because the data are usually normalized such that scaling problems are trivial and because fixed point arithmetic is usually faster.

Input data for the maxplane program consist of three hyperplane widths, a transformation matrix, and an associated V matrix. More will be said about the hyperplane widths later. The transformation matrix is a unit matrix if the input data are orthogonal and unrotated; otherwise the transformation matrix is that formed by the premaxplane computation used. The V matrix is the unrotated, orthogonal matrix, V_0. Checks are made on the input data to verify that the transformation matrix and the V matrix contain consistent numbers of rows and columns. It would be a sad waste of computer time—on some problems a great deal of time—to rotate data which was worthless because of an error in the preparation of the input data.

In the most general terms possible, a rotation cycle is as follows. A pair of factors is chosen. In the column of the V matrix associated with the factor to be rotated, the elements (i.e., the projections of the points on the factor to be rotated) which are within a given hyperplane width are counted and the elements for the "high salients" (Cattell & Baggaley, 1954) which are greater than a given maximum value are counted—these are the counts of the unrotated position of the factor. The factor in question is now rotated (i.e., the V matrices are calculated which reflect the rotated positions of the factor) in the plane of the factor pair to predetermined positions. The elements in the column of the new V matrices associated with the rotated factor which are within the hyperplane width are counted and those which are greater than the maximum value are counted for each rotated position, respectively. The rotated positions are chosen in a regular manner and so that they lie both toward the positive end and the negative end of the other factor of the pair. Note is taken of the position of the rotated factor (and this could be the unrotated position!) which produces the greatest sum which is formed by adding the count of elements which are within the hyperplane width to one quarter of the count of elements which are greater than the maximum value, for the factor pair in question. Another pair of factors is chosen and, using the unrotated V matrix (i.e., each set of factor pair rotations is calculated independently of all the rest), the process is repeated. This continues until every possible permutation of factors by pairs has been through the rotation process and note has been taken of the respective positions which gave the highest sum. A shift matrix is then formed such that the noted rotation for every permutation of factor pairs can be performed simultaneously.

The mathematics of a rotation are essentially those given by Thurstone (1947, p. 194–216). When a particular factor is to be rotated against another factor, a shift matrix is formed which has 1's on its main diagonal and has the tangent of the angle between the original position of the rotated factor and its rotated position as an entry in the shift matrix at a position whose column number is the number of the rotated factor and whose row number is the number of the co-plot. The tangent is positive for a rotation towards the positive end of the against factor from the unrotated position and negative for a rotation in the opposite direction.

Thurstone allows the main diagonal elements to be —1 as well as 1, but this implies a rotation of more than 90°. Such a large rotation is never allowed in this program in order to simplify the formation of the shift matrix. Since only one off-diagonal element of the shift matrix is filled for each rotation of one factor against another, the shift matrix will support the simultaneous rotation of every factor against every other factor—exactly the operation desired in the maxplane program. To form the new transformation matrix, λ_n, the shift matrix, S, is premultiplied by the last transformation matrix and the product matrix is normalized by dividing each column element by the square root of the sum of the squares of the elements in the column. The new V matrix, V_n, is formed by the operation: $V_n = V_o \lambda_n$.

If the shift matrix has its columns divided by the square root of the sum of the squares of the elements in the corresponding columns of the transformation matrix-shift matrix product, then a new matrix, H_n, is formed which has the property—according to Thurstone—that $V_n = V_{n-1}H_n$. It would be most convenient to use this type of operation—i.e., forming V_n directly from V_{n-1}—in the part of the rotation cycle which rotates one factor into various positions against another factor. Such an operation must be fast and efficient. It is used many times for each permutation of factors by pairs which exists in a problem, and even for relatively small problems the number of permutations is quite large. For just this reason, the H matrix is not convenient to use since its formation is long and complicated. However, an approximation to the H matrix would be adequate for reasons given below.

The effect that the rotation of other factors might have on the determination of the best rotation to be performed of one factor of a pair against the other is not considered in a single rotation cycle in spite of the fact that the calculation which forms the new V matrix, which reflects the factor rotations determined during the rotation cycle, rotates every factor against every other factor simultaneously. This simultaneous rotation probably causes more perturbations away from the simple structure desired in the new V matrix than the use of any reasonable approximation of an H type matrix to determine which of a number of trial rotations is to be used in the rotation of one factor of a pair against another. Even if some additional perturbations were caused by the use of an approximation to the H matrix, the iteration of the rotation cycle causes both the simultaneous rotation perturbations and any perturbations caused by the H matrix approximation to damp down until they are relatively small.

The approximation to the H matrix which is used, then, is that of dividing the elements of the columns of the shift matrix, S, by the square root of the sum of the squares of its own respective columns rather than those of the columns of the transformation matrix-shift matrix product (λS). A little thought will show that this approximation is simply performing the rotation calculations as if the V matrix vectors were referenced to an orthogonal system rather than to an oblique one. It might seem at first glance that this approximation saves nothing more than the handling of the transformation matrix and its multiplication by the shift matrix—if, indeed, this is "nothing more." However, an even greater savings can be attained—the elimination of the squaring, division, and square root operations. Since only a modified shift matrix is being manipulated now, for any rotation of one factor of a pair against the other, only two entries in the modified shift matrix need be determined—the main diagonal entry and another entry in the same column as the main diagonal entry. The column is determined by the rotating factor and the row of the non-main diagonal entry is determined by the co-plot. Further, if predetermined angles of rotation are to be used, the value of the modified shift matrix entries as well as their positions are known. Thus at any time that a rotation is desired, the modified shift matrix elements can be calculated from the angles of rotation and their positions can be determined from the factors being rotated.

Although mathematically clear this is not easy programming. The elements of the modified shift matrix are trigonometric functions of the angles of rotation, and these functions are normally time consuming to calculate. The way out of this for the maxplane program was to resort to a table look-up. There were not so many different angles of rotation to be used that a table of necessary element values for the modified shift matrix would be prohibitively long.

The actual rotation procedure used in the maxplane program, then, is as follows. When one factor of a factor pair is to be rotated against the other to determine which of a number of trial rotations should be performed in the final all factors against all other factors simultaneously rotation, the approximate H matrix—the modified shift matrix—is used. In this case only two of the elements of the modified shift matrix are non-zero or non-unity and they are looked up in a table. Multiplying the V matrix into this modified shift matrix changes only one column in the V matrix—the column of the rotated factor. This is all that is needed since this is the column of the V matrix whose elements determine the sum in question. Actually, the whole matrix multiplication is not even calculated. The same result is accomplished by multiplying an element in the column of the V matrix associated with the rotating factor by the element in the modified shift matrix (obtained by a table look-up) associated with the rotating factor and adding it to the number associated with the same row in the V matrix calculated similarly for the co-plot to obtain the new V matrix element for the row being calculated upon.

When the rotated position which produces the greatest sum for a factor pair is determined, the shift matrix element appropriate for that rotation is determined by a table look-up and placed into the position determined by the factor pair being used in the shift matrix used for the simultaneous rotations. As soon as the proper rotation of every permutation of factors by pairs has been determined—i.e., when the shift matrix is full (again a zero element is legitimate since it just indicates that no trial rotation of that factor of a pair against its appropriate factor gave a better sum than the unrotated position), the rotation cycle can be completed by the matrix manipulations which form the new transformation matrix for the cycle from the shift matrix in question. Then the new V matrix which is used in the next rotation cycle can be calculated from the new transformation matrix and V_0.

If a judicious choice of a factor pairs sequence is made, considerable information handling and storage time can be saved. In the above trial rotation calculations, for instance, the rotating factor number determines the number of the V matrix column to be associated with the rotating factor and the shift matrix column to be filled while the co-plot number determines the number of the V matrix column to be associated with the co-plot. Now if the sequence of factor pairs is chosen as follows: $F_1F_2, F_1F_3, \ldots, F_1F_{1n}$; $F_2F_1, F_2F_3, \ldots, F_2F_n$; $F_3F_1, F_3F_2, F_3F_4, \ldots, F_3F_n$; $F_nF_1, F_nF_2, \ldots, F_nF_{n-1}$, then the shift matrix can be filled one column at a time so that no transfer has to be made from working storage to back-up storage until the column is full and each column of the V matrix associated with the rotating axis need be transferred from back-up storage to working storage only once per rotation cycle.

It has been stated before that the determination of the best rotation of one factor of a pair against the other is to be accomplished independently of all others. This is not in the end exactly followed because of the manner in which the best rotation position is determined—namely by choosing the position at which a sum is greatest. Since there is some factor position in the plane of two factors where the sum in question is a maximum, and since at one time the first factor is rotated against the second and at a later time in the same rotation cycle the second factor is rotated against the first, it is almost a certainty, especially in an iterative procedure, for the two factors to both rotate to that same position where the sum is the greatest. This procedure can degenerate to where a third factor will rotate into the position where two have already collapsed and so on. For this reason an arbitrary limitation was put on the determination of the best rotation of one factor of a pair against the other such that if any trial rotation being considered brought the two factors closer than a specified angle that rotation position would not be used and if no trial rotation position could bring two factors from a too close position to one at least the specified angle apart the trial rotation position which brought the factors the farthest apart would be used no matter what the count was for any of the other positions.

The above procedure, of course, requires both the knowledge of the angle between each factor of a pair before any trial rotations are attempted and, as well, the angle of rotation that was determined best at the time a factor was rotated against the other factor of a pair against the time when the latter factor is to be rotated against the former. The angle between factors before any new factor pair rotations take place can be found from the correlation matrix, C, which is the transpose of the current transformation matrix times the current transformation matrix. The correlation matrix contains the cosines of the angles between factor pairs at the element positions whose row number is the number of one factor and whose column number is the number of the other factor. The angle of best rotation of a factor against another factor of a pair for use when the latter factor is to be rotated into the best position against the former is determined from the entry placed in the shift matrix by the rotation calculation already performed.

Again in this instance, a table look-up is used to save time. The correlation (C) matrix gives the cosines which must be converted to angles. However, the length of a cosine table is limited only by the accuracy of the angle desired. In the maxplane program it was decided that rotations of less than one degree would never be used

—thus one degree could be the limit of accuracy required in the determination of angles. Since the cosine is symmetrical about 0° and since rotations of more than 90° either direction are not required since it is not necessary to have factors cross one another, a cosine table of only 90 entries is necessary. To conserve memory space this table can be packed more than one entry per word depending on the word size of the computer.

The determination of the minimum angle between two factors is rather an arbitrary matter. It should be small enough to give the program as much freedom as possible and not so small as to allow the factors to tend to collapse as far as possible against one another. In the maxplane program a number of runs were made using test data, and 35° seemed a reasonable minimum angle. The only effect increasing the angle would have would be to restrict solutions which would like to have some factors closer than tne prescribed minimum angle.

In order to keep computation time within reason, the number of trial rotation positions used when determining the best rotated position of one factor of a pair against another must be kept small. At the same time, if the grid of trial rotation positions is not extensive enough to search out the best position, then the program may fail at its attempt to attain simple structure or at best require a time consuming large number of iterations of the rotation cycle. One way out is to use a minimum number of trial rotation positions but change from a large angular spacing between positions early in the number of rotation cycle iterations while the structure is still spread and must be hunted out over a large area to a small angular spacing later on in the number of rotation cycle iterations when a finer search of the proper position must be made but when the structure is not so spread as to require a large area to be searched through.

The above solution is used in the maxplane program with the specific details being as follows. If the input data is the unrotated, V_0, then on the first rotation cycle, five trial rotation positions are used on either side of the unrotated position (10 in all) all spaced 9° apart. The hyperplane width used for these trial rotations is the greatest of the three read in with the input data—it seems only logical to use a broader hyperplane width when the data are more scattered since then the hyperplane projections are not as well defined either. On the second and subsequent rotation cycle iterations for unrotated data and on the first and subsequent rotation cycle iterations for data prerotated before maxplane, five trial rotation positions are used on either side of the unrotated position all spaced 4° apart. The hyperplane width used is the middle of the three read in. If, during any rotation cycle, the number of projections for any rotating factor of a factor pair which are within the middle hyperplane width reaches 50 per cent of the total number of points, on the next rotation cycle and thence ever after for the rotating factor of that factor pair five trial rotation positions on either side of the unrotated position all spaced 2° apart and the least hyperplane width read in are used. Only those rotating factors of those factor pairs which have passed the 50 per cent criterion, but no matter when they have passed it as long as it is on the second rotation cycle or later, are allowed to use the finest trial rotation grid. This is an attempt to keep very fine rotations from being tried on data which are spread so much that the program may not be able to see the forest for the trees.

Since any rotating factor of a factor pair may pass the 50 per cent criterion at any time—or never, some track must be kept of the situation for every permutation of factors by pairs. This could take a considerable amount of storage space and require, again, quite a bit of information transfer during the rotation cycle. Since the situation has only three alternatives—the greatest, middle, or least grid spacing and hyperplane width—two binary digits per factor pair are all that is required. Thus both storage space and information transfer time could be saved by packing a computer word with information about a number of factor pairs. The packing should be done such that the information about each rotating factor is grouped since that is the way the factor pairs are picked—each factor is rotated against all others.

The constants used above were determined by running a number of test problems (the solutions of which were determined by hand rotations) using various angular spacings and numbers of trial rotations and picking those constants which gave the best, consistent results. The hyperplane widths were purposely left up to the program user since it was considered that there would be enough variation from data to data to make it hard for a fixed set of constants to be used and easy enough to tell from the data approximately what widths to use. The maximum value against which elements of the rotating factor column of the V matrix are to be compared is 0.50. This value is used throughout the program and was set on the basis of experience with hand rotations.

Since the data to be used in a routine such as maxplane are already normalized, fixed point arithmetic can be used for all calculations to take advantage of its speed. What little scaling needs to be done can be done on input and output and in the matrix multiplication subroutine used to form the new transformation matrix and the new V matrix.

The end test for the maxplane routine had to be a rather special one. Probably because of the perturbations caused by the simultaneous rotation of every factor against every other factor, the plot (as determined by test data runs) of the total number of elements in the V matrix within the least hyperplane width against the number of the rotation cycle being calculated, had a constantly increasing envelope but itself was oscillatory. This means that no end test which uses only the improvement from one rotation cycle to the next as a criterion would do. The end test finally used is one that counts the number of times that the improvement from one rotation cycle to succeeding ones is less than one per cent—where less than one per cent includes degeneration as well as scant improvement. The count is set to 0 each time improvement greater than one per cent is attained and the calculation terminates when the count in question reaches some fixed number. This end test essentially bridges the gap between one secondary maximum and the next until the gap is too wide to be worth the calculation time. The program of course, keeps track of the last best solution (i.e., the last solution at which the count was set to 0) and uses that one when the end test terminates the calculation. The maxplane program presently uses 10 as the fixed number used to terminate calculation. Making this number smaller allows the program to terminate sooner but, in general, gives a poorer solution than using a larger number which requires more time since it encourages more rotation cycles.

At the end of any rotation cycle the program user has three alternatives which are switch controlled. (1) The calculation can be allowed to go on to its ultimate end; (2) the number of the rotation cycle being performed and the number of elements in the V matrix which are within the least hyperplane width, by columns and for the whole V matrix both as raw numbers and as percentages of the total elements in a column and in the V matrix respectively, can be printed out; or (3) the calculation can be terminated as if the end test had terminated the problem at the last best solution attained. The second alternative is used to check the progress of a problem without disturbing the calculation. The third alternative is used if the problem is to be run in sections, and the time for the problem to be pulled is at hand. When the problem is re-entered on the computer, that version of the maxplane program should be used which is normally used for preprocessed data.

The final output from the maxplane program consists of (a) the output for the progress checking alternative above; (b) the C matrix of inter-vector cosines from the last best solution attained; (c) the three hyperplane widths and the last best transformation matrix, λ_n, attained (printed out in such a manner that they could be re-entered along with the V_0 matrix to continue the calculation, if it is desired); (d) the last best V matrix, V_n; and plots of every permutation of factors by pairs with all the point projections, factor numbers, and the appropriate C matrix element (inter-coordinate angle) for the factor pair recorded on each plot. These are the necessary materials for any further results one could require, e.g., the factor inter-correlations (by inversion), factor loadings, estimates of individual's factor scores, second-order factors, etc.

5. *Empirical Verdict on Maxplane*

The effectiveness of a rotation program can only be determined by application to either (a) a model in which the real structure is known because it was put in, and this model should have as many properties as possible of the psychological data on which the program is typically to be used; (b) psychological data which have been thoroughly overdetermined as to structure by two or more independent rotations carried to twenty-five or more over-all shifts.

The Laboratory of Personality Assessment and Group Behavior has used three physical models for testing various methodological factor analytic procedures: Thurstone's Box Problem (Thurstone, 1947) and two of our own called the Ball Problem (Cattell & Dickman, in press) and the Coffee Cups Problem. The second, consisting of measures of some 32 behaviors of 100 balls is a particularly suitable example, being "organic," containing a normal sprinkling of experimental error of measurement and having exactly four factors. Maxplane found all four factor positions (size, weight, elasticity, length of suspending string) about as well as oblimax, but not as well as the visual, personally-directed rotaplot. The machine time required on these examples was about the same as for visual "hand rotation," excluding the psychologist's time on the latter, but, of course, coming in a single block on one day, instead of in the 10 to 30 re-entries, extending over twice or thrice as many days, required for controlled visual shifting.

As an empirical psychological example, we took the 24-variable, 12-factor questionnaire data for eight-year-old children reported by Cattell and Coan (1958). Two "markers" exist for each factor, and the data have been repeatedly worked over in our laboratory to the same unique solution. Its only drawback for the present purpose is that it is among the clearest cut of psychological factorings and has a sharp distinction between the two markers for each factor and the rest of the variables. Maxplane quickly gave all but one of the 12 factors, yielding a hyperplane count of 74% in $\pm$ 10. The mean angle among the factors was cosine $\Theta = .10$, as against .19 for oblimax, there being a consistent tendency elsewhere also for oblimax angles to run higher, as stated above. Experiments with less clean cut psychological examples, i.e., with lower final hyperplane counts, are in progress.

These empirical proofs that maxplane works should not tempt us, however, into saying that it works better than oblimax or (hand) rotaplot. On a combination of criteria of "goodness" worked out for the above examples, it is about as good as oblimax and not quite so good as a long, patient, highly skilled visual rotation. Indeed, as it now stands, it is the beginning of the development of a first rote instrument, not the end. Its chief vice so far has been that if a hyperplane crystallizes quite late in rotation, at perhaps 30° or more away from where Maxplane is making its final, refined, "settling down" (on an inadequate hyperplane) it will not, as a human would, stretch out and shift to the entirely new hyperplane. Incidentally, as every experienced investigator knows, such late "coagulations" of a hyperplane 20° to 40° away from where one has hitherto been working, are inherent in the nature of the cumulative clearing up process. Maxplane would probably be improved by being programmed to make periodic wide (40°) sweeps at late stages in the settling down, i.e., when the hyperplane width has already converged to $\pm.10$. Actually, our best final results have been obtained by hitching Maxplane tandem to oblimax, so that essentially it tidies up from the known slightly erratic position reached by oblimax. In this it has worked best by graduating the three stages of convergence on 4°, 4°, 2° instead of 9°, 4°, 2°, i.e., by allowing its movements to be relatively fine from the beginning.

In stating that Maxplane is the beginning of a program we refer particularly to all the experimentation that can be done in trying out, singly and collectively in patterns, variations on the parameters (made readily adjustable by keying into the program) of (a) the weight given to high, salient variable count as compared to hyperplane count, (b) the size and succession of angles of sweep, and (c) the width in which the hyperplane is counted. With proper attention to, and discovery about, the effects of these insertable parameters we believe that the method has promise of becoming that fully effective and dependable automatic rotation program which factor analysts have long desired. The possibilities of such improvement are inherent in this kind of design, but it is not presented as having yet reached this satisfactory level.

REFERENCES

Bargmann, R. "Signifikanzuntersuchungen der Einfachen Struktur in der Faktoren Analyse." Mitt f Mathem. Statistik. Sonderdruck, Physica-Verlag., Wurzburg, 1954.

Carroll, J. B. "An Analytical Solution for Approximating Simple Structure in Factor Analysis." *Psychometrika,* XVIII (1953), 23–38.

Carroll, J. B. "Biquartimin Criterion for Rotation to Oblique Simple Structure in Factor Analysis." *Science,* CXXVI (1957), 1114–1115.

Carroll, J. B. "Solution of the Oblimin Criterion for Oblique Rotation in Factor Analysis." Unpublished manuscript, 1958.

Cattell, R. B. *Factor Analysis.* New York: Harper Brothers, 1952.

Cattell, R. B. *Personality and Motivation Structure and Measurement.* New York: World Book Company, 1957.

Cattell, R. B. "Extracting the Correct Number of Factors in Factor Analysis." EDUCATIONAL AND PSYCHOLOGICAL MEASUREMENT, XVIII (1958), 791–838.

Cattell, R. B. and Baggaley, A. R. "The Salient Variable Similarity Index—S—for Matching Factors." Adv. Publ. No. 4, 1954. Laboratory of Personality Assessment, University of Illinois, Urbana.

Cattell, R. B. and Cattell, A. K. S. "Factor Rotation for Proportional Profiles: Analytical Solution and an Example." *British Journal of Statistical Psychology,* VIII (1955), 83–92.

Cattell, R. B. and Coan, R. W. "Personality Dimensions in the Questionnaire Responses of Six and Seven Year Olds." *British Journal of Educational Psychology,* XXVIII (1958), 232–242.

Cattell, R. B. and Dickman, K. "Simple Structure: Proof of Immanence in Natural Data, Absence from Random Matrices and Correspondence to Real Entities." (In press)

Kaiser, H. F. "Computer Program for Varimax Rotation in Factor Analysis." EDUCATIONAL AND PSYCHOLOGICAL MEASUREMENT, XIX (1959), 413–420.

Kaiser, H. F. and Dickman, K. W. "Analytical Determination of Common Factors." Presented at American Psychological Association, Annual Convention, Cincinnati, September 9, 1959.

Landahl, H. D. "Centroid Orthogonal Transformations." *Psychometrika,* III (1938), 219–223.

Neuhaus, J. O. and Wrigley, C. "The Quartimax Method." *British Journal of Psychology,* VII (1954), 81–91.

Pinzka, C. and Saunders, D. R. "Analytic Rotation to Simple Structure: II. Extension to an Oblique Solution." *Educational Test Service Research Bulletin,* No. 54, 31, 1954.

Thurstone, L. L. *Multiple Factor Analysis.* Chicago: University of Chicago Press, 1947.

Tucker, L. R. "The Objective Definition of Simple Structure in Linear Factor Analysis." *Psychometrika,* XX (1955), 209–225.

13

Reprinted from *Psychometrika,* **31**(3), 313–323 (1966)

ROTATION FOR SIMPLE LOADINGS

R. I. JENNRICH

AND

P. F. SAMPSON

HEALTH SCIENCES COMPUTING FACILITY
UNIVERSITY OF CALIFORNIA, LOS ANGELES

Existing analytic oblique rotation schemes proceed by optimizing a simplicity function applied to the reference structure. This article suggests optimizing a simplicity function applied to primary loadings directly. The feasibility of the suggestion is demonstrated using the quartimin criterion. An algorithm to implement the optimization is derived and the existence of an admissible solution proved. Practical comparisons with the biquartimin method are made using Thurstone's Box Problem and Holzinger and Swineford's Twenty-Four Psychological Tests Problem.

Analytic rotation schemes in factor analysis are based on a number of simplicity criteria. In the case of orthogonal rotations the primary factor loading matrix is identical to the primary factor structure and rotation for simple primary loadings is equivalent to rotation for simple primary structure. In the case of oblique "rotations," however, the matrices are not identical and it appears a choice would have to be made between rotation for simple primary loadings or simple primary structure. In fact, neither of these alternatives are used. Following Thurstone [5] rotations are made instead for simple reference factor structure. The reference factors are defined as the set of variables of unit variance which lie in the common factor space and are biorthogonal to the primary factors. Since the reference factors seem more like mathematical abstractions than variables of primary interest, an interest in simple reference structure seems a little strange. The reference structure, however, is quite similar to the primary loading matrix (the columns of one are scalar multiples of the columns of the other) and it is apparently this similarity which motivates the interest in simple reference structure.

The purpose of this paper is to show that such an indirect method of obtaining simple loadings is not required.

There are a number of popular oblique rotation schemes—oblimax, quartimin, covarimin, biquartimin, and Kaiser-Dickman to name a few. Most of the simplicity criteria used in these schemes are sufficiently complex that it is difficult, if not impossible, to tell in terms of the reference structure or the primary loading matrix what "simplicity" really means. It will be shown here that by applying the fairly simple quartimin criterion to the loadings directly, very satisfactory results are obtained while avoiding the problem of rotating to singularity, which appears to plague some oblique rotation methods.

I. *Basic Concepts and Statement of the Problem*

Each of the n basic variables z_i in a factor analysis model is a sum

$$z_i = c_i + u_i$$

of two orthogonal variables c_i and u_i called the common and unique parts of z_i. Each common part is a linear combination

$$c_i = l_{i1}f_1 + \cdots + l_{im}f_m$$

of m variables $f_1, \cdots, f_m$ called the primary factors. These factors are assumed to be a basis of the space spanned by the common parts (called the common factor space) and to have unit variance. The model is called orthogonal if it is assumed in addition that the primary factors are uncorrelated. In matrix notation, the model may be written

$$z = Lf + u,$$

where $z = \{z_i\}$ is an $n \times 1$ vector of basic variables, $L = \{l_{ij}\}$ is an $n \times m$ matrix of primary factor loadings, $f = \{f_i\}$ is an $m \times 1$ vector of primary factors and $u = \{u_i\}$ is an $n \times 1$ vector of unique parts.

The selection of a primary factor basis for the common-factor space is called the rotation problem. The selection is made in such a way that the corresponding loading matrix L will be as simple as possible. Formal conditions for simplicity have been stated by Thurstone ([5], p. 335). Roughly speaking, L is called simple if it has many nearly zero elements and a few relatively large elements. The goal is to represent each variable with relatively large loadings on one or at most a few factors and nearly zero loadings on the remaining.

As mentioned earlier, the usual method for doing this is somewhat indirect. Let $g_1, \cdots, g_m$ be a set of variables of unit variance in the common-factor space which are bi-orthogonal to the primary factors. These variables, which are called reference factors, uniquely determine the primary factors and are uniquely determined by them. The covariance matrix

$$S = [\text{cov}(z_i, g_j)]$$

of the basic variables with the reference factors is called the reference factor structure. It is easy to show ([3], p. 279) that the reference structure is related to the primary loading matrix by the equation

$$S = LD,$$

where D is a diagonal matrix with diagonal elements $d_{ii} = \text{cov}(f_i, g_i)$. In other words, the columns of the reference structure S are simply scalar multiples of the columns of the primary loading matrix L. Roughly speaking, S will look simple if L does and conversely. The common analytic rotation schemes proceed by defining a function $F(S)$ which measures the simplicity or complexity of the reference structure S. The rotation problem is then solved by selecting the reference factors which maximize or minimize $F(S)$. In perhaps the simplest case (the quartimin method [1]) the function

$$F(S) = \sum_{p<q} \sum_{i} s_{ip}^2 s_{iq}^2 \tag{1}$$

is minimized.

The authors believe that the reference structure method is unnecessarily complicated and indirect. They suggest that a simplicity function $F(L)$ be applied directly to the loadings and that the rotation problem be solved by maximizing or minimizing $F(L)$. It is always possible (and customary) to begin with an initial set of primary factors which are orthogonal. If A denotes an initial primary loading matrix then L is an admissible primary loading matrix if and only if it can be written in the form $L = AT^{-1}$, where T is a nonsingular normalized matrix (i.e., the rows of T have length one). Similarly S is an admissible reference structure if and only if it can be written in the form $S = AT'$, where again T is a nonsingular normalized matrix. The simple loadings solution amounts to finding a nonsingular matrix T to

$$\text{minimize } F(AT^{-1}) \text{ under the condition diag } (TT') = I, \tag{2}$$

while the simple reference structure solution amounts to finding a nonsingular matrix T to

$$\text{minimize } F(AT') \text{ under the condition diag } (TT') = I. \tag{3}$$

Formally the only difference between the methods is that in the simple loadings method T^{-1} replaces T' in the argument of F.

The T's defined by (2) and (3) are in general different and give rise to different loading and primary factor correlation matrices. In the case of the simple loadings solution, the rotated loadings L and the covariance matrix C of the rotated primary factors are given by

$$L = AT^{-1} \quad \text{and} \quad C = TT'.$$

In the case of the simple structure solution, these formulas are a little more complicated

$$L = AT'D^{-1} \quad \text{and} \quad C = D(TT')^{-1}D$$

where

$$D = [\text{diag } (TT')^{-1}]^{-1/2}$$

It is difficult to understand why the simple loadings method has not been used. One explanation may be Thurstone's leadership in simple structure techniques. Other explanations may include a fear of the complexity of the mathematical details, a fear that the solution methods would behave poorly numerically or that the final solutions would be unsatisfactory. The next three sections will attempt to demonstrate that these fears are unfounded.

II. *Mathematical Details of Method*

The quartimin simplicity function F defined by (1) will be used. The reason for this choice is that both algebraically and conceptually it is about the simplest criterion available and it seems to give very satisfactory results.

The method proceeds by means of a sequence of elementary rotations. Let $f_1, \cdots, f_m$ denote the factors (i.e., primary factors) at an intermediate step, L the corresponding loading matrix, and $C = [c_{ij}]$ the corresponding factor correlation matrix. Choosing two factors, say f_1 and f_2, a simple rotation consists of rotating f_1 in the plane of f_1 and f_2 in such a way that the resulting loading matrix $\tilde{L}$ minimizes $F(\tilde{L})$. The rotated factor

$$\tilde{f}_1 = t_1 f_1 + t_2 f_2, \tag{4}$$

where t_1 and t_2 are chosen so that $\tilde{f}_1$ has unit length. This amounts to requiring that

$$t_1^2 + 2t_1t_2c_{12} + t_2^2 = 1. \tag{5}$$

Let $l_1, \cdots, l_m$ and $\tilde{l}_1, \cdots, \tilde{l}_m$ denote the columns of L and $\tilde{L}$. By equating common parts it is easy to see that

$$\tilde{l}_1 = \frac{1}{t_1} l_1, \qquad \tilde{l}_2 = -\frac{t_2}{t_1} l_1 + l_2 \quad \text{and} \quad \tilde{l}_i = l_i \quad \text{for} \quad i \neq 1, 2. \tag{6}$$

Letting xy denote the element-wise product of two arbitrary vectors x and y and letting

$$(x, y) = \sum_{i=1}^{n} x_i y_i,$$

the function $F(\tilde{L})$ can be written in the form

$$F(\tilde{L}) = \sum_{p<q} (\tilde{l}_p^2, \tilde{l}_q^2) = (\tilde{l}_1^2, \tilde{l}_2^2) + \left(\tilde{l}_1^2 + \tilde{l}_2^2, \sum_{q=3}^{m} l_q^2\right) + K,$$

where K is constant with respect to t_1 and t_2. Letting $w = \sum_{q=3}^{m} l_q^2$, the problem reduces to minimizing

$$f = \left(\left(\frac{1}{t_1} l_1\right)^2, \left(\frac{t_2}{t_1} l_1 - l_2\right)^2\right) + \left(\left(\frac{1}{t_1} l_1\right)^2 + \left(\frac{t_2}{t_1} l_1 - l_2\right)^2, w\right) \tag{7}$$

over all t_1 and t_2 satisfying restriction (5). Using the change of variable

$$\gamma = \frac{1}{t_1}, \qquad \delta = \frac{t_2}{t_1}, \tag{8}$$

(5) and (7) become

$$\gamma^2 = 1 + 2c_{12}\delta + \delta^2, \tag{9}$$

and

$$f = a + b\delta + c\delta^2 + d\delta^3 + e\delta^4, \tag{10}$$

where

$$\begin{aligned}
a &= (l_1^2, w) + (l_2^2, w) + (l_1^2, l_2^2),\\
b &= 2c_{12}(l_1^2, w) - 2(l_1^2, l_1l_2) - 2(l_1l_2, w) + 2c_{12}(l_1^2, l_2^2),\\
c &= (l_1^2, l_1^2) - 4c_{12}(l_1^2, l_1l_2) + (l_1^2, l_2^2) + 2(l_1^2, w),\\
d &= 2c_{12}(l_1^2, l_1^2) - 2(l_1^2, l_1l_2),\\
e &= (l_1^2, l_1^2).
\end{aligned}$$

Hence the problem reduces to minimizing, without restrictions, the fourth degree polynomial given in (10). The minimum may be found by choosing the root δ of the cubic equation

$$b + 2c\delta + 3d\delta^2 + 4e\delta^3 = 0 \tag{11}$$

which minimizes f. Then, γ and δ are found from (9) and (11) (the sign of γ is arbitrary) and t_1 and t_2 are obtained from (8). The new loadings matrix $\tilde{L}$ is obtained from (6) and the new correlation matrix $\tilde{C} = \{\tilde{c}_{ij}\}$ of the rotated factors from the equations

$$\begin{aligned}
\tilde{c}_{1j} &= t_1c_{1j} + t_2c_{2j}, & j &\neq 1;\\
\tilde{c}_{ij} &= c_{ij}, & i, j &\neq 1.
\end{aligned} \tag{12}$$

This completes a simple rotation. Rotations are performed stepping uniformly through all possible pairs of factors until $F(L)$ converges. The final values of L and C are the loading matrix and correlation matrix of the rotated factor solution.

Harry Harman has suggested that the method described in this section be called the direct quartimin method and that simple loading methods in general be called direct rotation methods. For example, methods based on minimizing the oblimin criterion applied directly to the loadings rather than to the reference structure would be called direct oblimin methods. A FORTRAN IV subroutine which implements the direct quartimin method described here and the direct oblimin methods in general has been written by the authors and may be obtained by writing to the program librarian of the Health Sciences Computing Facility, UCLA, Los Angeles, California 90024. The subroutine also exists as part of a factor analysis program BMDX72 which is available at the same address.

III. *Existence and Nonsingularity of Solution*

In this section we will prove that as the common factors approach linear dependence, the simplicity criterion $F(L)$ approaches infinity and also that there exists a loading matrix L which minimizes $F(L)$. While it may be argued that these are merely mathematical niceties, it is not clear that other rotation schemes enjoy them. This will be discussed in the next section.

LEMMA 1. *For any* l_{ij} ,

$$F(L) \geqq \left(\frac{l_{ij}^2 - |l_{ij}|}{m - 1}\right)^2.$$

PROOF.

$$\operatorname{cov}(c_i, f_j) = l_{i1} \operatorname{cov}(f_1, f_j) + \cdots + l_{im} \operatorname{cov}(f_m, f_j).$$

Thus, $|l_{ij}| \leq 1 + |l_{i1}| + \cdots + |l_{i,j-1}| + |l_{i,j+1}| + \cdots + |l_{im}|$ and thus for some

$$k \neq j, \qquad |l_{ik}| \geqq \frac{|l_{ij}| - 1}{m - 1}.$$

The required result follows from formula (1).

THEOREM 1. *If L is a factor loading matrix and C is the corresponding factor correlation matrix, then* $F(L) \rightarrow \infty$ *as* $\det(C) \rightarrow 0$.

PROOF. Since $L = AT^{-1}$ and $C = TT'$,

$$\det(L'L) = \frac{\det(A'A)}{\det(C)} \rightarrow \infty \quad \text{as} \quad \det(C) \rightarrow 0.$$

The required result follows from Lemma 1.

THEOREM 2. *There exists a loading matrix L which minimizes $F(L)$.*

PROOF. Let K be the greatest lower bound of $F(L)$ over all admissible loading matrices L. Since $L = AT^{-1}$ and $C = TT'$ it follows from Theorem 1 that $F(AT^{-1}) \rightarrow \infty$ as $\det(T) \rightarrow 0$. Since the set of all (not necessarily non-singular) normalized matrices T is closed and bounded it follows by continuity that there exists a non-singular normalized matrix T such that $F(AT^{-1}) = K$ and hence an admissible loading matrix $L = AT^{-1}$ which minimizes $F(L)$.

In all fairness it should be pointed out that it has not been proved that the algorithm presented in the previous section converges to a loading matrix L which minimizes $F(L)$ or even that it converges at all. No such proofs exist for any of the common analytic rotation schemes. For all methods the fact that they converge to a loading or structure matrix is a matter of experience and that the matrix optimizes the required criterion a matter of faith.

IV. *Some Comparisons with Other Methods*

The oblimin methods minimize the function

$$G(S) = \sum_{p \neq q} \left(\sum_i s_{ip}^2 s_{iq}^2 - \frac{\gamma}{n} \left(\sum_i s_{ip}^2\right)\left(\sum_i s_{iq}^2\right) \right), \qquad 0 \leqq \gamma \leqq 1$$

of the reference structure S. When $\gamma = 0, 1/2, 1$ these methods are called the quartimin [1], biquartimin [2], and covarimin methods [4]. In this section, a simple example will be used to raise some objections to these methods while in the next section more practical comparisons will be made.

For the example, we will begin with orthogonal factors and the loading matrix

$$\begin{bmatrix} .960 & .140 \\ .480 & .070 \\ -.560 & .300 \end{bmatrix}$$

The loading matrices of the simple loading and several oblimin solutions are

$$\underset{\text{Simple loading}}{\begin{bmatrix} .970 & .000 \\ .485 & .000 \\ .000 & .635 \end{bmatrix}}; \quad \underset{\gamma = 0}{\begin{bmatrix} .970 & .000 \\ .485 & .000 \\ .000 & .635 \end{bmatrix}}; \quad \underset{\gamma = 1/4}{\begin{bmatrix} .951 & -.028 \\ .475 & -.014 \\ -.117 & .547 \end{bmatrix}};$$

$$\underset{\gamma = 1/2}{\begin{bmatrix} .940 & -.051 \\ .470 & -.026 \\ -.217 & .486 \end{bmatrix}}; \quad \underset{\gamma = 3/4}{\begin{bmatrix} .933 & -.084 \\ .467 & -.042 \\ -.294 & .456 \end{bmatrix}}; \quad \underset{\gamma = 1}{\begin{bmatrix} .928 & -.144 \\ .464 & -.072 \\ -.350 & .459 \end{bmatrix}}.$$

A perfect cluster solution exists and the simple loading and quartimin methods produce it. The remaining oblimin methods fail to obtain the "perfect" solution. This observation suggests the use of the simple loadings or quartimin method but the latter method, and to an extent all the oblimin methods with small γ are plagued by a problem which is called the problem of rotating to singularity. More precisely stated, the problem is that in some cases, as the iterations employed in the solution proceed, the reference factor correlation matrix becomes singular. To the authors' knowledge it has never been shown for the oblimin methods that there exists a set of linearly independent reference factors whose structure S minimizes the criterion $G(S)$. It is not clear whether the problem of rotating to singularity is due to the non-existence of a solution S or simply to problems in numerical analysis.

Carroll has suggested two solutions to the problem. One is to use a high value of γ which tends to make the reference factor correlations smaller ([3], p. 324). The other is to put an arbitrary upper bound on the correlations. Either of these alternatives can make it impossible to rotate to a perfect cluster solution when one exists. They also introduce added complexity into the algebraic formulation of the criterion G and into its interpretation.

On the other hand, the simple loadings method uses an algebraically simple criterion, can produce perfect cluster solutions when they exist and, thanks to Theorem 1, will not rotate to singularity.

V. *Practical Comparisons with the Biquartimin Method*

It remains to be demonstrated that the simple loadings method given in Section II works for practical problems—that for real data the method in fact gives loadings that look simple, or at least loadings which look as simple as those of other analytic oblique rotation methods. To do this the simple loadings and biquartimin methods will be compared using two well-known problems from the literature—Thurstone's Box Problem ([5], p. 140) and Holzinger and Swineford's Twenty-Four Psychological Tests Problem ([3], p. 135).

For the purpose of making comparisons with the results in the literature, Kaiser communality normalizations ([3], p. 325) have been used for both the simple loadings and oblimin methods. In both examples the initial factors are orthogonal.

Table 1 contains the simple loadings and biquartimin solutions to the

TABLE 1

Thurstone's Box Problem

Initial Loadings			Simple Loadings			Biquartimin Loadings		
.659	-.736	.138	-.035	1.020	-.056	-.002	1.003	-.036
.725	.180	-.656	1.011	-.032	-.022	.980	.020	.029
.665	.537	.500	-.046	-.066	1.018	-.016	-.047	1.001
.869	-.209	-.443	.765	.456	-.045	.756	.488	.003
.834	.182	.508	-.073	.344	.879	-.032	.353	.870
.836	.519	.152	.364	-.071	.834	.377	-.035	.840
.856	-.452	-.269	.525	.727	-.079	.531	.742	-.037
.848	-.426	.320	-.045	.866	.363	-.004	.859	.373
.861	.416	-.299	.787	-.092	.452	.775	-.041	.484
.880	-.341	-.354	.648	.610	-.059	.648	.633	-.013
.889	-.147	.436	-.066	.656	.644	-.022	.656	.645
.875	.485	-.093	.611	-.092	.653	.611	-.046	.673
.667	-.725	.109	.000	1.006	-.066	.031	.990	-.045
.717	.246	-.619	.989	-.087	.044	.959	-.034	.091
.634	.501	.522	-.091	-.039	.997	-.059	-.023	.978
.936	.257	.165	.327	.222	.725	.347	.250	.736
.966	-.239	-.083	.450	.629	.244	.465	.648	.276
.625	-.720	.166	-.073	.998	-.041	-.039	.980	-.024
.702	.112	-.650	.977	.023	-.071	.947	.072	-.020
.664	.536	.488	-.036	-.068	1.009	-.005	-.049	.992

Primary Factor Correlations

1.000			1.000			1.000		
.000	1.000		.334	1.000		.242	1.000	
.000	.000	1.000	.337	.249	1.000	.248	.196	1.000

Box Problem. Included are the initial loadings, the rotated loadings, and the primary factor correlations. The rotated loadings look fairly similar. This similarity along with their common similarity to Thurstone's graphical solution ([5], p. 228) can be seen in Fig. 1 which is an extended vector representation on the second and third original factors. The primary factor correlations for the simple loadings method are a little higher than those for the biquartimin method.

The simple loadings and biquartimin solutions to the Twenty-Four Psychological Tests Problem are very similar as can be seen from Table 2. The maximum absolute difference in the loadings is .071 and the average absolute difference is .016. All of the loadings except one nearly zero loading have the same sign. Again, the correlations for the simple loadings method are a little higher than those for the biquartimin method.

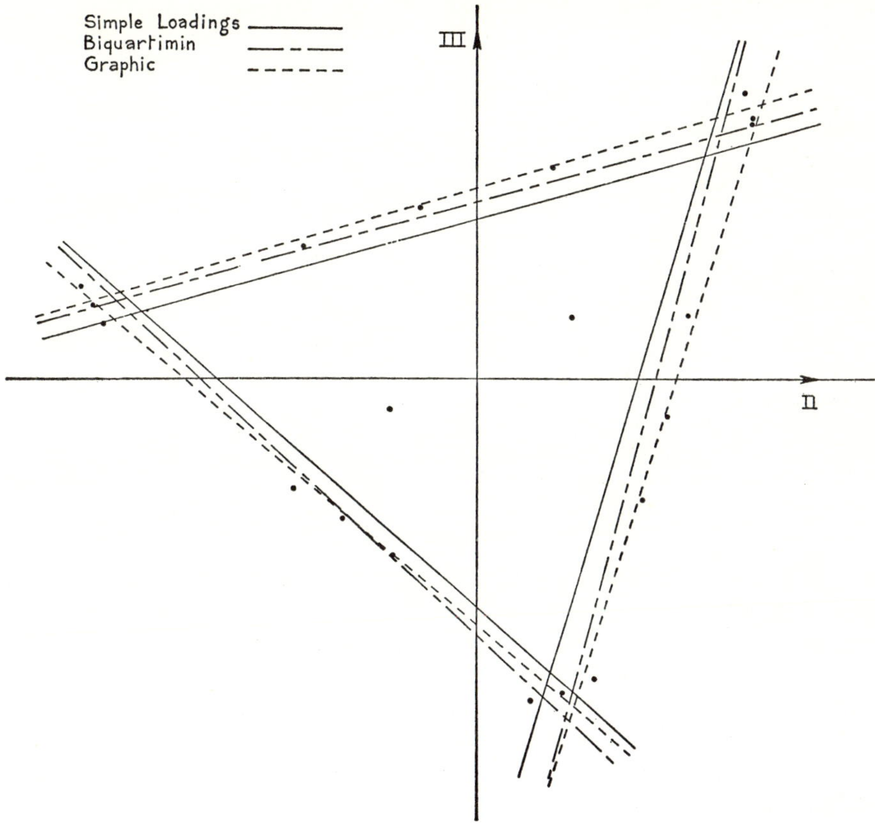

FIGURE 1

Comparison of Oblique Rotations, Extended Vector Representation

TABLE 2

Twenty-Four Psychological Tests Problem

Initial Loadings				Simple Loadings				Biquartimin Loadings			
.608	-.116	.300	-.250	.665	.116	-.004	.054	.666	.103	.015	.058
.372	-.119	.207	-.135	.435	.015	.019	.032	.436	.008	.031	.033
.427	-.220	.262	-.155	.554	-.049	.061	.000	.555	-.057	.074	.000
.477	-.211	.206	-.184	.540	.022	.107	-.030	.545	.012	.117	-.025
.668	-.306	-.344	.108	.024	.113	.779	-.033	.083	.120	.748	-.009
.661	-.337	-.258	.216	.039	-.031	.771	.079	.099	-.018	.742	.095
.652	-.396	-.384	.124	.013	.062	.872	-.096	.076	.069	.836	-.069
.662	-.225	-.153	-.060	.245	.176	.512	-.042	.285	.174	.497	-.021
.664	-.394	-.240	.308	.023	-.140	.831	.134	.087	-.122	.800	.146
.462	.455	-.365	-.136	-.238	.717	.097	.123	-.208	.711	.089	.149
.569	.397	-.208	-.063	-.091	.576	.085	.260	-.060	.574	.082	.276
.484	.360	-.149	-.388	.138	.721	-.087	-.039	.146	.701	-.080	-.012
.608	.130	-.099	-.402	.328	.594	.078	-.121	.341	.573	.082	-.092
.442	.199	-.013	.293	-.110	.059	.152	.501	-.076	.077	.149	.492
.407	.170	.146	.266	.035	-.037	.025	.525	.058	-.021	.032	.508
.523	.077	.300	.076	.357	-.013	-.053	.418	.368	-.008	-.035	.404
.492	.317	.082	.338	-.091	.066	.032	.668	-.059	.086	.037	.652
.547	.307	.248	.072	.229	.171	-.165	.531	.242	.176	-.144	.518
.452	.125	.129	.111	.150	.065	.031	.372	.168	.072	.039	.365
.612	-.174	.128	.004	.397	-.003	.263	.158	.421	-.002	.265	.161
.601	.114	.080	-.171	.339	.324	.032	.149	.353	.316	.042	.159
.608	-.144	.145	.136	.317	-.086	.276	.295	.246	-.070	.277	.291
.691	-.164	.129	-.116	.494	.109	.243	.087	.516	.104	.247	.095
.654	.151	-.150	-.003	.054	.371	.266	.223	.091	.372	.260	.237

Primary Factor Correlations

1.000				1.000				1.000			
.000	1.000			.313	1.000			.295	1.000		
.000	.000	1.000		.434	.313	1.000		.341	.261	1.000	
.000	.000	.000	1.000	.376	.412	.405	1.000	.329	.338	.337	1.000

Conclusion

It has been shown both from a mathematical and practical point of view that it is possible to solve the oblique rotation problem by applying a simplicity criterion directly to the primary factor loadings rather than to the reference factor structure. While it is not the intent of the authors to promote any particular criterion it turns out that the quartimin criterion in addition to being algebraically and conceptually simple, works quite well. The criterion produces perfect cluster loadings when they exist and cannot rotate to singularity, a problem which appears to plague the criterion when applied to the reference structure.

It would be interesting to investigate the behavior of other existing criteria applied to the primary loadings rather than the reference structure.

The authors, for example, have used other oblimin criteria with apparent success. Since some of the elements of the primary factor loading matrix become infinite as the factors become dependent (Lemma 1), while the elements of the reference factor structure matrix are always bounded by one, the problem of rotating to singularity seems less likely to plague a criterion applied to primary loadings than one using reference structure.

REFERENCES

[1] Carroll, J. B. An analytical solution for approximating simple structure in factor analysis. *Psychometrika*, 1953, **18**, 23–28.

[2] Carroll, J. B. Biquartimin criterion for rotation to oblique simple structure in factor analysis. *Science*, 1957, **126**, 1114–1115.

[3] Harman, H. H. *Modern factor analysis*. Chicago: Univ. Chicago Press, 1960.

[4] Kaiser, H. F. The varimax criterion for analytic rotation in factor analysis. *Psychometrika*, 1958, **23**, 187–200.

[5] Thurstone, L. L. *Multiple-factor analysis*. Chicago: Univ. Chicago Press, 1947.

Manuscript received 10/18/65
Revised manuscript received 1/15/66

14

Reprinted from *Psychometrika,* **32**(4), 381–401 (1967)

A COMPARISON OF FOUR METHODS OF CONSTRUCTING FACTOR SCORES*

RODERICK P. McDONALD AND E. J. BURR

UNIVERSITY OF NEW ENGLAND, N.S.W., AUSTRALIA

Four least-squares methods for constructing factor scores have been described in the literature. The formal properties of these scores are developed, and they are compared in terms of four generally desirable properties of constructed factor scores. In particular, it is shown that two of the methods yield scores that are conditionally unbiassed, and univocal in the sense of Guilford and Michael, though not orthogonal, while one of the other methods yields orthogonal scores.

It is shown that constructed factor scores cannot be simultaneously univocal and orthogonal, unless we choose the special basis in factor space given by Canonical Factor Analysis.

The general problem of choosing between the methods is discussed, on the basis of the theoretical relations obtained.

1. *Introduction*

In the past thirty years or so the theory of common factor analysis has undergone a tremendous amount of development. Almost all of this development has been concerned with methods for the determination of matrices of factor loadings, and methods for transforming these into an "acceptable" form for interpretation. Very little attention has been given to the correlative problem of determining factor scores.

From one point of view, this is not surprising, since the usual objectives of factor analysis are commonly considered to be attained when one has an interpretable factor structure. On the other hand, the theoretical problems connected with the indeterminacy of factor scores [Guttman, 1955] have led to misgivings about the usefulness of the common factor model, and some investigators [Guilford and Michael, 1948; Heermann, 1963] have been concerned to seek methods for constructing factor scores that may be used in a further phase of research, following on from the initial factor analysis itself. Further, Horst [1965] and Burkett [1964] show that reduced-rank approximations to a set of multivariate observations have useful properties in multiple regression problems. Also, the method of nonlinear factor analysis described by McDonald [1962, 1967] requires, loosely speaking, knowledge

*EDITOR'S NOTE: The reader will quickly discover that this article develops several of the generalizations given in the second part of the preceding article, "On Factors and Factor Scores." Independent development of the same generalizations is, of course, not a new phenomenon. Because the Presidential Address automatically is accepted for publication and given space in the December issue, it was decided that the only fair thing to do was to print this article in the same issue.

about factor scores, and it is not immediately evident whether the methods recommended for obtaining such knowledge are the most suitable for this purpose. The primary motive for the present paper was to examine this last question.

It has commonly been pointed out that factor scores cannot be determined precisely, but only "estimated", since the number of common and unique factor scores exceeds the number of observed variables. Two defects of the most commonly used "estimates" (due to Thurstone, 1935) are, firstly, that even when the "true" factor scores in the model are assumed mutually orthogonal, the "estimates" are typically correlated, and secondly, the estimated scores do not correlate only with the corresponding true factor scores, but they also have non-zero correlations with non-corresponding factors in the model.

Given a $m \times r$ matrix of factor loadings for m fixed tests, $m > r$, from an observed $m \times 1$ vector it is possible to construct a $r \times 1$ vector, such that this will be, in some sense, an approximation to the corresponding $r \times 1$ vector of factor scores in the model. For convenience we will refer to the components of the former as *constructed factor scores*, and the components of the latter as *true factor scores*. For the purposes of this paper it will suffice to consider orthogonal models only.

In the context of the orthogonal factor model, there are at least four properties of a vector of constructed factor scores that may be generally desirable. Let $\mathbf{x}_0' = [x_1^{(0)}, \cdots, x_r^{(0)}]$, be a vector of true factor scores, such that $\mathcal{E}\{\mathbf{x}_0\mathbf{x}_0'\}$ is a diagonal matrix, and $\mathbf{x}' = [x_1, \cdots, x_r]$, be a corresponding vector of constructed factor scores. Then, firstly, $\mathbf{x}$ should in some sense approximate $\mathbf{x}_0$ as closely as possible. Secondly, the constructed factor scores should also be orthogonal, i.e. $\mathcal{E}\{\mathbf{x}\mathbf{x}'\}$ should be a diagonal matrix. Thirdly, each constructed factor score should correlate zero with each non-corresponding true factor score, i.e. $\mathcal{E}\{\mathbf{x}_0\mathbf{x}'\}$ should be a diagonal matrix. Constructed factor scores having this property have been called *univocal* by Guilford and Michael [1948]. Lastly, we might desire the property that the constructed factor scores be conditionally unbiassed estimators of the corresponding true factor scores, that is $\mathcal{E}\{\mathbf{x} \mid \mathbf{x}_0\} = \mathbf{x}_0$ (the conditional expectation being taken over the subpopulation of "persons" whose true factor scores are $\mathbf{x}_0$).

By far the most widely known method for obtaining constructed factor scores is the least squares procedure due to Thurstone. It is well known that, in general, factor scores so constructed do not possess the last three of the four properties listed above. (It should be noted that Heermann [1963] has described procedures whereby one may transform a set of constructed factor scores in the sense of Thurstone, into either a set having the second property or a set having the third property.)

In addition to the Thurstone method, three other least squares procedures for obtaining constructed factor scores have been described in the literature. One of these has been characterized by Horst [1965] as "the" least-squares method. A similar, but distinct method was described by Bartlett [1937], and is given by Anderson and Rubin [1956] as the maximum likelihood estimator under the usual normality assumptions. A fourth method, given by Anderson and Rubin [1956], is designed to yield constructed factor scores which are mutually orthogonal, i.e. which have the second property listed above. The last two methods do not seem to have been widely noticed in the ten years since the appearance of Anderson and Rubin's paper, in spite of the fact that both maximum likelihood estimates and orthonormal estimates are highly desirable from many points of view.

The object of this paper is to examine the properties of the constructed factor scores given by the four methods indicated above. Section 2 contains a formal statement of the problem. In Section 3, the four methods are described and their main consequences developed. In particular, it is shown that the second and third methods yield, in general, constructed factor scores that are univocal and unbiassed, while only the fourth method yields orthogonal scores. In Section 4, the classical Spearman case is considered, in which we have just one common factor. In this case it is shown that three of the methods are equivalent, while the remaining method (the least squares method in the sense of Horst) in general yields a less precise estimate. In Section 5, a method is sought which will yield constructed factor scores that are both univocal and orthogonal, without regard for the extent to which they approximate the true factor scores. It is shown that such a method cannot exist for arbitrarily prescribed factor loadings, but only if we choose the special basis in common factor space given by Canonical Factor Analysis [Rao, 1955]. In Section 6, we examine the effects of rotation on the constructed factor scores. Two inequalities are established, one relating the Thurstone factor scores and the Bartlett factor scores, and another relating the latter and the least squares method as described by Horst. In the last section we examine the general problem of choosing between the methods, in the light of the preceding argument.

2. *The general problem*

Consider a matrix Z, of order $m \times n$, $m \leq n$, consisting of n observations on a vector variable $\mathbf{z}' = [z_1, \cdots, z_m]$. We wish to fit an orthogonal common factor model to the matrix Z. A number of forms of the model need to be distinguished.

In one form, Model I, we treat the n column-vectors $\mathbf{z}_i$, $i = 1, \cdots, n$, as n observations independently and identically distributed as the random vector $\mathbf{z}$. We may choose an origin and scale for each element of $\mathbf{z}$ such that $R = \mathcal{E}\{\mathbf{z}\mathbf{z}'\}$ is a correlation matrix. We then write the model in the form

(1) $$\mathbf{z} = [F\ U]\begin{bmatrix}\mathbf{x}_0\\ \mathbf{e}\end{bmatrix} = F\mathbf{x}_0 + U\mathbf{e},$$

where F is of order $m \times r$, and of rank $r \leq m$, U is an $m \times m$ diagonal matrix, and $\mathbf{x}_0' = [x_1^{(0)}, \cdots, x_r^{(0)}]$, and $\mathbf{e}' = [e_1, \cdots, e_m]$ are random vectors. The orthogonal model has the properties

(2) $$\mathcal{E}\{[\mathbf{x}_0'\ \mathbf{e}']\} = \mathbf{0}$$

$$\mathcal{E}\{\mathbf{e} \mid \mathbf{x}_0\} = \mathbf{0},$$

$$\mathcal{E}\left\{\begin{bmatrix}\mathbf{x}_0\\ \mathbf{e}\end{bmatrix}[\mathbf{x}_0'\mathbf{e}']\right\} = I_{(r+m)}\ .$$

We then have the well known result that

(3) $$R = \mathcal{E}\{\mathbf{z}\mathbf{z}'\} = FF' + U^2.$$

The parameters of the model are F and U^2.

In another form of the model, Model II, we regard the entire matrix Z as a sample of one member from an ensemble of matrices, such that

(4) $$Z = FX_0 + UE,$$

where the $r \times n$ matrix X_0 is fixed, and E is an $m \times n$ random matrix. We choose an origin and scale for each row of Z such that $R = \mathcal{E}\{(1/n)ZZ'\}$ is a correlation matrix. The model has the properties

(5a) $$\mathcal{E}\{E\} = \mathbf{0},$$

(5b) $$\mathcal{E}\{X_0E'\} = \mathbf{0},$$

(5c) $$X_0X_0' = nI_r\ ,$$

(5d) $$\mathcal{E}\{EE'\} = nI_m\ ,$$

and

(5e) $$X_0\mathbf{1}_n = \mathbf{0},$$

where $\mathbf{1}_n$ is the column-vector whose elements are all unity. We easily obtain

(6) $$R \equiv \mathcal{E}\left\{\frac{1}{n}ZZ'\right\} = FF' + U^2,$$

which is formally the same as (3).

The arguments in the sequel are based on Model I and Model II only. However, three further interpretations of the common factor model should be briefly noted first. In Model III, say, the m rows of Z are regarded as, in some sense, a sample of size m from a universe or domain of $(1 \times n)$ vectors which might have been observed. In Model IV, say, the unique factor scores $\mathbf{e}$ are represented as the sum of two orthogonal components, viz., specific factors, which might be random or fixed, plus random errors of replication. These two models will not be considered further here.

Lastly, in one interpretation of equation (4), we regard E as fixed, as well as X_0. In consequence, Z must also be regarded as fixed. Unfortunately, some accounts of factor analysis do not make it clear which of the models is intended, and in such accounts it usually seems as though this last situation is meant. In this case, Kestelman [1952] has shown that given Z, F and U, such that $n \geq m \geq r$, satisfying equation (4), there exist matrices X_0^* and E^* such that analogues of the conditions in (5) are satisfied. In outline, his argument is that, given Z, F, and U such that (4) holds, and

$$\frac{1}{n} ZZ' = FF' + U^2, \tag{7}$$

there exists an orthonormal $n \times n$ matrix Q such that

$$[F \; U \; \mathbf{0}] = \frac{1}{\sqrt{n}} ZQ, \tag{8}$$

where $\mathbf{0}$ is the $m \times (n - m - r)$ null matrix. Given such a matrix Q, we could then choose

$$[X_0'E'] = \sqrt{n}\, Q_1, \tag{9}$$

where Q_1 is the $n \times (n - r)$ matrix whose columns are the first $(m + r)$ columns of Q. (See also Heermann [1964] for a geometrical account of this situation.) This case is, of course, quite distinct from that of Model I or Model II.

The problem of estimating common factor scores has been dealt with by a number of investigators, by first estimating F and U^2 for some $r < m$, and then obtaining the factor score estimates by a least-squares procedure, treating the estimates of F and U^2 as though they were equal to the corresponding population values. In the sequel, we assume that F and U^2 are known. The results for Model I and Model II are formally identical, but require different interpretations. It is convenient to develop the argument in terms of Model I, merely indicating the Model II interpretation where necessary.

The general problem may be described as follows:—Let $\mathbf{x}_0$, $\mathbf{e}$, $\mathbf{z}$, be random vectors satisfying (1) and (2). Given an observation of $\mathbf{z}$, (a sample of size 1), we seek to determine a vector $\mathbf{x}' = [x_1, \cdots, x_r]$, in the form

$$\mathbf{x} = B'\mathbf{z}, \tag{10}$$

where B' depends only on F and U. We require that $\mathbf{x}$ be in some sense an approximation to $\mathbf{x}_0$. Three desirable properties of $\mathbf{x}$ are, (a) orthogonality, i.e. that

$$\mathcal{E}\{\mathbf{x}\mathbf{x}'\} = \Delta_1 , \tag{11}$$

where Δ_1 is a $r \times r$ diagonal matrix, (b) conditional unbiassedness, in the sense that

$$\mathcal{E}\{\mathbf{x} \mid \mathbf{x}_0\} = \mathbf{x}_0 , \tag{12}$$

for every $\mathbf{x}_0$, and (c), *univocality* in the sense that

$$\mathcal{E}\{\mathbf{x}_0\mathbf{x}'\} = \Delta_2 , \tag{13}$$

where Δ_2 is a $r \times r$ diagonal matrix.

We note that the condition (12) implies the condition (13) but not conversely, for by (1), (2) and (10) we have

$$\mathcal{E}\{\mathbf{x} \mid \mathbf{x}_0\} = B'F\mathbf{x}_0 , \tag{14}$$

so that (12) is satisfied if and only if

$$B'F = I, \tag{15}$$

in which case we also have

$$\mathcal{E}\{\mathbf{x}_0\mathbf{x}'\} = \mathcal{E}\{\mathbf{x}_0\mathbf{x}_0' + \mathbf{x}_0\mathbf{e}'UB\} = I, \tag{16}$$

by (1), (2) and (10).

Further, from (10) and (11) we have

$$\mathcal{E}\{\mathbf{x}\mathbf{x}'\} = B'RB = \Delta_1 . \tag{17}$$

We shall see in Section 5 that the conditions (15) and (17) cannot both be satisfied for arbitrary F, but only if F is that matrix of factor loadings which is given by Canonical Factor Analysis [Rao, 1955].

3. *The methods*

The four methods to be considered here are as follows:

Method I: Take $\mathbf{x} = \mathbf{x}_1 = B_1'\mathbf{z}$, where B_1 is chosen to minimize the quantity

$$\begin{aligned} \varphi_1 &= \mathrm{Tr}\,[\mathcal{E}\{(\mathbf{z} - F\mathbf{x})(\mathbf{z} - F\mathbf{x})'\}] \\ &= \mathrm{Tr}\,[\mathcal{E}\{(F(\mathbf{x}_0 - \mathbf{x}) + U\mathbf{e})(F(\mathbf{x}_0 - \mathbf{x}) + U\mathbf{e})'\}]. \end{aligned} \tag{18}$$

(Note that when $\mathbf{x} = \mathbf{x}_0$, $\varphi_1 = \mathrm{Tr}[U^2]$.) It is well known that the solution is given by

$$\mathbf{x} = \mathbf{x}_1 = B_1'\mathbf{z} = (F'F)^{-1}F'\mathbf{z}. \tag{19}$$

Such a choice of $\mathbf{x}$ has been characterized by Horst [1965] as the least squares solution. It appears to be well known and commonly employed.

Method II: Take $\mathbf{x} = \mathbf{x}_2 = B_2'\mathbf{z}$, where B_2 is chosen to minimize the quantity

$$\begin{aligned} \varphi_2 &= \mathrm{Tr}\,[\mathcal{E}\{U^{-1}(\mathbf{z} - F\mathbf{x})(\mathbf{z} - F\mathbf{x})'U^{-1}\}] \\ &= \mathrm{Tr}\,[\mathcal{E}\{[U^{-1}F(\mathbf{x}_0 - \mathbf{x}) + \mathbf{e}][U^{-1}F(\mathbf{x}_0 - \mathbf{x}) + \mathbf{e}]'\}]. \end{aligned} \tag{20}$$

(Note that when $\mathbf{x} = \mathbf{x}_0$, $\varphi_2 = \mathrm{Tr}[I] = m$.) The solution is given by

$$\mathbf{x} = \mathbf{x}_2 = B_2'\mathbf{z} = (F'U^{-2}F)^{-1}F'U^{-2}\mathbf{z}. \tag{21}$$

This method was given by Bartlett [1937], but does not seem to have been widely noted.

Method III: Take $\mathbf{x} = \mathbf{x}_3 = B_3'\mathbf{z}$, where B_3 is chosen to minimize the quantity

$$\varphi_3 = \mathrm{Tr}[\mathcal{E}\{(\mathbf{x}_0 - \mathbf{x})(\mathbf{x}_0 - \mathbf{x})'\}]. \tag{22}$$

The solution is given by

$$\mathbf{x} = \mathbf{x}_3 = B_3'\mathbf{z} = F'R^{-1}\mathbf{z}. \tag{23}$$

This method is due to Thurstone [1935] and is generally known as the regression method. It is certainly the most widely known of the methods.

Bartlett [1937] writes

$$\mathbf{e}_3 = UR^{-1}\mathbf{z}, \tag{24}$$

so that an analogue of (1) is satisfied identically. That is to say, we may write

$$\mathbf{z} = [F \ U]\begin{bmatrix} \mathbf{x}_3 \\ \mathbf{e}_3 \end{bmatrix} = [F \ U]\begin{bmatrix} F' \\ U' \end{bmatrix} R^{-1}\mathbf{z}. \tag{25}$$

However, it is easily verified that the orthogonality conditions in (2) are not satisfied by $\mathbf{x}_3$, $\mathbf{e}_3$.

Method IV: Take $\mathbf{x} = \mathbf{x}_4 = B_4'\mathbf{z}$, where B_4 is chosen to minimize the quantity given in (20), subject to the condition that

$$\mathcal{E}\{\mathbf{x}\mathbf{x}'\} = I. \tag{26}$$

The solution, given by Anderson and Rubin [1956], is

$$\mathbf{x} = \mathbf{x}_4 = B_4\mathbf{z} = A^{-1}F'U^{-2}\mathbf{z}, \tag{27}$$

where

$$A^2 = F'U^{-2}RU^{-2}F, \tag{28}$$

or, by (3),

$$A^2 = (F'U^{-2}F)^2 + F'U^{-2}F. \tag{29}$$

These, then, are the four methods to be discussed. It is convenient to define the matrices

$$G = F'U^{-2}F, \tag{30}$$

and

$$K = (G + I)G^{-1} = I + G^{-1}. \tag{31}$$

Equation (29) then becomes

(32) $$A^2 = G(G + I).$$

From the identity

(33) $$F'U^{-2}(FF' + U^2) = (F'U^{-2}F + I)F',$$

using (3) and (30) we have

(34) $$B_3 = F'R^{-1} = (G + I)^{-1}F'U^{-2}.$$

(This is the basis of Ledermann's well known short-cut procedure [Thomson, 1946] for Method III. One is thereby enabled to substitute the inversion of an $r \times r$ matrix, for the inversion of an $n \times n$ matrix, where in practice r is usually much smaller than n.) We may then rewrite (19), (21), (23), and (27), in the convenient forms

(35) $$\mathbf{x}_1 = B_1'\mathbf{z} = \mathbf{x}_0 + B_1'\mathbf{e},$$

where B_1 is given by (19);

(36) $$\mathbf{x}_2 = G^{-1}F'U^{-2}\mathbf{z} = \mathbf{x}_0 + G^{-1}F'U^{-2}\mathbf{e},$$

where G is given by (30);

(37) $$\mathbf{x}_3 = (G + I)^{-1}F'U^{-2}\mathbf{z} = K^{-1}\mathbf{x}_0 + (G + I)^{-1}F'U^{-2}\mathbf{e},$$

where G and K are given respectively by (30) and (31); and

(38) $$\mathbf{x}_4 = A^{-1}F'U^{-2}\mathbf{z} = A^{-1}G\mathbf{x}_0 + A^{-1}F'U^{-2}\mathbf{e},$$

where A and G are given respectively by (28) and (30). For Methods II, III and IV, we must have U positive definite. Given F and U^2 we may then readily obtain expressions for $\mathbf{x}_1$, $\mathbf{x}_2$, and $\mathbf{x}_3$ as functions of $\mathbf{z}$. To determine $\mathbf{x}_4$, we require the symmetric root of A^2, which may be found as follows. First obtain

(39) $$T'GT = D,$$

where T is an orthonormal matrix of order $r \times r$, and D is a diagonal matrix. Given U positive definite, we must have G positive definite, since F is of rank r. We then have by (32),

(40) $$T'A^2T = D^2 + D,$$

so that

(41) $$A^{-1} = T(D^2 + D)^{-1/2}T',$$

whence we may obtain $\mathbf{x}_4$ by (38).

We note that by (36), (37), and (38), we have

(42) $$G\mathbf{x}_2 = (G + I)\mathbf{x}_3 = A\mathbf{x}_4 = F'U^{-2}\mathbf{z},$$

and that

(43) $$\mathcal{E}\{\mathbf{x}_1 \mid \mathbf{x}_0\} = \mathcal{E}\{\mathbf{x}_2 \mid \mathbf{x}_0\} = \mathbf{x}_0 ,$$

i.e., $\mathbf{x}_1$ and $\mathbf{x}_2$ are conditionally unbiassed estimators of $\mathbf{x}_0$, whereas

(44) $$\mathcal{E}\{\mathbf{x}_3 \mid \mathbf{x}_0\} = K^{-1}\mathbf{x}_0$$

and

(45) $$\mathcal{E}\{\mathbf{x}_4 \mid \mathbf{x}_0\} = A^{-1}G\mathbf{x}_0 ,$$

i.e., $\mathbf{x}_3$ and $\mathbf{x}_4$ are, in general, conditionally *biassed* estimators of $\mathbf{x}_0$.

In the special case in which $U^2 = \theta I$, where θ is a scalar, (but not in general, otherwise), we have

(46) $$\mathbf{x}_2 = \theta(F'F)^{-1}F'\theta^{-1}\mathbf{z} = \mathbf{x}_1 .$$

We shall use the notation

(47a) $$C_{\alpha\beta} \equiv \mathcal{E}\{\mathbf{x}_\alpha \mathbf{x}'_\beta\},\ \alpha = 0, 1, 2, 3, 4;\ \beta = 0, 1, 2, 3, 4,$$

(47b) $$R_{\alpha\beta} \equiv [\text{Diag}\ \mathcal{E}\{\mathbf{x}_\alpha \mathbf{x}'_\alpha\}]^{-1/2}[\mathcal{E}\{\mathbf{x}_\alpha \mathbf{x}'_\beta\}][\text{Diag}\ \mathcal{E}\{\mathbf{x}_\beta \mathbf{x}'_\beta\}]^{-1/2},$$
$$\alpha = 0, 1, 2, 3, 4; \qquad \beta = 0, 1, 2, 3, 4,$$

and

(47c) $$S_\alpha \equiv \mathcal{E}\{(\mathbf{x}_\alpha - \mathbf{x}_0)(\mathbf{x}_\alpha - \mathbf{x}_0)'\}, \qquad \alpha = 1, 2, 3, 4,$$

so that $C_{\alpha\beta}$ is a covariance matrix, $R_{\alpha\beta}$ is a correlation matrix, since $\mathcal{E}\{\mathbf{x}_\alpha\} = \mathbf{0}$, $\alpha = 0, 1, 2, 3, 4$, and S_α may be thought of as a covariance matrix of errors.

Using the above relations we easily obtain expressions for the following covariance matrices:

(48a) $$C_{00} = I,$$

(48b) $$C_{01} = I,$$

(48c) $$C_{02} = I,$$

(48d) $$C_{03} = K^{-1},$$

(48e) $$C_{04} = GA^{-1},$$

(48f) $$C_{11} = (F'F)^{-1}F'RF(F'F)^{-1} = I + B_1'U^2B_1 ,$$

(48g) $$C_{12} = I + G^{-1} = K,$$

(48h) $$C_{13} = I,$$

(48i) $$C_{14} = (I + G)A^{-1},$$

(48j) $$C_{22} = I + G^{-1} = K,$$

(48k) $$C_{23} = I,$$

(48l) $$C_{24} = (I + G)A^{-1},$$

(48m) $$C_{33} = F'R^{-1}F = K^{-1},$$

(48n) $$C_{34} = GA^{-1},$$

(48o) $$C_{44} = I.$$

The corresponding correlation matrices are given by

(49a) $$R_{01} = [\text{Diag }\{I + B_1'U^2B_1\}]^{-1/2},$$

(49b) $$R_{02} = [\text{Diag }\{K\}]^{-1/2},$$

(49c) $$R_{03} = K^{-1}[\text{Diag }\{K^{-1}\}]^{-1/2},$$

(49d) $$R_{04} = GA^{-1},$$

(49e) $$R_{12} = [\text{Diag }\{I + B_1'U^2B_1\}]^{-1/2}[\text{Diag }\{K\}]^{-1/2}$$

(49f) $$R_{13} = [\text{Diag }\{I + B_1'U^2B_1\}]^{-1/2}[\text{Diag }\{K^{-1}\}]^{-1/2},$$

(49g) $$R_{14} = [\text{Diag }\{I + B_1'U^2B_1\}]^{-1/2}(G + I)A^{-1},$$

(49h) $$R_{23} = [\text{Diag }\{K\}]^{-1/2}[\text{Diag }\{K^{-1}\}]^{-1/2},$$

(49i) $$R_{24} = [\text{Diag }\{K\}]^{-1/2}(G + I)A^{-1},$$

and

(49j) $$R_{34} = [\text{Diag }\{K^{-1}\}]^{-1/2}GA^{-1},$$

and for the "error" covariance matrices we obtain

(50a) $$S_1 = B_1'U^2B_1 ,$$

(50b) $$S_2 = G^{-1},$$

(50c) $$S_3 = (G + I)^{-1},$$

and

(50d) $$S_4 = 2(I - A^{-1}G).$$

We note also that by (48j), (48k), and (48m)

(51) $$\mathbf{x}_2 = C_{33}^{-1}\mathbf{x}_3 \; ; \qquad \mathbf{x}_3 = C_{22}^{-1}\mathbf{x}_2 \, .$$

That is, $\mathbf{x}_2$ is a pseudo-inverse to $\mathbf{x}_3$ and conversely. We also see that

(52) $$\mathbf{x}_4 = \sqrt{C_{22}^{-1}}\, \mathbf{x}_2 = \sqrt{C_{33}^{-1}}\, \mathbf{x}_3 \, ,$$

where $\sqrt{Y}$ denotes the positive definite matrix whose square is Y, when Y is positive definite.

In the special case where G, and hence A and K, are diagonal matrices, R_{24}, R_{23}, and R_{34} reduce to the identity matrix, and each of R_{02}, R_{03}, and R_{04} is equal to $K^{-1/2}$. This case is treated in Section 5. But in general, neither G nor $B_1'U^2B_1$ is a diagonal matrix, thus in general only the Method IV constructed factor scores are orthogonal, in the sense of (11). However, these factor scores are not in general univocal in the sense of (13). On the other hand, Methods I and II both yield constructed factor scores that are unbiassed and univocal but not in general orthogonal. Method III yields none of these properties in general.

Formally identical results may be written for Model II, by a suitable redefinition of the quantities in (18), (20), and (22), and of $C_{\alpha\beta}$, $R_{\alpha\beta}$, and S_α in (47). Thus, we would write

$$\varphi_1 = \mathrm{Tr}\left[\frac{1}{n}\,\mathcal{E}\{(Z - FX)(Z - FX)'\}\right] \tag{53}$$

in place of (18) for example, and define

$$C_{\alpha\beta} = \frac{1}{n}\,\mathcal{E}\{X_\alpha X_\beta'\}, \qquad \alpha = 0, 1, 2, 3, 4; \qquad \beta = 1, 2, 3, 4, \tag{54}$$

and

$$C_{00} = \frac{1}{n}\,X_0X_0'$$

In this case, the matrices $C_{0\beta}$, $\beta = 1, 2, 3, 4$, are weighted first moment matrices, and the matrices $C_{\alpha\beta}$, $\alpha = 1, 2, 3, 4$; $\beta = 1, 2, 3, 4$, are second moment matrices, since X_0 is fixed.

4. *The Spearman case*

In the classical Spearman case, we have a single common factor, that is, we have $r = 1$. We may then write $F' = [f_1, \cdots, f_m]$, $U^2 = \mathrm{Diag}\,\{1 - f_i^2\}$. Corresponding respectively to (35) through (38), we easily obtain the scalar expressions for the four constructed factor scores,

$$x_1 = \left(1\Big/\sum_{i=1}^m f_i^2\right)\sum_{i=1}^m f_iz_i, \tag{55}$$

$$x_2 = \frac{1}{g}\sum_{i=1}^m \frac{f_i}{1 - f_i^2}\,z_i, \tag{56}$$

$$x_3 = \frac{1}{g+1}\sum_{i=1}^m \frac{f_i}{1 - f_i^2}\,z_i, \tag{57}$$

and

$$x_4 = \frac{1}{\sqrt{g(g+1)}}\sum_{i=1}^m \frac{f_i}{1 - f_i^2}\,z_i, \tag{58}$$

where G in (30) reduces to the scalar

$$g = \sum_{j=1}^{m} \{f_j^2/(1 - f_j^2)\}. \tag{59}$$

We note also that

$$gx_2 = (g + 1)x_3 = \sqrt{g(g + 1)}x_4 , \tag{60}$$

and that

$$v = \text{Var}\{x_1\} = 1 + \frac{\sum_{j=1}^{m} f_j^2(1 - f_j^2)}{\left\{\sum_{j=1}^{m} f_j^2\right\}^2}; \tag{61}$$

$$\text{Var}\{x_2\} = \frac{g + 1}{g}, \tag{62}$$

$$\text{Var}\{x_3\} = \frac{g}{g + 1}, \tag{63}$$

and of course

$$\text{Var}\{x_4\} = 1. \tag{64}$$

Corresponding to the correlation matrices given in (49), we readily obtain

$$r_{01} = 1/\sqrt{v}, \tag{65a}$$

$$r_{02} = r_{03} = r_{04} = \sqrt{g/(g + 1)}, \tag{65b}$$

$$r_{12} = r_{13} = r_{14} = \sqrt{(g + 1)/gv}, \tag{65c}$$

and

$$r_{23} = r_{24} = r_{34} = 1. \tag{65d}$$

We note the more special case where $f_j = \sqrt{\rho}$, $j = 1, \cdots, m$, i.e., where every off-diagonal element in the observed correlation matrix R is equal to a constant, ρ. This corresponds to the situation where we have m parallel tests as defined in classical true score theory. We then have $U^2 = (1 - \rho)I$, and hence $x_1 = x_2$, as in (46). Not surprisingly, we also have

$$r_{01}^2 = r_{02}^2 = r_{03}^2 = r_{04}^2 = \frac{m\rho}{(m - 1)\rho + 1}. \tag{66}$$

The expression on the right is the familiar Spearman–Brown formula for the effect of length on the reliability of a test [Gulliksen, 1950]. In this case the four constructed factor scores differ from each other and from $\sum_{j=1}^{m} z_j$ by a scale factor at most.

We now show that in the general Spearman case we must have

(67) $$r_{01} \leq r_{02} ,$$

or, an equivalent to this by (65a) and (65b),

(68) $$g(v - 1) \geq 1,$$

and that equality is attained if and only if $U^2 = \theta I$, where θ is a scalar.

We have by (59) and (61),

(69) $$\begin{aligned} g(v - 1) &= \left\{\sum_{j=1}^{m} \frac{f_j^2}{u_j^2}\right\}\left\{\sum_{j=1}^{m} f_j^2 u_j^2\right\} \bigg/ \left\{\sum_{j=1}^{m} f_j^2\right\}^2 \\ &= \left\{\sum_{j=1}^{m} \sum_{k=1}^{m} \frac{f_j^2}{u_j^2} f_k^2 u_k^2\right\} \bigg/ \left\{\sum_{j=1}^{m} f_j^2\right\}^2 \\ &= \left\{\sum_{j=1}^{m} f_j^4 + \sum_{j<k}\sum \left[f_j^2 f_k^2 \left(\frac{u_j^2}{u_k^2} + \frac{u_k^2}{u_j^2}\right)\right]\right\} \bigg/ \left\{\sum_{j=1}^{m} f_j^2\right\}^2. \end{aligned}$$

But, for any numbers u_j , u_k , we must have

(70) $$\frac{u_j^2}{u_k^2} + \frac{u_k^2}{u_j^2} \geq 2,$$

and equality is attained if and only if $u_j^2 = u_k^2$, so that

(71) $$g(v - 1) \geq \left\{\sum_{j=1}^{m} f_j^4 + 2 \sum_{j<k}\sum f_j^2 f_k^2\right\} \bigg/ \left\{\sum_{j=1}^{m} f_j^2\right\}^2 = 1,$$

and equality is attained if and only if $u_j^2 = u_k^2$, $j = 1, \cdots , m; k = 1, \cdots , m$. This completes the proof.

We see, then, that in the Spearman case Methods II, III, and IV yield constructed factor scores that differ at most by a scale factor, while Method I yields a constructed factor score which must have a lower correlation with the true factor score, except in the special case where $U^2 = \theta I$.

5. *The special case of canonical factor analysis*

Let us now consider the general problem described in Section 2 from a different point of view. Instead of seeking to construct factor scores that in some sense approximate the true factor scores, as in the methods of Section 3, we ask whether it is possible to construct factor scores satisfying simultaneously just the two conditions (15) and (17). It will be seen below that this cannot be done for arbitrarily prescribed F, but only if we choose the special basis in common factor space which is determined by Canonical Factor Analysis. To restate these conditions, we seek a $m \times r$ matrix B, of rank $r \leq m$, such that the constructed factor scores are univocal, which requires that

(15) $$B'F = I,$$

and such that the constructed factor scores are mutually orthogonal, which requires that

$$B'RB = D, \tag{17}$$

where D is an as yet unspecified diagonal matrix. By (15) we have

$$B'FF'B = I, \tag{72}$$

whence by (3) we obtain

$$B'U^2B = D - I, \tag{73}$$

so that $D - I$ must be positive definite. From (17) and (73), we then have

$$(D - I)^{-1/2}B'RB(D - I)^{-1/2} = (D - I)^{-1/2}D(D - I)^{-1/2} = \Lambda, \text{ say}, \tag{74}$$

and

$$(D - I)^{-1/2}B'U^2B(D - I)^{-1/2} = I. \tag{75}$$

Let

$$Q = UB(D - I)^{-1/2}, \tag{76}$$

so that (74) and (75) become respectively,

$$Q'U^{-1}RU^{-1}Q = \Lambda, \tag{77}$$

and

$$Q'Q = I. \tag{78}$$

But it is known [Rao, 1955] that if we are given R and U consistent with (3), then (77) and (78) determine the elements of Λ uniquely. These are the r latent roots of $U^{-1}RU^{-1}$ that are greater than unity. If they are all distinct, and arranged in descending order of magnitude, say, then Q is also determined uniquely. The pth column of Q is equal to the normalized latent vector of $U^{-1}RU^{-1}$ corresponding to the pth latent root λ_{pp} , p = 1, $\cdots$, r. We then have

$$D = \Lambda(\Lambda - I)^{-1}, \tag{79}$$

$$B' = (\Lambda - I)^{-1/2}Q'U^{-1}, \tag{80}$$

and by (15),

$$F = UQ(\Lambda - I)^{1/2}. \tag{81}$$

But (77), (78), and (81) provide one of several equivalent definitions of Rao's Canonical Factor Analysis model. Further, by (78), (80) and (81), we have

$$B' = (F'U^{-2}F)^{-1}F'U^{-2}. \tag{82}$$

That is, the transformation matrix B is precisely that obtained by the Method II least-squares procedure.

In this special case, (30), (31), and (32) become

$$G = F'U^{-2}F = \Lambda - I, \tag{83}$$

$$K = \Lambda(\Lambda - I)^{-1}, \tag{84}$$

$$A^2 = \Lambda(\Lambda - I), \tag{85}$$

and the correlation matrices given in (49) take the form

$$R_{02} = R_{03} = R_{04} = \Lambda^{-1/2}(\Lambda - I)^{1/2},$$

(86) and

$$R_{23} = R_{24} = R_{34} = I.$$

The constructed factor scores given by any of the Methods II, III, and IV are orthogonal and univocal as defined above, and they are related by at most a change of scale, that is (42) yields in this case

$$(\Lambda - I)\mathbf{x}_2 = \Lambda\mathbf{x}_3 = \Lambda^{1/2}(\Lambda - I)^{1/2}\mathbf{x}_4\,. \tag{87}$$

It should be noted that the sample analogue of the condition that G be a diagonal matrix was used by Lawley [1940] to fix the basis in the common factor space in his maximum likelihood method of factor analysis. This has sometimes been described as an arbitrary mathematical condition. It may now be seen that this basis has special properties in respect of the constructed factor scores.

For the sake of completeness, the corresponding behavior of the Method I factor scores may be noted briefly. Here, we find that (50a) takes the form

$$B_1'U^2B_1 = (\Lambda - I)^{-1/2}(Q'U^2Q)^{-1}Q'U^4Q(Q'U^2Q)^{-1}(\Lambda - I)^{-1/2}, \tag{88}$$

that is, the error covariance matrix for this method is not in general a diagonal matrix. Expressions for covariance and correlation matrices involving $\mathbf{x}_1$ may easily be written down, but clearly such expressions do not in general yield a simple explicit form, such as have been obtained for the other methods.

6. *Some further aspects of the general case*

Supposing that an investigator chooses the Lawley-Rao characterization of the common factor basis for the purpose of the initial factor analysis, it is nevertheless unlikely that he will retain this basis for the interpretation of the factors or the construction of factor scores. Much more commonly, he will choose some orthonormal transformation of the canonical factor loadings that is believed to yield interpretable factors. The problem still remains, then, of choosing a method of factor score construction in the general case where, we may suppose, the factor loadings obtained by a method such as canonical factor analysis, have thereafter been subjected to an arbitrary orthonormal transformation.

The object of this section is to develop some further results which could be of value in choosing between the methods in practice. It is suggested that one possible ground for choosing between two methods is the relative magnitude of the correlation between each constructed factor score and the corresponding true factor score. If all other considerations were indifferent, one would naturally choose the method yielding the higher correlation. With regard to the comparative merits of Methods I and II, both of which yield the unbiassedness property ((12) above), one might attempt to compare, in some sense, their error covariance matrices. It is possible to establish two useful inequalities, one with respect to the diagonal elements of R_{02} and R_{03}, as given by (49b) and (49c), and one with respect to the error covariance matrices S_1 and S_2 given by (50a) and (50b).

For the first of these results, it is sufficient, and convenient, to consider the squares of the diagonal elements of R_{02} and R_{03}. By (49b) and (49c) respectively, we may write

$$D^{(2)} \equiv [\text{Diag } \{R_{02}\}]^2 = [\text{Diag } \{K\}]^{-1}, \tag{89}$$

and

$$D^{(3)} \equiv [\text{Diag } \{R_{03}\}]^2 = \text{Diag } \{K^{-1}\}. \tag{90}$$

From (31) and (39), we obtain

$$D^{(2)} = [\text{Diag } \{T(I + D^{-1})T'\}]^{-1} \tag{91}$$

and

$$D^{(3)} = \text{Diag } \{T(I + D^{-1})^{-1}T'\}, \tag{92}$$

where T is an orthonormal matrix. We note immediately that

$$\text{Tr}\{D^{(3)}\} = \text{Tr}\{(I + D^{-1})^{-1}\}, \tag{93}$$

is independent of T, while in general $\text{Tr}\{D^{(2)}\}$ is not. That is, the sum of the squares of the r correlations between Method III factor scores and the corresponding true factor scores is invariant under orthogonal rotations in the common factor space, but this property does not obtain for Method II factor scores.

Let k^{pp} represent the pth diagonal element of K^{-1}. It is known that for any $r \times r$ symmetric matrix K, we must have

$$k^{pp}k_{pp} \geq 1, \; p = 1, \cdots, r, \tag{94}$$

with equality for a particular p if and only if $k_{pq} = 0 = k_{qp}$, for every $q \neq p$. Thence we must have

$$(95) \qquad d_{pp}^{(3)} \geq d_{pp}^{(2)}, \qquad p = 1, \cdots, r,$$

with equality for any p if and only if $k_{pq} = 0 = k_{qp}$, for every $q \neq p$, and further

$$(96) \qquad \mathrm{Tr}\{D^{(3)}\} \geq \mathrm{Tr}\{D^{(2)}\},$$

with equality if and only if $K = I + G^{-1}$ is a diagonal matrix, or, what is the same thing, G is a diagonal matrix.

It was shown in Section 4 that in the Spearman case, the correlation between the Method I factor score and the true factor score must be lower than that for the other three methods, with equality only in the special case where U^2 is a scalar matrix. In a comparison between Method I and Method II only, an equivalent statement is that the Method I error variance is larger than that of Method II, with the same condition for equality. In the more general case given by $r > 1$, a corresponding inequality holds between the generalized error variances given respectively by $|B_1'U^2B_1|$, and $|G^{-1}|$, from (50a) and (50b). That is, it may be shown that

$$(97a) \qquad |B_1'U^2B_1| \geq |G^{-1}|,$$

or, equivalently,

$$(97b) \qquad |(F'F)^{-1}F'U^2F(F'F)^{-1}| \geq |(F'U^{-2}F)^{-1}|,$$

and that equality holds in general only if $U^2 = \theta I$, where θ is a scalar. This result is stated more precisely in a theorem and corollary given in the appendix to this paper.

This result is formally similar to the well known result that in classical linear least squares theory [Kendall & Stuart, 1961, §§19.4–19.8] the least squares estimator of the (vector) population parameter has minimum generalized variance among all linear unbiased estimators. However, the resemblance is only formal since in (1), $\mathbf{x}_0$ is not a population parameter but a random variable.

8. *Discussion*

In the orthogonal, common factor model, as was pointed out above, we cannot, for a finite number of variables, determine the factor scores precisely. Each of the four methods considered above yields constructed factor scores that are in some sense a least squares approximation to the true factor scores.

Four desirable properties of constructed factor scores are: (a) The constructed factor scores for each factor will have high correlations with the corresponding true factor scores. (b) They will have zero correlations with non-corresponding true factor scores, i.e. they will be *univocal* in the sense of Guilford and Michael [1948]. (c) The constructed factor scores for different factors will have zero correlations, i.e. they will be mutually orthogonal.

(d) The constructed factor scores will be conditional by unbiassed estimates of the corresponding true factor scores.

We have seen that in general, all four methods may be said to satisfy condition (a), and Methods I and II, only, satisfy (b) and (d), while Method IV, only, satisfies (c). The generally popular Method III satisfies only condition (a).

We have also seen that for arbitrarily chosen axes in common factor space, conditions (b) and (c) are incompatible, using any method of construction, and may be jointly satisfied if and only if we choose the special basis in the common factor space that is given by Canonical Factor Analysis. In this case, Methods III and IV yield constructed factor scores equally satisfying conditions (a), (b), and (c), but only Method II meets all four conditions. Hence, when CFA axes are retained, the choice between the methods seems obvious.

It is possible that for some experimental purposes the combination of CFA and the Method II constructed factor scores may suffice. However, this would not be adopted in common practice, since the prevailing opinion is that "unrotated" factors do not lend themselves to the art of interpretation, a process which is commonly regarded as an important goal in factor analysis. We must, then, consider the grounds for choice between the methods in the general case.

In considering the conflicting conditions (b) and (c), it seems reasonable to suggest that condition (b) is in general the more important one, (though the opposite view is defensible). That is, it seems more important that each constructed factor score should be an approximate measure of the corresponding true factor score only, than that the constructed factor scores be mutually orthogonal, but not, loosely speaking, measuring the right quantity.

In general, Method III meets only condition (a). In the special case where the basis of the space is determined by CFA, it is not superior to Method II in this respect. More generally, however, Method III yields higher correlations between constructed factor scores and the corresponding true factor scores than does Method II. Thus in general there is a conflict between condition (a) on the one hand, and conditions (b) and (d) on the other. The choice here will depend in part on the aims of the investigator, and in part on just how large the difference is, in a particular application, between the correlations given by the two methods.

In the method of nonlinear factor analysis described by McDonald [1962, 1967] it is necessary to obtain estimates of certain joint moments of the distribution of the true factor scores. For this purpose, it is essential to obtain constructed factor scores satisfying condition (d), while conditions (b) and (c) are not essential. The likelihood of detecting nonlinearity in the model can be improved by minimizing the variances of the errors. McDonald's recommendations as given in [1962, 1967] were equivalent to combining

PFA with Method I. It would now appear that this procedure is permissible, but not the best possible. For this purpose, the choice lies between Method I and Method II. It has been shown above that the generalized variance of the errors in Method I must be greater than or equal to that of Method II. One would therefore expect to improve the efficiency of the procedures in nonlinear factor analysis by using Method II as against Method I. The extent of the improvement, in any application, will depend on the amount of variation in the unique variances. In the Spearman case, it is readily verified by trials with representative numerical values that the Method I error variance may quite commonly be as much as twice that given by Method II. In any situation where factor scores satisfying conditions (b) or (d) are desired, one can safely recommend the use of the Bartlett factor scores (Method II), rather than the seemingly more popular Methods I and III.

Appendix

Theorem Let F be a matrix of order $n \times r$ and rank r, with $r < n$; F' the transpose of F; $S = F'F$; and $U^2 = \text{diag}\{u_1^2, \cdots, u_n^2\}$, with $|U^2| > 0$. Let

$$W_1 = S^{-1}F'U^2FS^{-1}, \qquad W_2 = (F'U^{-2}F)^{-1}.$$

Then $|W_1| \geq |W_2|$, with equality if and only if the product $u_{i_1}^2 u_{i_2}^2 \cdots u_{i_r}^2$ is constant for all r-tuples $i_1, \cdots, i_r$ such that rows $i_1, i_2, \cdots, i_r$ of F are linearly independent.

Proof By a theorem on the determinant of a product of rectangular matrices [Hadley, 1965], we know that $|F'F|$ equals the sum of the squares of all $\binom{n}{r}$ minors of order $r \times r$ in F; that is,

$$|F'F| = \textstyle\sum_i (d_{i_1 i_2 \cdots i_r})^2 \tag{1a}$$

where $\sum_i$ denotes the sum over all r-tuples $i_1, \cdots, i_r$ satisfying $1 \leq i_1 < \cdots < i_r \leq n$, and $d_{i_1 i_2 \cdots i_r}$ denotes the determinant formed from rows $i_1, i_2, \cdots, i_r$ of F. Let $v_k = u_k^2$, $k = 1, \cdots, n$. By an obvious extension of (1a) we have

$$|F'U^2F| = \textstyle\sum_i (v_{i_1} v_{i_2} \cdots v_{i_r})(d_{i_1 i_2 \cdots i_r})^2,$$
$$|F'U^{-2}F| = \textstyle\sum_j (v_{j_1} v_{j_2} \cdots v_{j_r})^{-1}(d_{j_1 j_2 \cdots j_r})^2.$$

Now

$$\frac{|W_1|}{|W_2|} = \frac{|F'U^2F| \cdot |F'U^{-2}F|}{|F'F| \cdot |F'F|} = \frac{N}{D} \quad \text{say},$$

where the denominator is

$$D = \textstyle\sum_i \sum_j (d_{i_1 \cdots i_r})^2 (d_{j_1 \cdots j_r})^2 \tag{2a}$$

and the numerator is

$$N = \sum_i \sum_j (v_{i_1} \cdots v_{i_r})(v_{j_1} \cdots v_{j_r})^{-1}(d_{i_1 \cdots i_r})^2(d_{j_1 \cdots j_r})^2. \tag{3a}$$

Comparing each of the $\binom{n}{r}^2$ terms in (2a) with the corresponding term in (3a), we distinguish two types. In each term of type 1, $i_1 = j_1$, $i_2 = j_2$, $\cdots$, $i_r = j_r$, so that the product of the v's in such a term in (3a) reduces to unity and corresponding terms of this type in (2a) and (3a) are equal. In each term of type 2, we have $i_k \neq j_k$ for at least one k in $1 \leq k \leq r$. The number of such terms is $\binom{n}{r}^2 - \binom{n}{r}$, which is even and positive since $0 < r < n$. Now form an unordered pair associating the term in which

$$i_k = a_k, \qquad j_k = b_k, \qquad k = 1, \cdots, r$$

with the term in which

$$i_k = b_k, \qquad j_k = a_k, \qquad k = 1, \cdots, r,$$

for every $(2r)$-tuple $a_1, \cdots, a_r, b_1, \cdots, b_r$ having $1 \leq a_1 < \cdots < a_r \leq n$ and $1 \leq b_1 < \cdots < b_r \leq n$ and $a_k \neq b_k$ for at least one k in $1 \leq k \leq r$. The pairing of terms of type 2 is unique and exhaustive and no term is paired with itself, and two paired terms are either both positive or both zero. The ratio of the sum of any two positive paired terms in (3a) to the sum of the corresponding terms in (2a) is

$$\frac{1}{2}\left\{\frac{v_{a_1} \cdots v_{a_r}}{v_{b_1} \cdots v_{b_r}} + \frac{v_{b_1} \cdots v_{b_r}}{v_{a_1} \cdots v_{a_r}}\right\}$$

which has the form $\frac{1}{2}(x + x^{-1})$ where x is positive. Now for x positive, $\frac{1}{2}(x + x^{-1}) \geq 1$ with equality if and only if $x = 1$. Therefore $N \geq D$ and hence $|W_1| \geq |W_2|$, with equality if and only if

$$v_{a_1} \cdots v_{a_r} = v_{b_1} \cdots v_{b_r} \tag{4a}$$

whenever rows $a_1, \cdots, a_r$ of F are linearly independent and rows $b_1, \cdots, b_r$ of F are linearly independent. This completes the proof.

Corollary If among the n rows of F there are $r + 1$ rows no r of which are linearly dependent, then $|W_1| \geq |W_2|$ with equality if and only if u_i^2 is constant for all i such that row i of F is non-null.

Proof Let $\mathbf{f}_i$ denote row i of F, $i = 1, \cdots, n$. If the $\mathbf{f}_i$ and the u_i^2 are similarly permuted, $|W_1|$ and $|W_2|$ are unaltered. Hence we may assume without loss of generality that among $\mathbf{f}_1, \mathbf{f}_2, \cdots, \mathbf{f}_{r+1}$, no r are linearly dependent. To show that $|W_1| = |W_2|$ implies $u_i^2 = u_1^2$, $i = 2, 3, \cdots, r + 1$, choose the numbers $2, 3, \cdots, r + 1$ as subscripts in the left member of (4), and the numbers $1, 2, \cdots, i - 1, i + 1, \cdots, r + 1$ as subscripts in the right member. To complete the proof we must show that $|W_1| = |W_2|$ implies $u_i^2 = u_1^2$ when

$i > r + 1$ and $\mathbf{f}_i$ is non-null. Since $\mathbf{f}_1, \cdots, \mathbf{f}_r$ are linearly independent and $\mathbf{f}_i$ is a non-null linear combination of them, we know [Hadley, 1965] that there is at least one subscript k in $1 \leq k \leq r$ that can be replaced by i to give the new linearly independent set

$$\mathbf{f}_1, \cdots, \mathbf{f}_{k-1}, \mathbf{f}_i, \mathbf{f}_{k+1}, \cdots, \mathbf{f}_r. \tag{5a}$$

Set the r subscripts of (5a) in the left member of (4a) and the r subscripts $1, \cdots, r$ in the right member of (4a), giving $v_i = v_k = v_1$. This completes the proof.

REFERENCES

Anderson, T. W. and R. H. Statistical inference in factor analysis. *Proceedings of the third Berkeley Symposium on mathematical statistics and probability* Vol. V, 1956.

Bartlett, M. S. The statistical conception of mental factors. *British Journal of Psychology*, 1937, **28**, 97–104.

Burkett, G. R. A study of reduced rank models for multiple prediction. *Psychometric Monographs*, No. 12, 1964.

Guilford, J. P. and Michael, W. B. Approaches to univocal factor scores. *Psychometrika*, 1948, **13**, 1–22.

Gulliksen, H. *Theory of mental tests*. New York: John Wiley, 1950.

Guttman, L. The determinacy of factor score matrices with implications for five other basic problems of common-factor theory. *British Journal of Statistical Psychology*, 1955, **8**, 65–82.

Hadley, G. *Linear algebra*. Mass.: Addison–Wesley, 1965.

Heermann, E. F. Univocal or orthogonal estimators of orthogonal factors. *Psychometrika*, 1963, **28**, 161–172.

Heermann, E. F. The geometry of factorial indeterminacy. *Psychometrika*, 1964, **29**, 371–382.

Horst, P. *Factor analysis of data matrices*. New York: Holt, Rinehart & Winston, 1965.

Kendall, M. G. and S. A. *The advanced theory of statistics*. London: Griffin, 1961.

Kestelman, H. The fundamental equation of factor analysis. *British Journal of Psychology, Statistical Section*, 1952, **5**, 1–6.

Lawley, D. N. The estimation of factor loadings by the method of maxium likelihood. *Proceedings of the Royal Society of Edinburgh*, 1940, **60**, 64–82.

McDonald, R. P. A general approach to nonlinear factor analysis. *Psychometrika*, 1962, **27**, 397–415.

McDonald, R. P. Numerical methods for polynomial models in nonlinear factor analysis. *Psychometrika*, 1967, **32**, 77–112.

Rao, C. R. Estimation and tests of significance in factor analysis. *Psychometrika*, 1955, **20**, 93–112.

Thomson, G. H. *The factorial analysis of human ability*. (2nd ed.) New York: Houghton Mifflin, 1946.

Thurstone, L. L. *The vectors of mind*. Chicago: Univ. of Chicago Press, 1935.

Manuscript received 7/3/67

Revised manuscript received 9/18/67

15

Reprinted with permission from *Syst. Zool.*, **2**(2), 76–92 (1953)

An Application of Factor Analysis to the Systematics of *Kalotermes*

CLYDE P. STROUD

MULTIPLE factor analysis, as developed chiefly by Thurstone (1931, 1945*a*), is a general method which can be adapted to a wide variety of problems. Like other forms of multivariate analysis it constitutes a method for the examination of complex sets of phenomena, expressed either in terms of experimental or nonexperimental data, in an attempt to account for most of the variance of the observed values in terms of a restricted number of underlying parameters, factors, or explanatory concepts. It is particularly useful as an exploratory method for the investigation of certain kinds of complex domains, especially where one or more of the following conditions prevails: (1) the underlying causes are unknown; (2) the underlying causes cannot be controlled; (3) the underlying causes are very numerous; (4) the basic question of the investigation tacitly forbids any alteration of the underlying causes.

This method stands in an intermediate position between two extremes of multivariate analysis. On the one hand the object of analysis may be maximum parsimony of description in terms of wholly arbitrary parameters, such as the principal axis method as developed by Hotelling and others. At the other extreme is the method of path coefficients as developed by Wright (1918, 1921, 1923, 1932, 1934), which permits a great deal of freedom in the introduction of hypotheses based on knowledge additional to that included in the table of correlations of the variables.

Multiple factor analysis differs from purely statistical methods in that it does not always involve the same set of assumptions as do the usual statistical models, and in that there have not been developed, as yet, adequate methods of establishing degree of confidence in the findings obtained. Discoveries made by multiple factor analysis could be related to statements of reliability in either of two ways. In some cases findings could be subjected after discovery to more refined statistical techniques which can establish reliabilities. It is also possible that in many applications, assumptions basic to statistics could be made at the outset and appropriate sampling error criteria applied. It is hoped that a further relation of multiple factor analysis to statistics can be developed.

In favor of multiple factor analysis it must be said that it is a comparatively easy method of obtaining exploratory information of a semireliable nature about many kinds of complex phenomena. The addition of techniques to establish statements of reliability would doubtless greatly increase the work involved in an analysis. Whether they are to be applied directly with the factor analysis, applied afterwards to verify the results of the analysis, or not applied at all, is a judgement to be made in terms of the nature of the problem at hand. A characteristic of multiple factor analysis which makes it particularly useful for certain types of studies and less useful for other types is that it employs a standard set of simple assumptions in all applications, and derives its findings from the coefficients of correlation alone. Only in the interpretation of the findings are data extraneous to the correlations introduced.

Some facets of the central problem of systematic biology—which may be defined as the relation of organisms to each other —lend themselves remarkably well to investigation using a method such as multiple factor analysis. Four kinds of information are developed relative to this problem: (1) organisms are assigned to categories corresponding as closely as possible to their phylogenetic relationships; (2) characters are evaluated with regard to their relations to each other and their reliabilities as indices of closeness of kin; (3) phylogenies or historical lines of descent are discovered and related to the categories and characters of organisms; (4) the evolutionary processes are studied in relation to characters which define categories and to phylogenies which generate categories. Multiple factor analysis can make substantial contributions to the organization and integration of all these four related kinds of information.

Materials and Methods

The material studied in the present work consisted of specimens of more than fifty species of termites belonging to the genus *Kalotermes.* This is the most primitive living genus of the family Kalotermitidae, which contains eight other genera. The kalotermitids are a primitive family of termites remotely related to the more primitive family Mastotermitidae, represented today only by a single Australian species (Ahmad, 1950). The chief criterion of primitiveness here is the possession of blattoid characteristics, since the evidence for the descent of the termites from primitive cockroaches is very clear and probably irrefutable.

In addition to the possession of blattoid characteristics, the genus *Kalotermes* shows its primitive aspects in other ways. It is a group in which many species show such a wide range of individual variation that specific characters are often difficult to ascertain reliably. Characters involving serially replicated components are highly variable. The number of such elements is usually relatively large and variable, and they show relatively little differential specialization among themselves.

Measurements were made on soldiers (usually a single specimen) of each of 48 species of this genus, and on imagoes of 43 species. Because of the time consumed in making accurate measurements and in dealing with the additional analytical problems involved, the element of individual variation within a species was not studied by measurements of a large number of individuals of those species for which they were available. However, for two species, *K. snyderi* and *K. immigrans,* two soldier specimens of each species were measured, and the results give some hint of the relationship of individual variation to the study. Variation within the species should be investigated by a combination of the techniques of analysis of variance and of covariance with the factor analytical methods of the present study.

Each specimen was measured for an array of fourteen characters. The choice of characters was made in such a way that homologous dimensions of soldier and imago anatomy were measured. The measurements were scattered over the body as much as possible, rather than being concentrated in a particular region. Characters were also chosen in such a way that measurements were made on hard, sclerotized regions of the body, thereby minimizing errors due to handling or positioning of the specimens. The characters so selected are listed below:

1. Length of right mandible.
2. Length of second antennal segment.
3. Length of third antennal segment.
4. Length of third (metathoracic) tibia.
5. Width of third tibia.
6. Width of third femur.
7. Height of head.
8. Length of head.
9. Width of head.
10. Length of pronotum.
11. Width of pronotum.
12. Length of postmentum.
13. Width of postmentum.
14. Maximum diameter of eye.

The method of multiple factor analysis was introduced by Thurstone (1931). Previous to this, a number of other workers employed various types of multivariate analysis to data of different kinds. Since 1931 multiple factor analysis and related methods have been employed very extensively in psychology by Thurstone (1945*a*), Holzinger and Harman (1941), Kelley (1935), Thomson (1939), Burt (1941), and others. The methods have been applied to anthropometry by Burt (1938), Cohen (1938, 1939–1941), Hammond (1942), McCloy (1940), Moore and Hsu (1946), Rees and Eysenck (1945), and Thurstone (1945*b*, 1946).

It was expected that information of some interest might be gained by studying the measured characters of *Kalotermes* by factor analysis. Such a study might permit an evaluation of the characters as to their usefulness in describing subgroups within the genus, and might also make it possible to relate characters to evolutionary trends for the genus as a whole. Moreover, the results could serve as the basis for a comparison with some more specialized genus.

It was found that all the measured characters were distributed approximately normally both in the soldier caste and in the imago caste over the population of species measured. Most of the characters show some skewness, always in the same sense. There are a relatively small number of species having quite large values and a large number of species having moderately small values. Bimodality is suggested for some of the characters but is not demonstrable at a significant level with the relatively small number of measurements made.

Factor analysis makes the assumption of the additive combination of factors, whereas such factors as those of the present study may often combine more nearly as multiplying factors in the sense used by Wright (1921). As Wright pointed out, the error term in considering additive combination as an approximation of multiplicative combination is small unless the amount of variation in the causative factors is large in comparison with the mean values. The additive assumption of factor analysis is that for the uncorrelated common factors *a*,

$$s_{ij} = \Sigma_a x_{ia} u_{aj} + \delta$$

where δ is due to unique factor and error variance.

The skewness of the data suggests that a change to logarithmic scale might have improved the study, but it was thought best to abstain from the additional complication in this, the first factor analytical approach to a systematic problem.

Product-moment coefficients of correlation were computed relating each character to each other character. This was done separately for the imago and soldier caste studies.

It is assumed that the regression of one variable on another, if not linear, is at least monotonic so that the use of linear coefficients of correlation is approximately correct. Whether or not the factors have a Gaussian distribution over the population studied need not concern us at present (Thurstone 1945*a*, pp. 63–67). It would be interesting to compare the results of the analysis of these coefficients of correlation with those obtained on the logarithms of the measurements.

Each of these correlation matrices R was then factored by the complete centroid method (Thurstone 1945*a*, pp. 149–157). In this method the correlations are accounted for in terms of a number of parameters much smaller than the number of measurements. The projections of the character vectors on the axes of the "common factor space" are given as a matrix F whose rows represent the measurements and whose columns refer to the orthogonal factor axes. These projections are called the centroid factor loadings.

The length of a character vector is the square root of the corresponding "communality," h_j^2, or the proportion of the whole variance of the measurement which enters into the correlations.

Thus, the correlation between a pair of characters j and k,

$$r_{jk} = \sum_{a=1}^{b} f_{ja} f_{ka} + \epsilon$$

where the subscripts a refer to the orthogonal factors. The quantity ε is a residual which is vanishingly small.

The communality,

$$h_j^2 = \sum_{a=1}^{b} f_{ja}^2$$

In matrix notation

$$F \cdot F' \doteq R$$

a matrix whose non-diagonal entries are, approximately, the coefficients of correlation. The entries in the principal diagonal of R are the communalities.

Since the communalities were unknown at the outset of the study, these were estimated for a first factoring (Thurstone 1945*a*, pp. 299–300), then corrected for a second factoring. Those obtained from the second factoring did not differ substantially from the estimates based on the first factoring, so that further iteration of the factoring process was not necessary.

The coefficients of correlation and the communalities are so related to the angle ϕ_{jk} between pairs of character vectors that

$$r_{jk} = h_j h_k \cos \phi_{jk}$$

The position of the first centroid axis is located in such a way that it passes from the origin through the centroid of the termini of the measurement vectors. It thus lies approximately in the longest axis of the hyper-ellipsoidal region occupied by the measurement vectors. In cases such as these, where all the correlations are positive, this is strictly true and all the measurements have positive projections for this first centroid axis, as all the measurement vectors lie in the same half of the hyper-ellipsoid.

The "residual correlations"

$$r_{(-\mathrm{I})jk} = r_{jk} - f_{j\mathrm{I}} f_{k\mathrm{I}}$$

are computed, and the table of these is then analyzed to obtain positions for the second centroid axis. Strictly speaking, this second axis (and this is true of subsequent axes as well) does not pass through the centroid of the residual vectors. This centroid is near the origin since the sum of the residual correlations is near zero. Instead, the second axis passes from the origin through the centroid of a system of termini of vectors, some of which have been "reflected" so that all lie in the same half of the residual hyper-ellipsoidal space. It does not really matter much where the second axis is placed so long as it is orthogonal to the first and accounts for a large amount of the variance involved in the residual correlations. Arbitrarily, vectors are reflected one at a time starting with that vector the sum of the residual correlations of which has the greatest negative value, and continuing until no sum of the residual correlations of a vector, corrected for the vectors reflected, is negative. The projections of the original vectors on this axis are determined and a second set of residual correlations $r_{(-\mathrm{II})jk}$ are computed.

The process is continued until all entries $r_{(-b)jk}$ are vanishingly small. These are the "residuals" and are to be ascribed, largely, to errors of estimate of the correlations.

The first centroid axis may have some biological significance but the others are certainly no more than convenient arbitrary orthogonal axes for the description of the vectors in the common factor space. This space is of b dimensions, where b, the number of factors, is considerably less than the number of characters. It is important to devise a statistical means of determining the best value of b.

Many factor studies have been carried only to this point. In other cases the axes have been located by least-squares fitting instead of the simple centroid method. It was obviously desirable to try to find a simple hypothesis about the relation of factors to characters so that meaningful interpretations of factors could be made.

One such hypothesis is furnished by Thurstone's simple-structure concept. The hypothesis is that for any meaningful common factor (other than a general factor) one would expect to find a subset of the original characters on which the factor has no (or only a very small) influence. The vectors representing such characters would lie in (or very nearly in) a subspace of $(b-1)$ dimensions of the b-dimensional common factor space. The factor axis could be described as orthogonal to this null subspace. Similarly a whole class of b or more or less factors might be located in this way with the axis of each orthogonal to a $(b-1)$-dimensional subspace containing the vectors on which a given factor does not act.

A brief set-theoretical interpretation of the simple-structure concept seems helpful. Denoting by B_p the set of vectors on which factor P has no influence, a complete simple structure can be said to exist if there is a set B_p for each of the b factors P, and if

$$\begin{array}{rr} B_1 \neq B_2, B_1 \neq B_3, \ldots & B_1 \neq B_b \\ B_2 \neq B_3, \ldots & B_2 \neq B_b \\ \ldots & \ldots \\ & B_{b-1} \neq B_b \end{array}$$

that is, all intersections of pairs of sets contain fewer elements than the corresponding unions,

$$(B_p \cup B_q) - (B_p \cap B_q) \neq O$$

Thus in a given study a particular subset of the variables can determine one and only one group factor. If there are two (or more) group factors having the same null subset of variables, their presence can only be determined by additional studies in which supplementary variables are introduced so that this is no longer the case. The relativity of the unitary nature of factors is not necessarily unfortunate, especially in applications to domains in which the number of ultimate factors is legion and what are really sought are major groups of ultimate factors. Of course there is no point in reporting as separate, two group factors P and Q, whose null subsets B_p and B_q are really a pair of approximations to what is actually the same null subspace. In a case such as this, either $r_{pq} \doteq 1$ or $r_{pq} \doteq -1$.

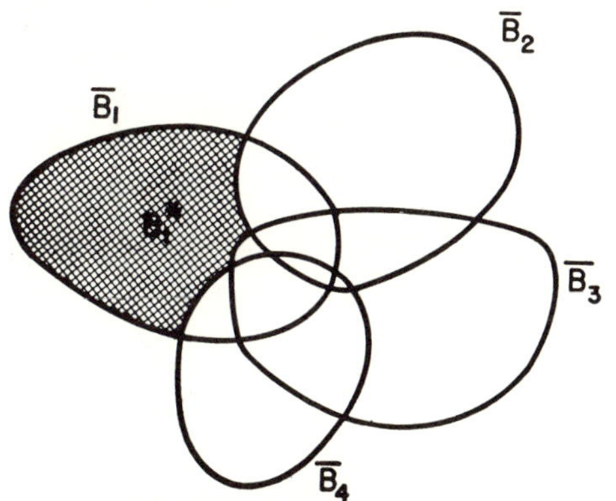

FIG. 1. The set B_p*.

A quantitative test could be based on determining the rank of the matrix R_{pq}, the entries of which are the correlations of the primary factors (with unity as the entries of the principal diagonal), and seeing that this is equal to its order. For studies with high communalities (which are common in morphological as opposed to psychological studies) the coefficients of correlation of the variables have smaller standard errors so that primary group factors which are relatively highly correlated with each other can be discerned.

The complement of B_p, denoted $\overline{B_p}$, is the set of characters which can be used to describe the factor P, as these are the measurements on which factor P exerts an influence. A set B_p*, which may be defined by a relation such as

$$B_p^* = \overline{B}_p - \overline{B}_p \cap (\overline{B}_1 \cup \overline{B}_2 \cup \ldots \cup \overline{B}_{p-1} \cup \overline{B}_{p+1} \cup \ldots \cup \overline{B}_b)$$

contains only characters which are influenced by no group factors other than P. The set B_1* is indicated as the shaded portion of set $\overline{B}_1$ in Figure 1 for a case in which $b = 4$. Obviously, the elements of B_p* describe the action of factor P in its purest form. In Figure 2, the projections of the termini of vectors are shown on a two-dimensional subspace including the origin O of the system. The two axes represent the factor axes P and Q and lie in this plane. The points representing the termini of character vectors can be assigned to sets $\overline{B}_q$, B_q, $\overline{B}_p$, and B_p.

The rotational process is one in which a search is made for a set of $\frac{b(b-1)}{2}$ such planes (for the case where all b common factors are group factors), each containing a pair of factor axes, by locating the "streaks" of projection points which define the null subspaces of factors as they appear in the two-dimensional projections of the space. These factors need no longer be orthogonal to each other. A rotational matrix Λ is found. Then

$$F \cdot \Lambda = V$$

where V is a new matrix in which the number of zero (or near zero) entries is maximized. When the columns of V are plotted against each other, a set of diagrams similar in general form to Figure 2 is obtained. These are characterized by the fact that a good number of the points lie on or near the lines representing zero values for the factors. The rotated factor matrix V is examined, and each factor is interpreted in terms of the characters for which it has high loadings.

It was of interest to compute approximate factor evaluations for the individuals measured in the study. The method used is based on the discussion by Thurstone (1945*a*, pp. 511–516), and elsewhere. First, the matrices S_{ij} of the standard values for each individual i for each measurement j were computed. Next, the reciprocal or inverse matrix R_{jk}^{-1} of each correlation matrix (with unit diagonal entries) was obtained. The method as given by Thurstone (1945*a*, pp. 46–48) after Tucker (1938) was satisfactory except that it was necessary to carry all numbers to four decimal places in order to have satisfactory accuracy as the rounding errors accumulate in the process. For each caste a matrix T_{ap} is needed to relate the rotated factors to the centroid axes. It is obtained by

$$T_{ap} = (D \cdot \Lambda^{-1})' = (\Lambda^{-1})' \cdot D$$

where D is a diagonal matrix, each of whose entries is the reciprocal of the square root of the corresponding entry in the principal diagonal of the matrix

$$M = C^{-1} = (\Lambda' \cdot \Lambda)^{-1}$$

It should be noted that C is the matrix of the cosines c_{pq} of the angles θ_{pq} between pairs of primary factor axes P and Q. Furthermore the matrix of the correlations of the primary factors

$$R_{pq} = D \cdot M \cdot D = D \cdot C^{-1} \cdot D$$

Then

$$W_{ja} = R_{jk}^{-1} \cdot F_{ka}$$

and

$$W_{jp} = W_{ja} \cdot T_{ap}$$

are matrices whose entries are the regression weights for the characters on the centroids and on the rotated factors respectively.

Finally these weights were applied to the standard values:

$$S_{ij} \cdot W_{ja} = X_{ia}$$
$$S_{ij} \cdot W_{jp} = X_{ip}$$

to obtain matrices X which are evaluations of the individuals i for the centroids a and the rotated factors p respectively. The evaluations x_{ia} and x_{ip} are subject to errors due to values δ attributed to unique factors and error variance as shown in the fundamental additive equation. The matrix W_{ja}, which relates S_{ij} to X_{ia}, the evaluations for the centroids, is simply related to the matrix U_{aj} of values u_{aj} of the fundamental additive aquation by:

$$W_{ja} = U_{aj}^{-1}$$

where the squares of the residuals δ due to unique factors and error variance have been minimized. Matrix X_{ia} gives the coordinates of points representing the measured individuals in a space of b dimensions, represented by the orthogonal axes a.

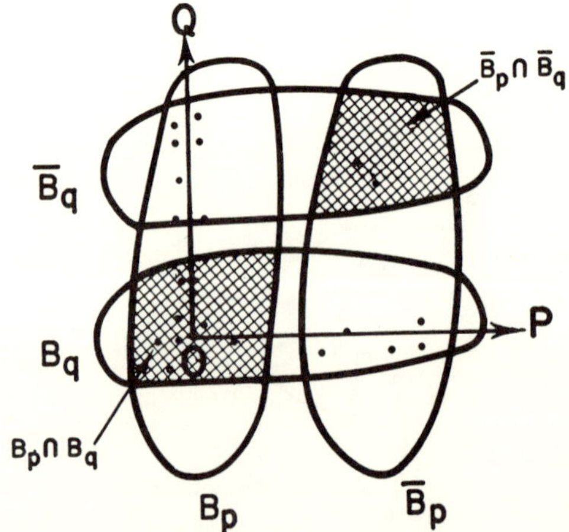

FIG. 2. Projections of termini of vectors on plane containing a pair of group-factor axes.

In addition to other sources of error, a species, which should be represented by a cluster of points, each determined by the evaluations for a single individual, is represented in this study by the evaluations of a single individual. There are two exceptions to this in the soldier caste. The relation of this individual variation to Centroid I (size) and to primitiveness should be studied further.

The matrix X_{ia} is a convenient starting point for an examination of the points to see if subsets corresponding to systematic categories can be discerned. The underlying assumption is that a group of species, all descended from a single branch of an early dichotomy, and represented by a single interbreeding population for some time before further dichotomy, would be expected to differ from other members of the genus by some combination of attributes. That combination of attributes for which the variance of the group is minimized is found by rotation. The points representing members of the group then lie approximately in a subspace of $(b-1)$ dimensions. It should be noted that only large subgroups can be found by this means. A subspace of $(b-1)$ dimensions can be fitted exactly to any set of b species; hence only groups containing considerably more than b species can be demonstrated.

The search for such groupings was made in a manner analogous to the single plane method of rotation (Thurstone, 1945*a*, pp. 216–224). Some species S_1 is chosen as the starting point in the search for each species-group. It is assumed that this species is to be a member of the group to be located. A first trial axis, α_1, is located, running from the origin through the point representing the species chosen. The direction cosines $\lambda_{\alpha_1 a}$ of this axis are obtained by normalizing the set of centroid evaluations $x_{S_1 a}$ of species S_1.

The entire set of species is then evaluated for projections on this axis α_1, and the resulting values are plotted as the ordinate, using each centroid axis in turn as the abscissa. These diagrams are then examined for subspaces of $(b-1)$ dimensions near S_1 as indicated by streaks of points including S_1. A line is drawn through S_1 in the direction of each streak. A parallel to this line is drawn from the origin to the intersection with the ordinate at the unity centroid value. The value of the ordinate of intersection is entered as s_{aa} in the equation

$$\Psi_{a_2 a} = \frac{\lambda_{a_1 a} - s_{aa}}{1 - s_{aa}\lambda_{a_1 a}}$$

The obtained values of $\Psi_{a_2 a}$ are then normalized by

$$\lambda_{a_2 a} = \frac{\Psi_{a_2 a}}{\sqrt{\sum_a \Psi^2_{a_2 a}}}$$

to obtain the direction cosines of a second approximation to the axis, α_2. The process is continued until a streak including S_1 is demonstrated or is shown not to exist.

Results

Table I lists the coefficients of correlation of pairs of characters in the imago caste and in the soldier caste respectively. The highest correlation for a pair of characters of the soldier caste is nearly 0.98 between head width and pronotum width. It is found that this especially high correlation is interpretable in terms of the factor analysis and is of great adaptive significance. Other very high correlations are found between the length of the tibia and the length of the pronotum and between the height and width of the head. Both these are also related to adaptive trends found by the factor analysis. The lowest correlations for the soldier caste are those between tibia width and the two antennal segment measurements, and the correlations of postmentum length with third antennal segment length and with eye diameter. All but the lowest two correlations are significant at the $p < .01$ level, and one of these is significantly different from zero at the $p < .05$ level.

For the imago caste none of the correlations is as high as the two greatest soldier correlations. The highest are the correlations relating the two measurements

TABLE I—COEFFICIENTS OF CORRELATION

SOLDIER CASTE—MATRIX $R_{♃}$—ABOVE PRINCIPAL DIAGONAL

	1	2	3	4	5	6	7	8	9	10	11	12	13	14
1	—	.685	.663	.857	.679	.741	.867	.802	.877	.865	.862	.682	.889	.657
2	.588	—	.761	.571	.350	.488	.560	.490	.535	.586	.537	.446	.548	.413
3	.455	.451	—	.528	.389	.598	.494	.446	.587	.539	.653	.231	.473	.572
4	.735	.655	.633	—	.788	.850	.815	.813	.823	.927	.827	.801	.892	.597
5	.809	.487	.417	.756	—	.862	.760	.756	.790	.807	.800	.668	.811	.500
6	.760	.529	.438	.790	.897	—	.728	.768	.790	.866	.812	.668	.800	.501
7	.765	.648	.542	.821	.792	.828	—	.865	.939	.819	.910	.714	.888	.606
8	.761	.551	.584	.780	.878	.902	.858	—	.863	.840	.845	.844	.814	.528
9β	.820	.570	.595	.805	.904	.894	.910	.937	—	.864	.979	.647	.884	.694
10	.827	.589	.505	.838	.892	.905	.895	.884	.930	—	.895	.765	.886	.612
11	.820	.610	.572	.864	.861	.876	.880	.878	.919	.946	—	.636	.874	.666
12	.679	.611	.616	.774	.724	.717	.759	.754	.802	.757	.743	—	.780	.357
13	.644	.518	.415	.815	.769	.778	.802	.756	.833	.834	.846	.680	—	.579
14	.610	.488	.433	.817	.684	.740	.641	.709	.784	.738	.742	.578	.573	—

IMAGO CASTE—MATRIX R_{im}—BELOW PRINCIPAL DIAGONAL

of the pronotum, the correlations of head width with head length and height, and with pronotum length and width, and between the width of the femur and the lengths of head and pronotum. The relation of these high correlations to the factors is very complex in the imago caste.

None of the correlations of imago characters is as low as the lowest correlations of the soldier caste. All the imago correlations are significant at the $p<.01$ level. These facts suggest that adaptive evolution has acted in different ways on the two castes. In the imago caste a single general pattern of characters seems to be favored for all species, whereas in the soldier caste different combinations of characters are especially adapted to ecological differences of the species. Hence the correlations of the imago characters do not range as high or as low as those for soldiers.

Table II gives the centroid matrices of the soldier and imago castes. It should be noted that although five factors suffice to account for the soldier correlations, six are involved in the correlations of imago characters.

TABLE II—CENTROID MATRICES

	SOLDIER CASTE—$F_{♃}$					IMAGO CASTE—F_{im}					
	I	II	III	IV	V	I	II	III	IV	V	VI
1	.93	—.16	—.11	.14	.10	.84	.06	.08	—.08	—.14	—.12
2	.66	—.49	—.24	—.11	—.17	.65	—.22	.12	.12	—.18	—.08
3	.65	—.60	.15	—.22	—.16	.60	—.31	.24	.04	.15	.19
4	.93	.10	—.14	—.11	.13	.92	—.30	—.23	.14	.05	.05
5	.83	.32	.20	—.19	.03	.90	.27	—.08	—.16	.10	—.17
6	.88	.13	.11	—.34	—.03	.91	.21	—.12	—.11	—.03	.03
7	.92	.09	.04	.25	—.08	.92	.06	.13	.15	—.09	.05
8	.89	.23	—.08	.12	—.22	.93	.12	.08	—.12	.10	.15
9	.95	.04	.22	.22	—.03	.97	.17	.18	—.05	.11	.05
10	.94	.10	—.08	—.11	.13	.96	.17	—.06	.08	—.13	—.02
11	.94	.07	.22	.12	.03	.96	.09	—.07	.10	—.07	.05
12	.77	.33	—.40	.03	—.12	.83	—.20	.19	.04	.10	—.14
13	.93	.16	—.08	.06	.13	.84	.12	—.13	.28	.16	—.03
14	.66	—.21	.20	.14	.18	.77	—.15	—.28	—.21	—.11	.18

Table III lists the communalities h_j^2 and uniquenesses u_j^2 for the two castes. In both castes head width has the highest communality. This means that practically all of its variance is discernible in its correlations with other characters measured, and it shows almost no variance due to unique factors influencing the head width but none of the other characters of the study. The only high uniquenesses for the soldier caste are those for eye diameter and for the length of the second antennal segment. At least in the case of eye diameter this suggests one or more additional factors, acting on eye diameter but not on any of the other characters measured. This seems a reasonable consequence of the convergent reduction of the soldier eye in many different lines. There are five reasonably high uniquenesses of the imago caste, the largest two involving the second and third antennal segments. The data suggest that there are unique factors acting on each of these two characters.

The distributions of residuals from the factoring are noted in Table IV. It is found that the two distributions are very similar, even though five factors have been found for the soldiers and six for the imagoes. These residuals may be due to errors of estimate of the correlations, rounding errors in the factoring, and to very small common factors not influencing the variances of the characters enough to be detected in this study. The residuals are all small and their distributions are approximately normal.

TABLE III—COMMUNALITIES AND UNIQUENESSES

	SOLDIER CASTE		IMAGO CASTE	
	COMMUNALITY	UNIQUENESS	COMMUNALITY	UNIQUENESS
CHARACTER	h_j^2	u_j^2	h_j^2	u_j^2
1	.93	.07	.76	.24
2	.77	.23	.54	.46
3	.88	.12	.57	.43
4	.92	.08	1.00	.00
5	.86	.14	.95	.05
6	.92	.08	.90	.10
7	.92	.08	.90	.10
8	.91	.09	.93	.07
9	1.00	.00	1.00	.00
10	.93	.07	.98	.02
11	.95	.05	.95	.05
12	.87	.13	.80	.20
13	.92	.08	.84	.16
14	.58	.42	.78	.22

TABLE IV—DISTRIBUTIONS OF RESIDUALS FROM FACTORING

SIZE OF RESIDUAL	NO. OF SOLDIER RESIDUALS	NO. OF IMAGO RESIDUALS
.00	29	20
±.01	33	40
±.02	20	22
±.03	7	6
±.04	2	3

TABLE V—ROTATED FACTOR MATRIX V_{4}

	SOLDIER CASTE				
	A	B	C	D	E
1	.36	.01	.30	.07	.10
2	.00	.06	.26	—.06	.57
3	.00	—.04	—.01	.09	.52
4	.19	—.05	.32	.22	.05
5	.04	.03	—.03	.40	—.12
6	—.06	—.01	.04	.46	.12
7	.25	.26	.04	—.11	—.04
8	.03	.38	.02	—.01	—.06
9	.30	.18	—.06	—.04	—.10
10	.20	—.05	.28	.24	.02
11	.28	.10	—.02	—.07	—.11
12	—.01	.28	.30	.04	.09
13	.28	.03	.25	.10	—.08
14	.40	—.11	.08	—.03	—.07

Table V is the rotated factor matrix for the soldier caste. If ±0.10 is taken as the null value for a loading, it can be said that factor *A* is determined by six characters on which it has no influence, *B* by nine characters, *C* by eight, *D* by nine, and *E* by nine. Thus all factors except *A* are rather well determined. Even *A* is determined better than the rule-of-thumb criterion suggested by Thompson (1939), namely, that the number of near-zero entries in each column should be as great as or greater than *b*, the number of factors in the study.

FIG. 3. Final positions, rotated soldier factors.

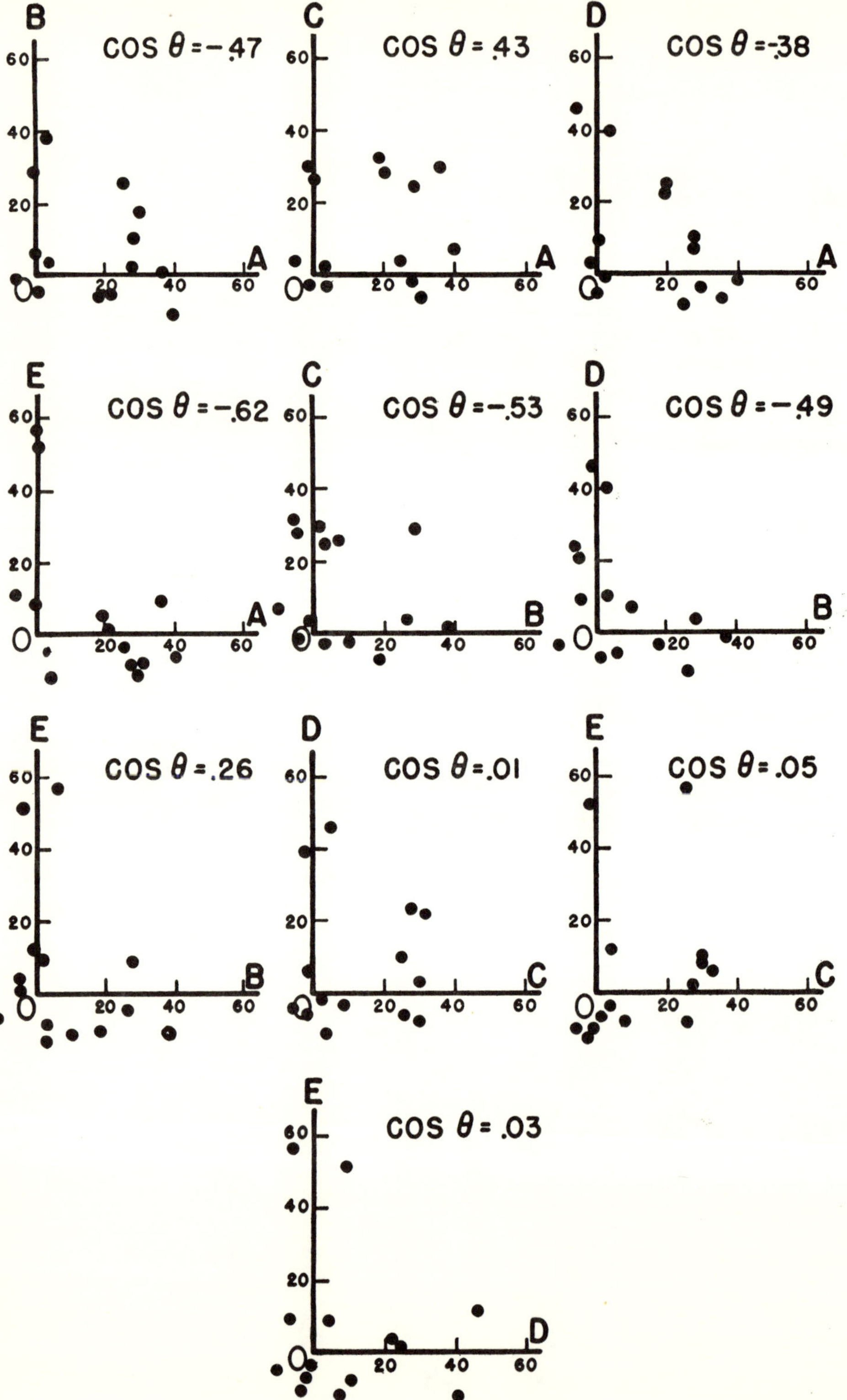
B
COS θ = −.47
60
40
20
0
20
40
60
A
C
COS θ = .43
D
COS θ = −.38
E
COS θ = −.62
COS θ = −.53
COS θ = −.49
COS θ = .26
COS θ = .01
COS θ = .05
COS θ = .03

Since the origin must lie in the subspace of $(b-1)$ dimensions which determines the factor, such a subspace can be found to fit exactly to any $(b-1)$ vectors. Hence the existence of such a subspace is suggested by numbers greater than or equal to b. In the soldier study b is five.

It should be noted that factors whose null set of characters B_p contain less than about $m-b$ elements are not discovered in a factorial analysis, where m is the number of measurements and b the number of factors. A primary factor for "general size," if it existed in the present study, would be of this sort, as there would be no characters whose variance it did not influence. If such a factor existed at the first-order level, then from a study based on b centroid axes, only $(b-1)$ linearly independent factors would be found by rotation. Since this was not the case in either of the present studies, it must be assumed that no such general size factor, having a relatively large degree of independence from the other primary factors, exists.

Table VI gives the rotated factor matrix for the imago caste. Here $b=6$, and the numbers of vectors defining each factor are: *A*, nine; *B*, nine; *C*, eleven; *D*, eleven; *E*, eleven; and *F*, nine.

Figures 3 and 4 show the "final positions" of the factor axes as seen in two-dimensional sections of the spaces. In the figures the angles θ_{pq} between each pair of factor axes are drawn as though they were right angles. The cosine of the actual angle c_{pq} between axes is noted with each section.

The matrices of rotation Λ relating each set of factor axes to its corresponding set of centroid axes are presented as Table VII.

The primary factors obtained by rotation are, typically, correlated with each other. The table of the correlations of these factors is given as Table VIII. Some of these correlations are remarkably high, but the linear independence of the factors

TABLE VI—ROTATED FACTOR MATRIX V_{im}

IMAGO CASTE						
	A	B	C	D	E	F
1	.15	.21	.01	.16	—.11	—.02
2	.29	.28	.00	.05	—.01	—.05
3	.26	—.04	.05	—.04	.03	.37
4	.26	—.01	.32	.02	.31	—.01
5	.02	—.04	.05	.28	.04	.01
6	—.08	.06	.15	.06	.00	.05
7	.05	.25	—.07	—.06	.02	.11
8	—.02	—.02	.08	.01	—.04	.30
9	.02	.03	—.07	—.07	—.03	.28
10	—.03	.23	.04	.00	.05	—.03
11	.00	.17	.09	—.04	.10	.03
12	.39	.04	—.04	.23	.08	.11
13	.00	.00	—.02	.00	.37	.01
14	.06	.00	.49	—.02	—.02	.04

TABLE VII—MATRICES OF ROTATION

SOLDIER CASTE Λ_{ψ}					
	A	B	C	D	E
I	.20	.08	.16	.11	.09
II	—.17	.23	—.14	.29	—.56
III	.07	—.06	—.76	.21	—.38
IV	.58	.40	—.07	—.90	—.42
V	.77	—.88	.61	.21	—.60

IMAGO CASTE Λ_{im}						
	A	B	C	D	E	F
I	.13	.09	.10	.08	.06	.09
II	—.75	.09	—.40	—.13	—.11	.01
III	.22	.27	—.72	.01	—.50	.51
IV	—.09	.33	—.45	—.42	.64	—.20
V	.11	—.90	—.08	.29	.55	.48
VI	—.60	—.03	.32	—.85	—.13	.68

TABLE VIII—CORRELATIONS OF PRIMARY FACTORS

SOLDIER CASTE					
	A	B	C	D	E
A	——				
B	.72	——			
C	—.27	.24	——		
D	.82	.84	.00	——	
E	.75	.34	—.40	.50	——

IMAGO CASTE						
	A	B	C	D	E	F
A	——					
B	—.62	——				
C	—.50	.77	——			
D	—.80	.88	.71	——		
E	—.56	.86	.66	.78	——	
F	—.57	.90	.72	.85	.82	——

FIG. 4. Final positions, rotated imago factors.

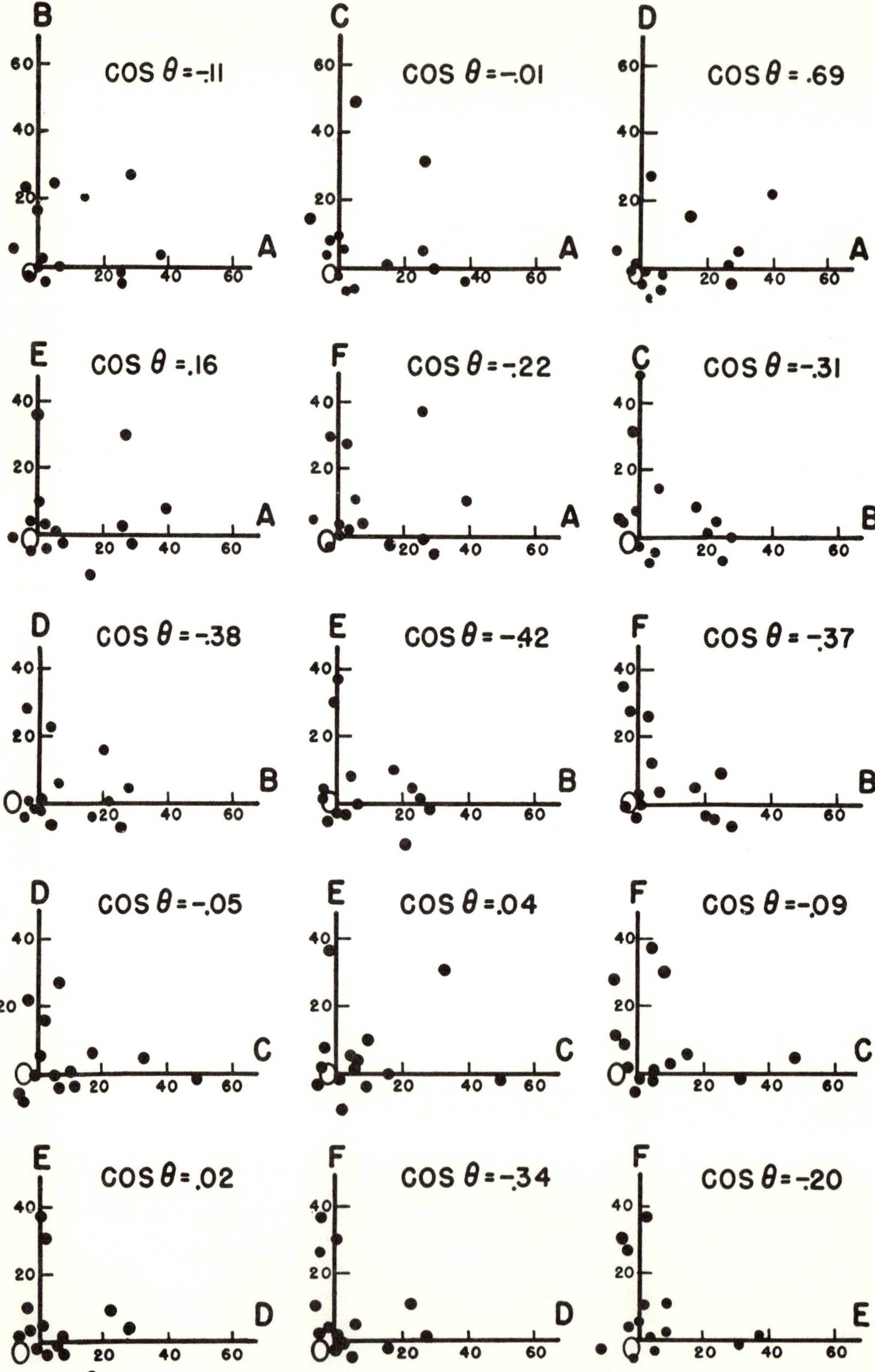
B
COS θ = -.11
A
C
COS θ = -.01
A
D
COS θ = .69
A
E
COS θ = .16
A
F
COS θ = -.22
A
C
COS θ = -.31
B
D
COS θ = -.38
B
E
COS θ = -.42
B
F
COS θ = -.37
B
D
COS θ = -.05
C
E
COS θ = .04
C
F
COS θ = -.09
C
E
COS θ = .02
D
F
COS θ = -.34
D
F
COS θ = -.20
E
0
20
40
60

Table IX—Second Order Domain

SOLDIER CASTE		
	I	II
A	.96	.02
B	.71	.64
C	—.23	.58
D	.85	.41
E	.78	—.39

IMAGO CASTE		
	I	II
A	—.72	.38
B	.96	.19
C	.77	.13
D	.95	—.25
E	.86	.17
F	.91	.21

seems clear enough. No serious change need be made if two of the factors should eventually be regarded as only one, but there would then be the possibility of a linearly independent general factor at the primary level, which does not seem, at present, to be the case in either study. These correlations were then factored by the centroid method to obtain the second-order centroid matrix for each caste. The results are shown in Table IX. When these loadings for the soldier caste are plotted against each other (Fig. 5), using one centroid axis as the abscissa and the other as the ordinate, no grouping of the primary soldier factors is evident. A comparable plot (Fig. 6) for the imago factors, however, indicates a group consisting of factors B, C, E, and F, which could

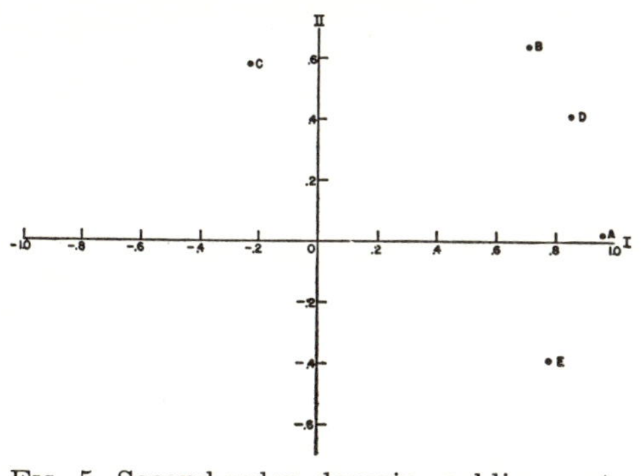

Fig. 5. Second-order domain, soldier caste.

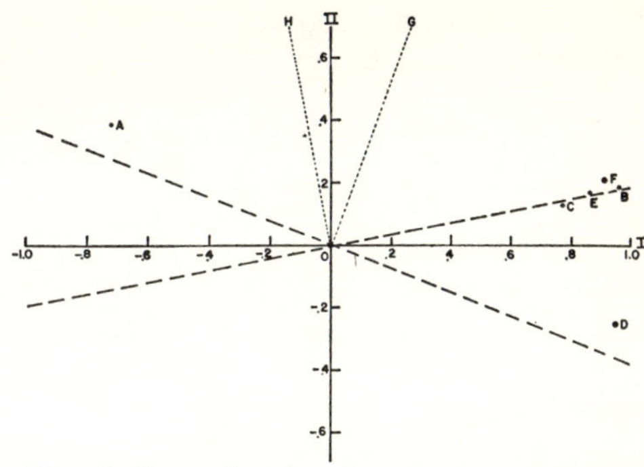

Fig. 6. Second-order domain, imago caste.

be assumed to lie in one subspace including the origin (in this two-dimensional example the subspace is a line). These second-order factors are used to account for the correlations among the primary factors.

The lack of structure corresponding to a group factor in the soldier study could be interpreted as indicating one or more general factors there. In the imago study what seem indicated are a bipolar second-order group factor relating the first-order factors A and D, and a general factor.

All second-order results in studies this small must be regarded as tentative and suggestive only. The addition of a few more characters might increase the number of first-order factors and alter the conclusions regarding the second-order domains.

Table X lists the evaluations of the *Proglyptotermes* individuals of the soldier

Table X—Alpha Projections of *Proglyptotermes* Soldiers

K. atratus	.79
K. banksiae	1.22
K. brouni	.99
K. castaneiceps	1.12
K. gracilidens	.73
K. hilli	.98
K. longiceps	.87
K. longus	1.41
K. pallidinotum	1.52
K. rufinotum	1.00
K. sordwanae	1.32
K. spoliator	1.22
K. tillyardi	1.19
K. umtate	1.15
K. n. sp. (Ihosy)	1.06

caste for the axis α discussed in the succeeding section of this paper.

Discussion and Conclusions

Factor-analytical methods can probably make a substantial contribution to the solution of problems of character evaluation. One advantage of their application is that of parsimony. A relatively large number of observed characters can be reduced, practically speaking, to a considerably smaller number of parameters. In the present study the original fourteen measurements were reduced, largely, to five parameters for the soldier caste and to six for the imagoes. Much of the error variance is eliminated along with unique factors affecting only one measurement. The use of centroid evaluations permits the individuals to be located as points in a space whose axes are orthogonal to each other and which represent statistically independent parameters. A much more significant addition to biological knowledge is gained, however, by an examination of the rotated factors. These can, in some cases, be subjected to interpretations as causative agencies or as trends accounting for the correlations of the characters.

It does not seem feasible in the present paper to discuss in great detail the results obtained. Instead the results will be very briefly summarized and sampled here. In the present study the soldier factors are more easily interpreted, on the whole, than are the imago factors. This is probably due to the well-known fact, true rather generally throughout the order Isoptera, that the soldier undergoes a variety of obviously adaptive changes, whereas the imago caste is more conservative.

Of the five factors discovered for the soldier caste, three are distinctly regional in nature: factor *B*, a head-volume factor; *D*, a leg factor; factor *E*, an antennal factor. The other soldier factors, *A* and *C*, are not limited in their influence to a regional group of characters, but operate on a variety of characters distributed over the body. Of six imago factors, only one, factor *F*, shows strong regionality in its operation. The other five operate on measurements on various parts of the body.

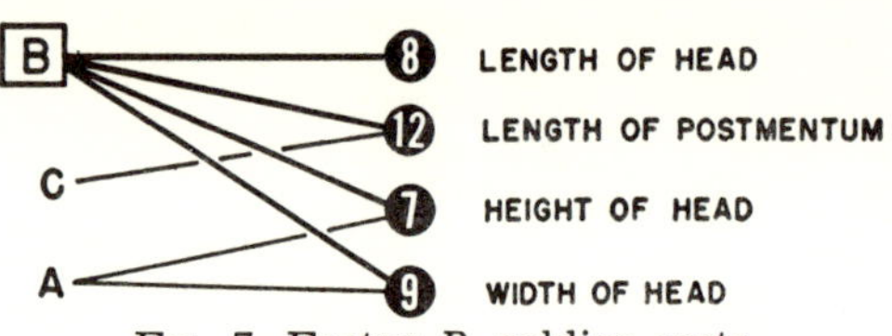

FIG. 7. Factor *B*, soldier caste.

The five soldier factors are all interpretable in terms of adaptive trends for the genus as a whole. Of the six imago factors only two seem probable as representing adaptive trends in the imago. The other four factors, and to some extent these two as well, are related in a very interesting way to the factors discovered for the soldier caste.

Factor *B* of the soldier caste affects the four measurements shown in Figure 7: the three dimensions of the head and the length of the postmentum, which runs along and constitutes a part of the floor of the head capsule. Thus it is a factor with a distinctly regional influence, affecting all dimensions of the head. Apparently, the width of the postmentum is independent of this factor. The existence of factor *B* may be interpreted in any of three ways which are not necessarily mutually exclusive: (1) it may be that a certain proportionality among the head measurements tends to be maintained in spite of any other changes in the body; (2) there may be a directional tendency toward increase in head volume for some or all the species; or (3) there may be a tendency for head-volume decrease for some or all the species. Another possibility is that some early dichotomy of the genus, through selective or random acquisition of different alleles in the two lines, resulted in two rather discrete groups of species whose principal difference is head volume. If this were the case one would expect a marked bimodality in the distribution of the individual evaluations for factor *B*. Since no such bimodality is evident this fourth possibility can be eliminated. It is not possible, as a result of the present study, or on the basis of other facts known about the genus, to distinguish clearly among the first three possibilities; but the third, head-volume reduction, seems most likely.

Factor *A* for the soldier study is an interesting example of a non-regional factor. It affects eight of the fourteen characters studied. The eye in the soldier caste is undergoing evolutionary regression, while in the imago caste, with the same genetics, the compound eyes are maintained as functional structures. The eye shows considerable unique variance, as suggested by the factor *U* in Figure 8, but 58 per cent of its variance is due to common factors included in this study. These findings strongly indicate that much of the regressive decrease in eye diameter in the soldier caste is attributable to the selection of alleles which happen to be deleterious to the soldier eye but which have pleiotropic effects of selective advantage in reducing the cross-sectional area of the soldier. The entire combination of characters involved in factor *A* amounts to a decrease in this cross-sectional area, fitting the soldier caste to more efficient function in small burrows of subcircular section.

Turning to imago factors, it is found that these are more difficult to interpret as functional adaptations of the imago caste, but that a number of them are closely related to the soldier factors. The closest such relationship is that between

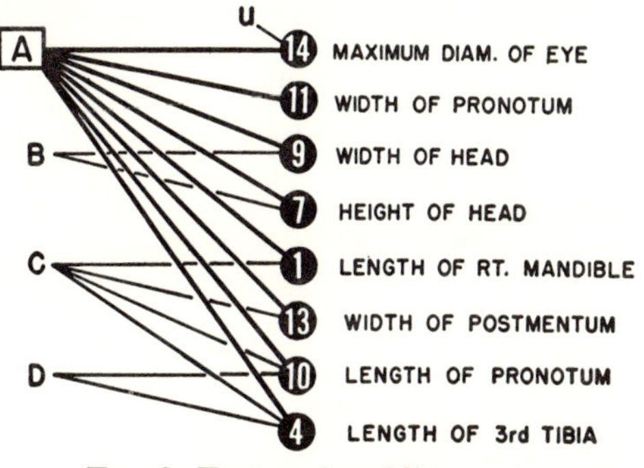

FIG. 8. Factor *A*, soldier caste.

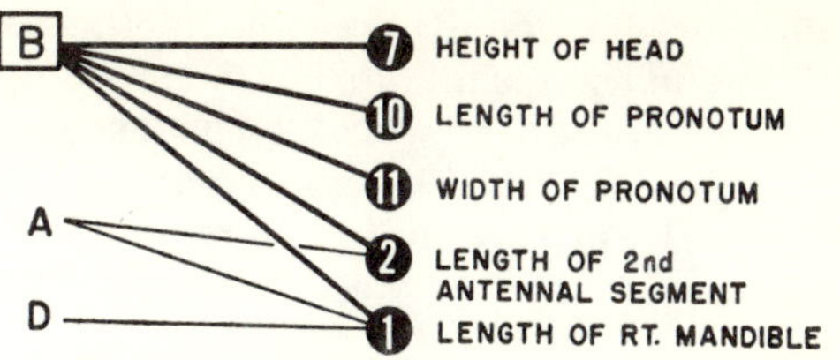

FIG. 9. Factor *B*, imago caste.

imago factor *B* and soldier factor *A*. The coefficient of correlation $r = .74 \pm .08$ for the 37 species in which both soldier and imago specimens were measured. Imago factor *B* (Fig. 9) affects four characters homologous to those affected by soldier factor *A*. It also affects the length of the second antennal segment, regionally almost contiguous with the head. The imago eye, head width, postmentum width, and tibia length are independent of imago factor *B* although their homologues are affected by soldier factor *A*.

The interpretation is that an obviously adaptive trend in the soldier caste is reflected by a trend in the imago caste affecting some but not all the homologous characters, and bringing in one additional character. Imago factor *B* may have some adaptive value in the imago caste but the relation is not clear. In this case, the alleles that have been selected seem to be those whose action is similar in sense in both castes, thus especially those alleles whose major contribution (other than eye reduction) to development is made prior to the time of soldier differentiation. In the case of a different relationship between factors of soldier and imago caste, the correlation is negative, suggesting the selection, primarily, of alleles acting in soldier differentiation.

This sample of the factors of the present study should suffice to indicate the extremely interesting possibilities which factor analysis offers for the interpretation and evaluation of characters.

A different application of multiple factor analysis lies in the identification of categories of organisms. A rotational search for groupings of species in the space described by the centroid evaluations was made. Since this is the first time such a technique has been applied in factor analysis, conclusions based on the findings must be viewed with some reserve. For this and other reasons, only the largest, most obvious streaks were examined.

In the soldier study an axis α was found on which a group of fifteen species had sub-equal projections. The difference between the mean for this species-group and for the rest of the genus is highly significant. This group of fifteen species, along with several others not in the soldier study, had previously been assigned to the genus *Proglyptotermes* by Emerson (Kirby, 1949) on the basis of a survey of the imago caste. Its members are, in general, smaller and darker than *Kalotermes* and always have an aroleum (some *Kalotermes* lack this structure). Although it had appeared in the literature its status was still uncertain. Kirby found corroborative evidence in the similarity of their intestinal flagellates. This study confirms, for the soldier caste, the existence of such a category as *Proglyptotermes,* but further study is needed before a final decision can be reached as to whether it should be designated as a genus or as a category of subgeneric rank within *Kalotermes.*

A comparable rotational search of the imago evaluations suggested the existence of the same group, but it is not so clearly defined here as in the soldiers. Apparently, additional, more critical characters must be used to find reliable criteria for distinguishing *Proglyptotermes* imagoes.

It should be stated that multiple factor analysis permits the identification of character correlations with factors or trends which can be considered the causative agents accounting for the correlations. It presupposes the additive combination of linearly related causes. Biological factors can hardly be supposed to have exactly these properties; but this must be approximately so, or interpretable results such as those reported here could not be found.

It is hoped that the method of factor analysis can supplement the more usual systematic techniques and permit a better description of some of the millions of evolutionary experiments in which nature has engaged.

Summary

By means of multiple factor analysis and related techniques the genus *Kalotermes* was examined in an attempt to elucidate its systematic structure. Factor analysis can be of great aid in evaluating characters. By its use five primary factors have been found for the soldier caste and six for the imago caste of *Kalotermes.* Interpretations of a sample of these are presented, showing their relation to evolutionary trends in the genus and to regressive evolution and pleiotropism.

The method of multiple factor analysis is discussed briefly as it has been applied to systematic problems. A brief set-theoretical account of Thurstone's simple-structure concept is developed. An extension of the methods of multiple factor analysis is employed in which a rotational search is made for groupings among individual evaluation points in the common-factor evaluation space. The existence of a previously suspected genus or subgenus, *Proglyptotermes,* is demonstrated in the soldier caste.

REFERENCES

Ahmad, M. 1950. The phylogeny of termite genera based on imago-worker mandibles. *Bull. Am. Mus. Nat. Hist., 95,* 37–86.

Burt, C. 1938. The analysis of temperament. IV. Physical signs of temperament. *Brit. J. Med. Psychology, 17,* 184–187.

——— 1941. The factors of mind. Macmillan Co., New York.

Cohen, J. 1938. Determinants of physique. *J. Mental Sci., 84,* 496–512.

——— 1939–1941. Physique, size, and proportions. *Brit. J. Med. Psychology, 18,* 323–327.

Hammond, W. H. 1942. An application of Burt's multiple general factor analysis to the delineation of physical types. *Man, 42,* 4–11.

Holzinger, K. J., and Harman, H. H. 1941. Factor analysis. Univ. of Chicago Press.

Kelley, T. L. 1935. Essential traits of mental life. Harvard Univ. Press, Cambridge, Mass.

KIRBY, H. 1949. Devescovinid flagellates of termites. V. The genus *Hyperdevescovina*, the genus *Bullanympha*, and undescribed or unrecorded species. *Univ. Cal. Publ. Zool., 45,* 319–422.

MCCLOY, C. H. 1940. An analysis for multiple factors of physical growth of different age levels. *Child Development, 11,* 249–277.

MOORE, T. V., and HSU, E. H. 1946. Factorial analyses of anthropological measurements in psychotic patients. *Human Biology, 18,* 133–157.

REES, W. L., and EYSENCK, H. J. 1945. A factorial study of some morphological and psychological aspects of constitution. *J. Mental Sci., 91,* 8–21.

THOMSON, G. H. 1939. The factorial analysis of human ability. Houghton Mifflin, New York.

THURSTONE, L. L. 1931. Multiple factor analysis. *Psychological Review, 38,* 406–427.

——— 1945*a*. Multiple factor analysis. Univ. of Chicago Press.

——— 1945*b*. Factor analysis and body types. *The Psychometric Lab.*, No. 24, 1–7.

——— 1946. Analysis of body measurements. *The Psychometric Lab.*, No. 29, 1–13.

TUCKER, L. 1938. *Psychometrika, 3,* 189–197.

WRIGHT, S. 1918. On the nature of size factors. *Genetics, 3,* 367–374.

——— 1921. Correlation and causation. *J. Agricultural Res., 20,* 557–585.

——— 1923. The theory of path coefficients—a reply to Nile's criticism. *Genetics, 8,* 239–255.

——— 1932. General, group, and special size factors. *Genetics, 17,* 603–619.

——— 1934. The method of path coefficients. *Annals Math. Statistics*. Charles Griffin, London.

Although the mathematical nature of this paper makes it slow reading, the technique of multiple factor analysis promises to be of wide utility in systematics for reducing complex sets of data to a relatively small number of causative factors. At the time of his death on October 8, 1952, **CLYDE P. STROUD** was Associate Biologist in the Biological and Medical Division of the Argonne National Laboratories, Chicago. This paper is a portion of the dissertation accepted for the degree of Doctor of Philosophy by the Department of Zoology, University of Chicago. Financial support for portions of the investigation was given by the Dr. Wallace C. and Clara A. Abbott Memorial Fund of the University of Chicago. The dissertation was written under the direction of Dr. Alfred E. Emerson who also edited the manuscript after the death of Dr. Stroud. Advice concerning the mathematical techniques was given by Dr. Louis L. Thurstone and Dr. Sewall Wright. This paper was presented at the symposium on "Statistical Problems in Animal Experimentation," December 27, 1951.

16

Reprinted from *J. Animal Ecol.*, **32**, 535–547 (1963)

MULTIVARIATE ANALYTICAL TREATMENT OF QUANTITATIVE SPECIES ASSOCIATIONS: AN EXAMPLE FROM PALAEOECOLOGY*

BY R. A. REYMENT

Department of Geology, University of Stockholm†

INTRODUCTION

A common observation made by micropalaeontologists in analysing series of borehole samples is that the proportions of species may fluctuate widely from sample to sample. It would not seem unreasonable to assume that the oscillations in numbers of individuals were caused by ecological factors. Although it seems hardly likely that the palaeoecologist will ever be in a position to identify such factors specifically, the statistical methods of principal component analysis and factor analysis, applied to frequency data, would seem to be suitable for supplying a picture of the order of complexity of the environment in which these species lived, and their reactions to it. Both methods attempt to divide up the total variation into entities, which one tries to provide with a meaningful interpretation. Both were originally devised for application to psychometric problems; factor analysis has been used for some 30 years in the analysis of biological variation, while principal component analysis has only more recently been applied to such problems.

Numerous examples of the taxonomic use of principal component analysis with respect to the ostracod carapace are given in Reyment (1963). A discussion of factor analysis as applied to an ecological problem in botany, comparable to that treated in the present paper, is to be found in Greig-Smith (1957, pp. 147–9, 159, 161–3). A general text on multivariate statistical analysis, which contains a detailed account of the statistical theory involved in the following, is that of Kendall (1957).

Ostracods are useful indicators in palaeoecology as they are amongst the most common animal fossils in borehole samples, they are benthonic (in, on, or close to the substratum) and, as opposed to their greatest potential rivals, the Foraminifera, they occur in a wide range of aquatic environments.

A note on the interpretation of the mathematical results might be in place. The correlation coefficients between pairs of variables (species), ordered in a correlation matrix, are subjected to the mathematical procedure known as a transformation (equation (1)). The new matrix **D** in equation (1) has all off-diagonal elements equal to zero and its diagonal elements add up to the same total as the diagonal elements of the original correlation matrix, in our example 17. This diagonalization process is of wide application in applied mathematics. It was used by Jacobi over a hundred years ago in his study of the orbits of the planets and his solution is the one used in most electronic computer programmes today. Many statistical procedures make use of the process, including the statistical treatment of certain taxonomical problems and a problem in population dynamics.

The element d_1 of matrix **D** is larger than any of the succeeding elements. We shall here refer to it as an *eigenvalue*; other terms in use for it are 'latent root' and 'characteristic root'. Each eigenvalue is associated with two *eigenvectors* (for an asymmetric

* Publication No. 6 of the Department of Geology, University of Ibadan, Nigeria.

† Present address: Department of Geology, University of Ibadan, Ibadan, Nigeria.

matrix there are two different eigenvectors, but in this paper we shall only be concerned with symmetric matrices, where the eigenvectors do not differ); other terms are 'latent vector' and 'characteristic vector'. The relationship is indicated in equation (2).

It is easy to see that the elements of **D** represent another distribution of the total variance of the original matrix (the sum of its diagonal elements—this is called the *spur* or the *trace* of a matrix; thus, spur **A** = spur **D** in equation (2)).

In the taxonomical analysis of, for example, ostracods, the first eigenvalue is usually many times greater than the second eigenvalue and we say that most of the variation is concentrated to the *first principal component*. This is given by the elements of the first eigenvector, which are employed as the coefficients in an equation. Thus, if the elements of the eigenvector are (b_1, b_2, b_3), and the dimensions measured, x_1, x_2, x_3, the first principal component may be written out as

$$U_1 = b_1x_1+b_2x_2+b_3x_3.$$

The variance of U_1 is given by d_1, the first element of **D**.

In our palaeoecological problem the x_i are species and the b_i denote the importance of species x_i in a particular principal component. We should therefore be able to recognize associations of species, yielded by the principal components, the relative significance of each species being indicated by the corresponding coefficient of the pertinent eigenvector (in statistical parlance this coefficient is sometimes called a 'weight' or 'weighting'; the corresponding psychometric term of factor analysis is 'loading').

It will usually be found that the first few eigenvalues account for almost all of the variance. The question is, then, whether any importance is attached to the remaining eigenvalues. This may be tested by seeing if the remaining eigenvalues are indistinguishable from each other (isotropic residue), which also implies that they are unimportant with respect to magnitude. Obviously, all eigenvalues must, in a sense, be significant (*cf.* Kendall 1957), unless we have a matrix, the rank of which is less than its order, which is a remote possibility in the kinds of biological problems we consider here. What the aforementioned test helps us to do is to decide the number of important eigenvalues; that is, in the ecological problem under review, the ones that have been most important in determining the frequencies in the various species associations, indicated by the elements of the corresponding eigenvalues. One would like to be able to test these coefficients for significance, but, as far as the author is aware, no such tests appear to be available. Consequently, there must be an element of subjectivity attached to decisions regarding the importance of any vector element.

Up to now the discussion has been in terms of principal component analysis. The ecological interpretation of the results of factor analysis is made in much the same way. However, the structure of factor analysis differs somewhat from principal component analysis, as indicated by equation (3), for it includes a term for random variation. The 'communalities' of factor analysis, discussed further on, are derived from the random matrix, as is indicated by equation (4). The method of factor analysis employed in the present paper would appear to be an improvement over previous methods, as it provides a more realistic assumption concerning the matrix of residual variances (equation (5)).

STATEMENT OF THE PROBLEM

We have, say, k samples from boreholes in which p different species occur (not necessarily all p species occur in each of the k samples). For each sample a certain arbitrary total number of individuals, N, are picked (randomly), this N being the same for all samples.

A matrix **T** of 'scores' (relative frequencies) results, each row of which consists of the score for each of the p species in the j-th sample ($j = 1, \ldots, k$).

$$\mathbf{T} = \begin{bmatrix} \nu_{11}s_1 & . & \nu_{12}s_2 & . & . & . & . & \nu_{1p}s_{1p} \\ . & . & . & . & . & . & . & . \\ \nu_{p1}s_1 & . & . & . & . & . & . & \nu_{pp}s_p \end{bmatrix}$$

here ν_{ik} represents the score for species s_k ($\nu_{ik} \geqslant 0$).

It is desired to use these observations on the frequencies of the species to ascertain the extent of relationship in their occurrence. That is, one would like to see if species that react in the same way to all factors in their environment can be marshalled into one group, and those that are adversely affected by some environmental facet can be made to show up in another group or groups.

METHOD 1

It would seem that a possible solution of the problem might be offered by the well-known linear algebraic procedure, whereby a matrix may be reduced by means of transformations to a simple form (*canonical form*), the rank of which is equal to the rank of the original matrix. The new matrix is a diagonal matrix. This procedure forms the basis of '*principal component analysis*', introduced by Hotelling (1933) for the study of psychometrical data.

For any symmetric ($k \times k$) matrix **A**, there exists an orthogonal matrix **B** such that

$$\mathbf{B'AB} = \mathbf{D} = \begin{bmatrix} d_1 & 0 & . & . & . & . & 0 \\ . & . & . & . & . & . & . \\ 0 & . & . & . & . & . & d_k \end{bmatrix} (d_1 > d_2 \ldots > d_k) \qquad (1)$$

If the matrix **A** is positive definite, then the $d_i > 0$. (In other words a positive definite symmetric matrix has only positive eigenvalues.)

Corresponding to each eigenvalue there will be an eigenvector, **c**, such that if d is any eigenvalue of **A**, **c** is a non-zero vector satisfying the equation

$$(\mathbf{A} - d\mathbf{I}) = \mathbf{0} \qquad (2)$$

These eigenvalues and eigenvectors provide the basis for the ecological analysis. Primarily we are interested in the components of the eigenvectors, which we may regard as *ecospectra*; i.e. the observational vectors are broken up into functionally connected spectra of ecological reaction. The same reasoning is applicable to the matrix of factor loadings of *factor analysis*, treated as method 2.

Computational procedure for method 1

1. Note the frequencies of each of the p species in k samples for a total of N individuals.
2. Arrange the data for each sample in rows the one above the other; the order of the rows with respect to each other should be random (important for palaeoecological data in order to eliminate the possibility of serial correlation).

	s_1	. . . .	s_p
Sample 1	ν_{11}	. . . .	ν_{1p}
.	.	. . . .	.
Sample k	ν_{p1}	. . . .	ν_{pp}

The ν_{ij} represents the score for the j-th species in the i-th sample. Because of the constant $\Sigma \nu_{ij} = N$, these scores are proportions which may be directly compared between samples. Logically, the greater one takes the value of N, the more likely are the components of the observational vectors to approach stability.

3. Compute all combinations of correlation coefficients between frequencies of species. It may be more suitable to express the data as decimal proportions at this stage.

4. The correlation coefficients are arranged into a matrix of correlations and this correlation matrix is reduced to its canonical form. This is best done on an electronic computer as the arithmetical labour involved for more than three or four variables is prohibitive.

Interpretation

Each eigenvalue gives a quantitative indication of the relative importance of the species combination represented by the components of the associated eigenvector. Particularly in large matrices, it may be found that only the first few eigenvalues will be meaningful; the remaining eigenvalues can be tested to see whether they differ significantly from an isotropic residue (Lawley 1956). The eigenvector may be written out in equation form as

$$z = a_1b_1 + a_2b_2 + \ldots\ldots + a_pb_p$$

The usual computer solution of an eigenvalue problem provides normalized vector components so that an a_i may receive any value between 0 and 1, depending on the importance of the contribution of the variable. Hence, the variables (= species) of the above equation that have large coefficients are said to be causing most of the variation represented in the eigenvalue. The reasoning is further elucidated in the worked example.

Example using method 1

The example considered here is palaeontological. It will be apparent, however, that the same reasoning may also be applied to data concerning relative abundances of living associations (*cf.* Greig-Smith 1957, p. 147).

The first 600 randomly selected individuals of twenty-eight borehole samples from western Nigeria, containing seventeen species of Paleocene (Cenozoic) ostracods were used to construct the correlation matrix given in Table 1. (Significant correlations printed in bold face type; these are 'total' correlations not taking the influences of the other k-2 variables into account.)

The species involved, in order of representation in the matrix, are *Cytherella sylvester-bradleyi* Reyment (x_1), *Bairdia ilaroensis* Reyment and Reyment (x_2), *Ovocytheridea pulchra* Reyment (x_3), *Iorubaella ologuni* Reyment (x_4), *Brachycythere ogboni* Reyment (x_5), *Dahomeya alata* Apostolescu (x_6), *Leguminocythereis bopaensis* (Apostolescu) (x_7), *L. lagaghiroboensis* Apostolescu (x_8), *Trachyleberis teiskotensis* (Apostolescu) (x_9), *Veenia warriensis* Reyment (x_{10}), *Buntonia* (*Buntonia*) *fortunata* Apostolescu (x_{11}),

Table 1. *Correlation matrix based on collections of seventeen species of fossil ostracods*

1·0000																
−0·3176	1·0000															
0·1363	−0·1902	1·0000														
0·1804	−0·1775	−0·3495	1·0000													
−0·2547	0·1606	−0·1285	−0·1838	1·0000												
−0·0267	−0·2299	−0·1599	0·2067	−0·0948	1·0000											
−0·0038	−0·1499	−0·2642	0·0481	−0·1279	**0·4711**	1·0000										
−0·1540	−0·1756	−0·2558	−0·2026	−0·1547	0·0639	0·0332	1·0000									
−0·3659	−0·1807	−0·1862	−0·0920	−0·2624	0·0435	0·0945	0·0308	1·0000								
−0·0509	−0·0028	0·2996	−0·3229	0·2043	−0·0899	0·3722	0·2368	−0·0499	1·0000							
−0·1097	−0·2288	−0·2509	−0·2021	0·3699	−0·1206	0·0400	−0·1809	−0·0083	0·0675	1·0000						
−0·2104	−0·1734	−0·1110	0·0320	**0·5888**	−0·1685	−0·2358	−0·2040	−0·1603	−0·1889	**0·7303**	1·0000					
−0·3085	−0·3166	0·3419	0·0981	−0·2349	0·0722	−0·2635	0·0903	0·2735	−0·1878	−0·2679	0·0075	1·0000				
−0·0213	−0·2386	0·2331	−0·0183	−0·1927	0·0792	−0·1421	−0·1107	−0·0871	−0·1408	−0·1212	−0·1286	0·2957	1·0000			
0·0130	−0·2420	−0·1388	−0·0855	−0·1567	0·3008	0·0942	−0·0079	0·3098	−0·0515	−0·0136	−0·2747	0·0277	−0·0811	1·0000		
0·1513	−0·1549	**−0·5050**	**0·4269**	−0·3088	0·2614	0·2442	−0·3362	0·3636	−0·2613	0·0118	−0·2447	−0·1761	−0·0099	**0·4261**	1·0000	
0·1389	−0·1714	**−0·4500**	0·1173	0·2005	0·2841	0·3086	**−0·4612**	0·0315	−0·1857	**0·4265**	0·3249	−0·3127	−0·1052	−0·0165	**0·4563**	1·0000

B. (*Buntonia*) *beninensis* Reyment (x_{12}), *B.* (*Buntonia*) *bopaensis* Apostolescu (x_{13}) *B.* (*Quasibuntonia*) *livida* Apostolescu (x_{14}), *Ruggieria tattami* Reyment (x_{15}), *Schizocythere* spp. (x_{16}), *Xestoleberis kekere* Reyment (x_{17}). Data from Reyment (1960, 1963) The results of the principal component analysis are shown in Table 2.

We note that all eigenvalues are positive, which is a result of the correlation matrix being symmetric and positive definite. A very interesting aspect of the analysis is that the first component only accounts for 18% of the total variation. Compared with applications of principal component analysis in taxonomic work this is low indeed (*cf.* Reyment 1963). The first three eigenvalues add up to 46·57% of the total variation and the first four to 57%. The slight differences between successive eigenvalues indicate the great computational difficulties that would have been experienced had the work been done manually. The differences between successive eigenvalues are smaller between the first and last few eigenvalues than between the intermediate ones. We note a further feature of the results, notably, that the elements of the first eigenvector are not all positive, which is a situation differing from that usually found in principal component analysis of biological materials (*cf.* Reyment 1963).

One interpretation that suggests itself is that the several eigenvalues of roughly the same magnitude may indicate the existence of more than one ecological factor and that all are of roughly the same importance.

Using Lawley's (1956) test of significance it was found that the first five eigenvalues are significantly different from each other, while the remaining twelve may be regarded as being isotropic.

Approximate 95% confidence intervals for the eigenvalues were found as follows:

The procedure is to choose l and u so that

$$\Pr[nl<\chi_n^2] = \sqrt{(1-\epsilon)};\ \Pr[\chi_n^2<nu] = \sqrt{(1-\epsilon)}.$$

$$\text{Then } \Pr\left[\frac{\lambda_m(\mathbf{S})}{u} < \text{all } \lambda\ (\mathbf{\Sigma}) < \frac{\lambda_M(\mathbf{S})}{l}\right] \geq 1-\epsilon.$$

(λ = the eigenvalues and $n = N-1$; λ_m = smallest eigenvalue and λ_M = largest eigenvalue)

These intervals are $0{\cdot}07 < \text{all } \lambda_i < 35{\cdot}08$, with a probability of 0·96.

A complication encountered when interpreting the meaning of the eigenvectors is that the standard deviations of the variables are frequently very different, which is an outcome of the nature of the data. In the present problem it might have been advisable to perform some sort of scaling operation on the input material. The vector of standard deviations is:

$\mathbf{s}'$ = (0·1384, 0·1593, 0·0609, 0·0475, 0·0247, 0·0123, 0·0425, 0·0714, 0·0829, 0·0213, 0·0514, 0·0850, 0·0733, 0·0808, 0·0431, 0·0421, 0·0101)

Dividing each component of the eigenvectors by the appropriate standard deviation, as is often done in principal component analysis based on a correlation matrix, did not lead to intelligible results.

A rough interpretation of the eigenvectors corresponding to the first five eigenvalues is that the components of the first vector are mainly suggestive of some environmental factor which influences species 16 and 17 in one direction and species 3 in another. The second eigenvector is mainly concerned with species 5, 11 and 12 and suggests the exist-

Table 2. *The principal component analysis*

Eigenvalue	Percentage of total variation and	diff. $\lambda_i - \lambda_{i+1}$	Eigenvectors
3·1052	18·27		(0·0832, −0·0973, −0·4252, 0·2438, 0·0065, 0·2300, 0·2360, −0·2753, 0·1080, −0·1759, 0·1954, 0·0852, −0.1990, −0·1003, 0·1690, 0·4427, 0·4435)
2·8655	16·86	1·41	(−0·1007, 0·1494, −0·0737, −0·1465, 0·4621, −0·2244, −0·1195, −0·1376, −0·2157, 0·0651, 0·3962, 0·4695, −0·2127, −0·1497, −0·2506, −0·2386, 0·1672)
1·9441	11·44	5·42	(−0·0323, 0·1795, −0·0531, −0·2826, 0·0705, 0·1092, 0·4530, 0·2376, 0·0294, 0·5075, −0·0079, −0·2717, −0·3912, −0·3246, 0·1172, −0·0308, −0·0144)
1·7764	10·45	0·99	(0·4703, 0·3122, −0·0128, 0·2283, −0·0938, −0·1424, −0·0681, −0·2084, −0·4765, −0·0761, −0·2868, −0·2290, −0·3481, −0·0224, −0·2487, 0·0270, −0·0126)
1·6070	9·45	1·00	(−0·4411, 0·5877, −0·3302, 0·0077, 0·0149, −0·1952, −0·2028, −0·0827, 0·2877, −0·2192, −0·1959, −0·1202, 0·0128, −0·2083, 0·0313, 0·1065, −0·1587)
1·1766	6·92	2·53	(−0·3485, 0·1197, −0·1118, 0·3392, 0·1588, 0·4946, 0·3246, 0·1046, −0·2307, 0·0051, −0·2213, 0·0641, 0·1974, 0·1327, −0·3858, −0·1939, 0·0337)
1·0167	5·98	0·94	(−0·1522, 0·2167, 0·1220, −0·4273, 0·0597, 0·0898, 0·0825, −0·4874, −0·0272, 0·1011, −0·0061, −0·1876, −0·0137, 0·6258, 0·0580, 0·0531, 0·1796)
0·8610	5·06	0·92	(0·0465, 0·0717, 0·0209, −0·1252, 0·3317, 0·4378, −0·2595, 0·0706, −0·3452, −0·2520, −0·1079, 0·0190, −0·0469, −0·0974, 0·6129, −0·0976, −0·1053)
0·6940	4·08	0·98	(0·0110, −0·0613, 0·2005, 0·3263, 0·3224, −0·1049, 0·0111 −0·5049, 0·0432, 0·4600, −0·2023, 0·0037, 0·3202, −0·2689, 0·1878, 0·0864, −0·0869)
0·6050	3·56	0·52	(−0·1580, −0·1108, −0·3080, 0·3368, 0·1484, −0·2197, −0·0507, 0·1772, −0·1719, 0·2609, 0·1767, −0·0202, −0·2246, 0·4767, 0·1991, 0·1872, −0·4107)
0·4302	2·53	1·03	(0·1081, −0·1690, 0·0023, −0·0274, 0·5267, 0·0105, −0·3173, 0·2839, 0·3200, 0·0632, −0·3887, −0·1422, −0·1655, 0·1482, −0·2202, 0·1739, 0·3072)
0·3077	1·81	0·72	(−0·2783, 0·2141, 0·5282, 0·1785, −0·2111, 0·2332, −0·3230, 0·1314, −0·1002, 0·1839, 0·2030, 0·0830, −0·1806, −0·0689, −0·0704, 0·4424, 0·1234)
0·2286	1·34	0·47	(0·2314, 0·0514, 0·0591, 0·1078, 0·0234, 0·4292, −0·0955, −0·2610, 0·4973, −0·0047, 0·1418, 0·1084, −0·3362, 0·0570, −0·1597, −0·1986, −0·4542)
0·2025	1·19	0·15	(0·2141, 0·1178, −0·4328, −0·0664, −0·1403, 0·2547, −0·4470, 0·0370, −0·0612, 0·4007, 0·2759, −0·2355, 0·3607, −0·0298, −0·1124, −0·0899, 0·1232)
0·1194	0·70	0·49	(−0·1768, −0·0193, 0·0062, 0·3237, −0·2712, −0·0804, −0·1662, −0·0321, 0·1293, 0·1536, −0·1420, 0·0268, −0·2809, 0·1001, 0·3031, −0·5810, 0·4188)
0·0403	0·24	0·46	(−0·1869, −0·1720, 0·1738, 0·1889, 0·2332, −0·0474, −0·0011, −0·0830, −0·0181, −0·2729, 0·4286, −0·7038, −0·0342, −0·1215, −0·0848, −0·1680, 0·0301)
0·0198	0·11	0·13	(0·3735, 0·5383, 0·1971, 0·2544, 0·1939, −0·1027, 0·2350, 0·2889, 0·2147, −0·0601, 0·2427, 0·0343, 0·2514, 0·2109, 0·1967, −0·0496, 0·1472)

ence of some environmental factor that influences all of these in the same way. The third eigenvector indicates the existence of an environmental factor that influences species 7 and 10 in one way and species 13 and 14 oppositely. The fourth vector suggests a factor that influenced species 1 and 2 in one direction and species 9 and 13 in another. The fifth component would appear to indicate an influence on species 1 and 5 in one direction and on species 2 in another.

Now, in principal component analysis it is assumed that the variation observed is mainly due to factors having characteristic effects on each of the variates. It is therefore open to question whether the foregoing analytic model is really the best way of attacking the problem. The related technique of factor analysis takes the possibility of extraneous variation into account. The $r_{ii} = 1$ terms of the correlation matrix are replaced by so-called 'communalities' that represent the proportion of the variation of each variate which is due to factors operating on some or all of the other variables, the rest being assumed due to chance. It should be pointed out that the communality concept of factor analysis is not accepted by all quantitative biologists as being valid. Also, in principal component analysis one is primarily interested in studying the *shape* of the scatter of the observations.

METHOD 2

The method of *factor analysis* here employed is that recently proposed by Jöreskog (1962). As it involves certain modifications it is given detailed discussion.

The fundamental postulate of factor analysis is

$$\mathbf{S} = \boldsymbol{\alpha\beta} + \boldsymbol{\epsilon} \tag{3}$$

where $\boldsymbol{\alpha} = (a_{it})$, $\boldsymbol{\beta} = (\beta_{tv})$ and $\boldsymbol{\epsilon} = (\epsilon_{iv})$. Here a_{it} and β_{tv} are non-random quantities, while matrix $\boldsymbol{\epsilon}$ is a set of random variables for which $E(\boldsymbol{\epsilon}) = \mathbf{0}$ and $E(\boldsymbol{\epsilon\epsilon}') = n\boldsymbol{\Delta}$, where $\boldsymbol{\Delta}$ is a positive diagonal matrix, the diagonal elements of which are the residual variances. We note that the a_{it} are termed factor loadings (= coefficients) of k common factors and the β_{tv} are the factor values of the individuals.

Factor analysis assumes the covariance matrix to be made up of two matrices:

$$\boldsymbol{\Sigma} = \boldsymbol{\alpha\alpha}' + \boldsymbol{\Delta} \tag{4}$$

The diagonal elements of $\boldsymbol{\Sigma} - \boldsymbol{\Delta}$ are called *communalities.*

Under the common assumption of factor analysis that $\boldsymbol{\Delta} = \sigma^2\mathbf{I}$, where σ^2 is a positive constant and $\mathbf{I}$ is the unit matrix, a test for significant eigenvalues may give a large number of common factors, since the $p-k$ eigenvalues of the population correlation matrix (which is what is used in most analyses in place of the covariance matrix) are not likely to be nearly isotropic unless k is large. In order to find the least number of common factors Jöreskog (1962) has proposed another assumption for $\boldsymbol{\Delta}$, namely, that

$$\boldsymbol{\Delta} = \theta(\text{diag } \boldsymbol{\Sigma}^{-1})^{-1} \tag{5}$$

where θ is a positive scalar and $\boldsymbol{\Sigma}$ is non-singular.

Computational procedure for method 2

We shall write $\mathbf{C}$ for the correlation matrix. Beginning with the sample correlation matrix the following steps are required:

1. Compute $\mathbf{D} = \text{diag } \mathbf{C}^{-1}$.

2. $\mathbf{C}^* = \mathbf{D}^{\frac{1}{2}}\mathbf{C}\mathbf{D}^{\frac{1}{2}}$.

3. $|\mathbf{C}^* - l^*\mathbf{I}|$, where the l^*_i are the eigenvalues of $\mathbf{C}^*$. At the same time the corresponding eigenvectors are found, $\mathbf{z}^*_i = (z^*_{1i}, z^*_{2i}, \ldots, z_{pi})'$.

4. The diagonal elements of $\mathbf{L}^*$ are tested for significance. $\mathbf{L}^*_k$ is the diagonal matrix of the first k eigenvalues of $\mathbf{L}^*$.

5. Then one computes

$$\mathbf{A}^* = \mathbf{Z}^*_k(\mathbf{L}^*_k - t_k\mathbf{I})^{\frac{1}{2}}$$

where t is defined as

$$t_k = \frac{1}{p-k} \sum_{i=k+1}^{p} l^*_i$$

and $\mathbf{A}$, the estimate of $\boldsymbol{\alpha}$, is $\mathbf{A} = \mathbf{D}^{-\frac{1}{2}}\mathbf{A}^*$.

6. The diagonal matrix of estimated *residual* variances $\mathbf{R}$ is

$$\mathbf{R} = t_k\mathbf{D}^{-1}$$

7. In order to find a lower confidence limit for k one first tests the hypothesis $k = 1$ against $k > 1$; if this hypothesis is rejected one tests $k = 2$ against $k > 2$, and so on. The number of factors for which the test does not show significance is determined. If the level of significance is 5% in each test, this method leads to a lower confidence limit for the number of common factors. The test criterion is (after Lawley 1956 and Whittle 1952)

$$c_k = -(n-1)[\log (l^*_{k+1} \ldots l^*_p) - (p-k) \log t]$$

where t is defined as in a foregoing equation and the l^*_i are the eigenvalues of $\mathbf{C}^*$. If the hypothesis is true and if n is large, c_k is approximately distributed as χ^2 with degrees of freedom

$$d_k = \tfrac{1}{2}(p-k+2)(p-k-1)$$

Example using method 2

We use the same correlation matrix as in the previous example, that is, the data of Table 1. We observe that the number of observational vectors, twenty-eight, is relatively small for an analysis in which five factors appear to be of some consequence. Furthermore, the observations deviate widely from normality, which is not a desirable situation in a factor analysis. Component analysis does not necessarily require normally distributed variables but the tests of significance for it as well as for factor analysis do. The results of the computations appear in order below.

1.

$$
\mathbf{D} = \begin{bmatrix}
9{\cdot}29 & . & . & . & . & . & . & . & . & . & . & . & . & . & . & . & . \\
 & 15{\cdot}94 & . & . & . & . & . & . & . & . & . & . & . & . & . & . & . \\
 & & 4{\cdot}90 & . & . & . & . & . & . & . & . & . & . & . & . & . & . \\
 & & & 5{\cdot}90 & . & . & . & . & . & . & . & . & . & . & . & . & . \\
 & & & & 5{\cdot}15 & . & . & . & . & . & . & . & . & . & . & . & . \\
 & & & & & 2{\cdot}55 & . & . & . & . & . & . & . & . & . & . & . \\
 & & & & & & 4{\cdot}93 & . & . & . & . & . & . & . & . & . & . \\
 & & & & & & & 5{\cdot}67 & . & . & . & . & . & . & . & . & . \\
 & & & & & & & & 4{\cdot}26 & . & . & . & . & . & . & . & . \\
 & & & & & & & & & 3{\cdot}81 & . & . & . & . & . & . & . \\
 & & & & & & & & & & 8{\cdot}90 & . & . & . & . & . & . \\
 & & & & & & & & & & & 12{\cdot}97 & . & . & . & . & . \\
 & & & & & & & & & & & & 5{\cdot}60 & . & . & . & . \\
 & & & & & & & & & & & & & 3{\cdot}75 & . & . & . \\
 & & & & & & & & & & & & & & 3{\cdot}95 & . & . \\
 & & & & & & & & & & & & & & & 4{\cdot}77 & . \\
 & & & & & & & & & & & & & & & & 4{\cdot}25
\end{bmatrix}
$$

2. Computing $\mathbf{C}^* = \mathbf{D}^{\frac{1}{2}}\mathbf{C}\mathbf{D}^{\frac{1}{2}}$ and extracting the eigenvalues gives the matrix

$$
\mathbf{L}^* = \begin{bmatrix}
23{\cdot}76 & . & . & . & . & . & . & . & . & . & . & . & . & . & . & . & . \\
 & 20{\cdot}35 & . & . & . & . & . & . & . & . & . & . & . & . & . & . & . \\
 & & 14{\cdot}62 & . & . & . & . & . & . & . & . & . & . & . & . & . & . \\
 & & & 11{\cdot}17 & . & . & . & . & . & . & . & . & . & . & . & . & . \\
 & & & & 9{\cdot}66 & . & . & . & . & . & . & . & . & . & . & . & . \\
 & & & & & 5{\cdot}79 & . & . & . & . & . & . & . & . & . & . & . \\
 & & & & & & 4{\cdot}80 & . & . & . & . & . & . & . & . & . & . \\
 & & & & & & & 3{\cdot}60 & . & . & . & . & . & . & . & . & . \\
 & & & & & & & & 3{\cdot}17 & . & . & . & . & . & . & . & . \\
 & & & & & & & & & 2{\cdot}73 & . & . & . & . & . & . & . \\
 & & & & & & & & & & 2{\cdot}32 & . & . & . & . & . & . \\
 & & & & & & & & & & & 1{\cdot}57 & . & . & . & . & . \\
 & & & & & & & & & & & & 1{\cdot}12 & . & . & . & . \\
 & & & & & & & & & & & & & 0{\cdot}90 & . & . & . \\
 & & & & & & & & & & & & & & 0{\cdot}57 & . & . \\
 & & & & & & & & & & & & & & & 0{\cdot}32 & . \\
 & & & & & & & & & & & & & & & & 0{\cdot}13
\end{bmatrix}
$$

Up to this point factor analysis does not differ essentially from principal component analysis. Compared with the results of the principal component analysis it is interesting to note that the eigenvalues are percentually of the same order of magnitude. More essential, however, is the fact that the $p-k$ smallest eigenvalues of $\mathbf{C}^*$ tend to be more like each other than do the $p-k$ smallest eigenvalues of $\mathbf{C}$.

The first eigenvalue represents 22·29% of the total variation, the second 19·09% and the third 13·72%. The first three eigenvalues add up to 55·1% of the total variation, and the first four to 65·6%; these figures indicate a somewhat greater concentration of variation to the first eigenvalues than was found for the correlation matrix alone.

3. The results of the tests of significance for the number of common factors are given in the following table. (If $d_k > 100$, c_k is a λ-value, otherwise c_k is a χ^2-value.)

k	t	d_k	c_k
1	5·18	135	5·35***
2	4·16	119	4·17***
3	3·42	104	3·27***
4	2·82	90	126·47**
5	2·25	77	95·67
6	1·93	65	79·26
7	1·64	54	64·69
8	1·43	44	54·30
9	1·21	35	44·04
10	0·99	27	33·84

The approximate 95 % confidence interval for the eigenvalues is, using the same formula as before, $0{\cdot}08 < \text{all } \lambda_i < 45{\cdot}73$, with a probability of 0·96.

Hence the smallest number of factors that may be considered to fit the model is $k = 5$, which agrees with the principal component analysis. However, since the sample size is small and the data not truly multivariate normal, it might be wise to consider a further five factors.

4. To find the unrotated factor loadings one computes

$$\mathbf{A} = \mathbf{D}^{\frac{1}{2}}\mathbf{A}^*$$

where $\mathbf{A}^* = \mathbf{Z}_k^*(\mathbf{L}_k^* - t_k\mathbf{I})^{\frac{1}{2}}$. It is common in psychometry to 'rotate' these vectors by some graphical device or other in order to present the data in 'standard' terms; rotation is not considered useful in most biological situations by the present writer. For $k = 10$, the matrix of factor loadings is

$$\mathbf{A} = \begin{bmatrix}
-0{\cdot}15 & -0{\cdot}47 & 0{\cdot}30 & -0{\cdot}72 & -0{\cdot}03 & 0{\cdot}09 & 0{\cdot}09 & 0{\cdot}01 & -0{\cdot}04 & -0{\cdot}06 \\
-0{\cdot}25 & 0{\cdot}90 & 0{\cdot}21 & -0{\cdot}02 & -0{\cdot}06 & 0{\cdot}03 & 0{\cdot}00 & -0{\cdot}03 & 0{\cdot}00 & -0{\cdot}01 \\
-0{\cdot}21 & -0{\cdot}12 & -0{\cdot}69 & -0{\cdot}36 & 0{\cdot}02 & 0{\cdot}13 & -0{\cdot}18 & 0{\cdot}03 & -0{\cdot}13 & -0{\cdot}05 \\
-0{\cdot}02 & -0{\cdot}30 & 0{\cdot}38 & 0{\cdot}07 & -0{\cdot}54 & -0{\cdot}48 & 0{\cdot}09 & 0{\cdot}00 & -0{\cdot}09 & 0{\cdot}20 \\
0{\cdot}59 & 0{\cdot}40 & -0{\cdot}06 & -0{\cdot}10 & 0{\cdot}10 & -0{\cdot}14 & -0{\cdot}11 & 0{\cdot}44 & 0{\cdot}02 & 0{\cdot}05 \\
-0{\cdot}10 & -0{\cdot}31 & 0{\cdot}21 & 0{\cdot}28 & 0{\cdot}12 & -0{\cdot}29 & -0{\cdot}25 & 0{\cdot}20 & 0{\cdot}32 & -0{\cdot}24 \\
-0{\cdot}09 & -0{\cdot}21 & 0{\cdot}35 & 0{\cdot}19 & 0{\cdot}56 & -0{\cdot}37 & -0{\cdot}31 & -0{\cdot}11 & -0{\cdot}06 & -0{\cdot}08 \\
-0{\cdot}25 & -0{\cdot}09 & -0{\cdot}52 & 0{\cdot}06 & 0{\cdot}36 & -0{\cdot}32 & 0{\cdot}41 & -0{\cdot}10 & 0{\cdot}22 & -0{\cdot}03 \\
-0{\cdot}09 & -0{\cdot}18 & -0{\cdot}01 & 0{\cdot}69 & 0{\cdot}12 & 0{\cdot}26 & 0{\cdot}17 & -0{\cdot}07 & -0{\cdot}27 & -0{\cdot}08 \\
-0{\cdot}09 & 0{\cdot}08 & -0{\cdot}21 & -0{\cdot}14 & 0{\cdot}66 & -0{\cdot}16 & -0{\cdot}21 & 0{\cdot}07 & -0{\cdot}31 & 0{\cdot}27 \\
0{\cdot}85 & -0{\cdot}02 & 0{\cdot}08 & 0{\cdot}02 & 0{\cdot}27 & 0{\cdot}18 & 0{\cdot}05 & -0{\cdot}17 & 0{\cdot}02 & 0{\cdot}10 \\
0{\cdot}93 & 0{\cdot}09 & -0{\cdot}13 & -0{\cdot}02 & -0{\cdot}17 & -0{\cdot}07 & 0{\cdot}04 & 0{\cdot}00 & -0{\cdot}01 & -0{\cdot}04 \\
-0{\cdot}11 & -0{\cdot}26 & -0{\cdot}57 & 0{\cdot}38 & -0{\cdot}44 & 0{\cdot}01 & -0{\cdot}14 & 0{\cdot}05 & -0{\cdot}14 & -0{\cdot}10 \\
-0{\cdot}11 & -0{\cdot}24 & -0{\cdot}23 & 0{\cdot}01 & -0{\cdot}26 & 0{\cdot}20 & -0{\cdot}49 & -0{\cdot}20 & 0{\cdot}37 & 0{\cdot}25 \\
-0{\cdot}16 & -0{\cdot}32 & 0{\cdot}16 & 0{\cdot}33 & 0{\cdot}22 & 0{\cdot}35 & 0{\cdot}17 & 0{\cdot}46 & 0{\cdot}15 & 0{\cdot}09 \\
-0{\cdot}10 & -0{\cdot}36 & 0{\cdot}69 & 0{\cdot}34 & -0{\cdot}06 & 0{\cdot}16 & 0{\cdot}05 & 0{\cdot}00 & -0{\cdot}02 & 0{\cdot}12 \\
0{\cdot}47 & -0{\cdot}18 & 0{\cdot}58 & 0{\cdot}05 & 0{\cdot}07 & 0{\cdot}03 & -0{\cdot}23 & -0{\cdot}02 & 0{\cdot}01 & -0{\cdot}29
\end{bmatrix}$$

2·47	1·93	2·54	1·70	1·66	0·95	0·84	0·56	0·54	0·40

The last row gives the contribution of each factor to the total test variance.

The results show clearly that these ten factors reproduce both the correlations and variances of the original matrix very well and it is quite possible that even a lesser number of factors would do this too. The communalities are found from finding diag **C**—**R**, where $\mathbf{R} = t\mathbf{D}^{-1}$. That is, (1—estimated residual variance). There are in order:

0·90, 0·94, 0·80, 0·84, 0·81, 0·62, 0·80, 0·83, 0·77, 0·74, 0·89,
0·93, 0·83, 0·74, 0·75, 0·80, 0·25.

The first factor indicates that species 5, 11, 12 and possibly 17 are affected in the same way by some environmental factor; the other elements of the vector do not differ significanlty from zero. The second factor suggests an environmental control that mainly influences species 2 but also to a lesser degree, in the same direction, species 5. Species 1, 15 and 16 are influenced in the opposite direction. The third factor seems to represent an environmental control that affects species 3, 8 and 13 in one direction and species 16 and 17 in another direction. The fourth vector would appear to indicate the influence of an environmental factor which rather strongly affects species 1 in one direction and about equally as strongly species 9 in the reverse direction; other affected species are 3, 13, 15 and 16. The fifth vector suggests relatively strong influences on species 4, 7 and 10 in opposite directions, and lesser influences on species 8 and 13. These five factors seem to be the most informative. The remaining five factors included in the matrix of factor loadings are remarkable in that none of them suggests a strong reaction of any environmental agent on any of the species.

Hence, it would seem that most of the variation in frequencies of the seventeen species may have been controlled by five environmental stimuli of some kind or other (for example, temperature, light, salinity, variation in chemical proportions of sea water, pH, redox).

If as a criterion of non-reactivity to environment we take small factor loadings it may be suggested that species 6, 14 and 15 are euryoic and this agrees extremely well with what is to be observed qualitatively in the material. Judging from occurrences in the borehole samples one gains the impression that species 1 and 9 are also euryoic. Both of these are strongly affected only by the fourth factor, which may indicate that this factor is an unimportant one with respect to actual distribution. None of the species appears to give the impression of being stenoöic.

It will be seen that the results of the factor analysis tend to differ somewhat in detail from the component analysis, although both give a strong impression of the operation of several approximately equally important environmental factors.

Inasmuch as the factor analysis model would seem to be more suitable for the kind of ecological data here treated (remembering we are not studying the shape of the distribution) the results obtained with its aid could be more descriptive of the actual situation.

ACKNOWLEDGMENTS

The computations for method 1 were performed on the computer FACIT of the Swedish Board for Computing Machinery, Stockholm, and those for method 2 were made on the IBM computer 1620 of the University of Uppsala. Thanks are due to these institutions for providing free computing time. The investigation was supported financially by the

Swedish Natural Science Research Council. Dr K. G. Jöreskog, Uppsala, and Fil. mag. K. Arle, Stockholm, gave invaluable assistance in processing the data.

SUMMARY

1. Variations in the relative frequencies of different species in samples may be interpreted in terms of the major environmental factors to which the organisms react. The number of these factors, if not their nature, may be deduced from a multivariate statistical analysis of the relative frequencies. These frequencies are set in a matrix of covariances or correlations and the number of factors whose interplay could have accounted for the observed matrix is computed by factor analysis or principal component analysis.

2. In the present investigation the results indicate the influence of five major environmental factors on seventeen species of fossil ostracods. Although the study is based on palaeoecological material, the reasoning can equally well be applied to other ecological situations on which insufficient chemical and physical data are available, or where it may not be exactly known which environmental factors are influencing the relative abundances of the species.

3. The procedures discussed, although unable to disclose whether a species is, for example, stenohaline or euryhaline, stenothermal or eurythermal, can indicate whether the species is stenoöic or euryoic. Although both principal component analysis and factor analysis are made use of here, it would seem that the underlying model of the latter method might be more suited to such problems.

REFERENCES

Greig-Smith, P. (1957). *Quantitative Plant Ecology*. London.

Hotelling, H. (1933). Analysis of a complex of variables into principal components. *J. Educ. Psychol.* **24,** 417–41, 498–520.

Jöreskog, K. G. (1962). On the statistical treatment of residuals in factor analysis. *Psychometrika*, **27,** 335–54.

Kendall, M. G. (1957). *A Course in Multivariate Analysis*. Griffin's Statistical Monographs, No. 2.

Lawley, D. N. (1956). Tests of significance for the latent roots of covariance and correlation matrices. *Biometrika*, **43,** 128–36.

Pearce, S. C. & Holland, D. A. (1961). Analyse des composantes, outil en recherche biométrique. *Biométrie-Praximétrie*, **2,** 159–77.

Rao, C. R. (1955). Estimation and tests of significance in factor analysis. *Psychometrika*, **20,** 93–111.

Reyment, R. A. (1960). Studies on Nigerian Upper Cretaceous and Lower Tertiary Ostracoda. Part I. Senonian and Maestrichtian Ostracoda. *Stockh. Contr. Geol.*, vol. 7.

Reyment, R. A. (1963). Studies on Nigerian Upper Cretaceous and Lower Tertiary Ostracoda. Part II. Danian, Paleocene and Eocene Ostracoda. *Stockh. Contr. Geol.*, vol. 10.

Whittle, P. (1952). On principal components and least squares methods of factor analysis. *Skand. Akt.-Tidskr.* **36,** 223–29.

17

Reprinted with permission from *Syst. Zool.*, **16**(2), 144–148 (1967)

FACTOR ANALYSIS IN MORPHOMETRIC TRAITS OF THE HOUSE MOUSE

James T. Wallace and Robert S. Bader

Abstract

A factor analysis with a Varimax rotation of the initial solution was applied to a system of 27 dental and cranial measurements in the house mouse. Five common "factors" were identified with respect to the 27 variables. These factors in decreasing order of their contribution to the total variation, were *width, anterior length, posterior length, skull* and *M3*. The proportion of the total variance accounted for by the total communality was 73.99%.

When a number of continuous variables are measured on a relatively large number of individuals, it is highly desirable to treat all the variables simultaneously and to reduce the large number of correlations and complex interactions to a relatively few effects and simple interactions. These, hopefully, can then be placed in a meaningful biological context. Among the various multivariate statistical techniques available one of the most powerful and useful is that of factor analysis, originally developed by Spearman in 1904. This technique has been progressively refined and improved over the past two decades, particularly by biometrically-inclined psychologists, and has begun to receive attention from biologists and anthropologists since the digital computer has become available to a large number of researchers. The techniques and procedures of numerical taxonomy are closely related to those of factor analysis. Adequate discussions of the computational procedures and associated techniques may be found in Thurstone (1947), Harman (1960) and Seal (1964). Particularly lucid presentations for the non-mathematician can be found in Adcock (1954) and Brown *et al.* (1965).

The basic data of the factor analysis are the coefficients of correlation among the several measurements or variables. These coefficients may be considered as effects and the aim is to achieve scientific parsimony, or economy of description, by reducing them to a few causes or "factors." Although the variables are known, the sets to which they belong are not. Thus, the determination of these sets of effects, or factors, and a search for their causes represent the essence of the factorial analytic technique. As opposed to regression analysis, the variables in factor analysis are treated symmetrically, i.e., no independent or dependent variables are assigned and no prior assumptions are necessary concerning the functional relationships among the variables (Brown *et al.*, 1965).

Two major types of factors exist, common and unique. The former includes general factors present in all variables (measurements) and group factors which are in more than one variable but not in all. The unique factors are exclusively associated with each particular variable and represent that portion of a variable not ascribable to its correlation with other variables in the set. The factor "loading" or coefficient represents the degree of association or correlation of the variable with the factor. The nature of the factor is primarily identified on the basis of the highest factor loadings. For any given dimensional measurement, the proportion of the total variance of the trait due to common factors (the total communality) is simply the sum of the squares of the factor loadings. The unique factor is the difference between the communality and one (unity). Thus, the entire variance of the trait is divided into common and unique factors.

After an initial "solution," or partition into factors, or major influences acting on

TABLE 1. COMMUNALITIES AND UNIQUE FACTORS FOR EACH VARIABLE.

Variables	Communalities (h^2)	(R – L)	Unique factor	Rank order of mean communalities
Palate	0.867		0.134	Mandible 0.981
L-Mandible	0.978		0.022	
R-Mandible	0.983	+0.005	0.017	Palate 0.867
L-M^1w	0.735		0.265	
R-M^1w	0.778	+0.043	0.222	M_3w 0.810
L-M^2w	0.734		0.266	
R-M^2w	0.745	+0.011	0.255	M^1L 0.783
L-M^3w	0.615		0.385	
R-M^3w	0.734	+0.119	0.266	M_2w 0.776
L-M^1L	0.789		0.211	
R-M^1L	0.777	–0.012	0.223	M_1w 0.768
L-M^2L	0.619		0.381	
R-M^2L	0.713	+0.094	0.287	M^1w 0.756
L-M^3L	0.486		0.514	
R-M^3L	0.567	+0.081	0.433	M^2w 0.740
L-M_1w	0.764		0.236	
R-M_1w	0.771	+0.007	0.229	M_1L 0.699
L-M_2w	0.757		0.243	
R-M_2w	0.795	+0.038	0.205	M_2L 0.698
L-M_3w	0.798		0.202	
R-M_3w	0.821	+0.023	0.179	M_3L 0.677
L-M_1L	0.635		0.365	
R-M_1L	0.764	+0.129	0.236	M^3w 0.675
L-M_2L	0.737		0.263	
R-M_2L	0.659	–0.078	0.341	M^2L 0.666
L-M_3L	0.746		0.254	
R-M_3L	0.609	–0.137	0.391	M^3L 0.527

Sum of Communalities 19.977

the system, it may be desirable to rotate the axes in order to obtain a more meaningful final solution that shows "simple structure." The main requirements of simple structure as given by Thurstone (1947) are that each variable should have at least one zero factor loading, that each factor should have a number (at least equal to the number of factors) of zero variable loadings, and that for each pair of factors there should be some variables with zero loadings for one factor but significant for the other (Brown *et al.*, 1965). Thus, appropriate rotation of the axes tends to maximize the number of zero coefficient loadings and allows one to more readily recognize the factors. In the present analysis the original principal factor solution was rotated to yield the final varimax solution (Kaiser, 1958).

ANALYSIS

Twenty-seven variables were considered in the present factorial analysis of a sample of 86 specimens (43 males, 43 females) of *Mus musculus* from salt marshes on San Francisco Bay, California. These included the right and left, length and width molar dimensions, the right and left mandibular length and the palatal length. The communalities and unique factors for each variable are shown in Table 1. The communalities range from 0.98 (mandibular length) to 0.49 (M^3L). The three skull values are all above the highest dental figure. The widths and the lowers tend to be higher than the lengths and uppers, respectively. The lowest values involve the M3. The relative values of the communalities generally parallel those of the individual correlations which were determined in an-

TABLE 2. VARIMAX ROTATED FACTOR MATRIX.

Variable	Factors I	II	III	IV	V
Palate	0.079	−0.035	−0.050	0.909**	0.174
Left Mandible	0.134	0.088	0.067	0.967**	0.116
Right Mandible	0.142	0.086	0.063	0.966**	0.134
Left M^1w	0.762**	0.269*	0.246	0.074	0.126
Right M^1w	0.731**	0.363**	0.304*	0.040	0.135
Left M^2w	0.748**	0.116	0.392**	0.049	0.070
Right M^2w	0.748**	0.122	0.402**	−0.016	0.093
Left M^3w	0.413**	−0.031	0.584**	−0.051	0.317*
Right M^3w	0.476**	−0.021	0.652**	−0.056	0.282*
Left M^1L	0.242	0.780**	0.222	0.048	0.264*
Right M^1L	0.285*	0.790**	0.160	−0.017	0.214
Left M^2L	0.141	0.356**	0.685**	0.017	−0.051
Right M^2L	0.077	0.303*	0.784**	0.019	0.022
Left M^3L	0.311*	0.148	0.521**	0.111	0.290*
Right M^3L	0.357**	0.046	0.587**	0.092	0.291*
Left M_1w	0.760**	0.399**	0.079	0.134	0.054
Right M_1w	0.779**	0.333*	0.071	0.196	0.096
Left M_2w	0.784**	0.271*	0.183	0.133	0.131
Right M_2w	0.791**	0.286*	0.205	0.149	0.152
Left M_3w	0.705**	0.162	0.253	0.102	0.449**
Right M_3w	0.683**	0.180	0.266*	0.116	0.488**
Left M_1L	0.535**	0.572**	0.094	0.067	0.087
Right M_1L	0.530**	0.684**	0.090	0.081	0.024
Left M_2L	0.413**	0.596**	0.442**	0.111	0.066
Right M_2L	0.497**	0.491**	0.399**	0.053	0.095
Left M_3L	0.151	0.158	0.125	0.213	0.798**
Right M_3L	0.145	0.211	0.193	0.233	0.672**

* Significant at 0.05.
** Significant at 0.01.

other study by the authors. The bilateral values of the communalities are very similar to one another, generally being in the neighborhood of a few hundredths of each other. The greatest differences involve the semi-vestigial M3.

The varimax solution factor matrix is presented in Table 2 for the bilateral measurements. Five common factors have been identified for the 27 variables. Almost all the loadings are positive and vary from near zero to +0.97. The bilateral loadings are very similar to one another as they were for the communalities to which they gave rise. The similarity of the bilateral coefficients or loadings reflects the high degree of symmetry in the system and may be considered a check on the method and the results produced. The greatest bilateral differences are in the M3, very probably another manifestation of the greater inter-side variance of this tooth.

The amount and per cent of the variance contributed by each factor is shown in Table 3. The statistical methodology is such that the first factor will contribute the greatest variance, the second factor the next largest variance, etc. The sum of the variances, or the total communality, is approximately 20. This is obtained by squaring the individual loadings for each factor and summing over all factors. The variance of each variable (dental measurement) is set equal to unity (one). Therefore, the total variance is equal to 27. The proportion of the total variance accounted for by the total communality is, then, 20/27 or 73.99%. The remainder is, of course, due to the unique factors.

For convenience in assimilating and in-

TABLE 3. PER CENT OF TOTAL VARIANCE PER FACTOR.

Factor	Variance	Variance per cent	Cumulative per cent
I	7.484	37.462	37.462
II	3.670	18.372	55.834
III	3.665	18.346	74.180
IV	2.980	14.917	89.097
V	2.178	10.903	100.000

Total communality 19.977

Per cent of total variance 73.997

TABLE 4. VARIMAX SOLUTION OF PRINCIPAL AXIS ANALYSIS. VALUES FOR RIGHT SIDE ONLY ARE GIVEN.

Variable	Factors I	II	III	IV	V
Palate	0.08	–0.03	–0.05	0.91**	0.17
Mandible	0.14	0.09	0.06	0.97**	0.13
M^1w	0.73**	0.36**	0.30*	0.04	0.14
M^2w	0.75**	0.12	0.40**	–0.02	0.09
M^3w	0.48**	–0.02	0.65**	–0.06	0.28*
M^1L	0.29*	0.79**	0.16	–0.02	0.21
M^2L	0.08	0.30*	0.78**	0.02	0.02
M^3L	0.36**	0.05	0.59**	0.09	0.29*
M_1w	0.78**	0.33*	0.07	0.20	0.10
M_2w	0.79**	0.29*	0.21	0.15	0.15
M_3w	0.68**	0.18	0.27*	0.12	0.49**
M_1L	0.53**	0.68**	0.09	0.08	0.02
M_2L	0.50**	0.49**	0.40**	0.05	0.09
M_3L	0.14	0.21	0.19	0.23	0.67**

* Significant at 0.05;
** Significant at 0.01.

terpreting, the loading coefficients are rounded to two decimal places and only the values of the right side are shown in Table 4. Only the coefficients greater than 0.257 are significant at the 0.05 level, and only those greater than 0.339 are significant at the 0.01 level. Of the 70 coefficients, eight are significant at $0.05 > P > 0.01$ and 22 are significant at $P < 0.01$.

The first factor has the greatest number of significant loadings (10), nine of which are significant at the 0.01 level. The five greatest loadings on factor I are all width dimensions and all widths are significant at the 0.01 level. Four of the lengths have significant, but relatively low, loadings on this factor which may be tentatively identified as a *width* factor.

The number of loadings significant at the 0.01 level has been reduced in factor II from nine to four. These include M^1L (0.79), M_1L (0.68) and M_2L (0.49) and M^1w (0.36). The M^2L has a significant coefficient of 0.30. In addition, the M_1w and M_2w have significant (0.05) loadings. This factor may be called an *anterior length* factor with somewhat greater strength in the lowers than in the uppers.

There are seven significant loadings on factor III, five of which are uppers and three of which are lengths. The very significant coefficients include M^2L (0.78), M^3w (0.65), M^3L (0.59), M_2L (0.40), and M^2w (0.40). This factor shall be called a *posterior length* factor with particular strength in the uppers and some strength in the widths as well as lengths. It should be noted that M_3L, which might be expected to have a relatively high loading here, does not.

Factor IV is the most clear-cut of the several factors. Very high coefficients of 0.91 and 0.97 are found for the palate and mandible respectively. The next highest value is 0.23 for M_3L. These skull dimensions do not approach significance for any other factor. This factor may be called a *skull* factor.

Factor V has two coefficients significant at 0.05 and two at 0.01, each involving one of the four M3 measures. The coefficients significant at 0.01 are for the mandibular tooth, the greatest being the M_3L. This is the only factor with significant loading on the M_3L while the other M3 measures all have at least one other significant loading (exclusively on factors I and III). It may be referred to, therefore, as the *M3* factor although it could also appropriately be called the M_3L factor.

DISCUSSION

What emerges from the analysis is a more general picture of the forces, or fields of influence, at work on the dentition than could be obtained from study of the individual correlation coefficients. The first factor, width, is the most powerful single determinant. It is of interest to note that the absolute amount of variance in the upper and lower dental widths is very similar for the three tooth types although the relative variation is greater as one moves posteriorly in the series. The next two factors emphasize the length component. The second factor has its greatest influence in the anterior region of the dentition while the third factor has its greatest influence more posteriorly. These factors may be thought of as gradients of opposite sign or direction. Grüneberg (1965) has recently identified mutants in the house mouse with increasing (or decreasing) dental effects as one moves from the first through the third molar. Since there are significant loadings for both lengths and widths in both the uppers and lowers on each of the first three factors, both the intra-tooth and occlusal correlations are included in these components but the primary emphasis is on the intra-jaw, unidimensional element, especially in factor one. The fourth factor reflects the relative independence of the skull and dental systems and is the most clear-cut of the five factors. The last factor is clearly a reflection of the segmentation of the total morphogenetic field into an anterior (M1–M2) and a posterior (M3) region.

SUMMARY

The multivariate approach of factor analysis was employed on 27 measurements (24 dental, 3 skull) of the house mouse, *Mus musculus*. This resulted in the recognition of five factors (in order of decreasing contribution to the total variance): width, anterior length, posterior length, skull and M3. The analysis yields a more general view of the forces, or morphogenetic fields, affecting tooth size and interrelationships than can be obtained from the individual correlation coefficients.

ACKNOWLEDGMENTS

Funds were made available for this research by the Department of Zoology, University of Illinois, Urbana, and National Science Foundation grant G-19197. Publication costs were shared by the Department of Zoology, University of Illinois, and Carleton College, Northfield, Minnesota. Appreciation is extended to the members of the Statistical Services Unit and the Department of Computer Science of the University of Illinois, Urbana, for their aid in the data processing. The authors also wish to thank Dr. Donald Breakey, Department of Biology, Willamette University, Salem, Oregon, for loan of the specimens.

LITERATURE CITED

ADCOCK, C. J. 1954. Factorial analysis for non-mathematicians. Cambridge University Press, London. 88 pp.

BROWN, T., M. J. BARRETT, AND J. N. DARROCK. 1965. Factor analysis in cephalometric research. Growth, 29:97–107.

GRÜNEBERG, H. 1965. Genes and genotypes affecting the teeth of the mouse. J. Embryol. Exp. Morph., 14:137–159.

HARMAN, H. H. 1960. Modern factor analysis. University of Chicago Press, Chicago. 469 pp.

Kaiser, H. F. 1958. The varimax criterion for analytic rotation in factor analysis. Psychometrica, 23:187–200.

SEAL, H. 1964. Multivariate statistical analysis for biologists. John Wiley and Sons, Inc., New York. 207 pp.

THURSTONE, L. L. 1947. Multiple factor analysis. University of Chicago Press, Chicago. 535 pp.

Department of Biology, Eastern Kentucky University, Richmond, and Department of Zoology, University of Illinois, Urbana.

18

Reprinted from *Amer. Naturalist,* **105**(945), 455–466 (1971)

COMPONENTS OF SEXUAL DIMORPHISM IN *CHIRONOMUS* LARVAE (DIPTERA: CHIRONOMIDAE)*

WILLIAM R. ATCHLEY

Department of Biology, Texas Technological University, Lubbock, Texas 79409

Morphometric studies of sexual differences present at least three interrelated problems. The first is whether sexual differences exist in a given set of attributes, the second how different the sexes are morphometrically. In general, most analyses of sexual differences have been directed toward these two problems. A recent example is a study by Atchley and Martin (1971) of sexual differences in larvae of five species of *Chironomus*. Seventeen characters of the larval head capsule were examined by univariate analyses of variance to determine sexual differences, and discriminant functions were computed to ascertain the extent of the sexual dimorphism in each of the five species. These analyses indicated that differences occurred in these species not only in the number of sexually dimorphic structures but also in the amount of phenetic overlap between sexes.

A third problem, which is seldom examined and is the subject of this paper, relates to the number of independent components of morphometric variation involved in the expression of sexual dimorphism. When this aspect of sexual dimorphism has been examined (e.g., Blackith and Roberts 1958; Blackith and Blackith 1969), attention has focused on the adult organism.

Morphometric variation in the "external" phenotype of a group of organisms can be partitioned into "size and shape" contrasts by multivariate statistical techniques. The general body size component, which accounts for most of the morphometric variation, is often highly sensitive to factors such as environmental fluctuations, nutrition, and rate of growth, while the shape components are much less affected by extrinsic factors and may more accurately reflect genetic differences in the group. Contrasts in shape can sometimes be correlated with differences in the biologies of the organisms being compared. If variation among species or higher-ranked taxa is studied, these shape contrasts may also reflect evolutionary differences.

The simplest form of sexual dimorphism is expressed only in the general size component. In this case, morphometric differences can be thought simply to reflect unequal rates of feeding, metabolism, or longevity at some stage of the life history. If dimorphic shape components are also found, the situation may be more complex, since size- and shape-variation patterns can range from being highly intercorrelated to being independent of each

* Contribution No. 1470 from the Department of Entomology, University of Kansas.

other (orthogonal). When sexual dimorphism in shape is orthogonal to size dimorphism, it may be inferred that the sexes are ecologically distinct; that is, they differ functionally with regard to a particular set of ecological parameters.

Sexual dimorphism in the immature stages of an insect might be due to developmental precursors of the adult dimorphism rather than to ecological differences between the immatures of sexes. Examples of such cases are several pupal characters in certain groups of nematocerous Diptera (e.g., Ceratopogonidae), in which the shape of the pupal head reflects the developing, highly dimorphic, adult organ. It is doubtful whether any ecological differences in the pupal stage could be correlated with this dimorphism in the pupal head. If, however, shape components exhibit sexually dimorphic characters for which a functional interpretation can be found, this might be taken as evidence of ecological differences between the sexes in the immature stages.

In order to investigate size and shape differences in sexually dimorphic immature insects, the data on chironomid larvae reported by Atchley and Martin (1971) are reanalyzed in this study. It is my aim to elucidate the patterns of morphometric variation and of sexual dimorphism in the larval head capsule of each species.

MATERIALS AND METHODS

Larvae of five species of *Chironomus, Ch. frommeri, Ch. tepperi, Ch. staegeri, Ch. cloacalis,* and *Ch. nepeanensis* had been measured for 17 characters of the head capsule. A list of these characters is given in table 1. Further details on the characters, sample sizes, means and standard deviations, and other relevant data for these five species can be found in Atchley and Martin (1971).

In addition to the 17 morphometric characters, the sex of each specimen, coded as 0 for males and 1 for females, was entered as the eighteenth variable. This was done to determine in which pattern of variation sex was most important and if this "sex" factor could be related to other axes of variation.

To restrict extrinsic causes of variation, only sclerotized head-capsule characters were used, and soft body measurements were omitted. The sex of each specimen had been previously determined by the genital anlagen, and, to eliminate any seasonal or geographic variation, only individuals from a single collection were included.

Those characters shown by Atchley and Martin (1971) to exhibit significant sexual differences ($P < .05$) for any one species are indicated by asterisks in table 1 following the pattern coefficients for factor I of that species. Sexual dimorphism in these characters is defined in a statistical sense—that is, sexually dimorphic characters are those that show a statistically significant difference in means between the sexes. When this definition is used the term "dimorphism" is probably a misnomer. I have, however, retained it for use in a broad sense and used it interchangeably with "sexual differences."

Complex morphometric phenomena, such as sexual dimorphism, generally require multivariate methods for their analysis. Even if a number of different characters are employed for univariate studies, it is possible to select variables that are highly correlated and, as a result, sample only a single plane of morphometric variation, for example, a "size" relationship. Computing bivariate correlation coefficients for the variables cannot adequately elucidate patterns of size and shape variation.

The use of discriminant functions to show size and shape patterns, as advocated by Blackith (1960) and others, is also inappropriate. In the main, this technique maximizes among-group variation, minimizes within-group variation, and therefore can be said to *discriminate* among groups of samples along possible size and shape axes. This method selects variables that best separate a series of groups, a procedure that is not necessarily related to outlining the patterns of size and shape variation *within* a taxon.

When computing discriminant functions, $n - 1$ vectors (where n is the number of groups) are produced. If we assume for the moment that each discriminant function does, in fact, reflect an interpretable shape contrast, the sexes of each species examined here would be expected to differ by only a single component. As will be shown later, the sexes of each species always differ on at least two axes.

Comparison of discriminant function results of these five species of *Chironomus* obtained by Atchley and Martin (1971) with those obtained by the factor analyses reported here reveals little similarity as far as the patterns of the coefficients are concerned. Since two groups are required to compute a discriminant function, it is obviously not possible to study size and shape variation within a single statistical population.

For studying the patterns of size and shape variation *within* a taxon, factor analysis, which partitions variance components within a sample, is the technique of choice. Factor analysis permits the isolation and interpretation of those morphogenetic forces that produce the observed patterns of size and shape variation. Cattell's "Maxplane" method of oblique factor analysis (Cattell and Muerle 1960) was used for these data. Factors were extracted representing eigenvalues greater than 1.0. In some instances (e.g., *Ch. nepeanensis* and *Ch. frommeri*) this gave useful results, since the factors with eigenvalues closer to 1.0 generally had more than one variable with a fairly large factor coefficient ($>.50$ as an arbitrary value), and, further, the variables were such that not very much error variance was possible. In other cases (e.g., *Ch. tepperi* and *Ch. cloacalis*) rotating all factors with eigenvalues above 1.0 was not done, since the vectors with eigenvalues nearer to 1.0 had only a single highly loaded variable, and, in the case of *tepperi* and *cloacalis*, these factors appeared to be "error factors." In *tepperi*, the only character with a high factor loading was the length of the fourth antennal segment, while in *cloacalis* it was the length of the second segment. The mean lengths of these two variables in the

TABLE 1
OBLIQUE FACTOR PATTERN MATRIX OF 18 CHARACTERS FOR LARVAE OF FIVE SPECIES OF *Chironomus*

	frommeri					*tepperi*		
	I	II	III	IV	V	I	II	III
Width of labial tooth C_1	05	−07	97	02	−06	07*	01	83
Width of central labial teeth C_1 and C_2	−06	−07	92	07	−14	−06*	01	83
Height of labial tooth C_1 from base of C_2	−08	−93	04	−19	01	06	−92	06
Height of labial tooth C_2	06	−54	01	58	15	−17	−72	−15
Height of labial tooth L_1	03	−76	13	15	07	−21	−96	−04
Height of central labial tooth from base of L_1	−03	−90	−15	01	06	−11	92	−05
Width between apices of C_2 teeth	−11	02	87	03	02	10*	13	100
No. pecten epipharyngeal teeth	−33	12	−06	−47	−02	−11	−22	18
Ventral head length	−73**	−12	−02	03	13	−83**	−44	−03
Length antennal segment 1	−81**	05	16	−16	−20	−76**	−22	16
Length antennal segment 1 to ring organ	−77*	10	11	−28	−02	−37	17	00
Length antennal segment 2	−58*	03	06	−19	36	−49*	−02	07
Length antennal segment 3	−24	−15	−17	−33	−06	−20	−04	07
Length antennal segment 4	23	−12	−07	20	43	−31	−11	00
Width of antenna at ring organ	01	19	−09	−19	89	−54**	−17	16
Mandible length	−59*	−06	−10	−01	19	−72**	−50	17
Frontoclypeus width	−11	02	−39	71	−13	−40*	−04	26
Sex	−60	18	−13	20	−02	−71	02	04
I	...	...	...	...	...	...	...	...
II	09	...	...	...	...	−08	...	...
III	−44	01	...	...	...	−38	−36	...
IV	−27	−02	19	...	...	...	...	...
V	−35	−12	23	09	...	...	...	...

populations studied were 12 μ and 30 μ, respectively. Because of the very small size of these more distal antennal segments, the possibility clearly exists that some of the variation might be error variation due to slide-mounting technique, particularly in the case of *tepperi*. In these two species, only the first three factors were extracted and rotated in the final analysis. The subject of "error factors" has been discussed by Cattell (1965).

Factor analyses were performed on data in which the sexes had been pooled. It is important, however, to examine the variation patterns within each sex. As will be seen in the next section, in the pooled analyses the coded sex variable always had the highest loading on what could be construed as a "head length" factor, that is, one with highest loadings on various length measurements such as ventral head length and antennal length.

To ascertain whether this factor was an expression of variation in head length or of sexual dimorphism, a separate factor analysis was performed on the 17 morphometric variables in each sex for each species. If this head length factor disappeared or the patterns of loadings became radically different, this factor in the pooled data was termed a "sex factor."

The degree of similarity in the patterns of factor loadings between the pooled and separated data for this head length factor was determined by

TABLE 1—*Continued*

staegeri				*cloacalis*			*nepeanensis*				
I	II	III	IV	I	II	III	I	II	III	IV	V
22*	00	100	−19	−16	−00	78	32**	−10	−04	04	76
−20	03	83	12	−12**	21	67	48**	04	05	29	74
−06**	84	02	−03	−64	06	11	11	88	02	08	−01
−15	59	−19	04	−82	16	−15	08	70	−01	−04	−18
−22	79	−06	09	−62*	12	−02	−07	89	02	−08	04
02	95	03	01	−83	02	15	03	80	10	−03	05
00	08	95	−01	24*	−19	91	26**	−07	08	00	74
03	19	07	−02	−05	06	−05	−16	14	−16	−46	−21
−89**	−07	−03	02	−26**	82	05	−71**	20	−01	08	22
−52**	−06	−07	−56	−03**	72	−03	116**	−01	06	89	22
08	−02	01	−91	09	26	−17	88	05	−10	115	24
−13	20	13	−50	09*	−14	01	−16	−20	63	−06	−29
−28	−21	−14	05	−03**	04	17	−07**	05	100	21	−10
03	−03	15	−15	−06	25	04	−12**	06	64	−05	−09
−58	−32	−14	00	01	06	06	17**	−07	02	−41	−08
−78**	01	−14	−03	−20**	51	18	70**	01	48	−01	−19
−57**	13	15	−03	03**	60	05	68**	−02	−06	−06	23
−77	−02	13	23	−11	55	22	83	−05	12	10	14
...	...	...	...	...	...	...	...	...	...	...	...
−17	...	...	...	05	...	...	−10	...	...	...	...
32	33	...	...	−31	02	...	39	−12	...	..	...
46	−17	−06	...	...	...	...	71	−74	33	...	...
...	...	...	...	...	...	...	−40	01	−51	−38	...

NOTE.—Decimal points have been removed. The diagonal matrices at bottom give correlation coefficients between primary factors. Asterisks following pattern coefficients for factor I of each species indicate significant sexual dimorphism determined by an analysis of variance test of the character in that row.

* $P < .05$.

** $P < .01$.

a correlation coefficient. If the patterns in each sex were highly correlated with those in the pooled data, it was assumed that a head length pattern of variation which was highly correlated with the major expression of sexual dimorphism did exist.

RESULTS

Factor pattern coefficients for each species are shown in table 1 together with the correlations between primary factors. They are discussed below, species by species.

Chironomus frommeri.—The five major components of variation included a head length factor I with highest factor loadings on ventral head length, total length of the first antennal segment, length of the segment to the ring organ, and mandible length. The coded sex variable loaded highest on this factor. Factor II was a "labial tooth height" axis, while factor III had highest loadings on measurements of labial tooth width. Factor IV represented a relationship between the height of labial tooth C_2 and the width of the frontoclypeus. Factor V was an antennal width component.

The geometrical relationships between factors were as follows: Factors II and III were uncorrelated, as were I and II. The relationships of I and III–V were of a more oblique nature, with I and III having the smallest angle.

As can be seen in the plot of the first three factors (fig. 1*a*), the best separation of the sexes is along the first, or head length, axis; there is some differentiation on the third, or tooth width, component; and there is no differentiation on the second, or tooth height, factor. The third factor is included in that all specimens with a factor score greater than 0.0 are represented by a circle. The value of 0.0 represents the break in factor scores whereby most of the male and female larvae were separated.

All of the sexually dimorphic variables had highest factor loadings on the head length axis, the same axis on which the coded sex variable loaded highest, indicating sexual differences in the head capsule of *frommeri* to be of the simplest type.

When the variation in each sex was examined by factor analysis, the above-mentioned patterns of morphometric variation persisted. Comparison of the head length components in each sex with those of the pooled data revealed highly correlated patterns of factor loadings ($r_{♀} = .852$, $r_{♂} = .790$; $P < .01$ for both).

Martin (1963) opined that, in a species of the related subgenus *Kiefferulus*, sexual differences are reflected by overall body size, a suggestion in agreement with the results presented here for *frommeri.*

Chironomus tepperi.—The three factors extracted for this species had the same interpretation as in the preceeding species, with head length, labial tooth height, and labial tooth width as the major size and shape components. Factor III showed rather strong covariation with the others, while I and II were uncorrelated. The coded sex variable again loaded on the general size factor, while the 10 sexually dimorphic variables were spread over four of the five major axes of variation. Only the tooth height factor lacked sexually dimorphic characters.

The plot of scores on the first three axes indicated that the sexes were most distinct on the first and third components, those relating to head length and tooth width (fig. 1*b*).

When the sexes were analyzed separately, the same components of variation were expressed, although not in the same order as in the pooled data. Correlations between the head length factors in the separate and pooled data were $r_{♀} = .877$ and $r_{♂} = .698$ ($P < .01$ for both). Thus a head length component occurs in each sex, although it is sexually very dimorphic.

Chironomus staegeri.—Three of the four factors extracted represented head length, labial tooth height, and labial tooth width; the fourth represented the lengths of the first two antennal segments. Only factors II and IV, the tooth height and antennal length components, were orthogonal. The remainder exhibited varying degrees of covariation, with the antennal

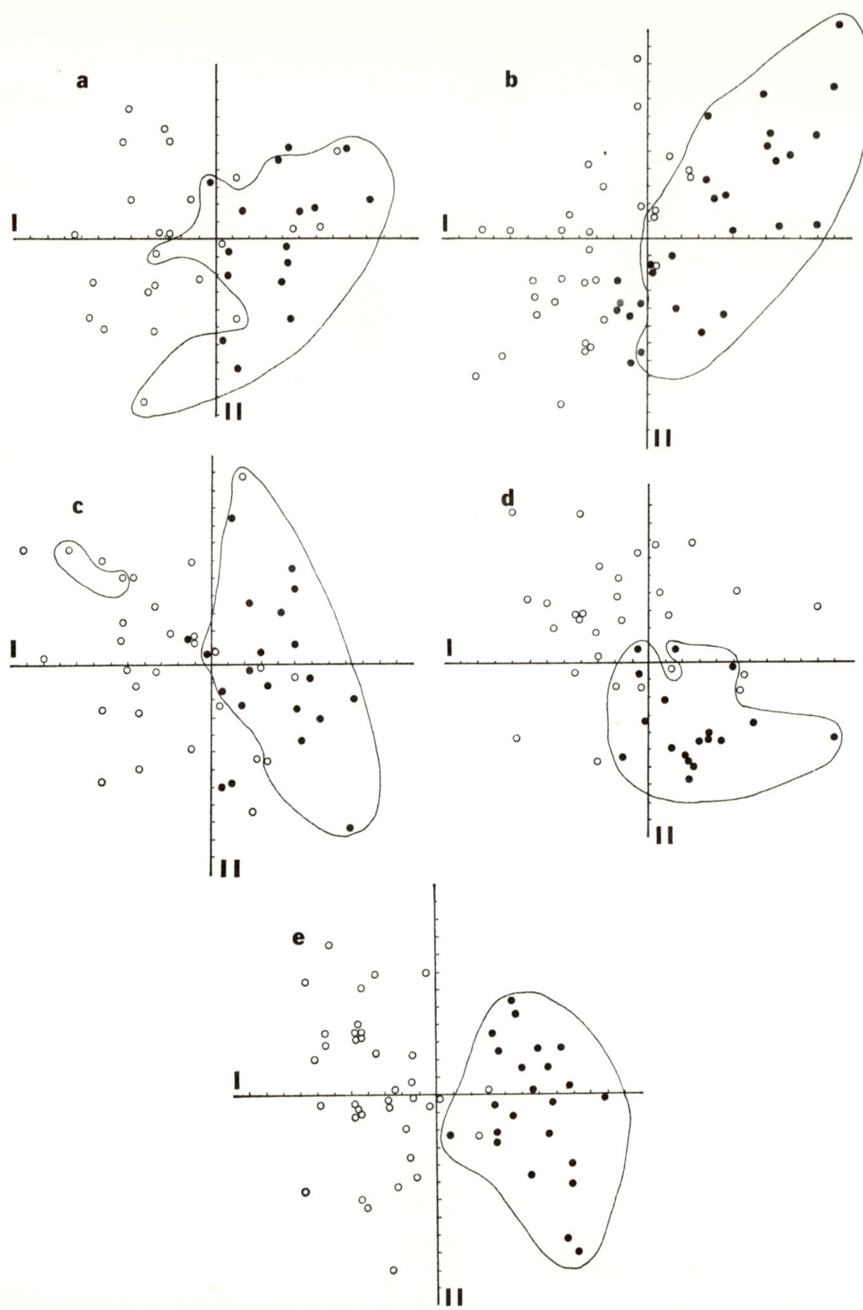

FIG. 1. Projection of standardized larval head capsule characters of five species onto oblique factors (*open circles* = females, *solid circles* = males). *a, Ch. frommeri* (*line* encloses individuals with factor scores >0.0 on factor III); *b, Ch. tepperi* (*line* encloses factor scores >−0.10 on III); *c, Ch. staegeri* (*line* encloses factor scores >0.0 on III); *d, Ch. cloacalis* (*line* encloses factor scores >−0.15 on III); *e, Ch. nepeanensis* (*line* encloses factor scores >−0.13 on III).

length and head length factors most closely related, as would be expected. Again the coded sex variable had highest loadings on the head length factor. Sexually dimorphic characters were found on three of the four factors; they are absent only from the antennal length component.

Once more, the plot of scores indicated distinctiveness of the sexes on the general size and tooth width components (fig. 1*c*).

A factor analysis performed on each sex showed that the labial tooth length and labial tooth width components persisted but that the head length component was changed. In the females, a head length vector which was highly correlated with the pooled data ($r_♀ = .696$, $P < .01$) was still evident; in the males, however, a head length component was found, but some of the variables that had loaded high on the pooled head length factor were now found on more than one axis. Ventral head length, for example, which is normally quite variable, did not have a loading $>.40$ on any of the first four factors. Correlation analysis of the pattern of factor coefficients for the head length component was not significant ($r_♂ = .506$, *ns*).

Chironomus cloacalis.—The three factors extracted represented the same components—head length, tooth height, and tooth width—which occurred in the other species. Sex again loaded with the head length factor, and the nine dimorphic characters were found on all three axes. The lengths of antennal segments 2 and 3, both of which exhibited sexual dimorphism, had highest loadings on factors which were not extracted in the final analysis. An examination of the geometrical relationships among factors revealed that factors I and III were not orthogonal with each other.

In this species the scattergram of factor scores indicated that the sexes are distinct on all three factors (fig. 1*d*).

In the analysis of each sex, similar loading patterns between the pooled data and the female larvae were found. The head length factor persisted in the females ($r_♀ = .828$, $P < .01$). In the males, one factor had high loadings on ventral head length as well as on other length measurements, but a marked inverse relationship occurred between ventral head length and the lengths of antennal segments 1 and 3. Because of this and other discrepancies, the patterns of factor loadings for the head length component were not significantly correlated with either the pooled data or that of the females ($r_♂ = .096$, *ns*).

Chironomus nepeanensis.—Five factors were extracted in *nepeanensis*. In addition to the head length/tooth height/tooth width relationship, factors were extracted that related to the length of the first antennal segment and to segments 2–4. The labial tooth height factor was more or less uncorrelated with head length; however, the other factors were correlated to varying degrees, with the first antennal segment's length component highly correlated with general size. The remaining relationships are depicted in table 1.

The coded sex variable loaded most heavily on the size axis, but the nine dimorphic characters were distributed over four of the five axes of variation. Only the tooth height component lacked sexually dimorphic variables.

The scattergram of factor scores (fig. 1*e*) showed good separation on the general size and tooth width components.

When the sexes were analyzed separately, the same morphometric contrasts were evident. The head length component persisted in each sex, and the patterns of factor loadings were highly correlated with the head length factor in the pooled analysis ($r_♀ = .805$, $P < .01$; and $r_♂ = .641$, $P < .05$).

DISCUSSION

Results presented here indicate considerable variation in the components of sexual dimorphism in larvae of closely related species of *Chironomus.* In each species, the coded sex variable always had highest loading on the major axis of variation, but in four of the five species other axes of variation were involved to varying degrees in the expression of sexual differences. In two species, the major component of variation appeared to be a head length factor in the pooled data, but, in actuality, this was an expression of sexual dimorphism, since it disappeared when the sexes were analyzed separately.

Chironomus frommeri exhibits the simplest form of sexual dimorphism, that is, a simple size difference between sexes, while *Ch. tepperi* shows a slightly more complex difference in that sexually dimorphic variables occur on two related factor axes. In *tepperi,* the dimorphic variables relate to head length and width of labial teeth, factors which exhibit considerable covariation. It is of interest that these two species are the simplest cytologically in that the latter form is cytologically monomorphic while the population from which *frommeri* was derived was heterozygous for a single chromosomal inversion (unpublished observations).

A third level of sexual differentiation is found in the remaining three species, in which dimorphic characters are found on at least three different factor axes, some of which are orthogonal. These three species are all cytologically highly polymorphic.

In *staegeri,* sexual differences occur on the head length, tooth width, tooth height, and antennal segment length factors, with the tooth height and antennal length components virtually orthogonal. In *cloacalis,* the head length, tooth width, and tooth height components are again represented, and only the two labial tooth factors are not orthogonal. Further, in *staegeri* and *cloacalis* the head length factor disappeared or was markedly changed when the males were analyzed alone. This indicated that what was thought to be an expression of head length in the analysis of pooled data was, in fact, an expression of sexual dimorphism.

With *nepeanensis,* the most dimorphic of the five forms, sex differences occur on four of the five factors. Some of these factors are orthogonal as far as the variational patterns are concerned.

In all cases, when the sexes were analyzed separately, the major components of variation became the tooth width or tooth height axes.

An examination of the larval biology might reveal ecological differences between the sexes to explain this separation of the sexes along the above-mentioned size and shape axes.

Chironomus larvae are aquatic and live in tubes or cases where they feed on organic detritus. Walshe (1951) has reported three types of feeding behavior in mud-inhabiting larvae. One type, characterized by *Ch. plumosus*, is generally referred to as a ''filter-feeding'' form. From the larval salivary glands, a net is secreted across the lumen of the tube. Undulatory movement of the body pulls water through the tube and catches organic material on the net; the net is then eaten and another secreted in its place by the salivary glands.

A second type of feeding behavior, found by Walshe in *Ch. ''dorsalis''* and *riparius,* can be characterized by the absence of filter feeding. (The correct taxonomic status of the species described by Walshe as *Ch. dorsalis* is unclear; consequently, it will be referred to here as *''dorsalis''*.) In these species, the larvae scrape up particles of mud from the surface with their mouthparts and anterior prolegs. Examination of the alimentary canal after carmine particles were scattered on the surface of the mud revealed large quantities of carmine, which indicated that the larvae were feeding by this method.

In a third type of behavior, found in *Ch. bathophilus,* the larvae feed deeper in the mud rather than on surface particles. In laboratory colonies, the larvae of this species were seen to feed well below the surface, and when carmine was added to the surface of the mud it was seldom found in the alimentary canal.

These studies indicate that, at least in the *''dorsalis''* and *bathophilus* modes of feeding, the mouthparts are being actively used in scraping and feeding. This has been further substantiated by Martin (personal communication), who found that differences in the wear on the labial teeth depended on whether the larvae were reared on sand and paper.

In blood-sucking Diptera, the nutritional reserves necessary for egg maturation and flight activities are obtained by a blood meal. Except for possibly obtaining carbohydrates in liquid form from plants, adults of *Chironomus* do not have an analog to a blood meal in the adult stage. Consequently, all nutritional reserves needed for egg maturation and other adult activities must be accumulated in the larva. The females, which are physiologically more active and longer lived than the males as adults, are often larger as larvae. This fact is reflected by the position of the female specimens along the primary axis of morphometric variation. This is not an isolated phenomenon found only in chironomids but rather occurs widely through at least the nematocerous Diptera.

One might wonder, therefore, how the females are able to attain this greater size and accumulate more reserves than the males during the larval stages. Several explanations are possible—for example, unequal longevity, differential metabolism, and different feeding rates.

Since, in general, the sexes emerge from the pupal stage at approximately the same time, unequal longevity in the larval instars can be dismissed as an explanation for the added reserves. Further, female larvae often become larger in the later stages of larval development, a fact not related to differential longevity. If we also reject a rather complicated differential rate of metabolism in each sex—for example, differences between sexes in mitotic rate (Mittwoch 1969)—the most plausible explanation is that one sex is better at exploiting the available food resources. This might be accomplished by faster and more active feeding or by unequal development of the mouthparts in the two sexes. With wider and longer labial teeth, the surface area for scraping would be increased. The sex with the better-developed mouthparts would, therefore, feed more efficiently in the "*dorsalis*" or *bathophilus* types of behavior.

As reported by Atchley and Martin (1971), the female larvae always have longer and wider labial teeth; the differences between males and females are often statistically significant. For example, in *nepeanensis,* the central labial tooth is about 12% wider and 10% longer in the females. The morphometric analyses reported here indicate that variation in the labial plate characters is only minimally related to size variation; therefore, any significant differences between the sexes in these structures probably reflect some type of ecological separation. It would seem most plausible that such ecological differences relate to feeding activities.

This size differential in mouthparts is sufficient to allow the female larvae to exploit more effectively the available food resources and thereby, in the same period of time as that spent by the males in the larval instars, to obtain food reserves required for adult females in the same time span needed by male larvae to complete developmnt.

Since detailed observations on larval biology in these five species are lacking, it is only possible to speculate on what type of feeding behavior might be exhibited by each form. It would be interesting, however, to see if *frommeri,* which exhibits dimorphism only in the characters related to head length, has the *plumosus* type of feeding behavior, in which a net is secreted and apparently less use is made of the labium and other mouthparts during feeding. In the case of *frommeri,* the females may be more active feeders than the males, which would explain the size differences. It is known that this species is a deepwater form living in muddy ooze at the bottom of lakes and, as such, probably does little scraping with the labial teeth. The remaining four species, which show sexual differences in size as well as in other growth components, may exhibit either the "*dorsalis*" or *bathophilus* type of behavior, in which the labium is apparently extensively used in feeding.

Some evidence exists for the latter hypothesis, at least in *cloacalis.* Martin (personal communication) has observed that, in laboratory colonies of this species, the larvae will readily leave the tubes in certain circumstances. Whether this can be taken as evidence of a "*dorsalis*" or *bathophilus* type of feeding behavior remains to be seen.

It is obvious that inferenees from morphometric results to modes of feeding behavior in these chironomids is highly speculative in view of our meager knowledege of the larval biology of these species. These results do, however, offer a plausible explanation for the observed size differences between the sexes in larvae of certain holometabolous insects as well as suggest a mechanism whereby differential nutritional reserves can be accumulated for use in the adult stage.

SUMMARY

The patterns of size and shape variation in the larval head capsules of five species of *Chironomus* were analyzed by a multiple factor analysis of 17 characters. Three major components of morphometric variation relate to head length, labial tooth width, and labial tooth height. Inclusion of a coded sex variable indicates that sexual differences in larvae of *Chironomus* are primarily reflected in size, but in four of the five species one or more shape components were found to be sexually dimorphic. It is suggested that the dimorphism along shape axes indicates ecological differences between sexes in the larval stage. The differences seem to relate to patterns of feeding and suggest a possible mechanism to account for the differential buildup of nutritional reserves needed in the adult stage.

ACKNOWLEDGMENTS

I am greatly indebted to Drs. Jon Martin and James E. Sublette for providing the data upon which this study was based. Further, I am indebted to Drs. Peter M. Neely and George W. Byers for reading and offering criticisms of the manuscript. These data were drawn from projects supported and financed, in part, by grants from the Australian Research Grants Commission (D67/16638) and the Federal Water Quality Administration, Department of the Interior, pursuant to the Federal Water Pollution Control Act (No. 18050 EOA). Computer facilities were provided by the University of Kansas Computation Center.

LITERATURE CITED

Atchley, W. R., and J. Martin. 1971. A morphometric analysis of differential sexual dimorphism in larvae of *Chironomus*. Can. Entomol. 103:319–327.

Blackith, R. E. 1960. A synthesis of multivariate techniques to distinguish patterns of growth in grasshoppers. Biometrics 16:28–40.

Blackith, R. E., and R. M. Blackith. 1969. Variation of shape and of discrete anatomical characters in the morabine grasshoppers. Australian J. Zool. 17:697–718.

Blackith, R. E., and M. I. Roberts. 1958. Farbenpolymorphisms bei einigen Feldheuschrecken. Z. Vererbungslehre 89:328–337.

Cattell, R. B. 1965. Factor analysis: an introduction to essentials. I. The purpose and underlying models. Biometrics 21:190–215.

Cattell, R. B., and J. L. Muerle. 1960. The "Maxplane" program for factor rotation to the oblique simple structure. Educ. Psychol. Measurement 20:569–590.

Martin, J. 1963. The cytology and larval morphology of the Victorian representatives of the subgenus *Kiefferulus* of the genus *Chironomus* (Diptera:Nematocera). Australian J. Zool. 11:301–322.

Mittwoch, U. 1969. Do genes determine sex? Nature 221:446–448.

Walshe, B. M. 1951. The feeding habits of certain chironomid larvae (subfamily Tendipedinae). Zool. Soc. London, Proc. 121:63–79.

Part III

MULTIVARIATE REGRESSION AND CORRELATION

Editors' Comments on Papers 19 Through 24

19 **BARTLETT**
On the Theory of Statistical Regression

20 **HOTELLING**
The Most Predictable Criterion

21 **GLAHN**
Canonical Correlation and Its Relationship to Discriminant Analysis and Multiple Regression

22 **POWER**
Evolutionary Implications of Wing and Size Variation in the Red-Winged Blackbird in Relation to Geographic and Climatic Factors: A Multiple Regression Analysis

23 **VOGT and JAMESON**
Chronological Correlation Between Change in Weather and Change in Morphology of the Pacific Tree Frog in Southern California

24 **LEE**
The Theory and Application of Canonical Trend Surfaces

Methods for studying functional relationships or degree of association between two or more variables are important procedures in all branches of science. In biology, statistical relationships among biological variables or between biological and ecological variables can provide information as to the underlying bases of biological variation and aid in the prediction of complex biological systems.

Correlation and regression methods were the first of the more complex statistical procedures to be developed. The least-squares hypothesis, which is basic to regression theory, was first mentioned in the literature by Legendre in 1805 and extensively developed by Gauss in 1809. The first biological applications of least-squares theory, however, were not made until almost the turn of the century.

The period of development of the correlation and regression

techniques in the late 1880s and 1890s signaled the beginning of the biometrical movement in biology. It was during this period that monumental contributions to statistical theory and applications were made by Francis Galton, W. F. R. Weldon, and Karl Pearson. The correlation-regression era, together with the "experimental design" phase that followed, has produced some of the most important tools in applied statistics.

The actual term *regression* was introduced by Francis Galton in connection with his early studies on inheritance in which he attempted to predict physical characteristics of children from those of the parents. Galton used the terms "reversion" or "regression" to refer to the tendency of the ideal mean filial type to depart from the parental type and revert to that of the ancestral form. Investigation into "reversion" or regression led to the rigorous mathematical interpretation of correlation and regression analysis, the latter meaning the least-squares predictive scheme, where a single criterion variable is estimated from one or more predictor variables.

In 1888 Galton published an important work in which he discussed co-relation or correlation of structure as a measure of the intensity of relationships between characters. He pointed out that co-relation must be the consequence of the variation of two structures being partly due to common causes rather than cause and effect. In 1892 F. Y. Edgeworth introduced the term "coefficient of correlation" and extended the idea of correlation to three and more variables. Although the ideas inherent to correlation and regression and their importance to biology were discussed by a number of statisticians and biologists during the early period of development, it was Karl Pearson (1896) who provided the definitive mathematical formulation for what we now know as multiple regression and correlation. Pearson's long treatise on regression, heredity, and panmixia synthesized much of the earlier work and produced the product-moment method for calculating the correlation coefficient. Although Pearson's 1896 paper is considered a classical work in the early developments of correlational methods, we have not included it in this series of papers because of its length. However, we have included a classical development of multivariate regression by M. S. Bartlett (Paper 19), in which he outlined much of the important statistical theory of modern analysis.

The linear model for multivariate regression is the common model of the form

$$\hat{Y} = b_0 + b_1X_1 + b_2X_2 + \cdots + b_pX_p$$

where the *b*'s are selected so as to maximize the correlation between *Y*, the actual observation, and $\hat{Y}$.

One limitation of the standard multivariate or multiple regression model is that only a single *Y* variable can be examined at a time. It was evident early in the history of biometrics that many problems exist in biology, as well as other fields, where several dependent variables are involved and the use of repeated multivariate regression analyses would not suffice. Thus, a generalized regression method was needed where the variation in one *set* of variables could be predicted by a second set of variables. Our original linear model should then be expanded to read

$$\hat{X} = b_1X_1 + b_2X_2 + \cdots + b_pX_p \qquad \hat{Y} = a_1Y_1 + a_1Y_1 + \cdots + a_qY_q$$

or, more simply,

$$\hat{X} = B'X \qquad \hat{Y} = A'Y$$

where the correlation is maximized between the two sets of variables, $\hat{X}$ and $\hat{Y}$.

In a classical pair of papers in multivariate statistics, Hotelling (Paper 20; 1936) resolved the problem of the generalized regression model and proposed his "most predictable criterion" to relate a battery of mental test scores to a set of physical measurement variables. This procedure Hotelling called *canonical correlation* and the linear function which maximized the correlation between the two sets of the original variables he referred to as a *canonical variate*. An unfortunate policy by the Trustees of Biometrika prevents our reproducing Hotelling's complete 1936 paper here. Thus, the more abbreviated 1935 paper must suffice.

The importance of Hotelling's contribution to multivariate statistical theory is considerable since canonical correlation is central to much of multivariate statistical methodology. Multiple regression, for example, can be considered simply a special case of canonical correlation analysis in which the number of variables in one set is equal to 1. The Fisherian discriminant function, which is also a multiple regression technique, can be generalized to multiple groups through canonical correlation analysis (Bartlett, 1938), in which one set of variables acts as dummy variables to be used as the discriminators. Glahn (Paper 21) has provided a recent account of the relationships among canonical correlation, discriminant analysis, and multiple regression. The relationships of canonical correlation to canonical factor analysis and scaling theory as well as multiple regression and discriminant analysis are discussed by McKeon (1966).

More recently, the canonical correlation technique itself has been shown to be inadequate in many instances because many experimental techniques exist where more than two sets of variables are involved. Horst (1961a), McKeon (1966), and others have provided methods whereby the relations among *m* sets of variables can be examined.

We have reprinted three papers in this volume in which multivariate regression and correlation methods have been applied in systematics and ecology. Power (Paper 22) has employed multiple regression analysis to study the adaptive significance of morphometric variation in blackbirds. This is a well-written and easily comprehended account of multiple regression methodology. Vogt and Jameson (Paper 23) used canonical correlation and stepwise multiple regression to relate morphological variation in the frog *Hyla regilla* to climatic variation. This paper was included because it permits comparison of both stepwise multiple regression and canonical correlation for the same sets of data.

The third paper by P. J. Lee (Paper 24) is a rather unique application of canonical correlation theory. Lee has combined canonical correlation with trend surface analysis to create a "canonical trend surface," which is a graphical summarization of areal variations expressed as polynomial functions of the canonical variates. Lee's treatment should prove to be of considerable utility. In addition, when this technique is combined with such data reduction methods as principal components or factor analysis, it will provide a powerful statistical tool for those interested in studying plant and animal variation.

Additional applications of multiple regression analysis in systematics and ecology include Sinha et al. (1969a), Atchley (1971), and Soulé and Kerfoot (1972). In spite of its obvious importance to statistical theory, it is difficult to find many papers in biology where canonical correlation techniques have been used. This is no doubt due to difficulty in interpreting canonical correlation results. Among those applications of canonical correlation that have involved biological data are Bronswijk and Sinha (1971), Calhoon and Jameson (1970), Jameson et al. (1973), Sinha et al. (1969a, 1969b), and Sinha et al. (1972).

BIBLIOGRAPHY

Atchley, W. R. 1971. A comparative study of the cause and significance of morphological variation in adults and pupae of *Culicoides:* A factor analysis and multiple regression study. *Evolution 25:*563–583.

Barkham, J. P., and J. M. Norris. 1970. Multivariate procedures in an investigation of vegetation and soil relations of two beech woodlands, Cotswold Hills, England. *Ecology 51:*630–639.

Barnett, V. D., and T. Lewis. 1963. A study of the relations between G.C.E. and degree results. *J. Roy. Stat. Soc. 126:*187–216.

Bartlett, M. S. 1938. Further aspects of the theory of multiple regression. *Cambridge Phil. Soc. Proc. 34:*33–40.

———. 1939. A note on tests of significance in multivariate analysis. *Proc. Cambridge Phil. Soc. 35:*180–185.

———. 1941. The statistical significance of canonical correlations. *Biometrika 32:*29–38.

———. 1947. The general canonical correlation distribution. *Ann. Math. Stat. 18:*1–17.

———. 1947. Multivariate analysis. *J. Roy. Stat. Soc. Suppl. 9:*176–197.

———. 1948. Internal and external factor analysis. *British J. Stat. Psychol. 1:*73–81.

Bronswijk, J. G. M. H., and R. N. Sinha. 1971. Interrelations among physical, biological and chemical variates in stored-grain ecosystems, a descriptive and multivariate study. *Ann. Entomol. Soc. Amer. 65:*789–803.

Burt, C. 1948. Factor analysis and canonical correlations. *British J. Stat. Psychol. 1:*95–106.

Calhoon, R. E., and D. L. Jameson. 1970. Canonical correlation between variation in weather and variation in size in the Pacific tree frog, *Hyla regilla,* in Southern California. *Copeia 1970:*124–134.

Chakravarti, I. M. 1954. Relation between canonical correlations and partially canonical correlations. *Calcutta Stat. Assoc. Bull. 5:*185–187.

Chen, C. W. 1971. On some problems in canonical correlation analysis. *Biometrika 58:*399–400.

Das, R. S. 1965. An application of factor and canonical analysis to multivariate data. *British J. Math. Stat. Psychol. 18:*57–69.

DeGroot, M. H., and C. C. Li. 1966. Correlations between similar sets of measurements. *Biometrics 22:*781–790.

Dempster, A. P. 1969. *Elements of Continuous Multivariate Analysis.* Addison-Wesley Publishing Company, Inc., Reading, Mass. 388 pp.

Edgeworth, F. Y. 1892. Correlated averages. *Phil. Mag. 34:*190–204.

Galton, E. 1888. Co-relations and their measurement, chiefly from anthropometric data. *Proc. Roy. Soc. London 45:*135–145.

Green, P. E., M. W. Halbert, and P. J. Robinson. 1966. Canonical analysis: an exposition and illustrative application. *J. Marketing Res. 3:*32–39.

Hannan, G. J. 1961. The general theory of canonical correlation and its relation to functional analysis. *J. Austral. Math. Soc. 2:*229–242.

Healy, M. J. R. 1957. A rotation method for computing canonical correlations. *Math. Tables and Aids to Comp. 11:*83–86.

Hooper, J. W. 1959. Simultaneous equations and canonical correlation theory. *Econometrika 27:*245–256.

Horst, P. 1961a. Generalized canonical correlations and their applications to experimental data. *J. Clin. Psychol. 17:*331–347.

———. 1961b. Relations among *m* sets of measures. *Psychometrika 26:*129–149.

———., and C. MacEwan. 1960. Predictor elimination techniques for determining multiple prediction batteries. *Psychol. Rept. 7:*19–50.

Hotelling, H. 1935. The most predictable criterion. *J. Educ. Psychol. 26:*139–142.

———. 1936. Relations between two sets of variates. *Biometrika 28:*321–377.

Hsu, P. L. 1941. Canonical reduction of the general regression problem. *Ann. Eugenies 11:*42.

———. 1941. On the limiting distribution of the canonical correlations. *Biometrika 32:*38.

Jameson, D. L., J. P. Mackey, and M. Anderson. 1973. Weather, climate and the external morphology of Pacific tree toads. *Evolution 27:*285–302.

Klatzky, S. R., and R. W. Hodge. 1971. A canonical analysis of occupational mobility. *J. Amer. Stat. Assoc. 66:*16.

Koons, P. B., Jr. 1962. Canonical analysis, Chap. 12 in *Computer Applications in the Behavioral Sciences* (H. Borko, ed.). Prentice-Hall, Inc., Englewood Cliffs, N.J.

Kshirsagar, A. M. 1962. A note on direction and collinearity factors in canonical analysis. *Biometrika 49:*255–259.

———, and R. P. Gupta. 1965. A note on the use of canonical analysis in factional experiments. *J. Indian Stat. Assoc. 3:*165–169.

Laha, R. G. 1954. On some problems in canonical correlations. *Sankhya 13:*61–66.

Lancaster, H. O. 1963. Canonical correlations and partitions of X^2. *Quart. J. Math. Oxford Second Ser. 14:*220–224.

———. 1966. Kolmosorov's remark on the Hotelling canonical correlations. *Biometrika 53:*585–588.

Lawley, D. N. 1959. Tests of significance in canonical analysis. *Biometrika 46:*59–66.

Lorge, I. 1940. The computation of the Hotelling canonical correlation (with discussion). *Proc. Educ. Res. Forum,* pp. 68–74.

Love, W. A., and D. K. Stewart. 1968. *Interpreting Canonical Correlations: Theory and Practice.* American Institutes for Research and School of Education, Palo Alto, Calif.

Marriott, F. H. C. 1952. Test of significance in canonical analysis. *Biometrika 39:*58–64.

McDonald, R. P. 1968. A unified treatment of the weighting problem. *Psychometrika 33:*351–381.

McKeon, J. J. 1966. Canonical analysis: some relations between canonical correlation, factor analysis, discriminant function analysis, and scaling theory. *Psychometric Monograph 13.*

Meredith, W. 1964. Canonical correlations with fallible data. *Psychometrika 29:*55–65.

Pearson, K. 1896. Mathematical contributions to the theory of evolution. III. Regression, heredity and panmixia. *Phil. Trans. Roy. Soc. London A187:*253–218.

Reyment, R. A., and B. Brannstrom. 1962. Certain aspects of the physiology of Cypridopsis (Ostracoda, Crustacea). *Geol. Foren. I Stockholm Forhandl. 9:*207–242.

Rozeboom, W. W. 1965. Linear correlations between sets of variables. *Psychometrika 30:*57–71.

Sinha, R. N., and F. Chebib. 1969. Canonical correlation analysis between acarine predators and prey in grain bulk ecosystems, in *Advances in Acarology,* Vol. 3 (J. Naegele, ed.). Cornell University Press, Ithaca, N.Y.

———., H. A. H. Wallace, and F. S. Chebib. 1969a. Canonical correlations between groups of acarine, fungal and environmental variables in bulk grain ecosystems. *Res. Population Ecol. 11:*92–104.

———., H. A. H. Wallace, and F. S. Chebib. 1969b. Canonical correlations of seed viability, seed-borne fungi and environment in bulk grain ecosystem. *Canadian J. Bot. 47:*27–34.

———, H. A. H. Wallace, and F. Chebib. 1969c. Multiple regression analysis of acarine, fungal and environmental variables in grain bulk ecosystems, in *Advances in Acarology,* Vol. 3 (J. Naegele, ed.). Cornell University Press, Ithaca, N.Y.

———., J. G. H. Van Bronswijk, and H. A. H. Wallace. 1972. Canonical correlation analysis of abiotic and biotic variables in insect-infested grain bulks. *Oecologia 8:*321–333.

Soulé, M., and W. C. Kerfoot. 1972. On the climatic determination of scale size in a lizard. *Syst. Zool. 21:*97–105.

Srikantan, K. S. 1970. Canonical association between nominal measurements. *J. Amer. Stat. Assoc. 65:*284.

Stewart, D. K., and W. A. Love. 1968. A general canonical correlation index. *Psychol. Bull. 70:*160–163.

Thomson, G. 1947. The maximum correlation of two weighted batteries. *British J. Stat. Psychol. 1:*27–34.

Waugh, F. V. 1942. Regressions between sets of variables. *Econometrica 10:*290–310.

Williams, G. J. 1967. The analysis of association among many variates. *J. Roy. Stat. Soc. B29:*199–228.

19

Reprinted from *Proc. Roy. Soc. Edinburgh,* **53,** 260–283 (1933)

On the Theory of Statistical Regression

M. S. Bartlett

(MS. received May 12, 1933. Read June 5, 1933.)

1. The product moment distribution in the general case of p normal variates, obtained in 1928 (**1**), and again in 1933 (**2**), has been awaiting further analysis. Some indication has already been given (Wishart, 1928) that new results might be expected from it; in the particular case of two variates obtained previously by Fisher (**3**), it has been used to deduce the distributions of the correlation coefficient (**3**), co-variance (**4**), and regression coefficient (**5**). In the general case, it has been used by Wilks (**6**) to furnish a proof of Fisher's distribution of the multiple correlation coefficient (**7**), and also in connection with his idea of a generalized variance (**8**). Further analysis appears to be most fruitful in studying statistical regression in general. It is shown in Part I of this paper that the product moment distribution can be split up into a chain of independent factors. Most of the known distributions related to regression or partial correlation are simply obtained, in a manner which clearly indicates the relations they bear to one another; the distribution of a partial regression coefficient of any order is also readily derived.

In Part II it is pointed out that the assumption of a normal system is not altogether necessary for some of the distributions to hold. A distribution which may be regarded as a further generalization of the product moment distribution, being the generalized partial product moment distribution, is obtained *ab initio*, in order to show what are the minimum assumptions about normality necessary for the various distributions obtained in I.

A note should be made here on the notation used. This follows Yule (**9**) with regard to the symbols for partial correlations, partial variates, etc. Further, the convention that Greek and English letters are to be used for true and estimated values respectively is employed as far as possible not only for parameters but also for variates. Thus if ξ denotes the value of a variate measured from its true mean, the estimate of ξ given by $\xi - \Sigma\xi/n$ (where Σ, unless otherwise specified, will always denote summation over the n observations in a sample) will be denoted provisionally by x (usually this has been called $x - \bar{x}$). We may write further, for interdependent variates ξ_1, ξ_2, ξ_3, assuming the regressions to be linear,

$$\begin{cases} \xi_{2\cdot 1} = \xi_2 - \beta_{21}\xi_1, \\ \xi_{3\cdot 21} = \xi_3 - \beta_{32\cdot 1}\xi_2 - \beta_{31\cdot 2}\xi_1, \end{cases}$$

where β_{21} is the true simple regression coefficient of ξ_2 on ξ_1, $\beta_{32\cdot 1}$ the true partial regression coefficient of ξ_3 on ξ_2 for constant ξ_1, etc. Similarly

$$\begin{cases} x_{2\cdot 1} = x_2 - b_{21}x_1, \\ x_{3\cdot 21} = x_3 - b_{32\cdot 1}x_2 - b_{31\cdot 2}x_1, \end{cases}$$

where b_{21}, $b_{32\cdot 1}$, etc., are the corresponding estimated regression coefficients from the sample.

The product sums $c_{\mu\nu}$ obtained from the sample are defined by the equation

$$c_{\mu\nu} = \Sigma(x_\mu x_\nu), \quad . \quad . \quad . \quad . \quad . \quad (1)$$

and correspondingly we write

$$\gamma_{\mu\nu} = \Sigma(\xi_\mu \xi_\nu). \quad . \quad . \quad . \quad . \quad . \quad (2)$$

As an extension of this notation, we write further

$$\begin{cases} c_{\mu\nu\cdot 1} = \Sigma(x_{\mu\cdot 1}x_{\nu\cdot 1}), \\ \gamma_{\mu\nu\cdot 1} = \Sigma(\xi_{\mu\cdot 1}\xi_{\nu\cdot 1}), \text{ etc.} \end{cases}$$

I. An Analysis of the Product Moment Distribution.

2. The distribution of p normal variates will be written

$$U_p(\xi_\mu) \equiv \pi^{-\frac{1}{2}p} \mid A \mid^{\frac{1}{2}} e^{-A(\xi,\xi)} \prod_{\mu=1}^{p} d\xi_\mu, \quad . \quad . \quad . \quad (3)$$

where $A(\xi, \xi)$ is the positive definite quadratic form of the ξ_μ, with matrix $A \equiv (a_\mu) = (\frac{1}{2}\Delta_{\mu\nu}/\sigma_\mu\sigma_\nu\Delta)$, Δ being the determinant of correlations $\rho_{\mu\nu}$, and $\Delta_{\mu\nu}$ the co-factor of $\rho_{\mu\nu}$ in Δ. The determinant of A is $\mid A \mid$. Similarly we write $C \equiv (c_{\mu\nu})$, $\mid C \mid \equiv \mid c_{\mu\nu} \mid$.

The distribution of product moments is more conveniently for the present purpose regarded as a distribution of product sums $c_{\mu\nu}$. It has been shown to be independent of the distribution of the means, and the complete distribution may be written

$$U_p(u_\mu)V_p(c_{\mu\nu}, n-1), \quad . \quad . \quad . \quad . \quad . \quad (4)$$

where $u_\mu = \Sigma\xi_\mu/\sqrt{n}$, and

$$V_p(c_{\mu\nu}, n-1) \equiv \frac{\mid A \mid^{\frac{1}{2}(n-1)} \mid C \mid^{\frac{1}{2}(n-p-2)} \exp\left(-\sum_{\mu,\nu=1}^{p} a_{\mu\nu}c_{\mu\nu}\right)}{\pi^{\frac{1}{4}p(p-1)} \prod_{r=1}^{p} \Gamma\frac{1}{2}(n \quad r)} \prod_{1 \leqslant \mu \leqslant \nu \leqslant p} dc_{\nu\mu}. \quad (5)$$

We write the distribution (5) $V_p(c_{\mu\nu}, n-1)$ to show that one degree of freedom has been lost by using the variates x_μ in place of ξ_μ, the distribution of $\gamma_{\mu\nu}$ being given by $V_p(\gamma_{\mu\nu}, n)$.

The complete distribution (4), obtained originally (**1**) by geometrical methods, has recently been obtained by another method (**2**), as the result of a certain multiple integral. The fact that this integral was evaluated (**10**) as a repeated integral has suggested that $V_p(c_{\mu\nu})$ can be split up into a product of independent distributions. This analysis of (5) is carried out in the case of two and three variates, and in the general case of p variates.

3. *Two Variates.*—In the case of two variates, we have

$$V_2(c_{\mu\nu}, n-1) \equiv \frac{|C|^{\frac{1}{2}(n-4)} e^{-\frac{1}{2(1-\rho_{12}^2)}\left(\frac{c_{11}}{\sigma_1^2} - \frac{2\rho_{12}c_{12}}{\sigma_1\sigma_2} + \frac{c_{22}}{\sigma_2^2}\right)}}{\pi^{\frac{1}{2}}\left[2\sigma_1\sigma_2\sqrt{(1-\rho_{12}^2)}\right]^{n-1}\Gamma\frac{1}{2}(n-1)\Gamma\frac{1}{2}(n-2)} dc_{11}dc_{12}dc_{22}. \tag{6}$$

Change the variates c_{11}, c_{12}, c_{22} to new variates c_{11}, $c_{22\cdot 1}$, $u_{2\cdot 1}$, where

$$c_{22\cdot 1} = \Sigma x_{2\cdot 1}^2 = \Sigma(x_2 - b_{21}x_1)^2 = |C|/c_{11},$$

since

$$b_{21} = \Sigma(x_2x_1)/\Sigma x_1^2,$$

and

$$u_{2\cdot 1} = \sqrt{c_{11}}(b_{21} - \beta_{21}).$$

Then

$$dc_{11}dc_{22\cdot 1}du_{2\cdot 1} = dc_{11}dc_{12}dc_{22}/\sqrt{c_{11}},$$

and

$$\frac{1}{(1-\rho_{12}^2)}\left(\frac{c_{11}}{\sigma_1^2} - \frac{2\rho_{12}c_{12}}{\sigma_1\sigma_2} + \frac{c_{22}}{\sigma_2^2}\right) = \frac{c_{11}}{\sigma_1^2} + \frac{\Sigma(x_2 - \beta_{21}x_1)^2}{\sigma_2^2(1-\rho_{12}^2)}$$

$$= \frac{c_{11}}{\sigma_1^2} + \frac{c_{22\cdot 1}}{\sigma_{2\cdot 1}^2} + \frac{u_{2\cdot 1}^2}{\sigma_{2\cdot 1}^2},$$

where $\sigma_{2\cdot 1}$ is the standard deviation of $\xi_{2\cdot 1}$. Hence

$$V_2(c_{\mu\nu}, n-1) = \left[\frac{c_{11}^{\frac{1}{2}(n-3)} e^{-\frac{1}{2}c_{11}/\sigma_1^2} dc_{11}}{(2\sigma_1^2)^{\frac{1}{2}(n-1)}\Gamma\frac{1}{2}(n-1)}\right]\left[\frac{c_{22\cdot 1}^{\frac{1}{2}(n-4)} e^{-\frac{1}{2}c_{22\cdot 1}/\sigma_{2\cdot 1}^2} dc_{22\cdot 1}}{(2\sigma_{2\cdot 1}^2)^{\frac{1}{2}(n-2)}\Gamma\frac{1}{2}(n-2)}\right]\left[\frac{e^{-\frac{1}{2}u_{2\cdot 1}^2/\sigma_{2\cdot 1}^2} du_{2\cdot 1}}{(2\pi\sigma_{2\cdot 1}^2)^{\frac{1}{2}}}\right], \tag{7}$$

$$= V_1(c_{11}, n-1)V_1(c_{22\cdot 1}, n-2)U_1(u_{2\cdot 1}).$$

We thus obtain the following results.

(i) $V_2(c_{\mu\nu})$ has been split up into three independent distributions. The particular function $u_{2\cdot 1} = \sqrt{c_{11}}(b_{21} - \beta_{21})$ is normally distributed with standard deviation $\sigma_{2\cdot 1}$. Further, while c_{11} is distributed like γ_{11}, but with one less degree of freedom (see, for example (**2**), p. 4), $c_{22\cdot 1}$ is distributed like $\gamma_{22\cdot 1}$, but with two less degrees of freedom. We may in fact consider

$$V_1(c_{22\cdot 1}, n-2)U_1(u_{2\cdot 1}),$$

write $c_{22\cdot 1}+u_{2\cdot 1}^2=y$, $u_{2\cdot 1}^2=yz$, say, and integrate for z. We then have the distribution of $y=\Sigma(x_2-\beta_{21}x_1)^2$ given by

$$V_1(y, n-1)\int_0^1 \frac{\Gamma\frac{1}{2}(n-1)}{\pi^{\frac{1}{2}}\Gamma\frac{1}{2}(n-2)}(1-z)^{\frac{1}{2}(n-4)}z^{-\frac{1}{2}}dz=V_1(y, n-1),$$

showing that the only effect of the substitution of b_{21} for β_{21} is that the distribution loses one degree of freedom. Our estimate of $\sigma_{2\cdot 1}^2$ is thus

$$s_{2\cdot 1}^2=c_{22\cdot 1}/(n-2).$$

(ii) If we consider

$$V_1(c_{11}, n-1)U_1(u_{2\cdot 1}),$$

write $u_{2\cdot 1}=\lambda\sqrt{c_{11}}$, and integrate out for c_{11}, we have $f(\lambda)d\lambda$, say,

$$=\int_0^{\infty}\frac{c_{11}^{\frac{1}{2}(n-)2}e^{-\frac{1}{2}c_{11}\left(\frac{1}{\sigma_1^2}+\frac{\lambda^2}{\sigma_{2\cdot 1}^2}\right)}d\lambda dc_{11}}{(2\sigma_1^2)^{\frac{1}{2}(n-1)}\left(2\pi\sigma_{2\cdot 1}^2\right)^{\frac{1}{2}}\Gamma\frac{1}{2}(n-1)}$$

$$=\frac{\Gamma\frac{1}{2}n}{\pi^{\frac{1}{2}}\Gamma\frac{1}{2}(n-1)}\frac{\sigma_1}{\sigma_{2\cdot 1}}\left(1+\frac{\lambda^2\sigma_1^2}{\sigma_{2\cdot 1}^2}\right)^{-\frac{1}{2}n}d\lambda, \quad . \quad . \quad . \quad (8)$$

or writing $\lambda=b_{21}-\beta_{21}$, and $\sigma_{2\cdot 1}^2=\sigma_2^2(1-\rho_{12}^2)$, we have finally for the distribution of the simple regression coefficient b_{21},

$$f(b_{21})db_{21}=\frac{\left(1-\rho_{12}^2\right)^{\frac{1}{2}(n-1)}\Gamma\frac{1}{2}n}{\pi^{\frac{1}{2}}\Gamma\frac{1}{2}(n-1)}\frac{\sigma_1}{\sigma_2}\left(1-2\rho_{12}\frac{\sigma_1}{\sigma_2}b_{21}+\frac{\sigma_1^2}{\sigma_2^2}b_{21}^2\right)^{-\frac{1}{2}n}db_{21}. \quad . \quad (9)$$

This distribution was first obtained independently by Romanovsky and Pearson (5).

(iii) If instead we again consider

$$V_1(c_{22\cdot 1}, n-2)U_1(u_{2\cdot 1}),$$

this being likely to lead to a more useful result, since $c_{22\cdot 1}$ and $u_{2\cdot 1}$ are both related to the same variate $\xi_{2\cdot 1}$, we have a normal variate $u_{2\cdot 1}$ with standard deviation $\sigma_{2\cdot 1}$, of which our independent estimate is $s_{2\cdot 1}=\sqrt{\{c_{22\cdot 1}/(n-2)\}}$.

It at once follows that the distribution of

$$t=u_{2\cdot 1}/s_{2\cdot 1}$$

is given by the well-known "t distribution," with $n-2$ degrees of freedom. This enables us to test the significance of an estimate b_{21} from any hypothetical value β_{21}. This result should be compared with that obtained by Fisher, who has shown (11) that exactly the same test is applicable whatever the distribution of ξ_1, provided ξ_2 is normal for each ξ_1, and we can suppose the set of values ξ_1 in the sample fixed from sample to sample. Any function of ξ_1 and ξ_2, provided it is a linear function of

ξ_2, is then, of course, normally distributed—for example, the regression coefficient b_{21} itself. In II the question of assumptions is reconsidered, and these two apparently distinct problems are shown to be special cases, the only condition necessary for the test to hold being the normality of $\xi_{2\cdot 1}$.

If we consider the particular case when β_{21} or ρ_{21} is zero, we have

$$t = r_{12}\sqrt{(n-2)}/\sqrt{(1-r_{12}^2)},$$

where r_{12} is our estimate of ρ_{12}. This explains why this particular function of r_{12} is distributed in a "t distribution," the test of significance of a correlation from zero being more fundamentally the test of significance of a regression from zero.

4. *Three Variates.*—We have in the case of three variates

$$V_3(c_{\mu\nu}, n-1) \equiv \frac{|C|^{\frac{1}{2}(n-5)} \exp\left\{-\frac{1}{2\Delta}\sum_{\mu,\nu=1}^{3}\frac{c_{\mu\nu}\Delta_{\mu\nu}}{\sigma_\mu\sigma_\nu}\right\} dc_{11}dc_{12}\ldots dc_{33}}{\pi^{\frac{3}{2}}(2^3\sigma_1^2\sigma_2^2\sigma_3^2\Delta)^{\frac{1}{2}(n-1)}\Gamma\frac{1}{2}(n-1)\Gamma\frac{1}{2}(n-2)\Gamma\frac{1}{2}(n-3)}. \quad (10)$$

First write

$$\begin{cases} c_{22\cdot 1} = \Sigma x_{2\cdot 1}^2 = \Sigma(x_2 - b_{21}x_1)^2 = C_{33}/c_{11}, \\ c_{23\cdot 1} = \Sigma(x_{2\cdot 1}x_{3\cdot 1}) = \Sigma(x_2 - b_{21}x_1)(x_3 - b_{31}x_1) = -C_{23}/c_{11}, \\ c_{33\cdot 1} = \Sigma x_{3\cdot 1}^2 = \Sigma(x_3 - b_{31}x_1)^2 = C_{22}/c_{11}, \end{cases}$$

and

$$u_{2\cdot 1} = \sqrt{c_{11}}(b_{21} - \beta_{21}), \qquad u_{3\cdot 1} = \sqrt{c_{11}}(b_{31} - \beta_{31}).$$

Then

$$dc_{11}dc_{12}\ldots dc_{33} = c_{11}du_{2\cdot 1}du_{3\cdot 1}dc_{11}dc_{22\cdot 1}dc_{23\cdot 1}dc_{33\cdot 1},$$

and

$$\sum_{\mu,\nu=1}^{3}\frac{c_{\mu\nu}\Delta_{\mu\nu}}{\sigma_\mu\sigma_\nu\Delta} = \frac{c_{11}}{\sigma_1^2} + \sum_{\mu,\nu=2}^{3}\left(\frac{\Delta_{\mu\nu}}{\sigma_\mu\sigma_\nu\Delta}\Sigma(x_\mu - \beta_{\mu 1}x_1)(x_\nu - \beta_{\nu 1}x_1)\right)$$

$$= \frac{c_{11}}{\sigma_1^2} + \sum_{\mu,\nu=2}^{3}\left(\frac{\Delta_{\mu\nu}}{\sigma_\mu\sigma_\nu\Delta}[c_{\mu\nu\cdot 1} + u_{\mu\cdot 1}u_{\nu\cdot 1}]\right).$$

For this identity depends on the coefficients of c_{12}, c_{13}, and c_{11} being respectively equal on each side. For the first two we must have

$$\Delta_{12} + \rho_{12}\Delta_{22} + \rho_{13}\Delta_{32} = 0,$$
$$\Delta_{13} + \rho_{13}\Delta_{33} + \rho_{12}\Delta_{23} = 0,$$

which (since $\rho_{11} = 1$) are obviously true. Multiply the first of these by ρ_{12}, the second by ρ_{13}, and add; we get

$$\Delta - \Delta_{11} + \rho_{12}^2\Delta_{22} + 2\rho_{12}\rho_{13}\Delta_{23} + \rho_{13}^2\Delta_{33} = 0,$$

which is the identity required for the coefficient of c_{11}. We have also *

$$|C| = c_{11}\begin{vmatrix} c_{22\cdot 1} & c_{23\cdot 1} \\ c_{23\cdot 1} & c_{33\cdot 1} \end{vmatrix}$$

* *Cf.* Ingham (10), p. 5.

It readily follows that

$$V_3(c_{\mu\nu}, n-1) = V_1(c_{11}, n-1)V_2(c_{\mu\nu\cdot 1}, n-2)U_2(u_{\mu\cdot 1}), \quad . \quad . \quad (11)$$

where μ, ν on the right-hand side can take the values 2 and 3.

Since $-\Delta_{32}/\sqrt{(\Delta_{22}\Delta_{33})} = \rho_{32\cdot 1}$—the correlation between $\xi_{3\cdot 1}$ and $\xi_{2\cdot 1}$, or the partial correlation between ξ_3 and ξ_2—and we have also $\Delta = (1-\rho^2_{32\cdot 1})\Delta_{22}\Delta_{33}$, we may write

$$V_2(c_{\mu\nu\cdot 1}, n-2) = \frac{|c_{\mu\nu\cdot 1}|^{\frac{1}{2}(n-5)} e^{-\frac{1}{2(1-\rho^2_{32\cdot 1})}\left(\frac{c_{22\cdot 1}}{\sigma^2_{2\cdot 1}} - \frac{2\rho_{32\cdot 1}c_{32\cdot 1}}{\sigma_{2\cdot 1}\sigma_{3\cdot 1}} + \frac{c_{33\cdot 1}}{\sigma^2_{3\cdot 1}}\right)}}{\pi^{\frac{1}{2}}\left[2\sigma_{2\cdot 1}\sigma_{3\cdot 1}\sqrt{(1-\rho^2_{32\cdot 1})}\right]^{n-2}\Gamma\frac{1}{2}(n-2)\Gamma\frac{1}{2}(n-3)} dc_{22\cdot 1}dc_{32\cdot 1}dc_{33\cdot 1}. \quad (12)$$

If we compare this result with (6), we see that the quantities $c_{22\cdot 1}$, $c_{32\cdot 1}$, $c_{33\cdot 1}$ are jointly distributed exactly like c_{22}, c_{23}, c_{33}, but with one less degree of freedom. We may thus define partial variances and co-variance by the equations

$$\begin{cases} v_{22\cdot 1} = s^2_{2\cdot 1} = \Sigma x^2_{2\cdot 1}/(n-2) \\ v_{32\cdot 1} = r_{32\cdot 1}s_{2\cdot 1}s_{3\cdot 1} = \Sigma(x_{3\cdot 1}x_{2\cdot 1})/(n-2) \\ v_{33\cdot 1} = s^2_{3\cdot 1} = \Sigma x^2_{3\cdot 1}/(n-2). \end{cases}$$

We thus have the further results that not only, as we saw in the case of two variates, is $v_{22\cdot 1}$ distributed in an "s^2 distribution" with $n-2$ degrees of freedom, but also

(iv) the partial co-variance $v_{32\cdot 1}$ will be distributed in the Bessel function distribution (4) and (**12**) obtained for v_{32} or v_{21}, and the partial correlation coefficient $r_{32\cdot 1}$ will be distributed exactly like a simple correlation coefficient r_{32} or r_{21}, each, however, with one less degree of freedom. Fisher has shown this for the important case of the partial correlation coefficient by means of a geometrical argument (**13**).

We should notice that $r_{32\cdot 1}$ is defined above as

$$r_{32\cdot 1} = \Sigma(x_{3\cdot 1}x_{2\cdot 1})/\sqrt{\left[\Sigma(x^2_{2\cdot 1})\Sigma(x^2_{3\cdot 1})\right]},$$

but this may be written

$$r_{32\cdot 1} = (r_{32} - r_{31}r_{21})/\sqrt{\left[(1-r^2_{31})(1-r^2_{21})\right]}.$$

To reduce (11) further, we now treat $V_2(c_{\mu\nu\cdot 1}, n-2)$ exactly as we did $V_2(c_{\mu\nu}, n-1)$ in the case of two variates. That is, write

$$\begin{aligned} c_{33\cdot 12} &= (c_{33\cdot 1}c_{22\cdot 1} - c^2_{23\cdot 1})/c_{22\cdot 1} = |C|/C_{33} \\ &= \Sigma x^2_{3\cdot 12} = \Sigma(x_3 - b_{32\cdot 1}x_2 - b_{31\cdot 2}x_1)^2, \\ u_{3\cdot 21} &= \sqrt{c_{22\cdot 1}}(b_{32\cdot 1} - \beta_{32\cdot 1}). \end{aligned}$$

We then have

$$\begin{aligned} V_3(c_{\mu\nu}, n-1) = {} & V_1(c_{11}, n-1)V_1(c_{22\cdot 1}, n-2)V_1(c_{33\cdot 12}, n-3) \\ & \times U_2(u_{2\cdot 1}, u_{3\cdot 1})U_1(u_{3\cdot 21}). \quad . \quad . \quad . \quad . \quad . \quad (13) \end{aligned}$$

Besides furnishing again all the results obtained for two variates, the case of three variates now gives the following.

(v) The quantity $c_{33\cdot 12}$ is distributed like $\gamma_{33\cdot 12}$, but with three less degrees of freedom. We may thus write $c_{33\cdot 12}=(n-3)v_{33\cdot 12}$, where $v_{33\cdot 12}$ is our estimate of the variance of $\xi_{3\cdot 12}$, that is, of ξ_3 from the regression plane

$$\xi_3-\beta_{32\cdot 1}\xi_2-\beta_{31\cdot 2}\xi_1=0.$$

Further, the quantities $u_{2\cdot 1}$, $u_{3\cdot 1}$, $u_{3\cdot 21}$ are normally distributed. Notice that since the correlation between $u_{2\cdot 1}$ and $u_{3\cdot 1}$ is $\rho_{32\cdot 1}$, we can write $w_1=u_{3\cdot 1}-\beta_{32\cdot 1}u_{2\cdot 1}=\sqrt{c_{11}}(b_{31}-\beta_{32\cdot 1}b_{21}-\beta_{31\cdot 2})$, $w_2=u_{3\cdot 21}$, and obtain three independent normal variates $u_{2\cdot 1}$, w_1, and w_2; moreover, w_1 and w_2 have the same standard deviation $\sigma_{3\cdot 21}$.

(vi) If we consider

$$V_1(c_{22\cdot 1},\ n-2)U_1(u_{3\cdot 21}),$$

we may obtain the distribution of the partial regression coefficient $b_{32\cdot 1}$, exactly as we found the distribution of the simple regression coefficient b_{21} by considering

$$V_1(c_{11},\ n-1)U_1(u_{2\cdot 1}).$$

The only point to notice is that one degree of freedom has been lost, so that if we write $u_{3\cdot 21}=\lambda\sqrt{c_{22\cdot 1}}$, and integrate out for $c_{22\cdot 1}$, we shall have

$$f(\lambda)d\lambda=\frac{\Gamma\frac{1}{2}(n-1)}{\pi\Gamma\frac{1}{2}(n-2)}\frac{\sigma_{2\cdot 1}}{\sigma_{3\cdot 21}}\left(1+\frac{\lambda^2\sigma^2_{2\cdot 1}}{\sigma^2_{3\cdot 21}}\right)^{-\frac{1}{2}(n-1)}d\lambda \quad . \quad . \quad (14)$$

analogously to (8), where here $\lambda=b_{32\cdot 1}-\beta_{32\cdot 1}$. If we write $\sigma^2_{3\cdot 21}=\sigma^2_{3\cdot 1}(1-\rho^2_{32\cdot 1})$, we have the result analogous to (9),

$$f(b_{32\cdot 1})db_{32\cdot 1}=\frac{(1-\rho^2_{32\cdot 1})^{\frac{1}{2}(n-2)}\Gamma\frac{1}{2}(n-1)}{\pi^{\frac{1}{2}}\Gamma\frac{1}{2}(n-2)}\frac{\sigma_{2\cdot 1}}{\sigma_{3\cdot 1}}\left(1-2\rho_{32\cdot 1}\frac{\sigma_{2\cdot 1}}{\sigma_{3\cdot 1}}b_{32\cdot 1}+\frac{\sigma^2_{2\cdot 1}}{\sigma^2_{3\cdot 1}}b^2_{32\cdot 1}\right)^{(n-1)}\cdot db_{32\cdot 1} \quad (15)$$

Similarly we obtain the distribution of $b_{31\cdot 2}$ by interchanging the suffixes 1 and 2, but the two distributions will not, of course, be independent.

(vii) Like the distribution of b_{21}, this distribution is of little value for testing the significance of regression coefficients, since it contains unknowns besides $\beta_{32\cdot 1}$ It is therefore more useful to consider

$$V_1(c_{33\cdot 12},\ n-3)U_1(u_{3\cdot 21}),$$

since now we have simply the normal variate $u_{3\cdot 21}$ with standard deviation $\sigma_{3\cdot 21}$, of which our independent estimate $s_{3\cdot 21}$ is given by

$$s^2_{3\cdot 21}=c_{33\cdot 12}/(n-3).$$

We may thus write

$$t=u_{3\cdot 21}/s_{3\cdot 21},$$

where t follows the "t distribution" with $n-3$ degrees of freedom. This enables us to test the significance of $b_{32\cdot1}$ from any hypothetical value $\beta_{32\cdot1}$, since $\beta_{32\cdot1}$ is the only unknown that occurs.

When $\beta_{32\cdot1}$ or $\rho_{32\cdot1}$ is supposed zero, we have

$$t=r_{32\cdot1}\sqrt{(n-3)}/\sqrt{(1-r^2_{32\cdot1})},$$

and we can test the significance of a partial correlation from zero.

(viii) We had not only $w_2=u_{3\cdot21}$, but also $w_1=u_{3\cdot1}-\beta_{32\cdot1}u_{2\cdot1}$ normally and independently distributed with standard deviation $\sigma_{3\cdot21}$, both being further independent of the distribution of $s^2_{3\cdot21}$. It follows that if we write

$$\begin{aligned}2s^2&=w_1{}^2+w_2{}^2\\&=(b_{31\cdot2}-2\beta_{31\cdot2})c_{13}+(b_{32\cdot1}-2\beta_{32\cdot1})c_{32}\\&\quad+\beta^2_{31\cdot2}c_{11}+2\beta_{32\cdot1}\beta_{31\cdot2}c_{12}+\beta^2_{32\cdot1}c_{22},\end{aligned}\qquad(16)$$

then $z=\frac{1}{2}\log(s^2/s^2_{3\cdot21})$ is distributed in Fisher's "z distribution" with $s^2_{3\cdot21}$ having $n-3$ degrees of freedom, and s^2 two degrees of freedom. This enables us to make a single test for the significance of $b_{32\cdot1}$ and $b_{31\cdot2}$ from hypothetical values $\beta_{32\cdot1}$ and $\beta_{31\cdot2}$. It is clearly a problem in the analysis of variance (see Fisher (**14**)).

If we suppose $\beta_{32\cdot1}=\beta_{31\cdot2}=0$, we have simply

$$2s^2=b_{32\cdot1}c_{32}+b_{31\cdot2}c_{31},$$

or since R^2, the estimated multiple correlation coefficient of ξ_3 with ξ_1 and ξ_2 is given by

$$R^2=(b_{32\cdot1}c_{32}+b_{31\cdot2}c_{31})/c_{33},$$

we have

$$2s^2/(n-3)s^2_{3\cdot21}=R^2/(1-R^2).$$

Thus, just as the test of significance of a partial regression coefficient $b_{32\cdot1}$ from zero becomes identical with the test of significance of the corresponding partial correlation coefficient $r_{32\cdot1}$ from zero, so the joint test of significance of $b_{32\cdot1}$ and $b_{31\cdot2}$ from zero must become identical with the test whether there is any significant multiple correlation.

5. *The General Case of* p *Variates.*—To obtain the various distributions above in their general form, we consider finally

$$V_p(c_{\mu\iota},\ n-1).$$

Write

$$\begin{cases}c_{\mu v\cdot1}=\Sigma(x_{\mu\cdot1}x_{v\cdot1})=c_{\mu v}-c_{\mu1}c_{v1}/c_{11},\\u_{\mu\cdot1}=\sqrt{c_{11}}(b_{\mu1}-\beta_{\mu1}),\end{cases}$$

where the convention is adopted that μ, v can take all values 1 to p except those occurring explicitly—*e.g.* μ cannot take the value 1 in $c_{\mu v\cdot1}$.

Then

$$\Pi dc_{\mu\nu} = c_{11}{}^{\frac{1}{2}p}\Pi dc_{\mu\nu\cdot1}\Pi du_{\mu\cdot1},$$

and

$$\sum_{\mu,\,\nu=1}^{p}\frac{c_{\mu\nu}\Delta_{\mu\nu}}{\sigma_\mu\sigma_\nu\Delta} = \frac{c_{11}}{\sigma_1{}^2} + \sum_{\mu,\,\nu=2}^{p}\left(\frac{\Delta_{\mu\nu}}{\sigma_\mu\sigma_\nu\Delta}\Sigma(x_\mu - \beta_{\mu1}x_1)(x_\nu - \beta_{\nu1}x_1)\right)$$

$$= \frac{c_{11}}{\sigma_1{}^2} + \sum_{\mu,\,\nu=2}^{p}\left(\frac{\Delta_{\mu\nu}}{\sigma_\mu\sigma_\nu\Delta}[c_{\mu\nu\cdot1} + u_{\mu\cdot1}u_{\nu\cdot1}]\right);$$

for similarly to the case of three variates, we have

$$\sum_{r=1}^{p}\rho_{1r}\Delta_{sr} = 0 \quad (s \neq 1),$$

and multiplying this equation by ρ_{1s}, and adding for all $s \neq 1$, we have

$$\Delta - \Delta_{11} + \sum_{r,\,s=2}^{p}\rho_{1r}\rho_{1s}\Delta_{sr} = 0,$$

and these identities are the equations required for the coefficients of $c_{1\mu}$ and c_{11}. We have also

$$|\,C\,| = c_{11}\,|\,c_{\mu\nu\cdot1}\,|.$$

It follows that

$$V_p(c_{\mu\nu},\, n-1) = V_1(c_{11},\, n-1)V_{p-1}(c_{\mu\nu\cdot1},\, n-2)U_{p-1}(u_{\mu\cdot1}).$$

Similarly

$$V_{p-1}(c_{\mu\nu\cdot1},\, n-2) = V_1(c_{22\cdot1},\, n-2)V_{p-2}(c_{\mu\nu\cdot12},\, n-3)U_{p-2}(u_{\mu\cdot21}),$$

$$\vdots$$

$$V_3(c_{\mu\nu\cdot1\,\ldots\,p-3},\, n-p+2) = V_1(c_{p-2,\,p-2\cdot1\,\ldots\,p-3},\, n-p+2)V_2(c_{\mu\nu\cdot1\,\ldots\,p-2},\, n-p+1) \times U_2(u_{\mu\cdot p-2,\,\ldots\,1}).$$

We may deduce from the distribution

$$V_2(c_{\mu\nu\cdot1\,\ldots\,p-2},\, n-p+1) \qquad (17)$$

results corresponding to those obtained from (12) in the case of three variates. That is, we may write

$$v_{\mu\nu\cdot1\,\ldots\,p-2} = c_{\mu\nu\cdot1\,\ldots\,p-2}/(n-p+1),$$

and obtain that

(ix) the distribution of the partial co-variance $v_{p,\,p-1\cdot1\,\ldots\,p-2}$ is like that of a simple co-variance, and the distribution of the partial correlation coefficient $r_{p,\,p-1\cdot1\,\ldots\,p-2}$ like that of a simple correlation coefficient, each distribution, however, with only $n-p+1$ degrees of freedom.

We now write further

$$V_2(c_{\mu\nu\cdot1\,\ldots\,p-2},\, n-p+1) = V_1(c_{p-1,\,p-1\cdot1\,\ldots\,p-2},\, n-p+1)V_1(c_{pp\cdot1\,\ldots\,p-1},\, n-p) \times U_1(u_{p\cdot p-1,\,\ldots\,1}),$$

and hence have

$$V_p(c_{\mu\nu},\, n-1) = V_1(c_{11},\, n-1)\prod_{r=1}^{p-1}\{V_1(c_{\mu\nu\cdot1\,\ldots\,r},\, n-r-1)U_{p-r}(u_{\mu\cdot r\,\ldots\,1})\}. \qquad (18)$$

We may therefore deduce in the general case the following results.

(x) The quantity $c_{pp\cdot 1 \ldots p-1}$ is distributed like $\gamma_{pp\cdot 1 \ldots p-1}$, but with p less degrees of freedom. We thus write

$$v_{pp\cdot 1 \ldots p-1} = c_{pp\cdot 1 \ldots p-1}/(n-p)$$

as our estimate of the variance of ξ_p from its regression "plane."

Further, the quantities $u_{\mu\cdot r \ldots 1}$ are all normally distributed. We may, moreover, write

$$\begin{cases} w_1 = u_{p\cdot 1} - \beta_{p2\cdot 13 \ldots p-1} u_{2\cdot 1} \cdots\cdots - \beta_{p,\,p-1\cdot 1 \ldots p-2} u_{p-1\cdot 1} \\ w_2 = u_{p\cdot 21} - \beta_{p3\cdot 124 \ldots p-1} u_{3\cdot 21} \cdots\cdots - \beta_{p,\,p-1\cdot 1 \ldots p-2} u_{p-1\cdot 21} \\ \vdots \\ w_{p-1} = u_{p\cdot p-1, \ldots 1}, \end{cases}$$

and obtain $p-1$ independent normal variates each with the same standard deviation. They may alternatively be written

$$\begin{cases} w_1 = \sqrt{c_{11}}(b_{p1} - \beta_{p,\,p-1\cdot 1 \ldots p-2} b_{p-1,\,1} \cdots\cdots - \beta_{p1\cdot 2 \ldots p-1}) \\ w_2 = \sqrt{c_{22\cdot 1}}(b_{p2\cdot 1} - \beta_{p,\,p-1\cdot 1 \ldots p-2} b_{p-1,\,2\cdot 1} \cdots\cdots - \beta_{p2\cdot 13 \ldots p-1}) \\ \vdots \\ w_{p-1} = \sqrt{c_{p-1,\,p-1\cdot 1 \ldots p-2}}(b_{p,\,p-1\cdot 1 \ldots p-2} - \beta_{p,\,p-1\cdot 1 \ldots p-2}), \end{cases} \quad (19)$$

showing that they contain only the unknowns $\beta_{p1\cdot 2 \ldots p-1}, \ldots\ldots, \beta_{p,\,p-1\cdot 1 \ldots p-2}$.

(xi) The distribution of the partial regression coefficient

$$b \equiv b_{p,\,p-1\cdot 1 \ldots p-2}$$

is obtained exactly as that of b_{21}, $b_{32\cdot 1}$, etc. We consider

$$V_1(c_{p-1,\,p-1\cdot 1 \ldots p-2},\, n-p+1)U_1(u_{p\cdot p-1, \ldots 1}),$$

and obtain

$$f(\lambda)d\lambda = \frac{\Gamma\frac{1}{2}(n-p+2)}{\pi^{\frac{1}{2}}\Gamma\frac{1}{2}(n-p+1)} \frac{\sigma_{p-1\cdot 1 \ldots p-2}}{\sigma_{p\cdot 1 \ldots p-1}} \left(1 + \frac{\lambda^2 \sigma^2_{p-1\cdot 1 \ldots p-2}}{\sigma^2_{p\cdot 1 \ldots p-1}}\right)^{-\frac{1}{2}(n-p+2)} d\lambda, \quad (20)$$

where

$$\lambda = b - \beta_{p,p-1\cdot 1 \ldots p-2},$$

or

$$f(b)db = \frac{(1-\rho^2)^{\frac{1}{2}(n-p+1)}\Gamma\frac{1}{2}(n-p+2)}{\pi^{\frac{1}{2}}\Gamma\frac{1}{2}(n-p+1)} \frac{\sigma'}{\sigma}\left(1 - 2\rho\frac{\sigma'}{\sigma}b + \left[\frac{\sigma'}{\sigma}b\right]^2\right)^{-\frac{1}{2}(n-p+2)} db, \quad (21)$$

where $\rho = \rho_{p,\,p-1\cdot 1 \ldots p-2}$, $\sigma = \sigma_{p\cdot 1 \ldots p-2}$, and $\sigma' = \sigma_{p-1\cdot 1 \ldots p-2}$.

(xii) As before, if we consider instead

$$V_1(c_{pp\cdot 1 \ldots p-1},\, n-p)U_1(u_{p\cdot p-1, \ldots 1}),$$

we may write

$$t = u_{p\cdot p-1, \ldots 1}/s_{p\cdot p-1, \ldots 1},$$

where $s^2_{p\cdot p-1, \ldots 1} = v_{pp\cdot p-1, \ldots 1} = c_{pp\cdot 1 \ldots p-1}/(n-p)$, and $u_{p\cdot p-1, \ldots 1}$ ($\equiv w_{p-1}$) is given in (19).

This is therefore the test in the general case for the significance of the partial regression coefficient b from any hypothetical value β. When we put $\beta=0$, we have

$$t=r\sqrt{(n-p)}/\sqrt{(1-r^2)},$$

where r is the estimated partial correlation coefficient between ξ_p and ξ_{p-1} when $\xi_1 \ldots \xi_{p-2}$ are eliminated, and we have therefore a test of significance of r from zero.

(xiii) If we write

$$(p-1)s^2 = w_1{}^2 + w_2{}^2 + \ldots w_{p-1}^2$$

$$= \sum_{r=1}^{p-1} (b_{pr\cdot 1 \ldots r-1, r+1, \ldots p-1} - 2\beta_{pr\cdot 1 \ldots r-1, r+1, \ldots p-1})c_{pr}$$

$$+ \sum_{r,s=1}^{p-1} \beta_{pr\cdot 1 \ldots r-1, r+1, \ldots p-1}\beta_{ps\cdot 1 \ldots s-1, s+1, \ldots p-1}c_{rs}, \quad (22)$$

then $z=\frac{1}{2}\log(s^2/s^2_{p\cdot 1 \ldots p-1})$ is distributed in Fisher's z distribution with $s^2_{p\cdot 1 \ldots p-1}$ having $n-p$ degrees of freedom, and s^2 $p-1$ degrees of freedom. This gives, therefore, a single test for the significance of $b_{p1\cdot 2 \ldots p-1}, \ldots, b_{p, p-1\cdot 1 \ldots p-2}$ from hypothetical values $\beta_{p1\cdot 2 \ldots p-1}, \ldots, \beta_{p, p-1\cdot 1 \ldots p-2}$.

If we suppose these values all zero, we have simply

$$(p-1)s^2 = \sum_{r=1}^{p-1} b_{pr\cdot 1 \ldots r-1, r+1, \ldots p-1}c_{pr}$$

$$= c_{pp}R^2,$$

where R^2 is the estimated multiple correlation coefficient of ξ_p with $\xi_1, \ldots \xi_{p-1}$; and

$$(p-1)s^2/(n-p)s^2_{p\cdot 1 \ldots p-1} = R^2/(1-R^2) = v,$$

say.

The distribution of $v=n_1s_1{}^2/n_2s_2{}^2$, where $s_1{}^2$ and $s_2{}^2$ are independent estimates of a variance σ^2, with n_1 and n_2 degrees of freedom respectively, has been shown by Fisher to be

$$f(v)dv = \frac{\Gamma\frac{1}{2}(n_1+n_2)}{\Gamma\frac{1}{2}n_1\Gamma\frac{1}{2}n_2}\frac{v^{\frac{1}{2}n_1-1}}{(1+v)^{\frac{1}{2}(n_1+n_2)}}dv,$$

from which the z distribution is derived.

Write $n_1=p-1$, $n_2=n-p$, and $v=R^2/(1-R^2)$, and we obtain the well-known distribution (15) of R^2 when there is no real multiple correlation,

$$g(R^2)d(R^2) = \frac{\Gamma\frac{1}{2}(n-1)}{\Gamma\frac{1}{2}(p-1)\Gamma\frac{1}{2}(n-p)}(R^2)^{\frac{1}{2}(p-3)}(1-R^2)^{\frac{1}{2}(n-p-2)}d(R^2). \quad . \quad (23)$$

The joint test of significance of the coefficients $b_{p1\cdot 2 \ldots p-1}, \ldots, b_{p, p-1\cdot 1 \ldots p-2}$ from zero thus becomes identical with the test of the multiple correlation being significant.

II. On the Assumption of Normality.

6. So far the notation has been used that ξ denotes the deviation of a variate from its true mean, and x the deviation from its estimated mean. This was convenient, since it allowed the estimate of $\xi_{2\cdot 1}$ to be written $x_{2\cdot 1}$, etc.

It may now be remarked, however, that if x is given its usual interpretation as the value of an observation, *e.g.*

$$x_1 = \xi_1 + m_1,$$

where m_1 is the true mean of x_1, the notation can be made quite complete by defining a quantity $x_0 \equiv 1$, and writing now $x_{1\cdot 0}$ as our estimate of ξ_1 ($\xi_{1\cdot 0} \equiv \xi_1$, since ξ_1 is already measured from its true mean). For we had previously the relations

$$\begin{cases} x_{\mu\cdot\nu} = x_\mu - x_\nu \Sigma(x_\mu x_\nu)/\Sigma x_\nu^2 \\ c_{\mu\mu\cdot\nu} = c_{\mu\mu} - c_{\mu\nu}^2/c_{\nu\nu} \\ u_{\mu\cdot\nu} = \sqrt{c_{\nu\nu}}(b_{\mu\nu} - \beta_{\mu\nu}), \end{cases}$$

and analogously we now have

$$\begin{cases} x_{1\cdot 0} = x_1 - \Sigma x_1/n = x_1 - \bar{x}_1 \\ c_{11\cdot 0} = c_{11} - (\Sigma x_1)^2/n = \Sigma(x_1 - \bar{x}_1)^2 \\ u_{1\cdot 0} = \sqrt{n}(\bar{x}_1 - m_1), \end{cases}$$

since

$$\Sigma(x_1 x_0) = \Sigma x_1 = n\bar{x}_1, \qquad \Sigma x_0^2 = n.$$

We may now write $c_{22\cdot 01}$ instead of $c_{22\cdot 1}$, etc., and in these expressions we shall now have the number of digits referring to variates eliminated giving the number of degrees of freedom lost. The joint distribution of the means and the product sums given by equation (4) will now be written

$$\mathrm{W}_{p,0} \equiv \mathrm{U}_p(u_{\mu\cdot 0})\mathrm{V}_p(c_{\mu\nu\cdot 0}, n-1). \qquad . \quad . \quad . \quad (24)$$

7. In the first part of this paper, the distributions arising out of the problem of regression were conveniently derived from the distribution $\mathrm{W}_{p,0}$, which holds only on the assumption that the system of variates ξ_μ is normal. We have seen, however (using now the complete distribution $\mathrm{W}_{p,0}$ given by (24), which includes the distribution of the means), that in the case of two variates we may write

$$\mathrm{W}_{2,0} = \{\mathrm{V}_1(c_{11\cdot 0}, n-1)\mathrm{U}_1(u_{1\cdot 0})\}\{\mathrm{V}_1(c_{22\cdot 01}, n-2)\mathrm{U}_1(u_{2\cdot 0} - \beta_{21}u_{1\cdot 0})\mathrm{U}_1(u_{2\cdot 10})\}, \quad (25)$$

where the variance of $u_{1\cdot 0}$ is σ_1^2, estimated by $c_{11\cdot 0}/(n-1)$, and the variance of $u_{2\cdot 0} - \beta_{21}u_{1\cdot 0}$ or $u_{2\cdot 10}$ is $\sigma_{2\cdot 1}^2$, estimated by $c_{22\cdot 01}/(n-2)$.

The original normal distribution $U_2(\xi_\mu)$ can of course be written

$$U_2(\xi_\mu) = U_1(\xi_1)U_1(\xi_2 - \beta_{21}\xi_1),$$

and equation (25) suggests that $W_{2,0}$, which depends on the normality of ξ_1 and ξ_2, can also be split up into two factors which depend respectively on the normality of ξ_1 and $\xi_{2\cdot 1}$. If we assume for the moment this to be true, we may observe that any distribution or test derived from both factors, such as the distribution of the regression coefficient, of the correlation coefficient, or of the co-variance, will therefore depend on the normality of ξ_1 and $\xi_{2\cdot 1}$, *i.e.* of ξ_1 and ξ_2. On the other hand, any distribution or test *derived from the second factor alone*, such as the distribution of the partial variance, or the test of significance of the regression coefficient, will depend simply on the normality of the partial or residual variate $\xi_{2\cdot 1}$.

For three variates we may write

$$W_{3,0} = \{V_1(c_{11\cdot 0}, n-1)U_1(v_{1\cdot 0})\}\{V_2(c_{\mu\nu\cdot 01}, n-2)U_2(u_{\mu\cdot 0} - \beta_{\mu 1}u_{1\cdot 0})U_2(u_{\mu\cdot 10})\}, \quad (26)$$

where we may expect the first factor to depend on ξ_1 being normal, the second on $\xi_{2\cdot 1}$, $\xi_{3\cdot 1}$ being normal, and then for the second factor we may write further

$$\{V_1(c_{22\cdot 01}, n-2)U_1(u_{2\cdot 0} - \beta_{21}u_{1\cdot 0})U_1(u_{2\cdot 1})\}$$
$$\times\{V_1(c_{33\cdot 012}, n-3)U_1(u_{3\cdot 0} - \beta_{31\cdot 2}u_{1\cdot 0} - \beta_{32\cdot 1}u_{2\cdot 0})U_1(u_{3\cdot 10} - \beta_{32\cdot 1}u_{2\cdot 10})U_1(u_{3\cdot 210})\}, \quad (27)$$

where in this expression we may expect the last complete factor to depend simply on $\xi_{3\cdot 21}$ being normal.

In order to give a formal proof of these results, which makes no assumptions that are not necessary, it is proposed to establish *ab initio* a general distribution which is an extension of the distribution $W_{p,0}$ obtained by Wishart, being the corresponding partial distribution when κ of the variates are eliminated. It will be written

$$W_{p,\kappa}.$$

The method is to find the moment-generating function of the quantities whose distribution we wish to find. This method was used in a recent paper (2) to obtain the distribution $W_{p,0}$, and it is there more fully explained. The work below gives perhaps the most general result obtainable in this way.

8. First, however, it is necessary to give a very brief discussion of moment-generating functions and their properties. The most fundamental definition of the moment-generating function of a quantity ϕ, which is some function of the pn observations in a sample in p variates ξ_μ, is perhaps

$$M(t) = E(e^{it\phi}), \quad . \quad . \quad . \quad . \quad . \quad (28)$$

where E denotes mathematical expectation—that is, $e^{it\phi}$ is to be averaged for all possible values of ϕ. Thus the moment of the rth order, which may be written $E(\phi^r)$, is the coefficient of $(it)^r/r!$ in the expansion of $M(t)$. We write it so that $M(t)$ is finite for all real t.

If $\phi = x_1 + x_2$, where x_1 and x_2 are any two independent quantities, we have

$$\begin{aligned} E(e^{it(x_1+x_2)}) &= E(e^{itx_1} \,.\, e^{itx_2}) \\ &= E(e^{itx_1})E(e^{itx_2}) \end{aligned}$$

or

$$M_{x_1+x_2} = M_{x_1} \,.\, M_{x_2}.$$

For n independent quantities $x_1 \ldots x_n$, we have similarly

$$M_{\Sigma x} = \Pi M_{x_r}. \qquad (29)$$

In particular, if $x_1 \ldots x_n$ are observations referring to the same variate x, we have

$$M_{\Sigma x} = M_x^{\,n}.$$

These results hold whether $x_1 \ldots x_n$ are continuous variates or not, and the definition (28) thus renders obvious in all cases the property of moment-generating functions given in (29), which, since this may be written

$$K_{\Sigma x} = \log M_{\Sigma x} = \Sigma K_x,$$

is sometimes called the additive property of semi-invariants.

If $\phi = x$, one of the observations, and x is a continuous variate with distribution $f(x)$, we have

$$E(e^{itx}) = \int_{-\infty}^{\infty} e^{itx} f(x) dx,$$

and similarly we have in general, when ϕ is continuous,

$$M(t) = E(e^{it\phi}) = \int_{-\infty}^{\infty} e^{it\phi} F(\phi) d\phi;$$

this equation is important for determining $F(\phi)$, the distribution of ϕ, when we know $M(t)$. In order to find $M(t)$, we must, however, average $e^{it\phi}$ in terms of the original observations, since its value is clearly the same whether we average for values of ϕ or for values of the observations of which ϕ is a function, and while we do not yet know the chance of a particular ϕ arising, given by $F(\phi)d\phi$, we do know the chance of particular values of the observations arising.

Notice, if $\phi = \phi(x_1, x_2)$, say,

$$E(e^{it\phi}) = E_{x_1} E_{x_2}(e^{it\phi}),$$

or symbolically

$$E = E_{x_1} E_{x_2},$$

where E_{x_1} denotes averaging with respect to x_1, etc., and if in particular $E_{x_2}(e^{it\phi})$ were independent of x_1, we should have simply

$$E(e^{it\phi}) = E_{x_2}(e^{it\phi}).$$

The above discussion deals only with the moment-generating function of a single quantity ϕ. For the present purpose we require the joint moment-generating function of q quantities ϕ_r, but this may be defined in exactly the same way by

$$M(t_r) = E\left(\exp \sum_{r=1}^{q} it_r\phi_r\right), \quad . \quad . \quad . \quad . \quad (30)$$

and all the corresponding properties hold.

9. We suppose we have two sets of variates $\xi_1 \ldots \xi_\kappa, \eta_{\kappa+1} \ldots \eta_p$, where the variates $\eta_\mu(\mu = \kappa + 1 \ldots p)$ are independent of the variates $\xi_m(m = 1 \ldots \kappa)$, and constitute a normal system

$$U_p(\eta_\mu) = \pi^{-\frac{1}{2}p} \mid A \mid^{\frac{1}{2}} e^{-A(\eta, \eta)} \prod_{\mu=\kappa+1}^{p} d\eta_\mu \quad . \quad . \quad . \quad (31)$$

The sample contains n sets of values of the p variates ξ_m, η_μ. The n sets of values of η_μ are of course assumed independent. No assumption is made, however, about the variates ξ_m.

The estimated variates $x_{1\cdot 0}$, etc., corresponding to ξ_1, etc., are as before. We consider the quantities

$$w_{\mu,m} = \Sigma\lambda_m\eta_\mu,$$

where

$$\lambda_m = \begin{cases} x_0/\sqrt{\Sigma x_0^2}, & (m = 0) \\ x_{m\cdot 0 \ldots m-1}/\sqrt{\Sigma x^2_{m\cdot 0 \ldots m-1}}, & (m = 1 \ldots \kappa); \end{cases}$$

and

$$\Sigma_{\mu v} = \Sigma\eta_\mu\eta_v, \quad (\mu, v = \kappa + 1 \ldots p).$$

The quantity x_0 we may subsequently put equal to unity.

The joint moment-generating function of $w_{\mu, m}$, $\Sigma_{\mu v}$ will be evaluated by averaging first with respect to the η, and then with respect to the ξ. Since the quantities $w_{\mu, m}$, $\Sigma_{\mu v}$ are each sums of n quantities $\lambda_m\eta_\mu$, $\eta_\mu\eta_v$ independent with respect to the η, we may write

$$M(2t_{\mu, m}, t_{\mu v}) = E_\xi \Pi E_\eta\left(\exp\left\{2i \sum_{\mu=\kappa+1}^{p} \tau_\mu\eta_\mu + iT(\eta, \eta)\right\}\right), \quad (32)$$

where Π denotes a product of n factors corresponding to the n sets of observations in the sample, $T \equiv (t_{\mu v})$, and

$$\tau_\mu = \sum_{m=0}^{\kappa} \lambda_m t_{\mu, m}.$$

Now we have from (31),

$$\mathrm{E}_\eta\left(\exp\left\{2i\sum_{\mu=\kappa+1}^{p}\tau_\mu\eta_\mu+i\mathrm{T}(\eta,\eta)\right\}\right)$$

$$=\int\dots\int_{-\infty}^{\infty}\pi^{-\frac{1}{2}p}|\,\mathrm{A}\,|^{\frac{1}{2}}\exp\left\{2i\sum_{\mu=\kappa+1}^{p}\tau_\mu\eta_\mu-(\mathrm{A}-i\mathrm{T})(\eta,\eta)\right\}\prod_{\mu=\kappa+1}^{p}d\eta_\mu.$$

Integrating * this expression, we therefore have from (32)

$$\mathrm{M}=\mathrm{E}_\xi\Pi(\{|\,\mathrm{A}\,|/|\,\mathrm{A}-i\mathrm{T}\,|\}^{\frac{1}{2}}e^{-\mathrm{B}(\tau,\tau)}),$$

where $\mathrm{B}\equiv(b_{\mu\nu})$ is the reciprocal of $(\mathrm{A}-i\mathrm{T})$; that is,

$$\mathrm{M}=\{|\,\mathrm{A}\,|/|\,\mathrm{A}-i\mathrm{T}\,|\}^{\frac{1}{2}n}\mathrm{E}_\xi(e^{-\Sigma\mathrm{B}(\tau,\tau)}).$$

Now, since †

$$\Sigma x_{m\cdot 0\,\dots\,m-1}x_{n\cdot 0\,\dots\,n-1}=0$$

if

$$n\neq m,$$

$$\Sigma\lambda_m\lambda_n=0$$

if

$$n\neq m,$$

and also obviously from the definition of λ_m,

$$\Sigma\lambda_m^2=1;$$

hence we have

$$\begin{aligned}\Sigma\mathrm{B}(\tau,\tau)&=\sum_{\mu,\nu=\kappa+1}^{p}b_{\mu\nu}(\Sigma\tau_\mu\tau_\nu)\\&=\sum_{\mu,\nu=\kappa+1}^{p}b_{\mu\nu}\left(\sum_{m,n=0}^{\kappa}t_{\mu,m}t_{\nu,n}\Sigma\lambda_m\lambda_n\right)\\&=\sum_{m=0}^{\kappa}\mathrm{B}(t_m,t_m).\end{aligned}$$

Now this expression is independent of the ξ; hence averaging with respect to ξ does not affect the result, and we have finally

$$\mathrm{M}=\{|\,\mathrm{A}\,|/|\,\mathrm{A}-i\mathrm{T}\,|\}^{\frac{1}{2}n}\exp\left\{-\sum_{m=0}^{\kappa}\mathrm{B}(t_m,t_m)\right\}.\qquad(33)$$

Now since, if $\mathrm{F}(w_{\mu,m},\Sigma_{\mu\nu})$ represents the joint distribution of $w_{\mu,m}$, $\Sigma_{\mu\nu}$, we may write

$$\mathrm{M}=\int\dots\int_{-\infty}^{\infty}\exp\left\{2i\sum_{\mu=\kappa+1}^{p}\sum_{m=0}^{\kappa}t_{\mu,m}w_{\mu,m}+i\sum_{\mu,\nu=\kappa+1}^{p}t_{\mu\nu}\Sigma_{\mu\nu}\right\}\mathrm{F}\,dw\,d\Sigma,$$

where

$$dw\equiv\prod_{\mu=\kappa+1}^{p}\prod_{m=0}^{\kappa}dw_{\mu,m},\qquad d\Sigma\equiv\prod_{\kappa+1\leqslant\mu\leqslant\nu\leqslant p}d\Sigma_{\mu\nu},$$

we have by a generalized form of Fourier's Integral Theorem (see (**2**)),

$$\mathrm{F}=(2\pi)^{-(p-\kappa)}\pi^{-\frac{1}{2}(p-\kappa)(p+\kappa+1)}\int\dots\int_{-\infty}^{\infty}\exp\left\{-2i\sum_{\mu=\kappa+1}^{p}\sum_{m=0}^{\kappa}t_{\mu,m}w_{\mu,m}-i\sum_{\mu,\nu=\kappa+1}^{p}t_{\mu\nu}\Sigma_{\mu\nu}\right\}$$

$$\times\mathrm{M}\prod_{m=0}^{\kappa}dt_m\,dt,$$

* *Cf.* Wishart and Bartlett (**2**), p. 2, equation (5).

† See Yule (**9**), p. 183.

where

$$dt_m \equiv \prod_{\mu=\kappa+1}^{p} dt_{\mu,m}, \qquad dt \equiv \prod_{\kappa+1 \leqslant \mu \leqslant \nu \leqslant p} dt_{\mu\nu},$$

and M is given by (33). Integrate * successively with respect to the t_m ($m=0, 1, \ldots \kappa$), and we have

$$F = \int \ldots \int_{-\infty}^{\infty} \frac{|A|^{\frac{1}{2}n} \exp\left\{-i \sum_{\mu,\nu=\kappa+1}^{p} t_{\mu\nu}\Sigma_{\mu\nu} + \sum_{m=0}^{\kappa} B^{-1}(w_m, w_m)\right\}}{(2\pi)^{-(p-\kappa)}\pi^{-\frac{1}{4}p(p-\kappa)} |A - iT|^{\frac{1}{2}n} |B|^{\frac{1}{2}\kappa}} dt,$$

or since

$$B^{-1}(w_m, w_m) = A(w_m, w_m) - iT(w_m, w_m),$$

and

$$|B|^{-\frac{1}{2}\kappa} = |A - iT|^{\frac{1}{2}\kappa},$$

we have

$$F = \prod_{m=0}^{\kappa} \{\pi^{-\frac{1}{2}(p-\kappa)} |A|^{\frac{1}{2}} e^{-A(w_m, w_m)}\} . \int \ldots \int_{-\infty}^{\infty} \frac{|A|^{\frac{1}{2}(n-\kappa-1)} \exp\left\{-i \sum_{\mu,\nu=\kappa+1}^{p} t_{\mu\nu}\Sigma_{\mu\nu}'\right\}}{(2\pi)^{(p-\kappa)}\pi^{\frac{1}{4}(p-\kappa)(p-\kappa-1)} |A - iT|^{\frac{1}{2}(n-\kappa-1)}} dt,$$

where

$$\Sigma_{\mu\nu}' = \Sigma_{\mu\nu} - \sum_{m=0}^{\kappa} w_{\mu,m} w_{\nu,m}.$$

The integral remaining is exactly similar to the integral obtained in the paper quoted (2) to obtain $V_p(c_{\mu\nu\cdot 0}, n-1)$, but with $n-k-1$ instead of $n-1$. It has been evaluated by Ingham (10), and using his result, changing the variates from $w_{\mu,m}$, $\Sigma_{\mu\nu}$ to $w_{\mu,m}$, $\Sigma_{\mu\nu}'$, and inserting the differentials, we have

$$F dw d\Sigma' = \prod_{m=0}^{\kappa} \{U_{p-\kappa}(w_{\mu,m})\} V_{p-\kappa}(\Sigma_{\mu\nu}', n-\kappa-1).$$

To interpret this result, we simply write

$$\eta_\mu = \xi_{\mu\cdot 1 \ldots \kappa},$$

so that in a set of variates $\xi_1 \ldots \xi_p$ we have assumed only that the $p-\kappa$ partial variates $\xi_{\mu\cdot 1 \ldots \kappa}$ are normal.

The quantities $w_{\mu,m}$ are readily expressible in terms of the observations $x_1 \ldots x_p$. They are in fact analogous to the $w_1 \ldots w_{p+1}$ of equation (19), except for $w_{\mu,0}$, which are functions of the means—which were not considered in I. They may be written

$$\begin{cases} w_{\mu,0} = \sqrt{n}(\bar{x}_\mu - m_\mu - \beta_{\mu\kappa\cdot 1 \ldots \kappa-1}[\bar{x}_\kappa - m_\kappa] \ \ldots \ - \beta_{\mu 1\cdot 2 \ldots \kappa}[\bar{x}_1 - m_1]) \\ w_{\mu,1} = \sqrt{c_{11\cdot 0}}(b_{\mu 1} - \beta_{\mu\kappa\cdot 1 \ldots \kappa-1} b_{\kappa 1} \ \ldots \ - \beta_{\mu 1\cdot 2 \ldots \kappa}) \\ w_{\mu,2} = \sqrt{c_{22\cdot 01}}(b_{\mu 2\cdot 1} - \beta_{\mu\kappa\cdot 1 \ldots \kappa-1} b_{\kappa 2\cdot 1} \ \ldots \ - \beta_{\mu 2\cdot 13 \ldots \kappa}) \\ \vdots \\ w_{\mu,\kappa} = \sqrt{c_{\kappa\kappa\cdot 0 \ldots \kappa-1}}(b_{\mu\kappa\cdot 1 \ldots \kappa-1} - \beta_{\mu\kappa\cdot 1 \ldots \kappa-1}) \end{cases} \quad (34)$$

* *Cf.* Wishart and Bartlett, *loc. cit.*

Just as $x_{\mu\cdot 0}$, $b_{\mu 1}$, etc. refer to the variates $\xi_1 \ldots \xi_p$, let $x_{\mu\cdot 0}'$, $b_{\mu 1}'$, etc. refer to the variates $\xi_1 \ldots \xi_{\kappa+1}, \eta_\kappa \ldots \eta_p$. Then we have

$$\Sigma\eta_\mu\eta_\nu - \sum_{m=0}^{\kappa} w_{\mu, m}w_{\nu, m} = \Sigma x_{\mu\cdot 0}'x_{\nu\cdot 0}' - \sum_{m=1}^{\kappa} w_{\mu, m}w_{\nu,}$$
$$= \Sigma(x_{\mu 0}' - b_{\mu 1}'x_{1\cdot 0})(x_{\nu\cdot 0}' - b_{\nu 1}'x_{1\cdot 0}) - \sum_{m=2}^{\kappa} w_{\mu, m}w_{\nu, m}$$
$$= \Sigma(x_{\mu\cdot 01}'x_{\nu\cdot 01}') - \sum_{m=2}^{\kappa} w_{\mu, m}w_{\nu, n}$$
$$= \ldots$$
$$= \Sigma x_{\mu\cdot 01}' \ldots {}_\kappa x_{\nu\cdot 01}' \ldots {}_\kappa.$$

But

$$x_{\mu\cdot 0}' \ldots {}_\kappa = x_{\mu\cdot 0} \ldots {}_\kappa - \beta_{\mu\kappa\cdot 1} \ldots {}_{\kappa-1}x_{\kappa\cdot 0} \ldots {}_\kappa \ldots - \beta_{\mu 1\cdot 2} \ldots {}_\kappa x_{1\cdot 0} \ldots {}_\kappa$$
$$= x_{\mu\cdot 0} \ldots {}^\kappa,$$

since by definition $x_{\kappa\cdot 0} \ldots {}_\kappa$, etc. $=0$. Hence

$$\Sigma_{\mu\nu}' = c_{\mu\nu\cdot 0} \ldots {}_\kappa.$$

Thus we may write the distribution finally

$$W_{p, \kappa} \equiv \prod_{m=0}^{\kappa} \{U_{p-\kappa}(w_{\mu, m})\}V_{p-\kappa}(c_{\mu\nu\cdot 0} \ldots {}_\kappa, n-\kappa-1). \tag{35}$$

It is suggested that the set of distributions represented by $W_{p, \kappa}$ is the more fundamental, at any rate for the study of regression or partial correlation, for though we saw in I that it may be deduced from Wishart's distribution, it is now seen to be true under more general conditions—namely, that the $p-\kappa$ variates $\xi_{\mu\cdot 1} \ldots {}_\kappa$ are normal—and Wishart's distribution may alternatively be obtained from (35) by writing $\kappa=0$. For the present purpose, the most important cases are given by $\kappa=p-2$, and $p-1$.

10. Thus, for $p=2$, $\kappa=0$ gives Fisher's distribution for two variates. When $\kappa=1$, we have

$$W_{2, 1} = U_1(w_0)U_1(w_1)V_1(c_{22\cdot 01}, n-2), \tag{36}$$

where

$$\begin{cases} w_0 \equiv w_{2, 0} = \sqrt{n}(\bar{x}_2 - m_2 - \beta_{21}[\bar{x}_1 - m_1]) \\ w_1 \equiv w_{2, 1} = \sqrt{c_{11\cdot 0}}(b_{21} - \beta_{21}). \end{cases}$$

This requires only that $\xi_{2\cdot 1}$ exists and is normal (this condition implies that the regression of ξ_2 on ξ_1 is linear, and that the variance of ξ_2 for each ξ_1 is constant).

Thus the quantity w_1 is normally distributed whatever the distribution of ξ_1, whether the observations ξ_1 can be regarded as fixed from sample to sample or not, or whether they are specially selected. The reason for this is that w_1, regarded as a function of $\xi_{2\cdot 1}$, is so weighted that the variance is independent of ξ_1. Since the infinite population of ξ_1

and $\xi_{2\cdot 1}$ can be split up into an infinite number of sub-populations for which the ξ_1 are fixed while $\xi_{2\cdot 1}$ vary (for $\xi_{2\cdot 1}$ and ξ_1 are assumed independent), the distribution of w_1 will be the sum of an infinite number of distributions each of which is normal, with the same variance and mean. Consequently, the distribution of w_1 will be similarly distributed, however ξ_1 varies.

In the case where ξ_1 is fixed from sample to sample the problem is simplified, since it is not so essential to consider the particular function w_1. Any linear function of $\xi_{2\cdot 1}$ will then be normal, but for any other function the variance will of course depend on ξ_1. There is no doubt that the condition that the set ξ_1 may be supposed fixed is commonly met with in practice, and this case was the one considered by Fisher (**11**), although he seems to suggest in conclusion that his test holds under somewhat wider conditions than he assumed.

It is important to notice that the test of significance of the mean $\bar{x}_2$ obtainable from $W_{2,1}$ is *only* valid when the set ξ_1 is fixed, for then $\bar{x}_1 \equiv m_1$, and w_0 becomes $\sqrt{n}(\bar{x}_2 - m_2)$. Otherwise, as for a random sample in two normal variates, we should have to consider $W_{1,0}$ for the variate ξ_2, given by $U_1(\sqrt{n}[\bar{x}_2 - m_2])V_1(c_{22\cdot 0}, n-1)$, with the corresponding assumption that ξ_2 is normal. The "t test" of significance of b_{21} from β_{21} is, however, valid, with no restrictions on ξ_1. Moreover, the special case of this, the test of significance of b_{21} or r_{21} from zero, is clearly to be regarded as a special case of the test of regression, not of correlation, for fewer assumptions are involved in the test of regression than in the test of correlation. Thus if we regarded the distribution of the correlation coefficient r_{21} when ρ_{21} is zero as a special case of the general distribution when $\rho_{21} \neq 0$, we might have supposed that ξ_2, ξ_1 must both be normal.

The simplest way to consider this distribution is to use Fisher's geometrical methods (**3**), and consider the chance of the two *radii vectores* representing in n dimensions the sample of ξ_1 and ξ_2 making an angle θ with each other when they are independent of each other—since $r_{21} = \cos\theta$. Clearly, however restricted the *radius vector* of ξ_1, say, is, if any direction of the *radius vector* of ξ_2 is equally likely, then the angle θ will be perfectly random, and the distribution of r_{21} when ρ_{21} is zero follows. The condition that any direction of the radius vector is equally likely is the condition of the normal law (compare Maxwell's proof of the normal law in the dynamical theory of gases). Hence the only condition for the distribution to hold is seen to be that ξ_2, say, is a normal variate (and the n observations ξ_2, of course, independent of each other).

In the case of three variates, put $p=3$, $\kappa=1$, and we have

$$W_{3,1} = U_2(w_{\mu,0})U_2(w'_{\mu,1})V_2(c_{\mu\nu\cdot 01}, n-2). \quad . \quad . \quad (37)$$

From $V_2(c_{\mu\nu\cdot01}, n-2)$ is obtained the distribution of the partial correlation coefficient $r_{32\cdot1}$, and this therefore depends only on $\xi_{3\cdot1}$ and $\xi_{2\cdot1}$ being normal, as has been pointed out by Fisher ((**14**), p. 163).

If instead we put $\kappa=2$, we have

$$W_{3,2}=U_1(w_0)U_1(w_1)U_1(w_2)V_1(c_{33\cdot012}, n-3), \quad . \quad . \quad (38)$$

where

$$\begin{cases} w_0 \equiv w_{3,0}=\sqrt{n}(\bar{x}_3-m_3-\beta_{32\cdot1}[\bar{x}_2-m_2]-\beta_{31\cdot2}[\bar{x}_1-m_1]) \\ w_1 \equiv w_{3,1}=\sqrt{c_{11\cdot0}}(b_{31}-\beta_{32\cdot1}b_{21}-\beta_{31\cdot2}) \\ w_2 = w_{3,2}=\sqrt{c_{22\cdot01}}(b_{32\cdot1}-\beta_{32\cdot1}), \end{cases}$$

provided only that $\xi_{3\cdot21}$ exists and is normal.

The test of the partial regression coefficient $b_{32\cdot1}$ from $\beta_{32\cdot1}$ thus depends only on this condition. Similarly for that of $b_{31\cdot2}$ from $\beta_{31\cdot2}$. This applies also to the joint test of regression using the "z distribution," and to the special case of this latter test when $\beta_{31\cdot2}$ and $\beta_{32\cdot1}$ are put equal to zero, when it coincides with the test of a significant multiple correlation.

The condition that $\xi_{3\cdot21}$ exists implies that the partial regressions with ξ_2 and ξ_1 are linear, since

$$\xi_{3\cdot21}=\xi_3-\beta_{32\cdot1}\xi_2-\beta_{31\cdot2}\xi_1,$$

but since we have seen that the distributions of ξ_2 and ξ_1 are immaterial, we may suppose them to be any required functions of the original observations, or what is more usual in practice if we have no prior knowledge of these functions, may consider the partial regressions with ξ_1, ξ_1^2, etc. We may, for example, by putting $\xi_2\equiv\xi_1^2$, interpret the above equation as representing the parabolic regression of a variate ξ_3 on ξ_1. Thus curvilinear regression and curvilinear partial regression provide no further theoretical difficulties. When a curved regression line is being fitted, it is often convenient to consider the partial regressions not with ξ_1, ξ_1^2, . . ., but with some orthogonal set such as $\xi_{1\cdot0}$, $\xi_{2\cdot01}$, . . ., where $\xi_2\equiv\xi_1^2$, etc.; the regression line may then be fitted term by term. This procedure is practically most important when the values of ξ_1 are separated by equidistant intervals.

11. The test of significance of $\bar{x}_1-m_1$ has been extended by Fisher to include the significance of a difference $\bar{x}_1-\bar{x}_1'$, where $\bar{x}_1$ and $\bar{x}_1'$ are the means from two independent samples S and S'. We may now regard the variate

$$w_{1,0}=\sqrt{n}(\bar{x}_1-m_1)$$

as a special case of the variates $w_{1,0}$, $w_{2,1}$, etc.; one, moreover, which is especially simple, as it is a function of x_1 and x_0, where the variate x_0 is not only fixed from sample to sample, but can take only the value unity. Thus since x_0 is fixed, it is not necessary to eliminate it to eliminate m_1, and we can not only consider

$$w_{1,0} - w_{1,0}' = \sqrt{n}(\bar{x}_1 - \bar{x}_1') \quad . \quad . \quad . \quad . \quad . \quad (39)$$

when the samples are equal in size, but

$$w_{1,0}/\sqrt{n} - w_{1,0}'/\sqrt{n'} = (\bar{x}_1 - \bar{x}_1'), \quad . \quad . \quad . \quad . \quad (40)$$

say, if they are unequal.

In the general case of the significance of any $w \sim w'$, the condition that the "independent variates" should be fixed from sample to sample for the test to be strictly applicable immediately becomes apparent.

Thus in the case of two variates, we only have

$$u \equiv w_{2,1} - w_{2,1}' = \sqrt{c_{11\cdot0}}(b_{21} - b_{21}') \quad . \quad . \quad . \quad . \quad (41)$$

provided that $\sqrt{c_{11\cdot0}'} = \sqrt{c_{11\cdot0}}$, that is, the samples are equal in size, and the same set of ξ_1 is contained in each sample. We may then consider the normal variate in (41), with its corresponding estimated variance

$$v = \left(c^2_{22\cdot01} + (c_{22\cdot01}')^2\right)/2(n-2),$$

and test the significance of the difference between b_{21} and b_{21}' by means of $t = u/\sqrt{v}$, which will follow the "t distribution" with $2(n-2)$ degrees of freedom.

If the samples are unequal in size, but the ξ_1 are fixed, we can weight $w_{2,1}$ and $w_{2,1}'$ in order to eliminate β_{21}, similarly to (40).

For three variates, when ξ_1 and ξ_2 are supposed fixed from sample to sample, we have similarly to (41) for two equal samples,

$$u_2 \equiv w_{3,2} - w_{3,2}' = \sqrt{c_{22\cdot01}}(b_{32\cdot1} - b_{32\cdot1}'). \quad . \quad . \quad . \quad (42)$$

Further, under the same conditions,

$$u_1 \equiv w_{3,1} - w_{3,1}' = \sqrt{c_{11\cdot0}}(b_{31} - b_{31}'), \quad . \quad . \quad . \quad (43)$$

and we have not only from (42) a normal variate by which we can test the significance of the difference between $b_{32\cdot1}$ and $b_{32\cdot1}'$, but from (43) a more sensitive test of significance between b_{31} and b_{31}' if ξ_2 appreciably affects ξ_3 and ξ_1. Again, since

$$2v_1 \equiv u_1^2 + u_2^2 = (c_{32\cdot0} - c_{32\cdot0}')(b_{32\cdot1} - b_{32\cdot1}') + (c_{31\cdot0} - c_{31\cdot0}')(b_{31\cdot2} - b_{31\cdot2}'), \quad . \quad (44)$$

we may test the joint significance between $b_{32\cdot1}$, $b_{31\cdot2}$, and $b_{32\cdot1}'$, $b_{31\cdot2}'$ respectively, by means of the "z distribution."

Analogous tests hold for three variates if the samples are unequal, provided ξ_1 and ξ_2 are still fixed for each sample; similarly for any number of variates.

12. To sum up the results of II, the following distributions depend simply on the existence and normality of the two partial or residual variates $\xi_{p\cdot1\ldots p-2}$, $\xi_{p-1\cdot1\ldots p-2}$:

(*a*) The distribution of the partial correlation coefficient $r_{p,\,p-1\cdot 1\ldots p-2}$.

(*b*) The distribution of the partial co-variance $v_{p,\,p-1\cdot 1\ldots p-2}$.

(*c*) The distribution of the partial regression coefficient $b_{p,\,p-1\cdot 1\ldots p-2}$.

On the other hand, the following distributions and tests of regression depend only on the existence and normality of the one residual, $\xi_{p\cdot 1\ldots p-1}$:

(*d*) The distribution of the partial variance $v_{pp\cdot 1\ldots p-1}$.

(*e*) The distribution of the normal and independent variates (defined in (34)), $w_{p,\,0}\ldots w_{p,\,p-1}$.

(*f*) The test of significance of $b_{p,\,p-1\cdot 1\ldots p-2} \sim \beta_{p,\,p-1\cdot 1\ldots p-2}$—including the particular case when $\rho_{p,\,p-1\cdot 1\ldots p-2}$ assumed zero.

(*g*) The joint test of significance of $b_{pr\cdot 1\ldots r-1,\,r+1\ldots p-1} \sim \beta_{pr\cdot 1\ldots r-1,\,r+1\ldots p-1}(r=1\ldots p-1)$—including the particular case when no real multiple correlation is assumed.

(*h*) In the problem of two samples, provided we can regard the set of variates $\xi_1,\ldots\xi_{p-1}$ as fixed, the test of significance of $b_{p,\,p-1\cdot 1\ldots p-2} \sim b_p{}',_{\,p-1\cdot 1\ldots p-2}$, or the joint test of significance of $b_{pr\cdot 1\ldots r-1,\,r+1\ldots p-1} \sim b_{pr\cdot 1}{}'_{\ldots r-1,\,r+1\ldots p-1}(r=1\ldots p-1)$.

Regression is seen to be a wider concept than correlation; for fewer assumptions are involved in the more important distributions and tests, which are moreover simpler.

Though all correlation tests—when we are concerned only with finding whether there is any correlation at all—reduce to regression tests, a distinction should obviously be made between problems of regression in general, and those where the idea of correlation may be usefully employed. In the case of two variates, for example, since the distribution of the correlation coefficient when ρ_{21} is not zero requires a random sample of two normal variates ξ_2 and ξ_1, a correlation coefficient has most meaning in a sample of this kind.

No mention has so far been made of the test of goodness of fit of regression lines (see Fisher (**16**)), since this depends on our having an array of values of ξ_2 for each value of ξ_1. The test is, however, simply another application of the analysis of variance, the "*z* distribution" being used to test, say, whether we have satisfactorily obtained a normal variate $\xi_{2\cdot 1}$ by assuming $\xi_{2\cdot 1}=\xi_2-\beta_{21}\xi_1$ and fitting a straight line. The variation within arrays is compared with the variation of the means of the arrays from the fitted regression line. The test is thus strictly possible only when several values of ξ_2 correspond to each ξ_1, but in practice when this condition is not fulfilled we can sometimes group the values of ξ_2 provided the grouping is sufficiently fine. Alternatively, we could fit the regression line first, and afterwards group the deviations from this line; we should then, however, as an approximation have to neglect the number of degrees of freedom lost in fitting, and assume the fitted regression line to be the true (linear) regression line.

An exact theoretical test for the goodness of fit of a straight line would be to fit the regression line, and then group coarsely, making an estimate of $\sigma_{2\cdot 1}^2$ by finding $c_{22\cdot 01}$ *for each group*. That is, if there were a groups, finding

$$(n-2a)v_1 = \sum_{r=1}^{a} (c_{22\cdot 01})_r,$$

$$2(a-1)v_2 = c_{22\cdot 01} - \sum_{r=1}^{a} (c_{22\cdot 01})_r,$$

and comparing the independent estimates of $\sigma_{2\cdot 1}^2$ given by v_1 and v_2. Even if we were testing the fit of a curved regression line, it would probably be sufficient to assume the regression in any group to be linear. But the awkwardness of the above test makes its practical importance limited; especially as the test of goodness of fit is supplementary to the specific tests of significance of regression coefficients—tests which are usually more sensitive.

Thus it might happen that a regression coefficient b_{21} is significant, although the goodness of fit seemed adequate when β_{21} was assumed zero; this is because in the test of significance of b_{21} the variance due to the linear regression is isolated as a single square. This particular criticism of the test of goodness of fit is general, of course, and applies also to the case where there are arrays of values of ξ_2 for each ξ_1.

In conclusion, we may perhaps recall that the calculation of all regression coefficients is a problem in the theory of least squares, and consequently their accuracy depends only on the assumptions necessary for this theory. We had, for example, the normal variate

$$w_{2,1} = \sqrt{c_{11\cdot 0}}(b_{21} - \beta_{21}) = \Sigma\lambda_1\xi_{2\cdot 1},$$

where

$$\lambda_1 = x_{1\cdot 0}/\sqrt{\Sigma x_{1\cdot 0}^2},$$

whether the values of ξ_1 are fixed or not, provided $\xi_{2\cdot 1}$ is normal. It is clear, however, that $w_{2,1}$ is approximately normal whatever the distribution of $\xi_{2\cdot 1}$, provided no values of $\xi_{2\cdot 1}$ (or λ_1) are abnormal, since a linear function of n variates goes to normality as n becomes large.

Hence it is likely that

$$\sqrt{c_{11\cdot 0}}\,|\,b_{21} - \beta_{21}\,| < 2\sigma_{2\cdot 1},$$

and for a reasonably sized sample, so that $\sqrt{c_{11\cdot 0}}$ is reasonably large, we may conclude that the discrepancy between b_{21} and β_{21} is likely to be small. But the exact test of significance of $b_{21} - \beta_{21}$ must, for any finite sample, involve the assumption that $\xi_{2\cdot1}$ is normal.

I should like to express my thanks to Dr J. Wishart, to whom I owe the suggestion that a more systematic and complete derivation might be possible of the various distributions and tests associated with regression than has perhaps hitherto been given. I am also indebted to Dr Wishart for advice and criticism while this paper was being prepared.

REFERENCES TO LITERATURE.

1. WISHART, J., 1928. *Biometrika*, vol. xx, A, pp. 32–52.
2. WISHART, J., and BARTLETT, M. S., 1933. *Proc. Camb. Phil. Soc.*, vol. xxix, pp. 260–270.
3. FISHER, R. A., 1915. *Biometrika*, vol. x, pp. 507–521.
4. PEARSON, K., JEFFERY, G. B., and ELDERTON, E. M., 1929. *Biometrika*, vol. xxi, pp. 164–193.
5. ROMANOVSKY, V., 1926. *Bulletin de l'Academie des Sciences de l'U.R.S.S.*, p. 646.
   PEARSON, K., 1926. *Proc. Roy. Soc.*, A, vol. cxii, pp. 1–14.
6. WILKS, S. S., 1932. *Annals Math. Statistics*, pp. 196–203.
7. FISHER, R. A., 1928. *Proc. Roy. Soc.*, A, vol. cxxi, pp. 654–673.
8. WILKS, S. S., 1932. *Biometrika*, vol. xxiv, pp. 471–494.
9. YULE, G. U., 1907. *Proc. Roy. Soc.*, A, vol. lxxix, pp. 182–193.
10. INGHAM, A. E., 1933. *Proc. Camb. Phil. Soc.*, vol. xxix, pp. 271–276.
11. FISHER, R. A., 1925. *Metron*, vol. v, No. 3, pp. 90–104.
12. WISHART, J., and BARTLETT, M. S., 1932. *Proc. Camb. Phil. Soc.*, vol. xxviii, pp. 455–459.
13. FISHER, R. A., 1924. *Metron*, vol. iii, Nos. 3–4, pp. 329–333.
14. FISHER, R. A., 1930. *Statistical Methods for Research Workers* (3rd ed.).
15. FISHER, R. A., 1924. *Phil. Trans.*, B, vol. ccxiii, pp. 89–142.
16. FISHER, R. A., 1922. *Journ. Roy. Stat. Soc.*, vol. lxxxv, pp. 597–612.

(*Issued separately September* 29, 1933.)

20

Reprinted from *J. Educ. Psychol.*, **26**, 139–142 (1935)

THE MOST PREDICTABLE CRITERION

HAROLD HOTELLING

Columbia University

Several measurable criteria may often be found which can be taken as estimates of a single non-measurable and often ill-defined variable, such as college success, or the general price level. When it is desired to predict the non-measurable variable by means of a second set of observable quantities, such as the examination and test grades recorded for applicants for college admission, no single regression equation can provide a fully adequate solution. Any combination of criteria may be used as the dependent variate in a regression equation. The principle basic to the combination will vary from case to case, according to circumstances; and in general not one but several regression equations must be used to give a proper picture. In some circumstances, particularly when the criterion variates are numerous, it may be well to use as the dependent variate in a regression equation that linear function of the criteria whose mean square correlation with them is a maximum. This function is the first of what I have elsewhere called the principal components of the set.[1] Sometimes two or more of the principal components may well be used as dependent variates. In other cases extraneous considerations will be the dominating ones in the choice of a quantity to be predicted; for example, college grades in some subjects may be deemed more suitable measures of success in what the college is trying to do than grades in other subjects, for reasons not determined by statistical calculations. In spite of this variety of grounds for choice of a variate to be predicted, the problem of finding a linear function of the criterion variates which can *most accurately* be predicted from given observations, in the sense of least squares, admits a definite solution, which we shall set forth.

If there are p variates which may be called "predicters," such as scores in college entrance examinations of various kinds, and q others which will be called the criteria, such as college grades, of which some function to be chosen is to be predicted, let us denote these variates by x_i and x_α respectively, with the understanding that Greek indices take all values from 1 to p, while Latin indices take values from $p + 1$ to $p + q$, inclusive. Thus Greek subscripts, and also the Greek super-

[1] "Analysis of a Complex of Statistical Variables into Principle Components." *Journal of Educational Psychology*, Vol. XXIV, September and October, 1933, pp. 417–441 and 498–520. Published separately as a pamphlet by Warwick and York, Baltimore.

scripts which will be used, refer to the criterion variates, while Latin subscripts and superscripts refer to the predicters. We shall, further, make use of the convention that the double occurrence of an index in a term, once as subscript and again as superscript, means that the term is to be summed with respect to this index. The summation will be over all the criterion variates if the index is Greek, but over the predicters if the index is Latin. An arbitrary linear function of the criteria may thus be written

$$z = a^{\alpha}x_{\alpha}.$$

Our problem is to choose the constants a^{α} in such a way that z will have the greatest possible multiple correlation with the predicters x_i; this is equivalent to the requirement that z be estimated from a regression equation with the smallest possible mean square error, provided the variance of z is held to a fixed value, for example unity.

If r_i denote the correlation of z with x_i, while r_{ij} is the correlation of x_i and x_j, the multiple correlation may be written

$$R^2 = r^{ij}r_ir_j, \tag{1}$$

where r^{ij} is the cofactor of r_{ij} in the determinant of the correlations among the predicters, divided by the determinant. This formula for the multiple correlation is not well known, but is an immediate consequence of the usual expression in terms of a ratio of determinants, obtainable by expanding the numerator determinant with respect to the row and column of correlations with the criterion z.

Taking the variance of each x_{α} as unity, that of z will be unity if

$$r_{\alpha\beta}a^{\alpha}a^{\beta} = 1, \tag{2}$$

where $r_{\alpha\beta}$ is the correlation of x_{α} and x_{β}, and a double sum over the predicters is indicated by the repetition of α and β. We then have for the correlation of z with x_i

$$r_i = r_{i\alpha}a^{\alpha}.$$

Substituting this in (1) we have

$$R^2 = r^{ij}r_{i\alpha}r_{j\beta}a^{\alpha}a^{\beta}, \tag{3}$$

which is to be made a maximum by choosing a set of a^{α} satisfying (2). The minimizing conditions are

$$(r^{ij}r_{i\alpha}r_{j\beta} - \lambda r_{\alpha\beta})a^{\alpha} = 0 \tag{4}$$

where λ is a Lagrange multiplier. Putting for brevity

$$w_{\alpha\beta} = r^{ij}r_{i\alpha}r_{j\beta},$$

these may be written

$$(w_{\alpha\beta} - \lambda r_{\alpha\beta})a^{\alpha} = 0. \tag{5}$$

This set of p homogeneous linear equations in the p unknowns a^1, a^2, . . . , a^p will have a non-trivial solution only if the determinant is zero; that is, if λ is a root of the determinantal equation

$$|w_{\alpha\beta} - \lambda r_{\alpha\beta}| = 0. \tag{6}$$

If any root of this equation is substituted in (5), values of the a^{α} may be determined satisfying the linear equations, and then multiplied by a constant factor so as to satisfy (2). We shall also have, upon multiplying (4) by a^{β}, summing with respect to β, and then reducing with the help of (2) and (3),

$$R^2 = \lambda.$$

The last equation shows that, among all the roots of (6), the one to be used is the *largest*, since the root will actually equal the squared multiple correlation which we are trying to maximize. The other roots correspond to minimum or minimax values of R^2.

For $p = q = 2$, so that x_1 and x_2 are the criteria and x_3 and x_4 the predicters, (6) reduces to

$$(1 - r_{12}^2)(1 - r_{34}^2)\lambda^2 - (W_{11} - 2r_{12}W_{12} + W_{22})\lambda + (r_{13}r_{24} - r_{14}r_{23})^2 = 0,$$

where

$$W_{\alpha\beta} = (1 - r_{12}^2)w_{\alpha\beta}.$$

The greater root of this equation is to be substituted in

$$(w_{11} - \lambda)a + (w_{12} - \lambda r_{12})b = 0. \tag{7}$$

We have also in this case

$$w_{11} = 1 - \frac{\Delta_{22}}{1 - r_{34}^2} = (R_{\cdot 134})^2,$$

$$w_{22} = 1 - \frac{\Delta_{11}}{1 - r_{34}^2} = (R_{2\cdot 34})^2,$$

$$w_{12} = r_{12} + \frac{\Delta_{12}}{1 - r_{34}^2}$$

where $\Delta_{\alpha\beta}$ is the cofactor of $r_{\alpha\beta}$ in the four-rowed determinant of correlations. The quantities a and b determined by (7), identical with those previously denoted by a^1 and a^2, are the weights in the expression $ax_1 + bx_2$ for the most predictable function.

With a larger number of variates the calculations directly from the preceding formulae may become somewhat heavy. But if, as a preliminary, an analysis into principal components, or any other uncorrelated components, has been performed upon the criteria, the remaining work may easily be accomplished by an iterative process similar to that used in the paper referred to for the determination of the first principal component. For if the x^α are such components, they are uncorrelated with one another, and (5) reduces to

$$w_{\alpha\beta}a^\alpha = \lambda a^\beta. \quad (\beta = 1, 2, \cdots, p)$$

Substitution of arbitrary quantities for the a^α in the left members of these equations yields new quantities whose ratios are improved estimates of those among the a^α, as shown in the previous paper. Upon substituting the new quantities, and continuing to repeat the process, the correct ratios are approached, as before, and in such a way that the ratio of two consecutive estimates of the same a approaches the greatest root λ.

It is entirely possible that the most predictable criterion thus defined will have little correlation with many of the original criteria, especially if they are numerous. But if, instead of using all the original criteria, or all their principal components, we use in the above process only the first few principal components, the mean square correlation with them, or with the original criteria of the function chosen for prediction will be increased.

In a space of dimensionality equal to the number of individuals measured, each variate may be represented by a point, or by a vector from the origin. The cosine of the angle between two such vectors is the correlation of the variates. p criterion variates thus determine p vectors from the origin, which of course must lie in a flat subspace of p dimensions. The q predicters lie in a flat sub-space of q dimensions. These two sub-spaces will in the most general case have no point in common other than the origin. The problem of the most predictable criterion is that of finding a line in the p-space making the least possible angle with the q-space. This minimum angle is also the least angle that a line in the q-space can make with the p-space. The geometry makes it evident that, if we interchange the roles of predicters and criteria, the maximum attainable value of R^2 remains the same. It also shows that the most predictable linear function of the criteria is identical with the regression function that most accurately predicts that function of the predicters which is most predictable in terms of the criteria; for these two functions are represented by the lines in the two flat spaces which are the sides of the minimum angle.

21

Reprinted with permission of the American Meteorological Society from *J. Atmospheric Sci.*, **25**(1), 23–31 (1968)

Canonical Correlation and Its Relationship to Discriminant Analysis and Multiple Regression

HARRY R. GLAHN[1]

Weather Bureau, ESSA, Silver Spring, Md.

(Manuscript received 26 June 1967)

ABSTRACT

Canonical correlation analysis is concerned with the determination of a linear combination of each of two sets of variables such that the correlation between the two functions is a maximum. Under certain conditions this analysis is equivalent to discriminant analysis and under other conditions it is equivalent to multiple regression. In this paper the relationships among these techniques are discussed, equations relating to prediction by canonical variates are derived, a generalized correlation coefficient is proposed, and an example of canonical correlation analysis is presented.

1. Introduction

Multiple regression has been used to relate a dependent variable (predictand) to a set of independent (predictor) variables for many years. Twenty years ago Hotelling (1936) introduced the concepts of canonical correlation and canonical variates for the analysis of relationships between two sets of variables. At about the same time Barnard (1935) and Fisher (1936) proposed discriminant analysis as a means of using one set of variables to discriminate between two categories of another variable; later, this analysis was extended to more than two predictand groups by Brown (1947), Bryan (1950) and others. Miller (1964) generalized to several predictand categories a specialized application of regression which had been used for two predictand categories by Mook[2] and Lund (1955).

It is not generally recognized by meteorologists that all of the above statistical techniques are embodied in Hotelling's canonical correlation analysis. Proofs of this exist but not in the meteorological literature. In this paper the relationships among these techniques are discussed, equations relating to prediction by canonical variates are derived, a generalized correlation coefficient is proposed, and an example of canonical correlation analysis is presented.

2. Canonical variates relationships

Suppose that there exist n observations of each of p variables X_i $(i=1, 2, \cdots, p)$ and of q variables Y_i $(i=1, 2, \cdots, q)$. These observations represent points in a $(p+q)$ dimensional space and can be arranged in the matrices ${}_n\mathbf{X}_p$ and ${}_n\mathbf{Y}_q$. The variables have means $\bar{X}_i$ and $\bar{Y}_i$, respectively, and deviations from the mean are given by $x_i=X_i-\bar{X}_i$ and $y_i=Y_i-\bar{Y}_i$. New variables ${}_n\mathbf{x}_p\mathbf{A}_i$ and ${}_n\mathbf{y}_q\mathbf{B}_i$ $(i=1, 2, \cdots, r)$, where r is $\leq$ the smaller of p and q, can be formed such that their means are zero and

$$ {}_r\mathbf{A}_p'\mathbf{x}_n'\mathbf{x}_p\mathbf{A}_r=n_r\mathbf{I}_r, \tag{1}$$

$$ {}_r\mathbf{B}_q'\mathbf{y}_n'\mathbf{y}_q\mathbf{B}_r=n_r\mathbf{I}_r, \tag{2}$$

$$ {}_r\mathbf{A}_p'\mathbf{x}_n'\mathbf{y}_q\mathbf{B}_r=n_r\mathbf{\Lambda}_r, \tag{3}$$

where $\mathbf{I}$ is the unit matrix,

$$ {}_r\mathbf{\Lambda}_r=\begin{bmatrix} \lambda_1 & & & \\ & \lambda_2 & 0 & \\ & 0 & \ddots & \\ & & & \lambda_r \end{bmatrix}, \tag{4}$$

and $\lambda_1 \geq \lambda_2 \geq \cdots \geq \lambda_r$.

Eqs. (1) and (2) state that the variance of each of the new variables is unity and each is uncorrelated with all others in its respective set. Eqs. (3) and (4), together with (1) and (2), state that each ${}_n\mathbf{x}_p\mathbf{A}_i$ is uncorrelated with each ${}_n\mathbf{y}_q\mathbf{B}_j$ except when $i=j$ and then the correlation is λ_i.

It can be shown [for instance, see Anderson (1958)] that the ${}_p\mathbf{A}_i$ $(i=1, 2, \cdots, r)$ can be found from

$$ ({}_p\mathbf{S}_{11p}{}^{-1}\mathbf{S}_{12q}\mathbf{S}_{22q}{}^{-1}\mathbf{S}_{21p}-\lambda_i^2\,{}_p\mathbf{I}_p)\,{}_p\mathbf{A}_i=\mathbf{0}, \tag{5}$$

(providing ${}_p\mathbf{S}_{11p}$ and ${}_q\mathbf{S}_{22q}$ are not singular), where the λ_i satisfy the determinantal equation

$$ |{}_p\mathbf{S}_{11p}{}^{-1}\mathbf{S}_{12q}\mathbf{S}_{22q}{}^{-1}\mathbf{S}_{21p}-\lambda^2\,{}_p\mathbf{I}_p|=0, \tag{6}$$

[1] A major portion of this work was accomplished while the author, employed by the Techniques Development Laboratory of the Weather Bureau, was on Reserve training with the Computer Application Division, U. S. Air Force.

[2] Mook, C. P., 1948: An objective method of forecasting thunderstorms for Washington, D. C., in May. Unpublished manuscript. (Copy in Atmospheric Sciences Library, ESSA, Washington, D. C.)

and where

$$_p\mathbf{S}_{11p}=\frac{1}{n}{}_p\mathbf{x}_n'\mathbf{x}_p, \qquad (7)$$

$$_p\mathbf{S}_{12q}={}_p\mathbf{S}_{21q}'=\frac{1}{n}{}_p\mathbf{x}_n'\mathbf{y}_q, \qquad (8)$$

$$_q\mathbf{S}_{22q}=\frac{1}{n}{}_q\mathbf{y}_n'\mathbf{y}_q, \qquad (9)$$

are the variance-covariance matrices. Then the $_q\mathbf{B}_i$ can be found from

$$_q\mathbf{B}_r={}_q\mathbf{S}_{22q}{}^{-1}\mathbf{S}_{21p}\mathbf{A}_r\mathbf{\Lambda}_r{}^{-1}. \qquad (10)$$

Alternatively, use can be made of

$$({}_q\mathbf{S}_{22q}{}^{-1}\mathbf{S}_{21p}\mathbf{S}_{11p}{}^{-1}\mathbf{S}_{12q}-\lambda_i{}^2{}_q\mathbf{I}_q){}_q\mathbf{B}_i=\mathbf{0}, \qquad (11)$$

$$|{}_q\mathbf{S}_{22q}{}^{-1}\mathbf{S}_{21p}\mathbf{S}_{11p}{}^{-1}\mathbf{S}_{12q}-\lambda^2{}_q\mathbf{I}_q|=0, \qquad (12)$$

$$_p\mathbf{A}_r={}_p\mathbf{S}_{11p}{}^{-1}\mathbf{S}_{12q}\mathbf{B}_r\mathbf{\Lambda}_r{}^{-1}. \qquad (13)$$

The latter equations are to be preferred if $q<p$ because the matrix which must be diagonalized is then of a lesser dimension.

The "first" pair of functions, defined by the first column of each of $_p\mathbf{A}_r$ and $_q\mathbf{B}_r$, have as large a correlation λ_1 as any other possible pair of functions, each composed of a linear combination of the original variables. Also, the "second" function pair have as large a correlation λ_2 as any other possible pair of functions, each being composed of a linear combination of the original variables *and* each being uncorrelated with both members of the first pair.

Either set of new variables can be predicted in a least-squares sense by the new variables in the other set. The prediction equations are

$$\widehat{{}_n\mathbf{y}_q\mathbf{B}_r}={}_n\mathbf{x}_p\mathbf{A}_r\mathbf{\Lambda}_r, \qquad (14)$$

and

$$\widehat{{}_n\mathbf{x}_p\mathbf{A}_r}={}_n\mathbf{y}_q\mathbf{B}_r\mathbf{\Lambda}_r. \qquad (15)$$

Also, the original variables in one set can be predicted in a least-squares sense by the new variables in the other set[3] by

$$_n\mathbf{y}_q={}_n\mathbf{x}_p\mathbf{A}_r\mathbf{\Lambda}_r\mathbf{B}'{}_q\mathbf{S}_{22q}. \qquad (16)$$

In the case that $r=q$, (16) can be written as

$$_n\hat{\mathbf{y}}_q={}_n\mathbf{x}_p\mathbf{A}_q\mathbf{\Lambda}_q\mathbf{B}_q{}^{-1}. \qquad (17)$$

Eq. (16) contains the prediction equation for each of the y_i in terms of all of the x_i. One may want to relate one set of variables to the other set but involve only a portion of the correlations λ_i, perhaps those k correlations that are judged to be significantly different from zero. An equation corresponding to Eq. (16) can be written as

$$_n\hat{\hat{\mathbf{y}}}_q={}_n\mathbf{x}_p\mathbf{A}_r\underline{\mathbf{\Lambda}}_r\mathbf{B}_q'\mathbf{S}_{22q}, \qquad (18)$$

where $_r\underline{\mathbf{\Lambda}}_r$ has only k non-zero elements, the others having been set to zero. Eq. (18) has the effect of including a contribution from only those k columns of $_p\mathbf{A}_r$ and k rows of $_r\mathbf{B}_q'$ corresponding to the k non-zero correlations.

An error matrix $_n\boldsymbol{\varepsilon}_q$ can be defined as

$$_n\boldsymbol{\varepsilon}_q={}_n\mathbf{y}_q-{}_n\hat{\mathbf{y}}_q. \qquad (19)$$

Then the total variance of each variable y_i is given by the corresponding diagonal element of

$$\frac{1}{n}({}_q\mathbf{y}_n'\mathbf{y}_q)=\frac{1}{n}({}_q\hat{\mathbf{y}}_n'+{}_q\boldsymbol{\varepsilon}_n')({}_n\mathbf{y}_q+{}_n\boldsymbol{\varepsilon}_q). \qquad (20)$$

Substituting from Eqs. (16) and (1), recognizing that the predictors are uncorrelated with the errors, and simplifying, yields

$$\frac{1}{n}({}_q\mathbf{y}_n'\mathbf{y}_q)={}_q\mathbf{S}_{22q}\mathbf{B}_r\mathbf{\Lambda}_r{}^2\mathbf{B}_q'\mathbf{S}_{22q}+\frac{1}{n}({}_q\boldsymbol{\varepsilon}_n'\boldsymbol{\varepsilon}_q). \qquad (21)$$

In the event that $r=q$, (21) can be written as

$$\frac{1}{n}({}_q\mathbf{y}_n'\mathbf{y}_q)={}_q(\mathbf{B}')_q{}^{-1}\mathbf{\Lambda}_q{}^2\mathbf{B}_q{}^{-1}+\frac{1}{n}({}_q\boldsymbol{\varepsilon}_n'\boldsymbol{\varepsilon}_q). \qquad (22)$$

Each ith diagonal element of the first term on the right is the amount of variance of the corresponding y_i explained by the predictors and each ith diagonal element of the last term is the amount of variance of the corresponding y_i unexplained by the predictors. Division of a diagonal element of the first term on the right by the corresponding element of the term on the left gives the fraction of variance of that y_i which is explained. These q values are the diagonal elements of

$$_q\mathbf{R}_q={}_q\boldsymbol{\sigma}_{22q}{}^{-1}\mathbf{S}_{22q}\mathbf{B}_r\mathbf{\Lambda}_r{}^2\mathbf{B}_q'\mathbf{S}_{22q}\boldsymbol{\sigma}_{22q}{}^{-1}, \qquad (23)$$

where $_q\boldsymbol{\sigma}_{22q}$ is a diagonal matrix composed of the corresponding positive square roots of the diagonal elements of $_q\mathbf{S}_{22q}$.

The total explained variance (EV) and the fractional part $R_{y\cdot x}{}^2$ of the total variance (TV) explained are obtained by taking the trace (tr) as follows:

$$\mathrm{EV}=\mathrm{tr}({}_q\mathbf{S}_{22q}\mathbf{B}_r\mathbf{\Lambda}_r{}^2\mathbf{B}_q'\mathbf{S}_{22q}), \qquad (24)$$

$$R_{y\cdot x}{}^2=\frac{\mathrm{tr}({}_q\mathbf{S}_{22q}\mathbf{B}_r\mathbf{\Lambda}_r{}^2\mathbf{B}_q'\mathbf{S}_{22q})}{\mathrm{tr}({}_q\mathbf{S}_{22q})}=\frac{\mathrm{tr}({}_q\boldsymbol{\sigma}_{22q}\mathbf{R}_q\boldsymbol{\sigma}_{22q})}{\mathrm{tr}({}_q\mathbf{S}_{22q})}. \qquad (25)$$

Eqs. (23), (24), and (25) can be evaluated as a sum of r terms. For instance, Eq. (24) becomes

[3] Hereafter in this paper only the equations which arise from considering the y_i to be the predictand set are presented.

$$\mathrm{EV}=\mathrm{tr}\left({}_{q}\mathbf{S}_{22q}\mathbf{B}_{r}\begin{bmatrix}\lambda_1^2 & & & \\ & 0 & 0 & \\ & & 0 & \\ & 0 & & \ddots \\ & & & & 0\end{bmatrix}{}_{r}\mathbf{B}_{q}'\mathbf{S}_{22q}\right)$$

$$+\mathrm{tr}\left({}_{q}\mathbf{S}_{22q}\mathbf{B}_{r}\begin{bmatrix}0 & & & \\ & \lambda_2^2 & 0 & \\ & & 0 & \\ & 0 & & \ddots \\ & & & & 0\end{bmatrix}{}_{r}\mathbf{B}_{q}'\mathbf{S}_{22q}\right)+\cdots$$

$$+\mathrm{tr}\left({}_{q}\mathbf{S}_{22q}\mathbf{B}_{r}\begin{bmatrix}0 & & & \\ & 0 & 0 & \\ & & \ddots & \\ & 0 & 0 & \\ & & & \lambda_r^2\end{bmatrix}{}_{r}\mathbf{B}_{q}'\mathbf{S}_{22q}\right). \quad (26)$$

The ith diagonal element in the jth term on the right before taking the trace, is the amount of variance of the corresponding y_i explained by the jth canonical function. Also, the jth trace on the right gives the total amount of variance of the y_i $(i=1, 2, \cdots, q)$ explained by the jth canonical function.

If Eq. (23) is expanded in a similar manner, the ith diagonal element in the jth term is the fractional amount of variance of the corresponding y_i explained by the jth canonical function. The jth term in the similar expansion of the right side of Eq. (25) is the fractional amount of the total variance of the y_i $(i=1, 2, \cdots, q)$ explained by the jth canonical function. It should be noted that ${}_q\mathbf{R}_q$ is invariant under linear transformations of scale of the predictors and predictands, and that EV and $R_{y\cdot x}^2$ are invariant under linear transformations of scale of the predictors but not of the predictands.

The prediction equations can be put in terms of the X_i and Y_i, if desired. For instance Eq. (17) becomes

$${}_n\hat{\mathbf{Y}}_q={}_n\mathbf{X}_p\mathbf{A}_q\mathbf{\Lambda}_q\mathbf{B}_q^{-1}-{}_n\bar{\mathbf{X}}_p\mathbf{A}_q\mathbf{\Lambda}_q\mathbf{B}_q^{-1}+{}_n\bar{\mathbf{Y}}_q. \quad (27)$$

3. Discriminant analysis formulation

Suppose that there exist n observations of each of p variables X_i $(i=1, 2, \cdots, p)$, and that these observations can be divided into G groups of sizes $n_1, n_2, \cdots, n_G$. The observations can be considered to represent points in a p-dimensional space where each is tagged with its respective group number. Also, the observations can be arranged to form the matrix ${}_n\mathbf{X}_p$. The p overall means are $\bar{X}_i$ and the deviations are given by $x_i=X_i-\bar{X}_i$. The individual group means of these deviations are given by $\bar{x}_{ij}$ $(i=1, 2, \cdots, p;\ j=1, 2, \cdots, G)$. New variables ${}_n\mathbf{x}_p\mathbf{V}_i$, $(i=1, 2, \cdots, r)$, where $r \leq$ the smaller of p and $G-1$, can be formed such that the values corresponding to a particular group tend to cluster together about a mean and also that that mean tends to be separated from other group means. The ${}_p\mathbf{V}_i$ are called discriminant functions.

Within groups, between groups, and total sum of squares matrices can be defined, respectively, as

$${}_p\mathbf{W}_p=({}_p\mathbf{x}_n'-{}_p\bar{\mathbf{x}}_n')({}_n\mathbf{x}_p-{}_n\bar{\mathbf{x}}_p), \quad (28)$$

$${}_p\mathbf{B}_p={}_p\bar{\mathbf{x}}_n'\bar{\mathbf{x}}_p, \quad (29)$$

$${}_p\mathbf{T}_p={}_p\mathbf{x}_n'\mathbf{x}_p={}_p\mathbf{W}_p+{}_p\mathbf{B}_p, \quad (30)$$

where the matrix ${}_n\bar{\mathbf{x}}_p$ is composed of the group means in the order that the groups are represented in ${}_n\mathbf{x}_p$.

The discriminant functions can be found by solving [see Bryan (1950)]

$$({}_p\mathbf{W}_p^{-1}\mathbf{B}_p-\mu_{ip}\mathbf{I}_p){}_p\mathbf{V}_i=\mathbf{0}, \quad (31)$$

in which the μ_i are solutions of

$$|{}_p\mathbf{W}_p^{-1}\mathbf{B}_p-\mu_p\mathbf{I}_p|=0. \quad (32)$$

Each μ_i is interpreted as the ratio of between-to-within-groups variance of the X_j (predictors) due to the ith discriminant function.

Now suppose $G-1\equiv q$ dummy variables Y_i $(i=1, 2, \cdots, q)$ are defined such that $Y_{ij}=1$ if the jth observation belongs to group i and $Y_{ij}=0$ if the jth observation does not belong to group i. A dummy variable corresponding to the Gth group is not defined since it would be redundant with the other $G-1$ groups and ${}_G\mathrm{S}_{22G}$ would be singular. Deviations from the mean are given by $y_{ij}=Y_{ij}-\bar{Y}_{ij}$ and can be put into the matrix ${}_n\mathbf{y}_q$. This predictand matrix and the predictor matrix ${}_n\mathbf{x}_p$ can now be used in the canonical correlation framework and all of the equations in Section 2 apply.

Tatsuoka (1955) and others have shown that the discriminant analysis solution [Eqs. (31) and (32)] is equivalent to the canonical correlation solution [Eqs. (5) and (6)], where

$$\mu_i=\frac{\lambda_i^2}{1-\lambda_i^2}, \quad (33)$$

and

$${}_p\mathbf{V}_i=C_p\mathbf{A}_i, \quad (34)$$

where C is an arbitrary constant which is necessary because if ${}_p\mathbf{V}_i$ is a solution of Eq. (31), $C_p\mathbf{V}_i$ is also, and no restriction was imposed on the variance of the discriminant functions. However, the discriminant analysis solution as usually defined stops with the computation of the ${}_p\mathbf{V}_i$ and μ_i and some other technique must be employed to find the actual estimates of group membership [for instance, see Miller (1962)], whereas Eq. (27) can be used to produce estimates of the $G-1$ binary predictands directly.

After estimates $\hat{Y}_{ij}$ $(i=1, 2, \cdots, G-1)$ are made, the estimate for the Gth group is

$$\hat{Y}_{Gj}=1-\sum_{i=1}^{G-1}\hat{Y}_{ij}. \quad (35)$$

Predictand group membership is designated by a "one" and non-membership is indicated by a "zero." Therefore, group estimates near "one" or estimates large with respect to the other estimates would indicate membership in that group.

Usually, $G-1=q$ is much less than p and it is more efficient to solve Eqs. (11), (12) and (13) rather than than Eqs. (5), (6) and (10) or Eqs. (31) and (32) unless special methods of solution are devised.

4. Multiple linear regression

Multiple linear regression is concerned with the estimation of one predictand by a linear combination of predictors and is a special case of canonical correlation. If $q=1$, the equations in Section 2 hold and are much simplified. In particular, Eq. (16), put in terms of the original variables X_i and Y, is the usual regression equation,

$$ {}_n\hat{\mathbf{Y}}_1 = {}_n\mathbf{X}_p\mathbf{S}_{11p}{}^{-1}\mathbf{S}_{12_1} - {}_n\bar{\mathbf{X}}_p\mathbf{S}_{11p}{}^{-1}\mathbf{S}_{12_1} + {}_n\bar{\mathbf{Y}}_1. \quad (36) $$

The regression estimates can be found separately for any number of predictands even though the predictands are linearly related. If G dummy variables denoting group membership are formed as discussed in Section 3 and used as predictands, the resulting regression equations are identical to those derived from the canonical correlation analysis and can be written

$$ {}_n\hat{\mathbf{Y}}_G = {}_n\mathbf{X}_p\mathbf{S}_{11p}{}^{-1}\mathbf{S}_{12_G} - {}_n\bar{\mathbf{X}}_p\mathbf{S}_{11p}{}^{-1}\mathbf{S}_{12_G} + {}_n\bar{\mathbf{Y}}_G. \quad (37) $$

It may be sufficiently accurate for some purposes to consider the estimates $\hat{Y}_i$ $(i=1, 2, \cdots, G)$ to be the mean values of Y_i for each observable combination of predictor values. In this way the concept of $\hat{Y}_i$ being an unbiased estimate of the probability of Y_i is introduced. Also, for any observable combination of predictor values the sum of the probability estimates is

$$ \sum_{i=1}^{G} \hat{Y}_{ij} = \sum_{i=1}^{G} ({}_1\mathbf{X}_p - {}_1\bar{\mathbf{X}}_p)_p\mathbf{S}_{11p}{}^{-1} \begin{bmatrix} \frac{1}{n}\sum_{j=1}^{n} x_{1j}y_{ij} \\ \frac{1}{n}\sum_{j=1}^{n} x_{2j}y_{ij} \\ \vdots \\ \frac{1}{n}\sum_{j=1}^{n} x_{pj}y_{ij} \end{bmatrix} + \sum_{i=1}^{G} \bar{Y}_{i\cdot} \quad (38) $$

$$ = \frac{1}{n}({}_1\mathbf{X}_p - {}_1\bar{\mathbf{X}}_p)_p\mathbf{S}_{11p}{}^{-1} \begin{bmatrix} \sum_{j=1}^{n} x_{1j} \sum_{i=1}^{G} y_{ij} \\ \sum_{j=1}^{n} x_{2j} \sum_{i=1}^{G} y_{ij} \\ \vdots \\ \sum_{j=1}^{n} x_{pj} \sum_{i=1}^{G} y_{ij} \end{bmatrix} + \sum_{i=1}^{G} \bar{Y}_{i\cdot} $$

Since

$$ \sum_{i=1}^{G} y_{ij} = \sum_{i=1}^{G} (Y_{ij} - \bar{Y}_{i\cdot}) = \sum_{i=1}^{G} Y_{ij} - \sum_{i=1}^{G} \bar{Y}_{i\cdot}, \quad (39) $$

and

$$ \sum_{i=1}^{G} Y_{ij} = \sum_{i=1}^{G} \bar{Y}_{i\cdot} = 1, \quad (40) $$

it is seen that the sum over all G groups of the coefficients of each predictor equals zero and that the sum of the probability estimates equals unity. However, this does not guarantee that the individual estimates $\hat{Y}_{ij}$ are bounded by zero and one [see Miller (1964)].

The fact that $\sum_{i=1}^{G} \hat{Y}_{ij}=1$ justifies Eq. (35). It is also interesting to note that the criterion for the determination of the regression equations, i.e.,

$$ \sum_{j=1}^{n} (Y_{ij} - \hat{Y}_{ij})^2 \text{ is a minimum} \quad (41) $$

also assures the minimum (best) P-Score [see Brier (1950)] on the dependent data obtainable by linear prediction equations and the predictors being used.

5. Measures of association between two sets of variables

Hooper (1959) discusses three measures of association between two sets of variables—Hotelling's (1936) vector alienation and vector-correlation coefficients and his proposed trace correlation. The following definitions can be made:

$$ \text{Vector correlation coefficient} = [\prod_{i=1}^{q} \lambda_i^2]^{\frac{1}{2}}, \quad (42) $$

$$ \text{Vector alienation coefficient} = [\prod_{i=1}^{q} (1-\lambda_i^2)]^{\frac{1}{2}}, \quad (43) $$

$$ \text{Trace correlation coefficient} \equiv \bar{r} = \left[\frac{1}{q}\sum_{i=1}^{q} \lambda_i^2\right]^{\frac{1}{2}} = \left[\frac{1}{q}\,\mathrm{tr}_q\Lambda_q^2\right]^{\frac{1}{2}}. \quad (44) $$

Hooper (1959) notes that the vector correlation coefficient has the undesirable property of being zero if there are not q non-zero canonical correlations. Also the vector correlation and vector alienation coefficients both tend to zero when q is large. He states that the trace correlation coefficient has "... none of these defects ..." and upon comparing

$$ \bar{r}^2 = \frac{1}{q}\sum_{i=1}^{q} \lambda_i^2 \quad (45) $$

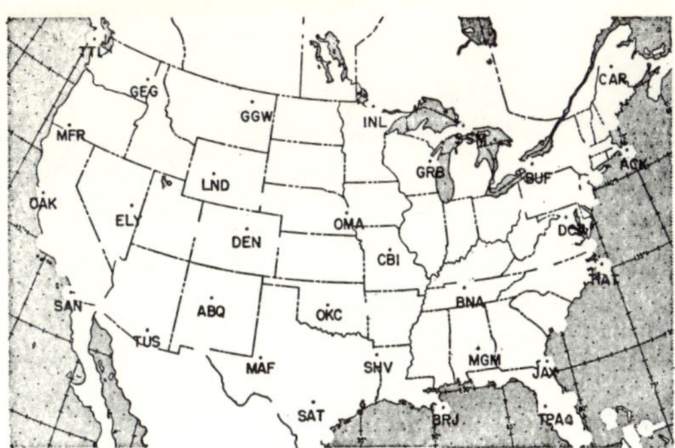

FIG. 1. The 30 stations used in the example.

with

$$1-\bar{r}^2=\frac{1}{q}\sum_{i=1}^{q}(1-\lambda_i^2), \tag{46}$$

concludes that "... $\bar{r}^2$ can be naturally interpreted as that part of the total variance of the jointly dependent variables that is accounted for by the systematic part of the reduced form, and $1-\bar{r}^2$ as the unexplained part ..." However, it has been shown in this paper that Eq. (25) gives the fractional part of the total variance of the dependent variables accounted for by the predictors. It is proposed, therefore, that if a single measure of association is desired, $R_{y\cdot x}{}^2$ defined in Eq. (25) is probably as good as any other even though it is no more nor less than

$$R_{y\cdot x}{}^2=\frac{\sum_{i=1}^{q}\sigma_i{}^2R^2{}_{y_i\cdot x_1\cdot x_2,\cdots,x_p}}{\sum_{i=1}^{q}\sigma_i{}^2}, \tag{47}$$

TABLE 1. The canonical correlations λ_i, canonical correlations squared λ_i^2, and fraction of total variance explained, EV_i/TV, by the most important 15 of the 30 canonical functions.

No. of function i	Canonical correlation λ_i	λ_i^2	EV_i/TV
1	0.94	0.88	0.231
2	0.93	0.86	0.173
3	0.91	0.82	0.097
4	0.90	0.81	0.094
5	0.87	0.76	0.066
6	0.82	0.67	0.023
7	0.79	0.62	0.020
8	0.76	0.58	0.012
9	0.72	0.51	0.013
10	0.67	0.44	0.004
11	0.64	0.41	0.005
12	0.59	0.34	0.002
13	0.55	0.30	0.002
14	0.49	0.24	0.003
15	0.46	0.21	0.001
$\sum_{i=1}^{6}$			0.683
$\sum_{i=1}^{30}$			0.749

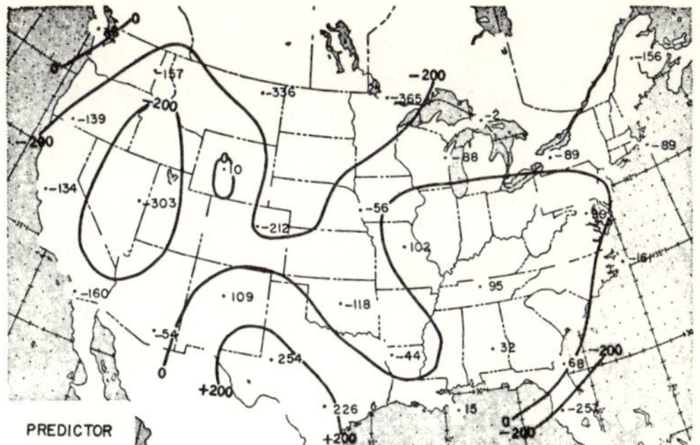

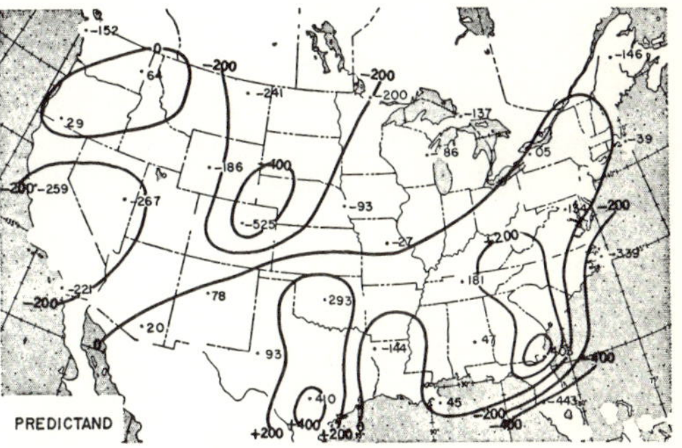

FIG. 2. The predictor and predictand functions $\times 10^5$ corresponding to λ_1.

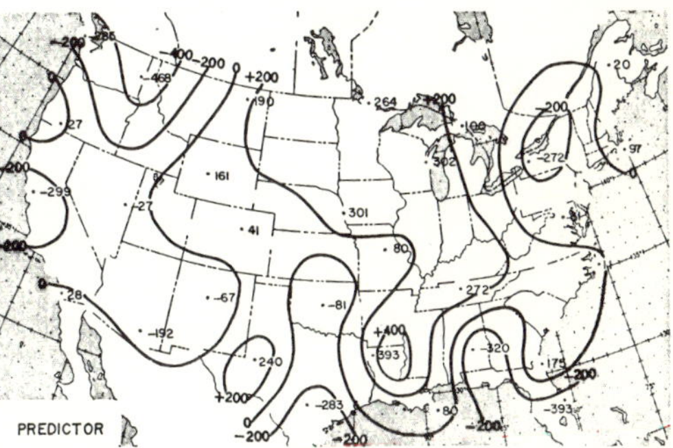

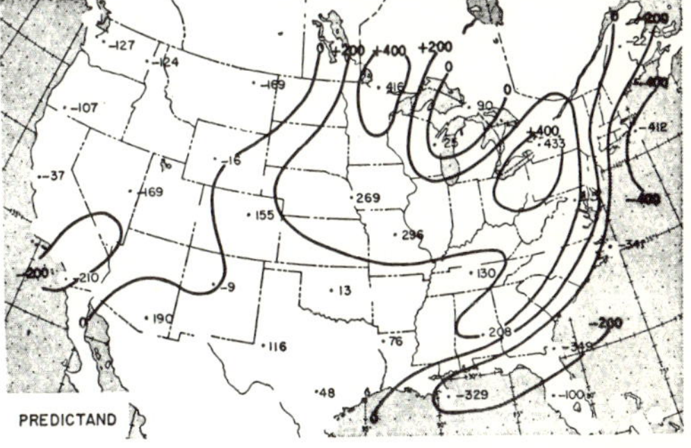

FIG. 3. The predictor and predictand functions $\times 10^5$ corresponding to λ_2.

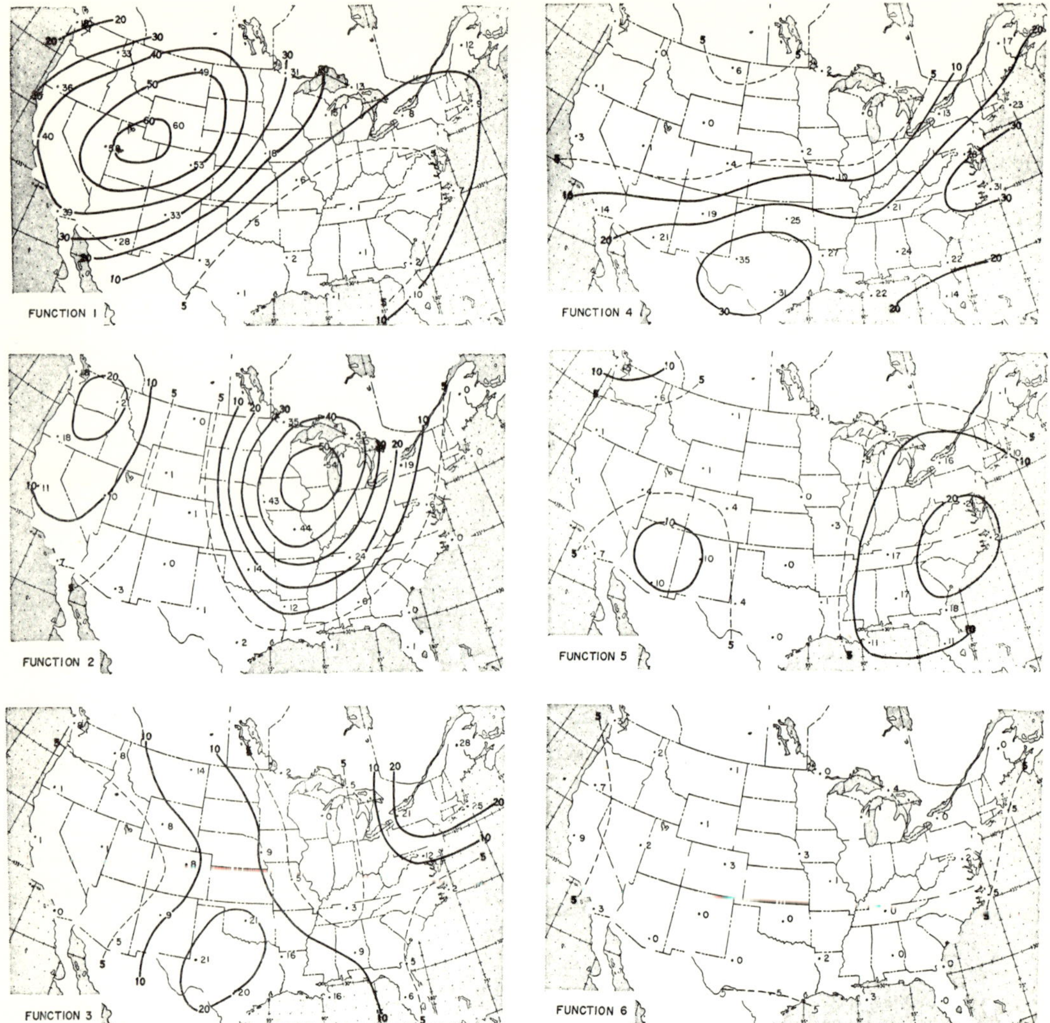

FIG. 4. The per cent of the variance of each predictand explained by each of the first 6 predictor functions.

where σ_i^2 is the variance of the ith predictand and $R^2_{y_i \cdot x_1 \cdot x_2, \cdots, x_p}$ is the reduction of variance of the ith predictand found by the usual multiple regression technique. (These reductions of variance and σ_i are also the diagonal elements of ${}_q\mathbf{R}_q$ and ${}_q\boldsymbol{\sigma}_{22q}$, respectively.) A suitable name for $R_{y \cdot x}$, the positive square root of $R_{y \cdot x}^2$, would be the "composite correlation coefficient."

As stated previously, $R_{y \cdot x}^2$ is not invariant to scale transformations of the predictands (unless the same linear transformation is imposed on each); however, this is not necessarily undesirable. If there is only a single predictand ($q=1$), then $R_{y \cdot x}^2$ is the square of the multiple correlation of that predictand with the p predictors.

6. An example

As an example of the canonical correlation technique, 5 years of 500-mb heights observed at 30 stations in the United States at 0000 GMT in June, July and August were related to the same variables 24 hr earlier. The stations used are shown in Fig. 1. The sample size was 455. The largest 15 of the resulting 30 canonical correlations are shown in Table 1.

The predictor and predictand functions corresponding to λ_1, and λ_2 are shown in Figs. 2 and 3; the coefficients are plotted at the station locations. These functions do not seem to correspond to any easily recognizable synoptic pattern, nor given the predictor functions

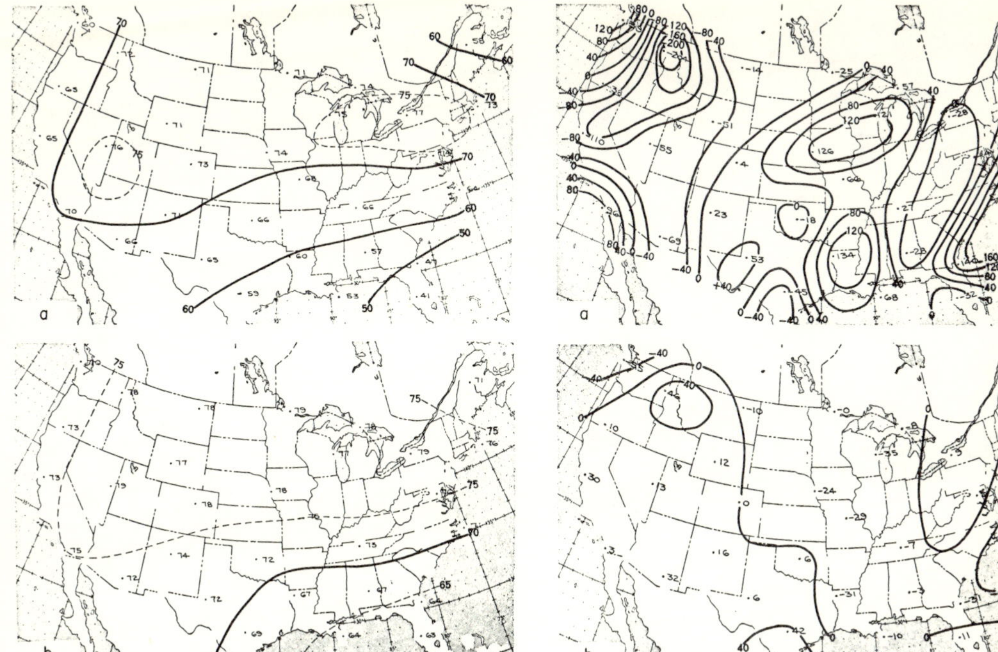

FIG. 5. The per cent of the variance of each predictand explained by the first 6 functions, a., and all 30 functions, b.

FIG. 6. The coefficients $\times 10^3$ in the regression equations for predicting the 500-mb height at Nashville, a., and Lander, b., derived from functions 2, 4 and 5.

would one readily expect the corresponding predictand functions.

The fractional part of the variance of each predictand explained by each of the first 6 predictor functions is shown in Fig. 4. These maps have well-defined patterns and it can be seen that a particular predictor function, even that corresponding to λ_1, is nearly useless for some predictands. The fractional part of the variance explained by the first 6 and all 30 predictor functions is depicted in Fig. 5. Fig. 5 shows that the first 6 functions explain most of the variance of which all 30 are capable.

In general, the heights at stations which have no "upstream" stations in the sample are less predictable than the others. Also, due to their less organized behavior, heights at southern stations are less predictable in terms of reduction of variance than those at northern stations. Finally, heights at those stations having few close neighbors are difficult to predict.

Figs. 4 and 5 show that functions 2, 4 and 5 together explain 62% of the variance of height at Nashville 24 hr later, while the other 27 functions increase this explained variance to only 73%. These same three functions explain only 2% of the variance of height at Lander. The coefficients in the regression equations derived from these functions for the predictands Nashville and Lander are shown in Fig. 6. The coefficients do not form a very smooth pattern although the larger positive values do tend to be near and slightly to the west of the predictand stations as would be expected.

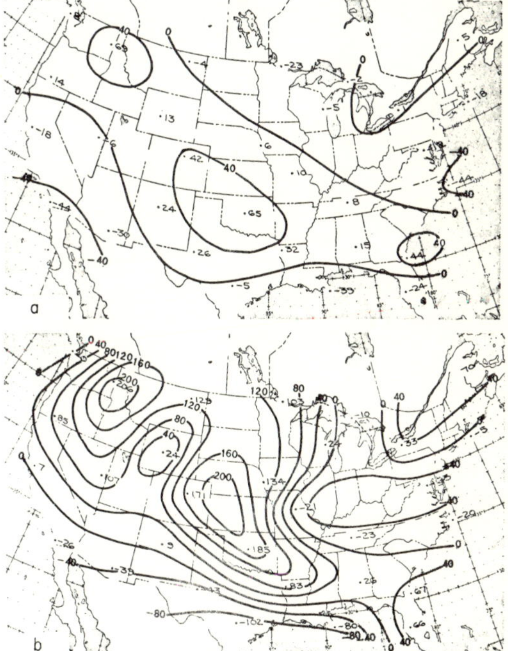

FIG. 7. The coefficients $\times 10^3$ in the regression equations for predicting the 500-mb height at Nashville, a., and Lander, b., derived from functions 1, 3 and 7.

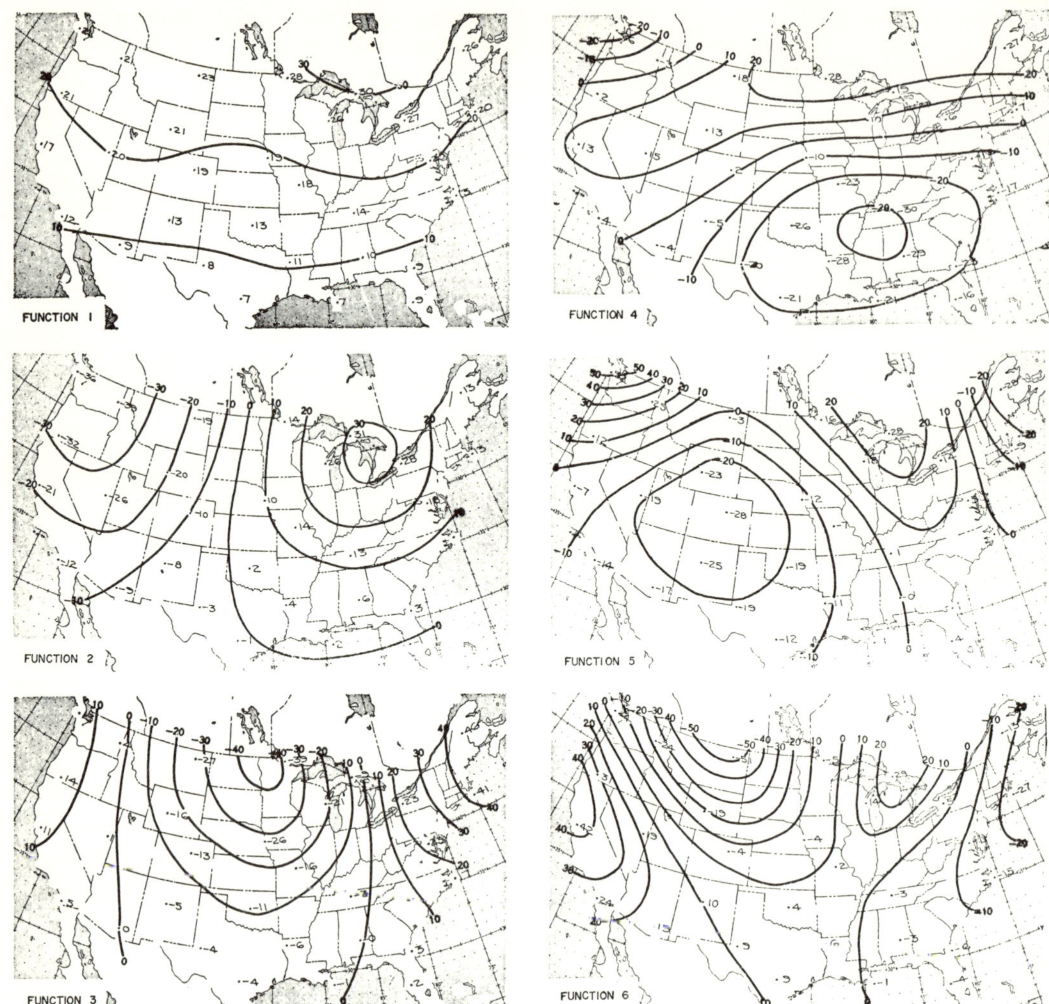

FIG. 8. The six most important principal components $\times 10^2$ of the predictand data.

The magnitudes of the coefficients are in general larger in the equation which explains a large portion of the variance of the predictand.

Functions 1, 3 and 7 together explain 71% of the variance of height at Lander, only 6% less than that afforded by all 30 functions, but they explain only 5% of the variance at Nashville. Again the magnitudes of the regression coefficients shown in Fig. 7 tend to be larger in the equation which explains a large portion of the variance of the predictand.

It is interesting to note that the predictor and predictand functions ${}_p\mathbf{A}_i$ and ${}_q\mathbf{B}_i$ (Figs. 2 and 3) do not exhibit the smooth patterns of the most important principal components of the predictand data which are shown in Fig. 8. (The principal components of predictor and predictand data are nearly the same in this example.) These principal components have the same characteristics as the empirical orthogonal functions (principal components) of sea-level pressure presented by Lorenz (1956).

Table 1 also contains λ_i^2, the fractional part of the variance of ${}_n\mathbf{Y}_q\mathbf{B}_i$ explained by ${}_n\mathbf{X}_p\mathbf{A}_i$ and vice versa, and $\mathrm{EV}_i/\mathrm{TV}$, the fractional part of the total variance of the predictands ${}_n\mathbf{Y}_q$ explained by the ith canonical predictor function. The canonical correlations decrease rather slowly as i increases but $\mathrm{EV}_i/\mathrm{TV}$ decreases rapidly. It is possible for $\mathrm{EV}_i/\mathrm{TV}$ to increase as i increases, as evidenced by the entries for $i=8$ and 9.

The data used here were highly correlated in time and 455 cases were evidently not sufficient to give very satisfactory results in terms of the patterns of the coefficients in the canonical functions and regression equations. (Existing significance tests give little guidance here since the assumptions underlying the tests are far from actuality. Lawley's (1959) test indicates the first 21 canonical correlations to be significant at the $2\frac{1}{2}\%$ level.) Given a much larger sample, one would expect the coefficients in the regression equations which explain a large portion of predictand variance to exhibit smoother patterns than those shown in Figs. 6 and 7. Since Fig. 6b and Fig. 7a represent regression equations which explain only a small portion of predictand variance and are probably the result of the random component in the data, one would not expect them to be meteorologically meaningful.

In many applications a predictor may occasionally be in error. If only a very small number of predictors are in the regression equation, the error may have a large and detrimental effect on the prediction. The more predictors there are in the equation the less an error in one of them will affect the prediction. In cases where there are several predictands, canonical correlation used very cautiously may give regression equations which are slightly better for operational use than those derived by the well known screening regression (Miller, 1962). Other factors, such as the problem of missing data, may overshadow this possible advantage.

Regression equations are not always to be preferred to discriminant (or canonical) functions. When the number of groups is large, analysis of the multi-dimensional discriminant space is difficult and involves additional assumptions. However if only 2, or perhaps 3, discriminant functions are important, hand analysis of scatter diagrams may yield probability estimates that surpass regression estimates; in this application, discriminant analysis will have reduced the problem from p dimensions to 2 or 3 dimensions.

REFERENCES

Anderson, T. W., 1958: *An Introduction to Multivariate Statistical Analysis.* New York, John Wiley & Sons, 374 pp.

Barnard, M., 1935: The secular variations of skull characters in four series of Egyptian skulls. *Ann. Eugenics*, **6**, 352–371.

Brier, G. W., 1950: Verification of forecasts expressed in terms of probability. *Mon. Wea. Rev.*, **79**, 1–3.

Brown, G. W., 1947: Discriminant functions. *Ann. Math. Statist.*, **18**, 514–528.

Bryan, J. G., 1950: A method for the exact determination of the characteristic equation and latent vectors of a matrix with applications to the discriminant function for more than two groups. Ed. D. Dissertation, Harvard University, 290 pp.

Fisher, R. A., 1936: The use of multiple measurements in taxonomic problems. *Ann. Eugenics*, **7**, Part II, 179–188.

Hooper, J. W., 1959: Simultaneous equations and canonical correlation theory. *Econometrika*, **27**, 245–256.

Hotelling, H., 1936: Relations between two sets of variates. *Biometrika*, **28**, 139–142.

Lawley, D. N., 1959: Tests of significance in canonical analysis. *Biometrika*, **46**, 59–66.

Lorenz, E. N., 1956: Empirical orthogonal functions and statistical weather prediction. Sci. Rept. No. 1, Statistical Forecasting Project, Massachusetts Institute of Technology, 49 pp.

Lund, I. A., 1955: Estimating the probability of a future event from dichotomously classified predictors. *Bull. Amer. Meteor. Soc.*, **36**, 325–328.

Miller, R. G., 1962: Statistical prediction by discriminant analysis. *Meteor. Monogr.*, **4**, No. 25, 54 pp.

——, 1964: Regression estimation of event probabilities. Tech. Rept. No. 1, Contract Cwb-10704, The Travelers Research Center, 153 pp.

Tatsuoka, M. M., 1955: The relationship between canonical correlation and discriminant analysis; and a proposal for utilizing qualitative data in discriminant analysis. Cambridge, Educational Research Corporation, 47 pp.

22

Reprinted with permission from *Syst. Zool.*, **18**(4), 363–373 (1969)

EVOLUTIONARY IMPLICATIONS OF WING AND SIZE VARIATION IN THE RED-WINGED BLACKBIRD IN RELATION TO GEOGRAPHIC AND CLIMATIC FACTORS: A MULTIPLE REGRESSION ANALYSIS

DENNIS M. POWER

Abstract

Power, Dennis M. (Royal Ontario Museum and University of Toronto, Toronto 5, Ont. Can.) 1969. Evolutionary implications of wing and size variation in the red-winged blackbird in relation to geographic and climatic factors: a multiple regression analysis. Syst. Zool., 18:363–373.—In red-winged blackbirds from central North America, geographic variation in wing length is a good indicator of geographic variation in body size. Multiple regression analysis is used to consider the relationship between geographic variation and eight geographic and climatic factors. Variation in males appears inversely related to two factors, latitude and July wet-bulb temperature, while variation in females is inversely related to only wet-bulb temperature. The statistical associations may indicate natural selection in terms of thermoregulatory efficiency. [Geographic variation. Thermoregulation. Multiple regression. *Agelaius*.]

Speculation on the adaptive signifiance of geographic variation is complicated by the fact that much of the genetic component of variation in a particular character may be the result of natural selection by more than one climatic, habitat, or biotic factor. Given a set of quantitative environmental factors which may be thought to relate in some way to variation in a character, a partial solution to the problem is in considering the statistical relationship between variation in the character and the enviromental factors by way of multiple regression analysis. The method yields, first, the multiple correlation coefficient which estimates the amount of variation in a character that is accounted for by the simultaneous effect of all of the environmental factors. Secondly, one obtains partial regression coefficients which may be used to estimate the degree of statistical relationship between the character and a single environmental variable while variation accounted for by the remaining environmental measures included in the test is held constant. In this way multiple regression analysis takes into account and largely removes the effect of correlations among the environmental factors and is thereby considerably more powerful a descriptive statistical procedure than would be a series of simple linear regression analyses.

The present study was designed to determine if variation in wing length and body size in red-winged blackbirds (*Agelaius phoeniceus*) is statistically associated with variation in certain geographic and climatic factors in the breeding season environment, and, if so, whether these relationships suggest what factors are most likely responsible for the evolution of wing length and body size in central North America. Multiple regression analysis was used in considering the possibliity of a selective influence by more than one factor. The study area (Figure 1) includes most of the Great Plains states of the U.S. and the prairie provinces of Canada. Eight geographic and climatic factors were of interest; these are, locality values for longitude, latitude, altitude, mean temperature during the breeding season (May to July), April minimum temperature, July maximum temperature, July wet-bulb temperature, and total precipitation during the breeding season. Only locality samples of birds collected during the breeding season were considered because specific wintering areas are unknown for birds of specific breeding sites; limiting locality samples to a single season does have the added advantage of reducing seasonal variation due to feather wear, molt, and fluctuations in fat content. Biological and physical factors encountered on the wintering grounds and in migration are likely to have a selective influence, but for the moment these factors and the extent of their influence may only be surmised.

It is useful to mention very briefly some of the more basic aspects of the biology of the species studied. Red-winged blackbirds are common and widespread throughout most of North America. In the study area they are migratory. The birds are sexually dimorphic, males being larger than females in overall size and other characters. Males, in the breeding season, have glossy black ("display") plumage with a characteristic red wing patch, while females are broadly striped with brown and off-white ("cryptic"). Generally, the birds are on the breeding grounds by late April and post–breeding flocks are formed in August. Males are polygamous, generally holding territories in which one to three females may nest and rear young. Nesting occurs primarily in the vicinity of water, with lake edge, marshes, irrigation ditches, and agricultural land comprising preferred habitat. During the late summer, autumn, and winter red-winged blackbirds, brown-headed cowbirds (*Molothrus ater*), and starlings (*Sturnus vulgaris*), along with certain other blackbirds, often occur in large, mixed flocks. In recent years red-winged blackbirds have occasionally been reported as responsible for extensive damage to corn and rice crops in midwestern and southeastern parts of the United States.

MATERIALS AND METHODS

Wing length was measured to the nearest tenth of a millimeter from the bend of the wing to the tip of the longest primary feather. Means were calculated for 54 samples of males and 38 samples of females distributed through the study area (Figure 1). Locality sample sizes ranged from 6 to 27, with a median sample size for males of 14 and a median sample size for females

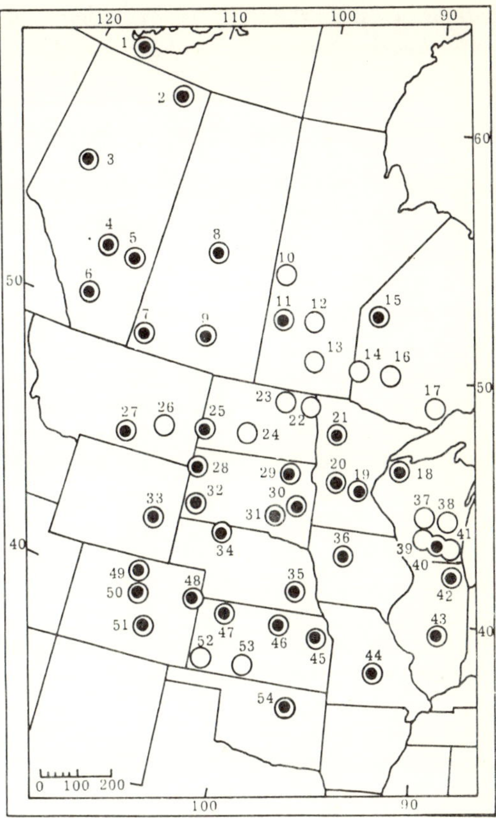

FIG. 1.—Map of the study area in central North America. Samples represented by open circles are of males only. Samples represented by dotted circles are of both males and females.

of 11. In some cases the use of museum specimens that had not been collected especially for studies in geographic variation made it necessary to lump several localities together in order to obtain reasonable sample sizes. Detailed locality descriptions and sample sizes have been given earlier (Power, 1969). Weights were recorded from fresh specimens on a triple beam balance to the nearest tenth of a gram, or were taken from specimen labels.

Longitude, latitude, and altitude were taken from maps for each locality or the center of each composite site. Climatic data for the United States are from Visher (1966) and the U.S. Agriculture Yearbook (1941), and for Canada are from climatic summaries published by the Canadian Department of Transport (1959, 1965). These sources provided summarized information that was accurate enough for this study. Mean breeding season temperature is the average of monthly mean temperatures for May, June and July. Breeding season precipitation is the total inches of rainfall for May, June, and July, averaged over 30 to 50 years. Mean April minimum temperature and mean July maximum temperature are also averaged over three to five decades and are chosen to represent the mean lowest temperature and mean highest temperature, respectively, most likely encountered on the breeding grounds. July wet-bulb temperature is the mean temperature to which the air may be cooled by evaporation during one of the hottest times of the year.

Multiple regression analysis.—The population model for a multiple regression analysis is

$$Y = \alpha + \beta_1 X_1 + \beta_2 X_2 + \ldots + \beta_k X_k + \beta_0 X_0 + e,$$

where Y represents estimated values of the dependent variable, α is a constant, β_i is the ith partial regression coefficient, X_i is the ith independent variable, k is the number of independent variables, the single term $\beta_0 X_0$ is the joint effect of all terms omitted from the model, and e is the "error term" or that part of Y which is distributed independently of the X_i's and X_0. In calculating the sample estimates b_i of β_i in

$$Y = a + b_1 X_1 + b_2 X_2 + \ldots + b_k X_k,$$

b_1, for example, is not an unbiased estimate of β_1, but of

$$\beta_1 + \beta_0{}^{b_{01 \cdot 2 \ldots k}},$$

where $b_{01 \cdot 2 \ldots k}$ is the sample partial regression coefficient of X_0 on X_1 while holding constant variation due to $X_2 \ldots X_k$. The same argument holds true for the remaining b_i's. As Snedecor and Cochran (1967:394) point out difficulties arise in the use of multiple regression because the investigator is never sure that there are not other independent variables related to the dependent variable that are not included in the analysis. Such variables may be thought to be unimportant, may not be feasible to measure or record, or may be unknown. It is difficult to judge the failure of estimating β_i since the amount of bias in b_i dependents on unmeasured variables. "If it is the purpose of the regression to find a formula that predicts Y then this bias is not a drawback; however in an analysis designed to discover the relative importance of X in relation to Y then serious misconceptions may develop" (Snedecor and Cochran, op. cit.).

From the foregoing it is clear that between Y and a given X the partial regression coefficients can change according to other X's included in the test. In the present study several preliminary multiple regressions were carried out with various numbers of independent variables. In one run locality values for isophane were included; isophanes (see Hopkins, 1938:9) are based on longitude, latitude, and altitude and reflect continent-wide changes in plant growing season. Partial regression coefficients on isophane were very large but not significant. This situation is due to the isophane correlating very highly with almost all of the geographic and climatic variables (all correlation coefficients are in the .80's except altitude where $r = .345$; in all cases $P < 0.01$). In other preliminary cases, where up to three independent variables were removed, the absolute values of the remaining standard partial regression coefficients differed from those given beyond, but were of the same sign and generally at the same probability level of significance. It is also worth noting that multiple correlation coefficients tend to increase with the number of independent variables, but that the critical R at a given probability level of significance also increases. The maximum number of independent variables is limited; as it approaches sample size the b's become unstable and R approaches unity even

TABLE 1. PRODUCT-MOMENT CORRELATION COEFFICIENTS BETWEEN INDEPENDENT VARIABLES OVER ALL LOCALITIES.[1]

	Lat.	Alt.	Mntemp.	Mitemp.	Mxtemp.	Wbtemp.	Ppt.
Long.	.471**	.533**	–.439**	–.513**	–.434**	–.710**	–.736**
Lat.		–.256	–.842**	–.728**	–.872**	–.592**	–.602**
Alt.			–.094	–.290*	–.018	–.536**	–.470**
Mntemp.				.920**	.964**	.805**	.717**
Mitemp.					.885**	.848**	.796**
Mxtemp.						.773**	.654**
Wbtemp.							.809**

[1] In order, the variables are longitude, latitude, altitude, mean temperature during breeding season (May to July), April minimum temperature, July maximum temperature, July wet-bulb temperature, and precipitation during the breeding season.
* $.01 < P < .05$
** $P < .01$

when the variates are uncorrelated (Morrison, 1967:104–105).

In a multiple regression analysis one considers a regression plane in an m–dimensional hyperspace, where m is the number of independent variables plus one (for the dependent axis). In order to show the partial regression of a character on a given independent variable in a simple two-dimensional diagram it is necessary to adjust the dependent variable for variation in all remaining independent variables. For example, take the sample problem

$$Y = a + b_1X_1 + b_2X_2 + b_3X_3,$$

where Y is made up of values of the dependent variable and X_i is made up of corresponding values of the ith independent variable, and say that it is desirable to show a two–dimensional scatter–diagram of values of Y, adjusted for variation in X_2 and X_3, regressed on X_1 with a slope of b_1. The adjusted values of Y are then calculated by

$$Y_{adj} = Y - (b_2X_2 + b_3X_3),$$

where here Y is made up of the original data of the dependent variable. A simple linear regression analysis of Y_{adj} on X_1 will then have a slope of b_1. Continuing, the values for Y adjusted for variation in X_1 and X_3, which would then be regressed on X_2 giving a slope of b_2, would be calculated by

$$Y_{adj} = Y - (b_1X_1 + b_3X_3),$$

and so on.

Finally, the magnitude of the partial regression coefficients is not independent of the scale in which the independent variables were recorded. To remove the scaling effect the standard partial regression coefficients (b'_i) may be calculated by initially converting the dependent and independent variables to their standardized form, or by calculating b'_i as follows:

$$b'_i = b_i(s_i/s_y),$$

where s_i is the standard deviation of ith independent variable and s_y is the standard deviation of the dependent variable.

RESULTS

Correlations among climatic factors.—A matrix of product–moment correlation coefficients between the eight geographic and climatic factors (the independent variables) helps to describe the climate of the study area (Table 1). To the west and north across the region climate becomes cooler and drier. As longitude and latitude increase, precipitation and all measurements of temperature decrease. Mean, minimum, and maximum temperatures are more highly correlated with latitude than with longitude, while wet-bulb temperature and precipitation find the highest correlation with longitude rather than with latitude. The small, but statistically significant correlation between longitude and latitude is due to the asymmetrical distribution of localities, primarily to the position of the far northwestern sites. Altitude

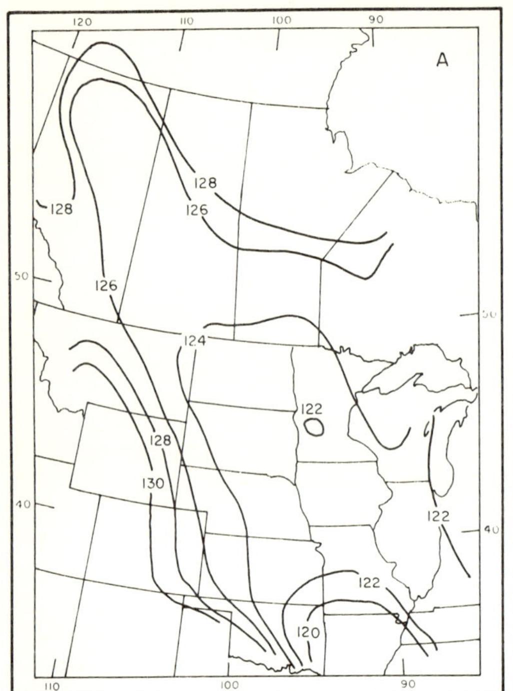

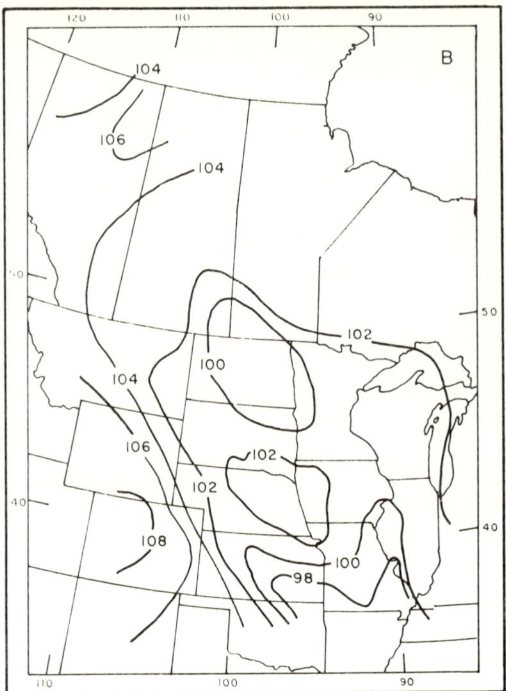

Fig. 2.—Contours (in mm.) of sixth-degree polynomial surfaces (trend surface analysis) for variation in wing length in males (A) and females (B).

ranges from 600 to 5800 feet over the study area and tends to increase in the western plains. Expected high negative correlations between temperature and altitude are lacking seemingly because of the overriding correlation between temperature and latitude. Areas of higher elevation tend to have less precipitation but this trend is weak. Finally, all temperature measures and precipitation are highly intercorrelated, indicating that as ambient temperatures increase, generally to the south and east, the amount of rainfall and the temperature to which the air may be cooled by evaporation also increase. Summarizing the important points, it may be said that during the breeding season the northern plains are relatively cool and dry, the western plains are relatively hot and dry, and the southeastern plains are relatively hot, humid, and wet.

Intralocality and interlocality correlations between wing length and weight.—Body weight is generally considered the best indicator of overall size or mass. The correlation coefficient (r) between wing length and weight indicates the degree of association between variation in these two variables. Weight data were available for ten locality samples of males and seven locality samples of females. Intralocality correlation coefficients between wing length and the cube root of body weight (Table 2) are, with one exception, not statistically significant. Thus, at the population level wing length and size are not correlated. However, interlocality correlation coefficients using locality sample means of wing length and means of cube root of body weight are highly significant, indicating that geographic variation in wing length is positively correlated with geographic variation in body size.

Pattern of geographic variation.—The patterns of geographic variation in wing length for males and females are sum-

TABLE 2. INTRALOCALITY AND INTERLOCALITY SAMPLE CORRELATION COEFFICIENTS (r) BETWEEN WING LENGTH AND THE CUBE ROOT OF BODY WEIGHT.

males			females		
loc.[a]	n	r	loc.[a]	n	r
1	12	–.124	1	9	–.477
3	14	.062	3	12	–.229
5	8	.286	5	7	.689
9	16	.088	9	17	.396
21	12	.067	21	8	–.324
37	36	.064	40	8	–.240
38	52	.085	54	12	.445
41	14	–.371			
52	8	.743*			
54	13	–.019			
means	10	.874**	means	7	915**

* $.01 < P < .05$; ** $P < .01$
[a] locality numbers refer to those in Figure 1.

marized by contour patterns of sixth–degree polynomial surfaces (trend surface analysis; O'Leary et al., 1966) fitted to the locality sample means (Figure 2). In both sexes wing lengths are largest in the western plains states of Colorado, Wyoming, and Montana, are slightly smaller in Northwest Territories and Alberta, decrease gradually through central Canada and the eastern and central plains states of the U.S., and are smallest in the southeastern plains region of eastern Kansas, Oklahoma, and Missouri. Variation in mean wing length for males is highly and significantly correlated with mean wing length for females ($r = .917$; $P < .01$). Since mean wing length has been shown above to be a good indicator of body size it is immediately obvious that variation in body size in red-winged blackbirds, at least for breeding localities, does not conform to variation predicted by Bergmann's ecogeographic rule (*viz*, in polytypic species of endotherms body size tends to be larger in cooler parts of the range and smaller in warmer parts).

Multiple regression analysis.—The results of the multiple regression analysis of mean wing length on eight geographic and climatic factors for 54 male locality samples and 38 female locality samples are as follows:

$$Y_{\text{males}} = 164.846 - 0.295\,(\text{long.}) - 1.011^{**}\,(\text{lat.}) - 0.454\,(\text{alt.}) - 0.262\,(\text{mean temp.}) - 0.356\,(\text{min. temp.}) + 0.184\,(\text{max. temp.}) - 0.848^{***}\,(\text{wet-bulb temp.}) - 0.233\,(\text{ppt.})$$

$$Y_{\text{females}} = 102.607 - 0.110\,(\text{long.}) + 0.376\,(\text{lat.}) + 0.226\,(\text{alt.}) + 0.090\,(\text{mean temp.}) - 0.663\,(\text{min. temp.}) + 0.763\,(\text{max. temp.}) - 0.992^{*}\,(\text{wet-bulb temp.}) - 0.423\,(\text{ppt.})$$

Values associated with the independent variables are the standard partial regression coefficients ($* = .01 < P < .05$; $** = .001 < P < .01$; $*** = P < .001$). For males and females the multiple correlation coefficients are .865 and .834, respectively, which are highly significant ($P < .005$).

The statistically significant standard partial regression coefficients for males are those associated with latitude and with wet-bulb temperature. The only significant coefficient for females is that with wet-bulb temperature. It is easier to visualize these trends if they are graphed. In order to graph the regression of wing length on latitude while holding constant variation due to all other factors, mean wing lengths are adjusted for all factors other than latitude and then regressed on latitude. This has been done in Figure 3 for males and females. The regression coefficients in these graphs are the partial regression coefficients rather than the standard partial regression coefficients, as given above, as they were calculated using raw data rather than standardized data. Only the inverse regression for males shows a significant departure from zero, indicating a tendency for a decrease in mean wing length and mean body size of males with an increase in latitude. This seemingly accounts in part for the slight decrease in mean wing length from the western plains to the northwest. Since it is known that mean wing

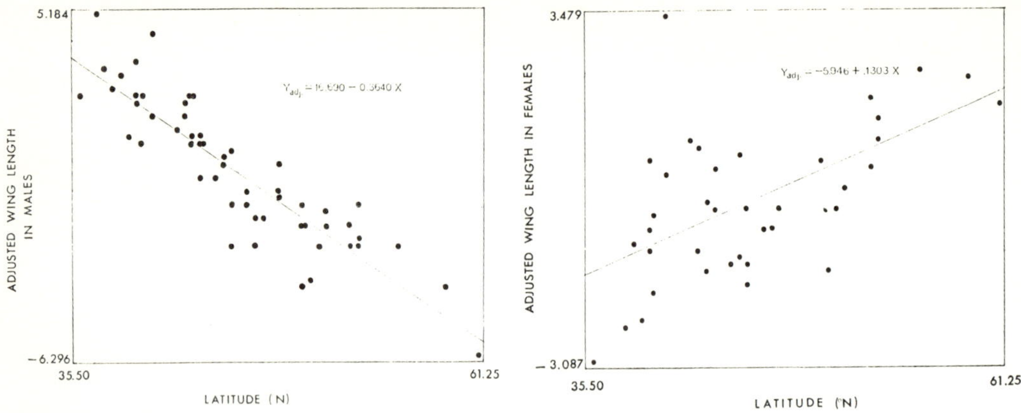

FIG. 3.—Partial regression of adjusted wing length on latitude for males (A) and females (B). The regression for females is not significant at the .05 level.

length in males decreases only slightly to the north and more extremely to the southeast the regression may be interpreted as indicating primarily a depressing effect by some factor correlated wtih latitude (but not included in the analysis). In other words wing length in the absence of influence by this factor would be expected to be larger than it actually is. For July wet-bulb temperature (Figure 4) the partial regressions of mean wing length, adjusted for all factors other than wet-bulb temperature, indicate that as wet-bulb temperature increases wing length and body size decrease.

Local differences in feather wear.—Among–locality differences in the density or the abrasive quality of the vegetation can result in local differences in feather wear and could thereby introduce a source of error in wing length measurements. If varying degrees of these qualities of the vegetation occur at random throughout the study area, then there is not a significant problem in biasing the results of the multiple regression analysis or in detailing patterns of geographic variation. If, on the other hand, these qualities vary clinally throughout the study area then a bias may be introduced. If this is the case, it is likely that variation in density or abrasiveness probably would be correlated with one of the climatic factors considered here, most likely precipitation. Variation in wear, then, would be accounted for by precipitation, if this is the important factor, and should not effect to any serious degree the statistical relationships found between wing length and July wet-bulb temperature.

DISCUSSION

Latitude.—For males, the significant negative partial regression coefficient of wing length on latitude may indicate selection for smaller size, or a depressing effect on size, with increasing latitude. This is not a temperature effect because temperature data are included in the analysis and any relation to temperature by either wing length or latitude would be accounted for. Since during the breeding season the number of hours of daylight increases with latitude, it may be that a smaller bird would absorb less solar radiation than a larger one and at high latitudes would have an advantage in reducing the heat load acquired during a period of activity in the sun. This interpretation, although highly speculative, is consistent with the fact that the relationship to latitude is seen only in males. Males, being black, would absorb more solar radiation than the brown and white females. Hamilton and Heppner (1967) and Lustick (1969) have shown

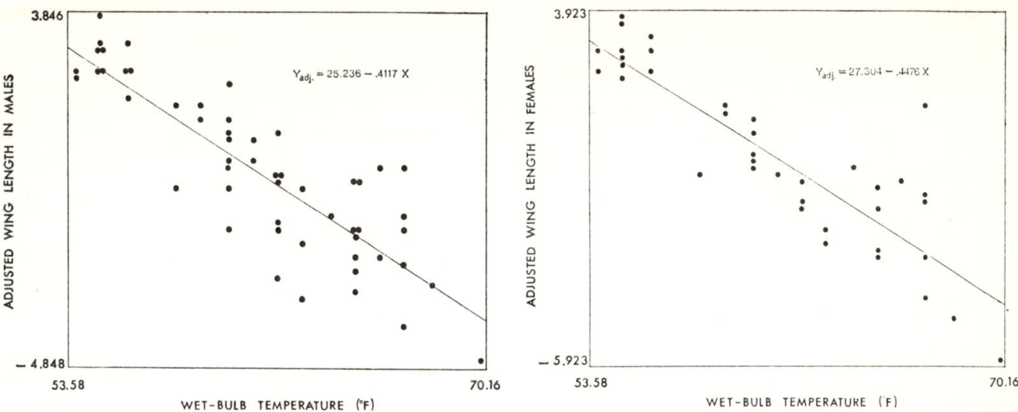

FIG. 4.—Partial regression of adjusted wing length on July wet-bulb temperature for males (A) and females (B).

that a black or heavily melanic bird absorbs more solar radiation than a white bird and that this heat can have a metabolic effect.

Wet-bulb temperature.—The significant negative partial regression coefficient of wing length on July wet-bulb temperature in both sexes may be interpreted as a relationship to wing length itself or as a relationship to wing length merely as it is positively correlated with body size. If the first alternative is taken, i.e., the significant coefficients indicate that wet-bulb temperature may exert a selective influence on wing length independent of body size, one interpretation is that the length of flight feather characters is related to such factors as sparsity of vegetation or distance between patches of feeding or breeding habitat, which are in turn inversely correlated with July wet-bulb temperature. It is not known if such a relationship exists. Hamilton (1961) and others have discussed the idea that a short, blunt wing facilitates flight in dense vegetation and that long wings may be favored in regions of sparse vegetation. A long wing may also be an advantage with increased time spent flying due to greater distances between feeding and breeding sites.

The second alternative is that the significant partial regression coefficients with wet-bulb temperature can be interpreted as relationships to body size or mass. As has been pointed out there is a high correlation between mean wing length and mean cube root of body weight.

July wet-bulb temperature is the average temperature to which the air may be cooled by evaporation during one of the hottest times of the year. Wet-bulb temperature depends on both dry-bulb temperature and vapor pressure (Figure 5). As dry-bulb temperature increases the need for heat dissipation increases, and as the vapor pressure increases the efficiency of evaporative cooling is diminished. Evolutionary

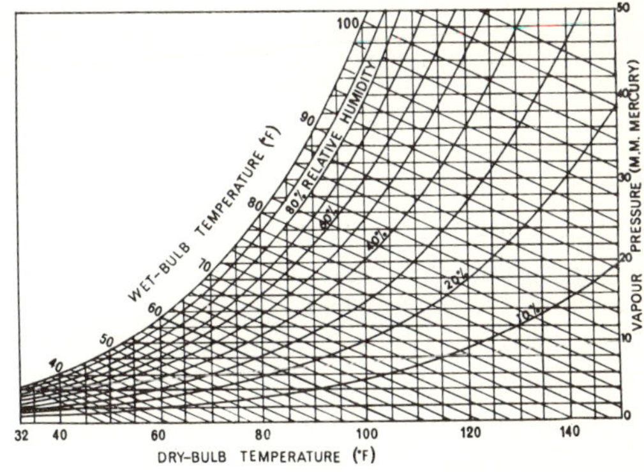

FIG. 5.—Relation between wet-bulb temperature, dry-bulb temperature, vapor pressure, and relative humidity. (Redrawn from Belding, 1967).

adaptations in regions characterized by high wet-bulb temperatures are likely to relate to the demands of metabolic heat dissipation. Birds maintain a body temperature (39–43° C for passerine species; King and Farner, 1960) nearer the margin of heat death than the margin of cold death and moderate motor activity even at relatively low ambient temperatures usually requires that a bird dissipate excess metabolic heat (King and Farner, 1964).

Pertinent aspects of the physiology of birds in maintaining homeothermy in various environments have been discussed by Bartholomew (1964), Dawson and Schmidt-Nielsen (1964), Kendeigh (1934), King and Farner (1960, 1964), Salt and Zeuthen (1960), and others. As ambient temperature and the heat produced by metabolic and motor activity increase, the evaporation of water from the respiratory surfaces becomes progressively more important in dissipating heat and in maintaining a relatively constant body temperature. As King and Farner (1960:262) point out, "Evaporative cooling provides essentially the only route for dissipation of heat from the body when ambient temperature approaches and exceeds the skin temperature." Evaporative cooling occurs when air-flow is increased over the surfaces of the mouth, pharynx, trachea, bronchi, lungs, and air sacs. An increase in blood temperature is detected in the dorsal wall of the midbrain, which in turn results in an increase in the respiratory rate and a decrease in the tidal volume of air, thereby increasing the "minute volume" and the rate of evaporation of water from the respiratory passages. This is generally accomplished by panting and, in some birds, gular fluttering. There is an advantage to increased air-flow over non-respiratory parts of the system (mouth, pharynx, trachea, and air sacs) in avoiding excess loss of CO_2 from tissues and the subsequent occurrence of alkalosis. The effectiveness of panting is limited, however, since as ambient temperature approaches body temperature only about one-half to one-fourth of the excess heat can be dissipated by evaporation. The efficiency of evaporative cooling is a function of the difference in partial pressure of water vapor between the surface of the respiratory system (*sensu lato*) and the inhaled environmental air. Respiratory evaporation is less efficient in cooling and the occurrence of hyperthermy is accelerated in a humid environment more than in a dry environment at the same ambient temperature.

Now it remains to be determined if and how geographic variation in body size in an endorthermic organism relates to fitness in terms of variation in mean wet-bulb temperature during one of the hottest times of the year. If there is a slight increase in respiratory surface per unit of mass in a smaller animal (I am aware of no evidence of this) it could be argued that in having a relatively greater surface area for evaporative cooling, an advantage is imparted for survival and development in a hot humid environment. There is some evidence that in desert birds, larger species and individuals expire relatively less respiratory moisture than smaller birds (Bartholomew and Dawson, 1953), but this may be due to their lower metabolic rates. In attempting to explain an inverse relation between body size and humidity in warm climates Hamilton (1961:184) speculated that " . . . the increase in wing length in warm, arid regions is an indication of size increases that facilitate conservation of metabolic water, and that the decrease of wing length in warm, humid regions is indicative of size decreases that facilitate effective heat dissipation." He suggested further that a decrease in body size might favor development and survival in warm, humid climates due to ". . . a relative increase in internal respiratory surfaces, along with slight increases in metabolic activities (e.g., respiratory rates?) which might facilitate heat dissipation." Hamilton's hypothesis may well explain the geographic variation in body size of red-winged blackbirds.

Concerning the advantages of large size

in dry, hot climates, it should be noted that even though conservation of metabolic water is not likely to be a critical factor for red-winged blackbirds, because the birds generally nest and feed during the breeding season in the immediate vicinity of water (marshes, lake edge, irrigation ditches, etc.), there may be some advantage in increased efficiency of water utilization in hot, dry climates. Less time spent drinking would also mean increased avoidance of terrestrial predators.

Small size in hot, humid climates may be an even more important adaptation. Perhaps significant here is the fact that birds and other endotherms may often employ behavioral means such as rest and seeking shade to aid in preventing the build-up of a heat load, in dissipating excess heat, and in recovering from hyperthermy. Thus, a plausible interpretation of the decrease in size and the negative statistical relationship between body size and July wet-bulb temperature is that in a hot, humid climate smaller individuals by virtue of their increased surface/volume ratios, their high respiratory and metabolic rates, and, perhaps, relative to size, their greater respiratory surface for evaporative cooling, by behavioural means such as rest and seeking shade may be able to effect a more rapid and efficient recovery from hyperthermy or a heat load than would a larger individual in the same situation.

SUMMARY

Geographic variation in wing length was examined for breeding season samples of red-winged blackbirds (*Agelaius phoeniceus*) in central North America. Within-locality sample variation in wing length is not correlated with body size. However, locality sample means of wing length are highly correlated with mean cube root of body weight suggesting that geographic variation in wing length is a good indicator of geographic variation in size. It was of interest to see if variation in wing length and body size is statistically associated with variation in one or more selected geographic and climatic factors, and to discuss the evolutionary implications of these associations. The environmental factors considered were longitude, latitude, altitude, mean breeding season temperature, April minimum temperature, July maximum temperature, July wet-bulb temperature, and breeding season precipitation. Since more than one enviromental factor may have a selective influence on a character, and since enviromental factors may be correlated with one another to some degree, a multiple regression analysis was used to (1) estimate the total amount of variation in wing length accounted for by the simultaneous effect of all environmental factors, and (2) to determine the amount of variation accounted for by a single environmental factor while holding constant the effect of all other factors in the analysis.

Wing length and body size in both sexes are largest in the western plains, slightly smaller in the northwest, decrease gradually through central Canada and the central U.S., and are smallest in the southeastern plains. Although the sexes are highly correlated in their patterns of geographic variation, variation in males has significant inverse partial regression coefficients on two enviromental factors, latitude and wet-bulb temperature, while variation in females has a significant inverse partial regression on only one, wet-bulb temperature. The latitude effect in males indicates a tendency for depression of body size with increasing latitude, independent of temperature. It is postulated that since during the breeding season the number of hours of daylight increases with latitude a smaller bird would absorb less solar radiation than a large one and at high latitudes would have an advantage in reducing the heat load acquired during a period of activity in the sun. This is highly speculative but is consistent with the observation that the relationship is found only with males, which are almost entirely black, rather than females, which are brown and white.

Significant negative partial regression coefficients on wet-bulb temperature during one of the hottest times of the year indicate that body size is larger in hot, dry climates and smaller in hot, humid climates. Size increases in arid regions may facilitate conservation of metabolic water and size decreases in humid regions may facilitate heat dissipation. In the latter case, a smaller endotherm by virtue of a relatively greater surface/volume ratio, higher respiratory and metabolic rate, and, perhaps, greater respiratory surface for evaporative cooling, by behavioural means such as rest and seeking shade may be able to recover from hyperthermy or a heat load more rapidly and efficiently than a larger endotherm in the same situation.

ACKNOWLEDGMENTS

I am grateful to Drs. J. C. Barlow, T. H. Hamilton, J. W. Hardy, R. F. Johnston, E. Mayr, L. J. Szijj, and J. R. Tamsitt for reading and commenting on these ideas. Gathering specimens and specimen data was assisted by National Science Foundation Grant GB-4446X (principal investigator, Dr. W. A. Clemens, University of Kansas). Computer time was made available by the Institute of Computer Science, University of Toronto.

REFERENCES

BARTHOLOMEW, G. A. 1964. The roles of physiology and behavior in the maintenance of homeostasis in the desert environment, p. 7–29. *In* Homeostasis and feedback mechanisms. Symposia of the Soc. Exper. Biol., no. 18. Oxford Univ. Press, Cambridge.

BARTHOLOMEW, G. A. AND W. R. DAWSON. 1953. Respiratory water loss in some birds of the southwestern United States. Physiol. Zool., 26: 162–166.

BELDING, H. S. 1967. Resistance to heat in man and other homeothermic animals, p. 479–510. *In* A. H. Rose, [ed.], Thermobiology. Academic Press, New York.

CANADIAN DEPARTMENT OF TRANSPORT (METEOROLOGICAL DIVISION). 1959. Climatic summaries for selected meteorological stations in Canada. Vol. II: Humidity and wind. Toronto, Ontario.

CANADIAN DEPARTMENT OF TRANPORT (METEOROLOGICAL DIVISION). 1965. Climatic summaries for selected meteorological stations in Canada. Temperature and precipitation normals. Mimeographed. Toronto, Ontario.

DAWSON, W. R. AND K. SCHMIDT-NIELSEN. 1964. Terrestrial animals in dry heat: desert birds, p. 481–492. *In* Handbook of physiology; section 4: adaptation to the environment. Amer. Physiol. Soc., Washington, D. C.

HAMILTON, T. H. 1961. The adaptive significance of intraspecific trends of variation in wing length and body size among bird species. Evolution, 19:355–367.

HAMILTON, W. J., III, AND F. HEPPNER. 1967. Radiant solar energy and the function of black homeotherm pigmentation. Science, 155:196–197.

KENDEIGH, S. C. 1934. The role of environment in the life of birds. Ecol. Monogr., 4: 299–417.

KING, J. R. AND D. S. FARNER. 1960. Energy metabolism, thermo-regulation and body temperature, p. 215–288. *In* A. J. Marshall, [ed.], Biology and comparative physiology of birds, vol. 2. Academic Press, New York.

KING, J. R. AND D. S. FARNER. 1964. Terrestrial animals in humid heat: birds, p. 603–624. *In* Handbook of physiology; section 4: adaptation to the environment. Amer. Physiol. Soc., Washington, D.C.

LUSTICK, S. 1969. Bird energetics: effects of artifical radiation. Science, 163:387–390.

MORRISON, D. F. 1967. Multivariate statistical methods. McGraw-Hill, New York.

O'LEARY, M., R. H. LIPPERT, AND O. T. SPITZ. 1966. FORTRAN IV and map program for computation and plotting of trend surfaces for degrees 1 through 6. Computer contribution 3, State Geological Survey, Univ. Kansas, Lawrence.

POWER, D. M. 1969. Geographic variation in red-winged blackbirds from central North America. Univ. Kansas Publ. Mus. Nat. Hist., *in press.*

SALT, G. W. AND E. ZEUTHEN. 1960. The respiratory system, p. 363–409. *In* A. J. Marshall, ed., Biology and comparative physiology of birds, vol. 1. Academic Press, New York.

SNEDECOR, G. W. AND W. G. COCHRAN. 1967. Statistical methods. 6th edition, Iowa State Univ. Press, Ames.

UNITED STATES DEPARTMENT OF AGRICULTURE. 1941. Climate and man. Washington, D.C.

VISHER, S. S. 1966. Climatic atlas of the United States. Harvard Univ. Press, Cambridge.

Department of Ornithology, Royal Ontario Museum and Department of Zoology, University of Toronto, Toronto, Ontario, Canada.

23

Reprinted from *Copeia*, No. 1, 135–144 (1970)

Chronological Correlation Between Change in Weather and Change in Morphology of the Pacific Tree Frog in Southern California

THOMAS VOGT AND DAVID L. JAMESON

Morphological variation in a single population of *Hyla regilla* was studied over a period of three years, and this variation was analyzed and correlated with environmental changes using discriminant analysis, principal component analysis and canonical correlation. A significant difference was found to exist between samples captured at different times and these differences were found to be strongly correlated with rainfall during development and high and low temperatures in the fall prior to breeding. The first year of the study was the driest on record, and the frogs taken during that year were smaller and had relatively longer fingers and shorter toes than the frogs obtained in two subsequent years. Frogs may differ allometrically at different times of the year and also, to a lesser degree, from year to year in response to fluctuations of environment. The rapidity of response in the population may indicate that initially some of the variation is non-genetic followed later by periods of genetic stabilization. The exact nature of the selective mechanism is not clear, but direct effects of heat and moisture on the frogs' abilities to retain water and dissipate heat may be important.

INTRODUCTION

THE response of a population to shifts in adaptive peaks resulting from environmental fluctuations is a part of the constant microevolution that occurs in natural populations. To understand these responses as representative of the entire phenomenon of biological evolution many workers have studied changes in gene frequencies in plants and animals and have tried to relate these changes to environmental action. Both geographic and chronological aspects of the problem have been attacked. The work of Dobzhansky and co-workers (Dobzhansky, 1943; Dobzhansky and Levene, 1955; Wright and Dobzhansky, 1946, *et al.*) on changes in the frequencies of chromosome types in *Drosophila pseudoobscura* are perhaps the best known of these studies.

This study attempts to correlate phenotypic changes with environmental changes in a single population of Pacific tree frogs measured for a period of three years. When significant morphological changes are found, the problem then becomes one of attempting to suggest the physiological basis for such a response.

The Pacific tree frog is distributed widely over western northern America from Mexico to British Columbia and from the Rocky Mountains to the Pacific Ocean in a vast variety of climatological conditions ranging from desert to mountain to coastal lowland, from very wet to extremely dry environments. Perhaps the single most significant and consistent trend found by Test (1898) and Jameson *et al.* (1966) is that moist areas have larger frogs, while the drier areas have smaller ones. Blair and Littlejohn (1960) cilted three anuran complexes which have their largest animals in the most xeric area and the smallest in moist environments. They suggested that larger animals have less surface area in relation to mass and presumably have a selective advantage in terms of water conservation in the drier areas. Allee *et al.* (1950) wrote that cold-blooded terrestrial animals tend to have their larger species and individuals in warmer rather than cold areas.

Thorson (1955) found that small amphibians have greater resistance to water loss than large frogs because they have a proportionately greater body water content. However, the smaller amphibian has a more rapid rate of loss due to the relatively greater body surface exposed. Thorson concluded that an optimum rate of water loss exists which differs between species but that resistance to desiccation is not an important factor in the appearance of terrestrialism in amphibians.

Blair (1955) and others suggested that size differences may be acting as ethological isolating mechanisms; closely related anuran species may have their greatest size differences in sympatric localities. Jameson (1966) found that the rate of weight loss under stress conditions of temperature and humidity may be the direct result of differences in genetic constitutions rather than simple differences in physiological adjustments to environment. Therefore, genetic variability is expected in the tolerance of stressful environmental conditions, and the population is thus capable of responding genetically and morphologically to changes in environment. Studying years in which environment is near stress levels may enable identification of these responses in the form of size and shape variations and the relation of the variations to specific environmental factors.

Morphological responses to environmental pressure represents the net effect of a number of interacting and even conflicting forces at work. A single measurement will be influenced not only by selective forces directly affecting it, but also by forces working on other measures. A general tendency toward an increased or decreased body size will affect any other characteristics measured regardless of the selective forces working on it separately. Just what happens if the body tends to grow larger while at the same time smaller shank lengths or finger lengths are also favored? The result will be a balance, the resultant vector of all individual forces acting, and the resolution of this vector into its component parts is an extremely difficult task. Perhaps some reasonably substantiated hypotheses can be formulated.

Because it is easier to study, geographical variation has been the subject of far more research than chronological variation. Nevertheless, the principles elucidated from either type of study should be applicable to the other. The following statement by Hamilton (1961:180) might just as well concern chronological as geographical, or intraspecific as interspecific variation.

> "Geographic variations of biological characters are generally not random, but orderly, for wide-ranging, continuously distributed sedentary species. It follows then that selection forces—admittedly operating through unknown mechanisms—are responsible for the non-random nature of ecogeographic variation. A corollary of this statement is that the characters involved are genotypic adjustments to geographically differential selection pressures. However, whether the biological characters manifested as clines are genotypic or phenotypic is not a problem that necessarily strikes at the foundation of the theory of ecogeographic adaptation . . . the theory of initial phenotypic adjustment followed by genotypic control is one explanation (the Baldwin effect) for the occurrence of climatic adaptation at the intraspecific level."

Materials and Methods

Frogs were collected in 1961, 1962, and 1963 from a small semi-permanent stream in La Mesa, California at an elevation of 530 ft.

The collection area, a few miles east of San Diego, California, has a Mediterranean climate. Rain occurs mostly from October to May: summer rain is rare. Average rainfall is from 10 to 12 inches per year. The frequency and amount of rain has a very direct effect on survival of frogs to the next generation. The period of collection was unusually dry—1961 in fact being the driest year on record—and thus represents a situation during which some sort of adaptive changes might be expected.

Seven different morphological measurements were made on each animal by the senior author within a few days of capture. Only males were used in the analysis. Snout–vent length (SV) is the distance from the tip of the snout to the anus. Head width (HW) was measured at the position of greatest protuberance of the jaw. Shank length (SL) is the distance from knee to ankle (*i.e.*, the length of the tibio-fibula). The lengths of the fourth or longest toe (TL) and the third or longest finger (FL) were also measured. The adhesive pad of the third finger was measured with respect to both length (PL) and width (PW). The first five measurements were made with a vernier caliper, the last two under a dissecting microscope with an ocular micrometer.

The samples were obtained over a period of 35 months, and the frogs grouped into samples according to the month of captures as indicated in Table 1.

These seven samples were then compared using four different multivariate statistical procedures: discriminant analysis, principal component analysis, multiple regression analysis and canonical correlation. Analysis was done on the IBM 1620 computer at San Diego State College and the SDS Sigma 7 at the University of Houston.

Twenty-four different environmental factors were tested in the canonical correlation.

TABLE 1. THE MEANS, GRAND MEANS, AND SUM OF THE ROOTS (λ_i) OF THE VARIANCE-COVARIANCE MATRIX OF MORPHOLOGICAL MEASUREMENTS IN MM.

Sample	Capture (date)	Size of Sample	SV	SL	HW	TL	FL	PW	PL	$\Sigma\lambda_i$	Σ_r
1	Mar 1961	76	27.78	14.96	10.92	6.24	6.20	1.30	1.01	6.19	2.5
2	Apr 1961	33	28.64	15.21	9.86	6.36	6.56	1.15	.93	13.64	4.9
3	Jan 1962	18	32.03	15.89	10.74	8.08	6.01	1.47	1.21	9.50	4.9
4	Feb 1962	49	33.38	16.62	11.70	8.43	6.50	1.44	1.07	14.89	4.9
5	Mar 1962	57	33.11	16.70	12.00	8.07	5.96	1.45	1.08	18.71	5.0
6	June 1962	12	33.38	17.12	10.92	8.30	7.00	1.35	.98	8.82	3.4
7	June 1963	30	31.16	15.49	10.49	8.23	6.01	1.25	.93	16.47	5.3
Grand means			31.39	16.03	10.83	7.69	6.19	1.35	1.03		

These included average high temperatures and precipitation for the months of February, March and April the year prior to capture, average high temperatures, average low temperatures, maximum temperatures, minimum temperatures and precipitation for the fall months prior to capture; and total precipitation for each of the two years preceding capture and also for the same year as the sample was taken. These were selected in an attempt to determine those periods which were most critical in terms of selection and to discover which particular factors are most strongly correlated with morphological changes.

Precipitation and average high temperatures from the previous winter were used because that is the time that the breeding frogs went through development and metamorphosis, a period which may be of great significance in selection. Of the 24 factors tested six showed significant intercorrelations with morphological changes. Environmental data were gathered from official records of the U. S. Weather Bureau. The weather station from which the data were obtained is a little more than a mile east of the collection site and at approximately the same elevation.

The mathematical theory behind multiple discriminant analysis, introduced by Fisher (1936), can be found in two papers, one by Bryan (1951) and the other by Rulon (1951). Jolicoeur (1963), Jameson, *et al.* (1966), and Snyder and Jameson (1965) have used this technique to describe morphological variation. Clear and reasonably brief explanations of the techniques are also in Cooley and Lohnes (1962).

Geometrically, the principal components reveal the directions and lengths of the principal axes of multidimensional ellipsoids circumscribing the data in a multivariate scatter diagram. Biologically, it shows what combinations of characters are most variable as well as uncorrelated with each other (Jolicoeur, 1963) and partitions the total measured variation into size and shape components (Jolicoeur and Mosimann, 1960). The first axis accounts for the greatest amount of variance and is essentially a size component; the signs of its direction cosines are all the same. Other axes have mixed signs and indicate shape difference between animals. Cattel, (1965) reviewed the general theory. For specific applications see Jameson *et al.* (1966) and Jolicoeur (1963).

Deviations from isometry are determined by calculating the magnitude of deviation from isometry in the equation

$$\cos \phi = U_1V_1 + U_2V_2 + \ldots U_iV_i,$$

where V_i = the isometric value of $\sqrt{1/7}$; U_i = the observed direction cosine value, and 7 is the number of measurements made (Jolicoeur, 1963).

Canonical correlation, developed by Hotelling (1935, 1936) permits a determination of the correlation between two distinct types of variates. Significant relationships between predictor variables (in the case of this paper, environmental data) and criterion variables (*e.g.*, morphological measurements on frogs) permit a reasonable estimate of the latter when the former are known. Essentially, this involves the calculation of a regression equation for predicting the value of some morphological characteristic in a given population with known significantly intercorrelated environmental data. The coefficients are termed standard partial regression coefficients and are influenced by the intercorrelations between predictors.

Initially the intercorrelation matrix is partitioned into four sub-matrices: R_{11} the in-

Table 2. Discriminant Analysis of Seven Measurements on 296 Male Frogs.

Discriminant Axis	Variance	% of Total	d.f.	χ^2	P[1]
1	1209.4	72	42	1683.6	< .01
2	308.3	18	30	474.2	< .01
3	91.8	5	20	166.0	< .01
4	46.9	2	12	74.2	< .01
5	26.2	1	6	27.3	< .01

[1] The probability that the variation between groups in that axis and all axes with smaller variances is greater than can be explained by chance.

tercorrelations among p predictor; R_{22}, the intercorrelations among q criteria; R_{12}, the intercorrelations of predictors with criteria; and R_{21}, the transposition of R_{12}. Next the latent roots (λ_i) are extracted. With $q < p$ there are q possible roots (seven in this paper). A χ^2 test for the lambda distribution tests the null hypothesis that predictors and criteria are unrelated (pq degrees of freedom). If this is rejected, the first or largest root is removed and the χ^2 analysis run again. The process is continued until no significant intercorrelations remain.

Thus canonical correlation allows a determination of those factors of one type which can be successively used to predict those of another. A very lucid explanation of canonical correlation can be found in Cooley and Lohnes (1962). A similar study by Calhoon and Jameson (1970) gives the computational formula.

Discriminant Analysis

The variation and significance of the discriminant axes are shown in Table 2. The total Chi Square value was 1683.6 with 42 degrees of freedom ($P \leqslant .001$). Of this variation 71% can be accounted for on the first (D_1) discriminant axis, 90% on the first two axes. These two are plotted against each other in Fig. 1. Fig. 2 shows the remaining 5% on the D_3 axis, 2% on the D_4 axis.

The plots of the first and second axes indicate an immediate and significant difference between the frogs collected in 1961 and those taken the following years. The most readily apparent differences between the 1961 samples and later ones is in the relatively longer finger lengths and shorter toe lengths, and also the shorter snout–vent length, a measure representing absolute size of the animal (Table 1). The standard deviation vectors (Fig. 1) represent directions of variation of the indicated measure when that measure has been adjusted by all others. Head width and pad width differ between the 1962 and 1963 samples but not between those and the 1961 frogs. In 1962 and 1963 there was a gradual increase in head width followed by a gradual decrease. The 9.2 inches of rain in 1962 was below average, but still considerably greater than the 5.2 inches in 1961, the driest year on record (Fig. 3).

Comparison of frogs from this population with frogs from a moist mountain environment showed that the finger length/toe length ratios were higher in the more xeric habitat, the same pattern followed within this single population when rainfall was low. The mountain frogs came from Julian, California, 40 miles east of San Diego at an elevation of 4200 ft. The annual rainfall at that site is about 35 inches. The finger length/toe length ratio in this area was .907 as opposed to a value of .959 in an overall survey of the La Mesa population.

The values for this ratio vary in the La Mesa population from a high of 1.03 in 1961 to a low of .725 in sample 6 captured in June 1962. The values given in the comparative study included frogs captured during all three of the years in both populations.

The plots of the third ($\chi^2 = 166$, df $= 20$, 5%) and fourth ($\chi^2 = 74.2$, df $= 12$, 2%) discriminant axes (Fig. 2) indicate that the June, 1962 sample (6) is considerably distinct from the other populations. This is the only group of frogs captured late in the season some time after the cessation of the major rains. Possibly these are the frogs which metamorphosed at the close of the rainy season the previous year under considerably more xeric conditions than the other samples. More likely this sample represents older frogs that have simply grown larger. Both shank length and snout–vent length are greatest in this sample, and the plot indicates that the shank length difference is significant despite the fact that the third and fourth axes contain only 7% of the total. All of the samples except those captured in 1961 lie more than two standard deviations of snout–vent length from the grand mean. This is more evidence that some significant change occurred in the population following that year.

Principal Component Analysis

The sum of the roots ($\Sigma\lambda_i$) of the variance-convariance matrices represents an estimate

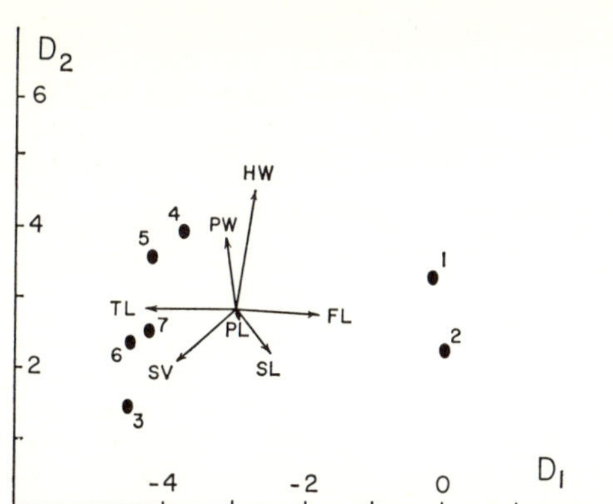

Fig. 1. Discriminant analysis of seven samples of *Hyla regilla* captured between 1961 and 1963. Mean of each sample is shown on first ($D_1 = 71\%$) and second ($D_2 = 19\%$) discriminant axes. Vectors emanating from grand mean show direction of original coordinates, lengths being representative of one standard deviation of the measurements. Sample numbers are in Table 1, Symbols are in text.

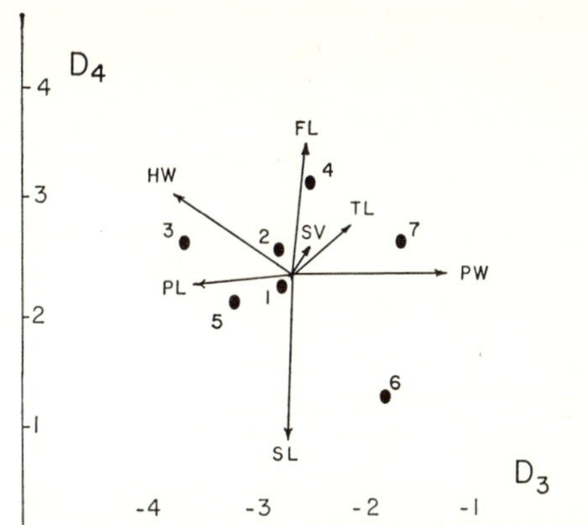

Fig. 2. Means and standard deviation vectors of third ($D_3 = 5\%$) and fourth ($D_4 = 2\%$) discriminant axes. Symbols and sample numbers are given in Fig. 1.

of the total variability in a sample. The differences in the sum of the roots suggests that the factors underlying this variability differ in the different samples (Table 1).

If each measurement was perfectly correlated with all others the sums of the correlation coefficients for that measure with respect to the six others would be equal to $1 \times 7 = 7$. Table 1 includes the results of such summation (Σr_i). Sample 1, captured in March, 1961, shows unusually low correlation values, significantly lower than all other samples when compared by means of a sign test. Sample 7 has some correlations—toe length and head width—which are significantly higher than all other samples.

Generally, the samples seem to be most like those taken in the same months of different years, a fact which suggests that a part of the sample differences may be due to a change in allometry as the frogs grow older. When true vectors are compared with an isometric growth vector the angles of deviation vary from 41° to 48° with the smallest $\chi^2 = 8 \times$

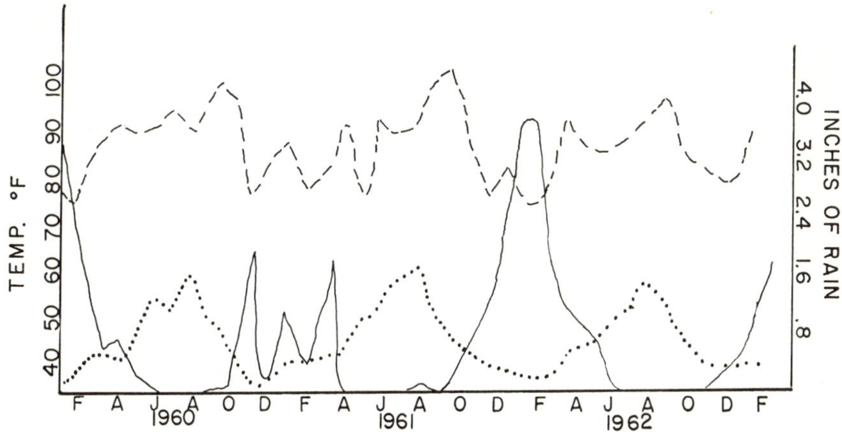

Fig. 3. Maximum (dashes) and minimum (dots) temperatures and precipitation (solid line) for each month from 1960 to 1963 at the site of collection. Temperature data represent highest or lowest temperature for that month.

TABLE 3. TRUE VECTORS FOR EACH SMPLE AND THE ANGLE OF DEVIATION FROM AN HYPOTHETICAL ISOMETRIC VECTOR. THE CHI SQUARE VALUES ARE GIVEN.

Sample	SV	SL	HW	TL	FL	PW	PL	χ^2	Angle of Deviation*	
1	.867	.419	.200	.119	.125	.047	.004	5×10^4	47°	48′
2	.839	.449	.255	.100	.134	.041	.017	2×10^4	46°	15′
3	.894	.319	.261	.108	.132	.033	.026	8×10^3	48°	4′
4	.804	.441	.265	.234	.179	.041	.024	2×10^4	41°	31′
5	.829	.416	.271	.217	.137	.030	.021	3×10^4	43°	37′
6	.763	.544	.250	.216	.103	.045	.014	1×10^9	43°	10′
7	.831	.370	.269	.266	.166	.042	.025	2×10^4	42°	7′

* The isometry vector is .377 for every measurement.

10^3 (Table 3). The change in direction cosines could be indicative of a number of phenomena. There may be different sets of genes working under the different conditions of environment, or perhaps a single set of genes are expressed differently with a changing age in the animals. This last possibility is certainly supported by the difference in shape indicated by the vectors of sample 6. This group has the largest size probably because it was collected in June and represents the oldest animals. Growth may have slowed not only with age, but possibly also in preparation for estivation. Finally it may be that the animals collected at different times belong to different sets. That is, selection removes different types of animals under varying conditions which would result in differential samples depending on the conditions preceding collection.

CANONICAL CORRELATION

The matrix of correlations between environmental factors and morphological data and the intercorrelations between the two sets of measurements, are available from the authors. All morphological measures are significantly correlated with each other with coefficients ranging from .359 to .885.

A Chi Square test for successive latent roots of the matrix of intercorrelations (Table 4) indicates that the analysis is highly significant, having $\chi^2 = 429.6$ with 84 degrees of freedom. Moreover, the successive removal of latent roots indicates that intercorrelations are significant at the 5% level on five of the seven vectors.

The first canonical vector ($\chi^2 = 768.2$, 53.8%) shows that as precipitation during growth and metamorphosis decreases (–.841 and –.463), snout–vent (.396) and toe length (.704) increase and finger length (–.503) decreases. On the second vector ($\chi^2 = 358.6$, 25%) we find snout–vent (.428) and shank length (.339) increasing, and head width (–.537) and pad size (–.481 length and –.350 width) decreasing as precipitation during growth and metamorphosis decreases; shank

TABLE 4. CANONICAL CORRELATIONS BETWEEN PREDICTORS AND CRITERIA FOR ALL FACTORS SHOWING SIGNIFICANT INTERCORRELATIONS ON EACH OF THE FIVE VECTORS CONTAINING SIGNIFICANT INFORMATION.

			PREDICTORS (Yr previous to capture)						CRITERIA						
Vector Roots Removed	χ^2	P	Fall low Temp.	Fall high Temp.	Feb precipitation	Mar precipitation	Apr precipitatiotn	Winter high Temp.	Snout–vent	Shank length	Head width	Toe length	Finger length	Pad width	Pad length
0	768.2	< .01				–.463	–.841		.396			.704	–.503		
1	358.6	< .01				–.774	–.568		.428	.339	–.537			–.350	–.481
2	181.5	< .01	–.321	–.741			.475		–.323		.864				–.310
3	76.6	< .01				–.679	–.672			.804	–.292	–.276	–.432		
4	37.5	< .05	.556	–.323	.537	.295	.275	.278	.686			–.386		–.486	–.330

NOTE: The chi square values for each of the vectors are also given. Correlations greater than .258 are significant at the 5% level, those greater than .287 are significant at the 1% level.

TABLE 5. STEPWISE MULTIPLE REGRESSION ANALYSIS OF EACH MORPHOLOGICAL MEASURE WITH THE SET OF ENVIRONMENTAL MEASURES. THE INTERCEPT AND SIGNIFICANT B COEFFICIENTS ARE GIVEN. F LEVEL FOR ENTRY AND REMOVAL WAS 2.0.

		B Coefficients								
Measures	Intercept	Winter High Temp	Winter Low Temp	Winter Rainfall	Fall Low Temp	Fall Rainfall	Feb Rainfall	Mar Rainfall	Apr Rainfall	Multiple R
SV	–11.95	1.94						6.783		.643
SL	14.01							1.537		.441
HW	6.47			.078	.078		.070	.767		.648
TL	7.22				.025				–2.49	.738
FL	2.17		.077	–.092	.029					.300
PL	1.49				.012	.057		.308		.525
PW	10.7	.132				–0.044				.535

length again has this relation (.804) in the fourth canonical vector (5.4%). The third vector (12.7%) shows that increase in precipitation during metamorphosis (.475) and high temperature prior to capture (.741) are correlated with increased head width (.864) and decreased relative length (–.323). This suggests that milder weather is correlated with rounder frogs while decreased rainfall is correlated with more elongate animals.

Toe length shows the strongest correlation with environmental factors. The change in this character was noted in the Results of Discriminant Analysis. Apparently toe length is the measure which responded most rapidly to changes in precipitation.

This analysis points out at least two critical selective periods through which the frogs must pass before breeding. The factor in the first is the amount of rain during growth and metamorphosis; in the second, the extremes of high temperature in the fall. There may, of course, be others.

Multiple regression coefficients (B) and the multiple R are given in Table 5 for each measure. These were determined by the stepwise method which selects the largest measures and tests each measure until all significant measures have been accepted and all non-significant measures rejected in stepwise fashion. The F level used for entry and removal of measures was 2.0. Eight of the environmental measures gave significant multiple correlation coefficients. Each morphological measure was significantly correlated with at least one environmental measure and head width was correlated with four measures. February, March and April rainfall, fall low temperature and winter high temperature provide significant correlation coefficients on both stepwise multiple regression and canonical correlation analysis. Fall high temperature is found to be significant by canonical correlation analysis while winter rainfall, fall rainfall and winter low temperature are unique to the stepwise multiple regression analysis. The canonical correlation analysis deals with sets of data and adjusts for both the intercorrelations between the environmental measures and the intercorrelations between the morphological measures. Other comparisons of the two methods may be found in Calhoon and Jameson (1969).

DISCUSSION

Change in size of a frog may affect survival value in at least four ways: (1) susceptibility to disease and predation, (2) the ability to obtain sufficient food, (3) susceptibility to heat or cold, and (4) susceptibility to moisture or the lack of it. Deciding in which direction selective pressures will push the population is a very complex and often impossible task. A small animal loses heat faster than a large one; it will also pick up heat faster, but during the many warm hours of a summer day this factor appears unimportant except for overheating. Thus the smaller animal might have an advantage in a warm climate. However, the smaller animal also loses moisture more rapidly than a larger frog. The same arguments hold as well for retention of heat. Still, extremes of both heat and cold are correlated with morphological change.

Lewontin (1965) pointed out that in a changing environment a population can be either genetically flexible or stable. The question as to what is "best" for the populations as a whole and what mechanisms exist for populations to adopt the "best strategy", given that one can be defined, must first be considered. The answer may not always be the same. Misra and Reeve (1964) in studying clines in two similar species of *Drosophila* found two different gene complexes controlling different sets of morphological characteristics. One set varies in the same direction for both species as environment changes, but the other moves in exactly opposite directions, a fact which suggests that perhaps these two species of *Drosophila* have different answers to the same problem. Thus, it is reasonable to assume that, while the species as a whole shows its larger individuals in mountain and more northern areas, selection within a single population might move in an opposite direction. Blair and Littlejohn (1960) found larger frogs in warmer environments in three anuran complexes. Blair (1955) has described a case in which size differences of two species of *Microhyla* are apparently part of a complex of isolating mechanisms to prevent interbreeding where sympatry exists. Here are just three of the possible responses that a population can make, each having a selective advantage but moving in response to different factors.

Animals captured only a month apart show significant differences. How much of this very rapid change is genetic cannot be immediately determined. Anderson (1966) analyzed six initially identical populations of *Drosophila pseudoobscura* kept at three different temperatures for seven years. After 1½ years he found significant differences in morphology that were not genetic, while after six years there were striking genetic differences in his populations. Some of these morphological changes must be the result of the plasticity of already present gene sets responding to environmental change. The extremely high selective pressure on the factors measured in the population would certainly produce genetic changes when the phenotype was the result of any appreciable heritability. Where one phenotype is favored in one area and another phenotype in a neighboring area, a single gene in question may be expected to show a frequency cline in the area of the boundary (Haldane, 1948). We may also expect a cline to occur where the boundary is chronological rather than geographic. Between 1961, the driest year, and 1962 there are large morphological changes, and following these more gradual change. Such a phenomenon could represent a purely, or mostly, non-genetic morphological response followed by genetic assimilation.

The results of the canonical correlation indicate that a number of environmental factors are significantly correlated with morphology. Those factors having the highest correlation suggest that there are at least two periods which are critical in terms of selection of parents for the next generation. The adults in the population developed and metamorphosed the previous March and April, and there are extremely high correlations between rainfall in these months and morphological characteristics the following year. Large numbers of metamorphosing individuals die at this time, any resistance to desiccation is only one of many possible modes of selection. Less rain means smaller pools and streams with a consequent result of crowding. This crowding could be the critical factor. Also in a dry year there is less food. Finally, there may remain any number of undetermined factors. The rainfall at this period appears critical; its method of action remains in doubt.

Other significant factors which may be critical are the high and low temperatures the fall preceding capture. In the area of collection September and October are generally the warmest months of the year, while late October and November often bring the year's coldest temperatures. Average temperatures do not appear to affect the population as much as a few extremely hot or cold days. Jameson (1966) suggested that the rate of weight loss in *Hyla regilla* under stress conditions may be due to differences in genetic constitutions rather than simple physiological adjustments to environment. If this is true, the few days that were hot enough or cold enough to cause stress may also have resulted in very high levels of selection for those animals with higher stress thresholds.

During one 16-month period June 1960 to December 1961, extremes of heat, cold and dryness occurred, all of which give significant correlations and thus appear to be critical factors. The most striking difference is in finger length and toe length as shown by the table of means (Table 1) and the plots of the first two discriminant axes (Fig. 1). In comparing the 1961 frogs with later samples it can be seen that the larger animals have slightly smaller finger lengths. The change is more accurately reflected by the finger length/toe length ratio which is 1.03 in 1961 and .725 in June, 1962. The cause of such a change is not at once apparent. Both heat and moisture regulation may be important. The two factors must strike some sort of balance, for the animal that loses heat the fastest also loses moisture the fastest.

The vectors of Table 3 indicate that allometry has not remained constant in the population. The time of year and consequent age of the population may be more closely related to these changes, however, than the fluctuations in environment. All samples captured during February and March regardless of year have similar vectors, while those taken in January and February differ only slightly. The June sample, representing older frogs that have survived an entire breeding season, differs substantially from all others. Growth of overall body size in relation to other characteristics has slowed while relative shank length has increased. If the changes in relative growth rates reflected shifts in environment, a difference in the ratios of the 1961 frogs might be expected corresponding to those noted in the other methods of analysis. Toe length is again the greatest exception. Certainly some of the changes in growth ratios may be ascribed to environmental variation, but it is probably overlaying a recurrent seasonal variation in allometry.

Calhoon and Jameson (1969), using canonical correlation to test members from populations of *H. regilla* in widely diverging environments against environmental data, revealed that rainfall during the months of March and July were major selective factors. Toe length and snout-vent length were positively correlated with rainfall while head width was negatively correlated on a geographic basis. These results are comparable to those obtained in this study.

An attempt to determine the nature of the morphological change involves several problems. First, can the population respond genetically in such a short period? The initial response between 1961 and 1962 suggests a large non-genetic component. What of later developments? Three years are equivalent to three generations in this species. Is this time enough? Many workers have described seasonal genetic variation in *Drosophila* which is apparently in response to changes in environment, and genetic selection in agricultural organisms is certainly apparent in three generations. Misra and Reeve (1964) noted that in intermediate latitudes the winter members of a population of *Drosophila subobscura* resemble more northern types, while the summer members resemble the southern animals. This is a case of regular and rapid response to changes in environment. If the pressure is strong enough, as it may be during this unusually dry, warm period, there is no reason to suppose that *H. regilla* would be any less capable of rapid genetic response.

Second, the nature of the selective agents involved is difficult to elucidate. At what point in the time continuum are these animals, and in which directions are they moving? Within the period of study there is no way to determine if the animals reacted initially to the extreme conditions of 1961 and were moving back toward the old adaptive peaks the following two years or if they were continuing to move genetically in the direction that they had moved morphologically in that first year. The several years preceding the study were drier than normal, though none approached the severity of that year. Misra and Reeve (1964) pointed out that the fact that different types or species react differently to the same selective agencies in *Drosophila* suggests that the selective agent itself may not be temperature, but some factor correlated with temperature such as available food. Even though very high correlation exists between precipitation during metamorphosis and changes in morphology, and between temperature extremes the previous fall and changes in morphology, it cannot automatically be assumed that these factors are the direct agents of the selection.

A third difficult problem is the determination of just how much of the changes are genetic in origin. Breeding tests would be required to resolve this question, and *H. regilla* is a very difficult animal to raise to sexual maturity. We can say that the samples are statistically different, and that because this difference extends between years, that all of it is not due to the fact that animals captured in different months are of different ages. We also suspect that a part of the difference between months is the result of the operation of selection and the result of age differential. Finally, those differences which are due to selection, whether within or between years, are strongly correlated with environmental factors of rainfall and temperature. Snout–

vent length and finger and toe lengths distingush the two 1961 samples from later ones (Fig. 1). The exact selective significance of these characteristics is not clear, but it appears that rounder frogs are correlated with increased rainfall, and drier weather is correlated with relatively more elongate animals.

ACKNOWLEDGMENTS

Drs. A. Romano and F. Awbrey have read the manuscript. The computer programs were developed by D. L. Jameson with considerable assistance from the San Diego State College Computer Center and special thanks are due E. Bauer and R. Bacon. The stepwise multiple regression was done on the University of Houston SDS Sigma with a program modified from Effroymson by Dr. Walt Sadlick. This paper is dedicated to Dr. Th. Dobzhansky.

LITERATURE CITED

ALLEE, W. C. *et al.* 1950. Principles of animal ecology. W. B. Saunders Co., Philadelphia.

ANDERSON, W. W. 1966. Genetic divergence in M. Vetukhiv's experimental populations of *Drosophila pseudoobscura.* Genet. Res. Camb. 7:255–266.

BLAIR, W. F. 1955. Size differences as a possible isolation mechanism in *Microhyla.* Am. Nat. 84:297–302.

——— AND M. J. LITTLEJOHN. 1960. Stage of speciation of two allopatric populations of chorus frogs (*Pseudacris*). Evolution 14:82–87.

BRYAN, J. G. 1951. The generalized discriminant function: mathematical foundation and computational routine. Harvard Ed. Rev. 21 (2):90–95.

CALHOON, R. E. AND D. L. JAMESON. 1970. Canonical correlation between variation in weather and variation in size of the Pacific tree frog, *Hyla regilla,* in southern California. Copeia 1970(1):124–134.

CATTELL, R. B. 1965. Factor analysis: an introduction to essentials. Biometrics 21(2):190–210, 405–435.

COOLEY, W. W. AND P. R. LOHNES. 1962. Multivariate procedures for the behavioral sciences. John Wiley and Sons, New York.

DOBZHANSKY, T. H. 1943. Genetics of natural populations. IX. Temporal changes in the composition of populations of *Drosophila pseudoobscura.* Genetics 28:162–186.

——— AND H. LEVENE. 1955. Genetics of natural populations. XXIV. Homeostasis in natural populations of *Drosophila pseudoobscura.* Genetics 40:797–808.

FISHER, R. A. 1936. The use of multiple measurements in taxonomic problems. Ann. Eug. 7:179–188.

HALDANE, J. B. S. 1948. The theory of a cline. J. Genet. 48:277–284.

HAMILTON, T. H. 1961. The adaptive significance of intraspecific trends of variation in wing length and body size among bird species. Evolution. 5(2):180–195.

HOTELLING, H. The most predictable criterion. J. Ed. Psych. 26:139–142.

———. 1936. Relations between two sets of variates. Biometrika 28:321–377.

JAMESON, D. L. 1966. Rate of weight loss of tree frogs at various temperatures and humidities. Ecology 47:605–613.

———, JAMES P. MACKEY, AND ROLLIN C. RICHMOND. 1966. The systematics of the Pacific tree frog, *Hyla regilla.* Proc. Calif. Acad. Sci. 33(19):551–620.

JOLICOEUR, P. 1963. The degree of generality of robustness in *Martes americana.* Growth 27:1–27.

——— AND J. E. MOSIMANN. 1960. Size and shape variation in the painted turtle. A principal component analysis. Growth 24:339–354.

LEWONTIN, R. C. 1965. Selection in and of populations. *In*: Ideas in modern biology. John A. Moore, ed., pp. 297–311. The Natural History Press, Garden City, N. Y.

MISRA, R. K. AND E. C. R. REEVE. 1964. Clines in body dimensions in populations of *Drosophila subobscura.* Genet. Res. Camb. 5:240–256.

RULON, P. 1951. Distinctions between discriminant and regression analyses and a geometric interpretation of the discriminant function. Harvard Ed. Rev. 21(2):80–90.

SNYDER, W. F. AND D. L. JAMESON. 1965. Multivariate geographic variation of mating call in populations of the Pacific tree frog (*Hyla regilla*). Copeia 1965(2):129–142.

TEST, F. C. 1898. A contribution to the knowledge of the variation of the tree frog, *Hyla regilla.* Proc. U.S. Nat. Mus. 21(1156):477–492.

THORSON, T. B. 1955. The relationship of water economy to terrestrialism in amphibians. Ecology 36:100–116.

WRIGHT, S. AND T. H. DOBZHANSKY. 1946. Genetics of natural populations. XII. Experimental reproduction of some of the changes caused by natural selection in certain populations of *Drosophila pseudoobscura.* Genetics 31:125–156.

MEDICAL CENTER, UNIVERSITY OF CALIFORNIA, SAN FRANCISCO 94122, AND DEPARTMENT OF BIOLOGY, UNIVERSITY OF HOUSTON, HOUSTON, TEXAS 77004.

24

Reprinted from *J. Geol.*, 77, 303–318 (1969)

THE THEORY AND APPLICATION OF CANONICAL TREND SURFACES[1]

P. J. LEE[2]

Department of Geology, McMaster University, Hamilton, Ontario, Canada

ABSTRACT

The theory of canonical correlation analysis has been combined with that of trend-surface analysis in order to construct a multivariate trend surface that is called a canonical trend surface. A canonical trend surface is a parsimonious summarization of areal variations of a set of geologic variates. This trend has a property of maximum correlation between variates and geographic coordinates. It does not show the absolute value of each variate, but it shows the nature of variation of a linear function of the variates. By use of this type of trend, it is possible to reveal the underlying pattern of geographic variation common to a set of variates. The Permian system in western Kansas and eastern Colorado was studied as a numerical example to illustrate the general procedures in solving practical problems and also to demonstrate the validity of this technique.

INTRODUCTION

REVIEW OF MULTIVARIATE TREND-SURFACE ANALYSIS

Investigations may be performed to analyze a set of geologic variates measured on a stratigraphic unit or a rock body regionally or locally. What geologists would like to do is to filter the error variance from systematic areal variance in order to evaluate the trend of each variate. The technique (trend analysis) that has been used frequently in geology is basically the fitting of a polynomial surface to the observed data of a single variate, by the principle of least squares. In geology, where most problems are beset with a highly multiple determination of events, conventional trend analysis is essentially a univariate technique, which is not adequate to handle the inherently multivariate data.

The concept of a composite end member in facies analysis was demonstrated by Krumbein (1955). Three predetermined end members were used to construct a facies triangle. Selection of a particular point within this facies triangle was considered as an optimum combination of end members. Contour lines around this point were used to measure the deviation of facies from the optimum facies. The deviation, called the distance function, as Forgotson (1960) pointed out, does not distinguish end members or give information on the absolute values of end members. It does not indicate the nature of change in composition from the optimum facies.

An entropy function derived from information theory was applied in multicomponent system mapping (Pelto, 1954) to express the degree of intermixing of end members and to define facies quantitatively with one set of contours. This technique does not distinguish between end members and does not show change of each individual variate.

The D-function (Pelto, 1954) divides a system into classes based on difference in amounts of components. It provides information on the relative proportion of a specific end member within its own class, but does not distinguish between end members or provide information on the absolute value of components.

One approach to spatially distributed multivariate data is to analyze by principal component analysis and then map each component by trend analysis. This technique provides information on the areal distribution of the principal component which is composed of a set of geologic variates with different weights. Two methods may be used in this approach.

[1] Manuscript received June 5, 1968; revised September 9, 1968.

[2] Present address: Fisheries Research Board of Canada, Freshwater Institute, Winnipeg 19, Manitoba, Canada.

1. A factor analysis (for example, maximum likelihood solution; Lawley and Maxwell, 1963; Jöreskog, 1967) or principal component analysis is performed first on outcrop or well data, and factor measurements or principal components are calculated for each sampling point. Then trend surfaces are fitted to factor measurements or principal components over all sampling points. The result is a factor measurement or principal component map. If a factor model is used, the factor map would probably imply lithologic or mineralogic associations resulting from underlying environmental factors. If based on the concept of principal component analysis, the first principal component map indicates the spatial variation of the first principal component, which is the component that explains the most variance of a set of geologic variates. In both models, the factor or principal component may or may not have a geologic interpretation.

2. A factor-vector map is obtained from a Q-mode factor analysis as suggested by Imbrie (1963) and Krumbein and Imbrie (1963). Reference wells are chosen for each factor-vector map, and all other wells are expressed in terms of similarity to the reference well for that particular map. The geologic meaning of each map can be obtained only by looking at the particular lithologic association of each reference well. The concept of a reference well is analogous to the concept of an optimum facies in the distance-function technique (Krumbein, 1955). This type of factor-vector map expresses the intergradation between the reference wells, in which respect it is similar to the concept of an entropy function map (Pelto, 1954).

PREVIOUS WORK ON CANONICAL CORRELATION

Suppose we have a number of rock specimens taken from a rock body and that the abundance of different types of minerals, major elements, trace elements, fossils, and so on, are estimated from each specimen. It may be determined whether relationships are present between the major and trace elements and clay minerals, or between sediment types and fossils.

For example, a simple correlation coefficient indicates a possible relation between one trace element and one clay mineral. If there are five clay minerals and ten trace elements, it will take fifty simple correlation coefficients to show all possible relationships. The relationship indicated by fifty coefficients is not easily understood; therefore, a technique is needed which will express the relationship in a more concise form.

Canonical correlation analysis was introduced by Hotelling in 1936 (the paper was read in 1935) as a technique to understand the relationship between two sets of variables. Relationships between the two sets of variables are summarized and expressed by a simple index. For example, the clay mineral and trace element abundances are weighted and combined into two linear functions such as $U = a_1$ (kaolinite) $+ a_2$ (illite) $+ a_3$ (chlorite) $+ \ldots$ and $V = b_1$ (vanadium) $+ b_2$ (zirconium) $+ b_3$ (titanium) $+ \ldots$, where $a_1, a_2, a_3, \ldots$ and $b_1, b_2, b_3, \ldots$ are weights to be determined such that U and V have a maximum simple correlation coefficient that is called the canonical correlation coefficient or canonical root. The strength of the link between the sets is measured by the canonical root, whereas the nature of the link is indicated by those trace elements and clay minerals having larger weights.

The principle of canonical correlation analysis was used immediately as a procedure for discriminatory analysis in several populations (Bartlett, 1938; Wilks, 1963, p. 576–581). Canonical correlation was generalized for three or more sets of variables and partial canonical correlation also was proposed by Roy (1957, p. 26). Canonical factor analysis (Rao, 1955) was developed by making use of the canonical correlation principle. Horst (1961*a*, *b*) has applied the same principle to the problem of matching two or more factor patterns. The probability function of the canonical roots has been studied by a number of mathematical statisticians (Bartlett, 1941, 1947*a;* Hsu, 1941; Constantine and James, 1958). Uses of this technique have been suggested in different fields: psychology (Bartlett, 1947*b*, 1948; Thomson, 1947; Burt, 1948; Hotelling, 1957; Beech and Maxwell, 1958; Kendall, 1961, p. 75–85; Maxwell, 1961; Meredith, 1964; Das, 1965; Dunteman, 1967), biology (Pearce and Holland, 1960; Kshirsagar, 1962; Seal, 1964, p. 123–152; Bartlett, 1965, p. 201–224; Cassie and Michael, 1968), economics (Waugh, 1942; Tintner, 1946; Morrison, 1967, p. 207–220), sociology (Cooley and Lohnes, 1962, p. 31–49; Koons, 1962), and meteorology (Glahn, 1968). In geology, Reyment (1963, 1966) and Buzas (1966) followed Bartlett's idea to use the principle of canonical correlation to solve discriminatory problems in several populations. Comparison of canonical correlation analysis and principal component analysis was discussed thoroughly with geologic data by Lee (1968). Other applications of canonical correlation in geology have been explained with illustrative geologic examples (Lee, 1968); namely, matching two factor patterns, Q-technique canonical correlation, and discriminatory analysis.

CONCEPT OF CANONICAL TREND

Suppose that a number of rock specimens have been collected from an area and that each specimen has its own geographic locality. After estimation of trace elements, for instance, we would like to know the areal variation of each trace element over the area studied. If we had ten trace elements, then we will have ten polynomial surfaces or trend maps indicating variations of trace elements. It may be asked whether it is possible to evaluate a single trend that is common to all or some of the trace elements. Again, we encounter the same problem as in the previous example; that is, a summarization of the ten maps is needed.

A canonical trend is a polynomial surface that shows the variations of many variables over an area simultaneously by making use of the principle of canonical correlation analysis. One set of variables consists of geologic variables such as trace elements, whereas the second set of variables is composed of location of specimens, that is, (X,Y)-coordinates. The canonical trend is a succinct summarization of areal variations of many geologic variables (Lee and Middleton, 1967; Middleton and Lee, 1967).

A quadratic surface which is a homogeneous expression of a second degree is reduced to a linear function of squares only, the cross-product terms being eliminated. A form of this type is said to be canonical form. This reduction process is also called a canonical analysis in chemical engineering (Hill and Hunter, 1966). This canonical reduction of chemical engineers is algebraically equivalent to the canonical analysis of Hotelling's method, but the underlying purposes, assumptions, and implications of the canonical reduction are completely different from that of the canonical trend analysis discussed in this paper.

THEORY OF CANONICAL TREND SURFACES

GENERAL STATEMENT

Suppose we have a sample from a p-dimensional space; then the purpose of canonical correlation analysis (Hotelling, 1936) is to find a linear function of the first p_1-variates and a linear function of the last p_2-variates ($p_1 + p_2 = p$) so that these two linear functions have the highest possible correlation coefficient. If the canonical correlation is zero, these two sets are completely independent, and it is useless to predict the dependent variates by means of the independent variates. If the canonical correlation is unity, this means that the dependent variates would be predicted perfectly by means of the independent variates based on the particular functions.

The geometrical meaning of the canonical correlation can be stated as follows: In a p-dimensional space, a sample of $p_1 + p_2$ variates determines one hyperplane of p_1 and one of p_2 dimensions, intersecting at the origin and containing a swarm of points representing the two sets. Linear transformations are developed for the first p_1 coordinate axes and also for the p_2 coordinate axes such that these two hyperplanes are as parallel as possible in a new p-dimensional space. The cosine of the angle between these two hyperplanes is defined as the canonical correlation coefficient.

The assumption is that the observed variates are linear functions of the canonical variates. Furthermore, it is assumed that the observed variates are distributed normally in order to derive the probability distribution of the canonical correlation coefficients and to make a statistical inference on the dependence between two sets.

CANONICAL CORRELATION AND VARIATES

Suppose (z_{ij}, $i = 1, 2, \ldots, p$; $j = 1, 2, \ldots, N$, $N > p$) is a random sample of size N from a p-dimensional distribution and has sample covariance matrix $\mathbf{S}$, the unbiased estimate of population covariance matrix, which is known to be a positive definite, real symmetric matrix. Let $\mathbf{Z}$ be geologic variates and (X,Y)-coordinates (plus cross-product and high-order terms); without loss of generality we may suppose that $\mathbf{Z}_i$ has zero mean, that is, $E[\mathbf{Z}_i] = 0$.

We partition $\mathbf{Z}$ into two subvectors, $\mathbf{Z}_1$ and $\mathbf{Z}_2$, in which $\mathbf{Z}_1$ is a matrix of p_1 geologic variates and $\mathbf{Z}_2$ is a matrix of geographic coordinates, or vice versa:

$$\mathbf{Z} = \begin{bmatrix} \mathbf{Z}_1 \\ \mathbf{Z}_2 \end{bmatrix},$$

$$\text{where } \mathbf{Z}_1' = (z_1, \ldots, z_{p_1}), \qquad \mathbf{Z}_2' = (z_{p_1+1} \ldots, z_p). \tag{1}$$

For convenience we shall assume $p_1 \le p_2$. The covariance matrix is partitioned into matrices as follows:

$$\mathbf{S} = \left[\begin{array}{c|c} \mathbf{S}_{11} & \mathbf{S}_{12} \\ \hline \mathbf{S}_{21} & \mathbf{S}_{22} \end{array}\right], \tag{2}$$

where $\mathbf{S}_{11}$ is the covariance matrix for $\mathbf{Z}_1$, $\mathbf{S}_{22}$ for $\mathbf{Z}_2$, and $\mathbf{S}_{12} = \mathbf{S}_{21}'$ is the covariance matrix between $\mathbf{Z}_1$ and $\mathbf{Z}_2$. The canonical variates U and V are defined as:

$$U = \mathbf{A}'\mathbf{Z}_1, \qquad V = \mathbf{B}'\mathbf{Z}_2, \tag{3}$$

where

$$\mathbf{A}' = (a_1, \ldots, a_{p_1}), \qquad \mathbf{B}' = (b_1, \ldots, b_{p_2}). \tag{4}$$

We require $\mathbf{A}$ and $\mathbf{B}$ to be such that U and V have unit variance, that is,

$$1 = E[U^2] = E[\mathbf{A}'\mathbf{Z}_1\mathbf{Z}_1'\mathbf{A}] = \mathbf{A}'\mathbf{S}_{11}\mathbf{A}, \tag{5}$$

$$1 = E[V^2] = E[\mathbf{B}'\mathbf{Z}_2\mathbf{Z}_2'\mathbf{B}] = \mathbf{B}'\mathbf{S}_{22}\mathbf{B}, \tag{6}$$

$$E[UV] = COV[UV] + E[U]E[V] = \mathbf{A}'\mathbf{S}_{12}\mathbf{B}. \tag{7}$$

Thus, the problem is to find $\mathbf{A}$ and $\mathbf{B}$ to maximize (7) or $E[UV]$ subject to (5) and (6). Let

$$\phi = \mathbf{A}'\mathbf{S}_{12}\mathbf{B} - \tfrac{1}{2}\mu(\mathbf{A}'\mathbf{S}_{11}\mathbf{A} - 1) - \tfrac{1}{2}\lambda(\mathbf{B}'\mathbf{S}_{22}\mathbf{B} - 1), \tag{8}$$

where μ and λ are Lagrange multipliers. We differentiate ϕ with respect to the variables of $\mathbf{A}$ and $\mathbf{B}$. The vectors of derivatives set equal to zero and manipulations of matrix algebra will show that

$$\lambda = \mu = \mathbf{A}'\mathbf{S}_{12}\mathbf{B} \tag{9}$$

and

$$(\mathbf{S}_{11}^{-1}\mathbf{S}_{12}\mathbf{S}_{22}^{-1}\mathbf{S}_{21} - \Lambda^2 I)\mathbf{A} = 0. \tag{10}$$

The solution involves finding latent roots, λ_i^2, of the equation

$$|\mathbf{S}_{11}^{-1}\mathbf{S}_{12}\mathbf{S}_{22}^{-1}\mathbf{S}_{21} - \Lambda^2 I| = 0. \tag{11}$$

The matrix, $\mathbf{S}_{11}^{-1}\mathbf{S}_{12}\mathbf{S}_{22}^{-1}\mathbf{S}_{21}$, is $p_1 \times p_1$ in dimension. Thus, Λ^2 is a $p_1 \times p_1$ diagonal matrix.

From (9) we see that $\lambda = \mathbf{A}'\mathbf{S}_{12}\mathbf{B}$ is the correlation between U and V. Thus, the elements, λ_i, of Λ were called the canonical roots or canonical correlation coefficients by Hotelling (1936). The value of λ_i ranges from zero to $+1$. A negative relation between two sets cannot be realized by just looking at the canonical root, but it will be shown by the signs of coefficients of the canonical variates. The value of λ_i^2 is the variance (of a linear function of geologic variates with unit variance) explained by corresponding linear function, which is the polynomial surface. Values of $\mathbf{A}_1$ in equation

(3) are the eigenvectors associated with λ_1^2. Solving for B_1 we have

$$B_1 = S_{22}^{-1}S_{21}A_1/\lambda_1(\lambda_1 \neq 0) . \quad (12)$$

We now consider finding the second linear functions of Z_1 and Z_2, respectively, such that each of these two linear functions has a maximum correlation and is uncorrelated with the first linear functions. This procedure is continued. At the rth step we have obtained linear functions $U_1 = A_1'Z_1$, $V_1 = B_1'Z_2, \ldots, U_r = A_r'Z_1$, $V_r = B_r'Z_2$, with corresponding roots $\lambda_1, \ldots, \lambda_r$ subject to the conditions:

$$0 = E[U_iU_j] = E[A_i'Z_1Z_1'A_j] = A_i'S_{11}A_j , \quad (13)$$

$$0 = E[V_jV_i] = B_j'S_{22}B_i , \quad (14)$$

$$0 = E[V_jU_i] = B_j'S_{21}A_i , \quad (15)$$

for any i and j $(i, j = 1,2, \ldots, r, i \neq j)$.

Consider approximating U by a multiple of V, say KV; then the mean square error of approximation is

$$E[(U - KV)^2] = s_u^2(1 - r) + (Ks_v - rs_u)^2 , \quad (16)$$

where the s_u^2 and s_v^2 are sample variances of U and V, respectively, and r is the simple correlation coefficient between U and V. This is minimized by taking $K = s_ur/s_v$. We can consider KV as a linear prediction of U from V; then $s_u^2(1 - r)$ is the mean square error of prediction.

Detailed discussion of canonical correlation is given by Hotelling (1936), Anderson (1958, p. 288–306), and Wilks (1963, p. 587–590).

A criterion which is useful in detecting the simultaneous departure of several roots λ_i^2 from zero was suggested by Bartlett (1938, 1941). Bartlett's statistic

$$\chi^2 = -[N - \tfrac{1}{2}(p_1 + p_2 + 1)] \log_e L , \quad (17)$$

where

$$L = \prod_{i=r}^{p_1} (1 - \lambda_i^2)$$

follows approximately a χ^2 distribution with $(p_1 - r + 1)$ $(p_2 - r + 1)$ degrees of freedom. The assumption is made that Z_1 and Z_2 follow a multivariate normal distribution with zero means.

A slightly different statistic, which was suggested by Lawley (1959) for the test of significance of residual canonical roots, is that the multiplying factor of (17) is replaced by taking the factor as

$$N - r - \tfrac{1}{2}(p_1 + p_2+1) + \sum_{i=1}^{r}(1/\lambda_i^2) . \quad (18)$$

If λ_i's are equal to 1, Bartlett's criterion and Lawley's criterion are identical. If the sample size N is large, these two criteria are approximately equal. In the present study, Bartlett's criterion is used for reference but not as the sole basis for making critical decisions.

COMPUTATION PROCEDURES

Step 1. Transformation of geologic variates, if necessary.—Transformations may be carried out in order to stabilize variance of geologic variates. Bartlett (1947*c*) has summarized those transformations appropriate to a particular situation and also suggested a general transformation.

There are three alternatives allowed in the computer program prepared (Lee, 1968): (1) arc sine square root transformation; (2) logarithm transformation; or (3) no transformation.

Step 2. Standardization of geologic variates.—A correlation coefficient is a dimensionless value; therefore a trend-surface equation derived from a correlation matrix is also scale-independent. To have a scale-dependent trend-surface equation, the covariance matrix for the (X,Y)-coordinates and geologic variates, z_i, should be used. On the other hand, geologic variates may be measured in non-comparable units. Thus, we should deal with the correlation matrix rather than the covariance matrix. In order to satisfy both of the two requirements, the geologic variates are standardized according to equation (19), and then the covariance matrix is computed for z_i and the (X,Y)-coordinates.

The equation for standardization of the geologic variates is as follows:

$$Z_{ij} = \frac{z_{ij} - \bar{z}_j}{\sqrt{s_j^2}}, \quad (19)$$

where s_j^2 = sample variance of variate z_j, $\bar{z}_j$ = sample mean of variate z_j, and z_{ij} = the ith observation on the jth variate.

Step 3. Generation of high power and cross-product terms of the (X,Y)-coordinates.—The high power and cross-product terms of the (X,Y)-coordinates are generated within the computer based on the input raw data (X,Y) as follows:

2	X	Y					
5	X^2	XY	Y^2				
9	X^3	X^2Y	XY^2	Y^3			
14	X^4	X^3Y	X^2Y^2	XY^3	Y^4		
20	X^5	X^4Y	X^3Y^2	X^2Y^3	XY^4	Y^5	
27	X^6	X^5Y	X^4Y^2	X^3Y^3	X^2Y^4	XY^5	Y^6.

Step 4. Computation of covariance matrix.—The covariance between ith and jth variates is calculated by the standard formula which is defined as

$$S_{ij}^2 = \frac{N\sum_{K=1}^{N}(Z_{iK}Z_{jK}) - \sum_{K=1}^{N}Z_{iK}\sum_{K=1}^{N}Z_{jK}}{N(N-1)}, \quad i \neq j. \quad (20)$$

If $i = j$, S_{ii}^2 is the variance of the ith variate.

The covariance matrix is partitioned into four parts, as in equation (2). In equation (2), the order of $\mathbf{S}_{11}$ is less than that of $\mathbf{S}_{22}$; $\mathbf{S}_{11}$ and $\mathbf{S}_{22}$ are covariance matrices of geologic variates and polynomial terms, respectively, if $p_1 \leq p_2$. The contents of $\mathbf{S}_{11}$ and $\mathbf{S}_{22}$ should be exchanged if $p_1 > p_2$.

Step 5. Calculation of canonical roots and variates.—The canonical roots λ_i's and their associated canonical variates (A_i) for variates in matrix $\mathbf{S}_{11}$ are, respectively, the eigenvalues and eigenvectors of the matrix equation (11).

The matrix on the left side of (11) is non-symmetric $(p_1 \times p_1)$. A Jacobi-like method (Eberlein, 1962) was used to solve for the p_1 roots, λ_i^2's. The eigenvectors are $\mathbf{A}_i$, each of which is the canonical variate for one set of variates, the canonical variate either for the geologic variates or for the (X,Y)-coordinates. The $\mathbf{A}_i$ are normalized, that is, $\mathbf{A}_i'\mathbf{A}_i = 1$.

Step 6. Determination of degree of polynomial surface.—We begin with a linear polynomial and evaluate the degree through canonical correlation analysis. If the canonical root was greater than 0.95 or if the difference between two successive roots is less than 0.05 (the initial canonical root is set at zero), then the canonical root corresponding to the degree of the polynomial is used as the canonical trend surface. If not, the degree of the polynomial is increased by 1, and steps 5 and 6 are repeated until the iterative process reaches a suitable degree.

It should be kept in mind that the sample size should always be greater than the sum of the number of the geologic variates and the number of polynomial terms. Suppose we have a sample of size 30 with 4 variates; the highest order of polynomial that may be fitted to this data is 5, because a sextic polynomial has 27 terms. In this example, for a sextic, the sum (=31) of the number of variates (=4) and the number of polynomial terms (=27) is greater than the sample size.

Step 7. Calculation of the canonical variate of the second set of variates.—The canonical variate of the second set is computed as equation (12). The $\mathbf{B}_i$ is normalized, that is, $\mathbf{B}_i'\mathbf{B}_i = 1$.

Step 8. Calculation of residuals.—Consider approximating U (a linear function of geologic variates) by a multiple of V (a linear function of the $[X,Y]$-coordinates) plus a constant, say $C + kV$; then k and c are defined as follows:

$$U = c + kV, \quad (21)$$

where

$$K = \frac{N\sum_{i=1}^{N}(V_iU_i) - \sum_{i=1}^{N}V_i\sum_{i=1}^{N}U_i}{N\sum_{i=1}^{N}V_i^2 - \left(\sum_{i=1}^{N}V_i\right)^2} \text{ and}$$

$$C = \frac{\sum_{i=1}^{N}V_i^2\sum_{i=1}^{N}U_i - \sum_{i=1}^{N}(V_iU_i)}{N\sum_{i=1}^{N}V_i^2 - \left(\sum_{i=1}^{N}V_i\right)^2} . \qquad (22)$$

The residual of canonical trend surface for the jth sample and the ith root is defined as

$$\begin{aligned}\text{Residual}_j &= U_j - (kV_j + c) \\ &= A_i'Z_j - (kV_j + c) ,\end{aligned} \qquad (23)$$

where Z_j is column matrix for standardized geologic variates; thus, $A_i'Z_j$ is called observed value, whereas $c + kV_j$ is called calculated value for U_j. The U and V should be exchanged; that is, A_i is replaced by B_i if U is a linear function of the (X,Y)-coordinates, while V is a linear function of geologic variates.

Step 9. Contouring of canonical trend surface.—In the canonical trend-surface map, the X-axis is the abscissa, whereas the Y-axis is the ordinate. In the computer printout, the length along the X-axis is greater than or equal to that of the Y-axis. The length of the X-axis is divided into 50 units, whereas the length of the Y-axis will be assigned a certain unit proportional to its relative length. Each point (having integer coordinates) of the new coordinate system is substituted into the polynomial equation $V = f(X,Y)$, so we have a set of V_{ij}, where $i, j = 1, 2, \ldots, 50, j \leq i$. The values, V_{ij}, are scaled according to equation $c + kV_{ij}$.

Suppose U is a linear function of geologic variates; then the calculated value for U is $KV + c$, as mentioned above. The difference between U_{max} and U_{min} (of all sampling points), ΔU, is divided into ten parts, and the values so defined are used for contours. The purpose of using ΔU rather than $\Delta V = V_{max} - V_{min}$ is to avoid extreme values introduced from a high order polynomial.

TWO-DIMENSIONAL CANONICAL TREND SURFACES

CANONICAL TREND SURFACES OF HYPOTHETICAL DATA

An illustration of some of the physical meaning of the canonical trend which results from the computations described in the previous sections can be given by making use of hypothetical examples. One hundred and fifty sampling points were spread over a rectangular area, the points being located by a stratified sampling method (Cochran, 1963, p. 87–88). Geographic coordinates of each sampling point within a grid were defined by a pair of random numbers. Values for the hypothetical variates at the sampling points were computed from equations incorporating an independent, random, normally distributed error term with zero mean and any desired variance.

1. The canonical trend surface achieves a parsimonious summarization of a set of trend surfaces showing distributions of variates from a single population.

Let two hypothetical variates be

$$z_1 = 0.707X + 0.707Y + E_1 \qquad (24)$$

and

$$z_2 = -0.707X - 0.707Y + E_2 , \qquad (25)$$

where E_1 and E_2 are independent variates with normal distribution $N(0,1)$. The canonical variates were that

$$U = -0.652z_1 + 0.758z_2 \qquad (26)$$

and

$$V = -0.693X - 0.721Y . \qquad (27)$$

The canonical root was 0.9834. Equation (27) is the polynomial of the canonical trend surface. Equation (26) represents the linear function of variates z_1 and z_2. The square values, -0.652^2 and 0.758^2 ($-0.652^2 + 0.758^2 = 1.000$), are the variances contributed by z_1 and z_2 to the trend, respectively. The variates z_1 and z_2 have approximately equal loadings but are in opposite signs, so that these two variates vary in approximately opposite directions. Equations (26) and (27) show that variate z_1 increases toward the northeast, while variate z_2 increases toward the southwest. This interpretation agrees with equations (24) and (25).

TABLE 1

RELATIONSHIP BETWEEN CANONICAL ROOT AND RANDOM ERROR WITH $N(0,\sigma^2)$

λ_i	σ^2
1.0000	0.0
0.9995	0.1
0.9873	0.5
0.9649	1.0
0.9222	1.5
0.8673	2.0
0.8126	2.5
0.7396	3.0
0.6751	3.5
0.6817	4.0

2. The coefficients of geologic variates are a function of the strength of the trend.

Let two hypothetical variates be

$$z_1 = 0.707X + 0.707Y + E_1 \quad (28)$$

and

$$z_2 = E_2 , \quad (29)$$

where E_1 was $N(0,4)$ and E_2 was $N(0,1)$. The canonical variates were that

$$U = 0.986z_1 - 0.196z_2 \quad (30)$$

and

$$V = 0.954X - 0.284Y - 0.014X^2 + 0.025XY + 0.088Y^2 . \quad (31)$$

The canonical root was 0.6967. It is obvious that variate z_2 contributes very little to the trend.

3. If the information (areal variations of the variates) cannot be obtained from the first canonical root alone, the second canonical root will supply part of the remaining information.

Let five constructed variates be

$$z_1 = X + E_1 , \quad (32)$$

$$z_2 = Y + E_2 , \quad (33)$$

$$z_3 = -X + E_3 , \quad (34)$$

$$z_4 = -Y + E_4 , \quad (35)$$

and

$$z_5 = E_5 , \quad (36)$$

where E_i ($i = 1,2,3,4,5$) were $N(0,1)$. The first canonical root was 0.9887; its associated canonical variates were that

$$U = 0.622z_1 - 0.163z_2 - 0.763z_3 + 0.048z_4 + 0.034z_5 \quad (37)$$

and

$$V = 0.974X - 0.228Y . \quad (38)$$

The second canonical root was 0.9701; its associated canonical variates were that

$$U = -0.098z_1 + 0.527z_2 - 0.272z_3 - 0.799z_4 - 0.005z_5 \quad (39)$$

and

$$V = 0.101X + 0.995Y . \quad (40)$$

The first canonical root explains the variations of variates z_1 and z_3, whereas the second canonical root explains the variations of variates z_2 and z_4. The variate z_5, does not contribute to the trend.

4. The value of the canonical root is a function of the magnitude of random error.

Suppose a hypothetical variate has a variation according to equation (28), where E_1 has a zero mean and variance σ^2 which ranges from 0.0 to 4.0. The relationship between canonical root for a linear trend and σ^2 is listed in table 1.

PERMIAN SYSTEM IN WESTERN KANSAS AND EASTERN COLORADO

Geologic setting.—The Permian system in western Kansas and eastern Colorado, which is above the Stone Corral dolomite, is composed of the upper part of the Lower Permian and lower part of the Upper Permian series. The rock sequence consists of alternating sandstone and shale with thin dolomite and evaporite. The stratigraphic sequence studied is equivalent to the upper part of intervals *B* and *C–D* in the central midcontinent region (Mudge, 1967).

A total of thirty-one wells were selected from this area and analyzed by W. T. Fox (see Krumbein, 1962) using conventional stratigraphic maps. Integration of stratigraphic maps shows that the four lithologic components—sandstone, shale, carbonate, and evaporite—thicken toward a center; that is, the area shows the characteristics of a sedimentary basin.

The Permian basin was a broad, shallow, fairly stable restricted marine basin. The basin was bounded on the west by the Front Range and Wet Mountains, from which feldspathic and quartz grains were derived. On the north and east the basin was bounded by low-lying land areas. There was a restricted connection with open sea to the south. The raw open data used in the analysis are from Krumbein (1962, table VI). This analysis is considered to be an illustrative example rather than a stratigraphic study.

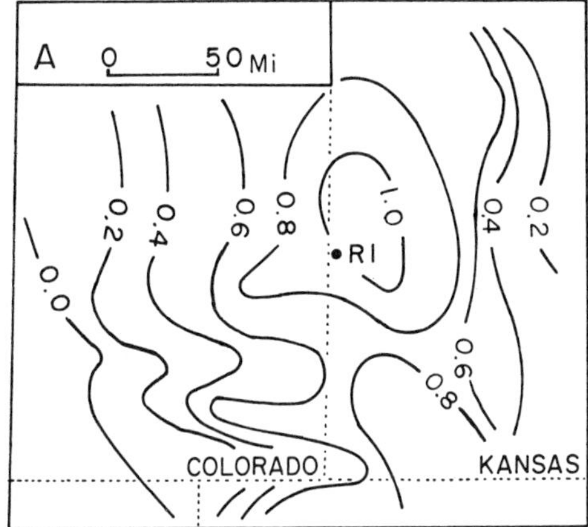

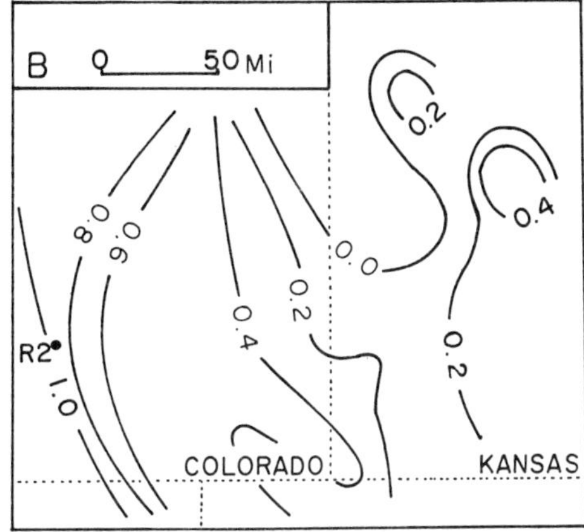

FIG. 1.—Stratigraphic factor-vector maps of the Permian system of western Kansas and eastern Colorado. *A*, factor 1. *B*, factor 2. (Data after Imbrie, 1963; redrawn by author.)

Stratigraphic factor-vector map analysis.—The thirty-one wells with six components—total thickness, non-clastic thickness, and thickness of the four components mentioned above—were subjected to factor-vector analysis. Three stratigraphic factor-vector maps were constructed (Imbrie, 1963), one for each of the first three factors. Geologic implications of the factor-vector analysis can be clarified by examination of the lithologic association of each reference well, which is defined as a well with an identity factor loading in a particular factor. The reference well of the first factor-vector map, which has maximum evaporite thickness and minimum carbonate thickness (fig. 1*A*), suggests an evaporite basin. The reference well of the second factor-vector map (fig. 1*B*), which has minimum thickness, shale content, and evaporite content, indicates distribution of shelf sedimentation. The reference well of the third factor-vector map, which has minimum sandstone and carbonate content and contains considerable shale and some evaporite, shows distribution of an offshore sedimentation of fine detrital material adjacent to the evaporite basin.

The fundamental approach of stratigraphic factor-vector map analysis is to visualize geologic implications of the distribution of a reference well through interpretations of lithologic associations of the reference well.

Canonical trend-surface analysis.—The hypothesis we expect to establish is that the area shows the characteristics of a sedimentary basin. A sedimentary basin, in general, should show that some of the sediments thicken toward the center and that the relative amount of shale and evaporite predominates over that of sand and carbonate, respectively, in the central part of the basin.

A sample of size 30 (excluding one well which did not contain any evaporite) was subjected to canonical trend analysis on two sets of different geologic variates. Variates of the first set are thickness of sand, shale, carbonate, and evaporite, whereas the variates of the second set are total thickness, sand:shale ratio, and carbonate:evaporite ratio. The order of polynomial fitted to these two sets of variates cannot be greater than 5.

TABLE 2

ROTATED AND NORMALIZED EIGENVECTORS OF SEDIMENT THICKNESS OF THE PERMIAN SYSTEM

VARIATE	PRINCIPAL COMPONENT		
	1	2	3
Sand	0.03	0.99	0.12
Carbonate	0.06	−0.12	−0.99
Shale	−0.99	−0.10	−0.10
Evaporite	−0.97	0.06	0.23

Principal components were obtained from a covariance matrix of the thickness of sand, shale, carbonate, and evaporite. The first principal component contributes 78.8 per cent of total variance, the second one contributes 20.6 per cent, and the third contributes 0.6 per cent. The rotated and normalized eigenvectors are listed in table 2.

The first principal component indicates a shale-evaporite association which is equivalent to the first factor obtained by Imbrie (1963). The second principal component explains the variance contributed by the thickness of sand, whereas the third principal component explains the variance contributed by the thickness of carbonate. The result from the principal component analysis indicates that the carbonate behaves differently from the evaporite and the sand behaves differently from shale and evaporite. Thus, the clastic ratio (=[sand + shale]/[carbonate + evaporite]) should not be included in the analysis, because its geologic meaning is ambiguous.

The first set of four variates—thickness of sand, shale, carbonate, and evaporite—was treated by canonical trend-surface analysis. After three iterations, the highest canonical root obtained was 0.9464, which indicates that a cubic polynomial is the most predictable response surface (table 3).

The canonical variate for the geologic variates was that

$$U = 0.516\ (\text{sand}) + 0.408\ (\text{shale}) - 0.104\ (\text{carbonate}) + 0.746\ (\text{evaporite})\,. \quad (41)$$

The canonical variate for the polynomial was that

$$V = -0.407X - 0.914Y + 0.002X^2 - 0.001XY - 0.001Y^2\,. \quad (42)$$

The canonical trend, equation (41), indicates the variation in thickness of evaporite, sand, and shale. The variance of the linear function (41), explained by the linear function (42), is equal to 91 per cent, the square of 0.9464.

TABLE 3

RECORD OF SUCCESSIVE EVALUATION OF ORDER OF POLYNOMIAL FOR FOUR VARIATES

Canonical Root	Order of Polynomial
0.7804	Linear
0.8810	Quadratic
0.9464	Cubic

The second canonical root of the third-order canonical trend surface was 0.8676. The associated canonical variates were as follows:

$$U = 0.757\ (\text{sand}) - 0.111\ (\text{shale}) + 0.617\ (\text{carbonate}) - 0.185\ (\text{evaporite}) \quad (43)$$

and

$$V = 0.183X - 0.983Y - 0.001X^2 + 0.001XY - 0.002Y^2\,. \quad (44)$$

The canonical trend, equation (43), summarizes variations in the thickness of sand and carbonate. The variance of linear function (43), explained by the linear function (44), is equal to 74 per cent.

Judging from figure 2*A* and equations (41) and (42), it is concluded that sand, shale, and evaporite thicken toward the southwest corner of Kansas. It is interesting to compare this canonical trend surface with net thickness maps of the same data. The trend of the net thickness maps of sand (fig. 3*A*), shale (fig. 3*B*), carbonate (fig. 3*C*), evaporite (fig. 3*D*), and isopach (fig. 4*B*) display patterns analogous to that shown by the canonical trend surface. Figure 2*B* displays a trend pattern similar in trend to the thickness of the carbonate and sand, whereas figure 2*A* displays a trend almost identical with the thickness of evaporite, shale, sand, and total isopach. The area having maximum thickness of carbonate and direction of thinning of the carbonate was also shown by figure 2*B*.

The maximum canonical root obtained was 0.9612, indicating that a cubic polynomial is an adequate fitting surface. The canonical variate for the geologic variate was that

$$\begin{aligned} U = {} & 0.803 \text{ (total thickness)} \\ & - 0.254 \text{ (sand : shale)} \\ & - 0.539 \text{ (carbonate : evaporite)} . \end{aligned} \tag{45}$$

The canonical variate for the polynomial was that

$$\begin{aligned} V = {} & -0.335X + 0.843Y + 0.233X^2 \\ & - 0.088XY + 0.335Y^2 \\ & - 0.026X^3 + 0.018X^2Y \\ & + 0.009XY^2 + 0.041Y^3 . \end{aligned} \tag{46}$$

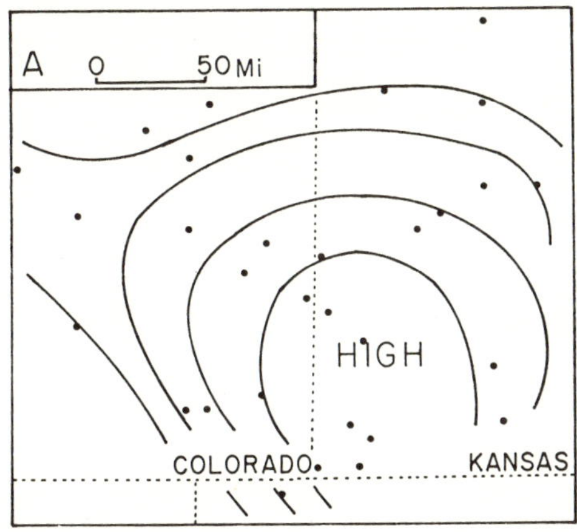

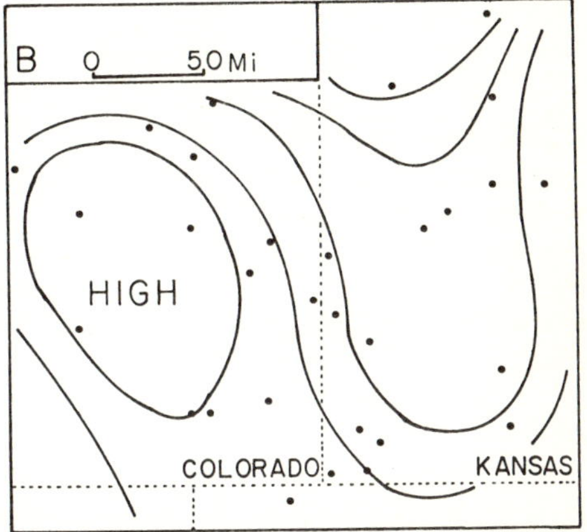

Fig. 2.—Canonical trend surfaces of the Permian system of western Kansas and eastern Colorado. *A*, Third-order canonical trend associated with the first canonical root showing variation in thickness of evaporite, sand, and shale. *B*, Third-order canonical trend associated with second canonical root showing variation in thickness of sand and carbonate.

Smaller canonical roots show different linear functions of the four variates. These linear functions are contradictory to the first two. The first one or first two canonical roots have a higher probability of approximating real trends, whereas the interpretation of the smaller canonical roots may be obscured by local variation ("noise"). Thus, the other functions are not discussed.

The second set of the three variates—total thickness, sand:shale ratio, and carbonate:evaporite ratio—were analyzed in the same manner. The record of evaluation of the trend surface is listed in table 4.

TABLE 4

Record of Successive Evaluation of Order of Polynomial for Three Variates

Canonical Root	Order of Polynomial
0.7829	Linear
0.9189	Quadratic
0.9612	Cubic

The negative sign for the ratio indicates that the high values in the map should be the low value for the ratios. Figure 4*A* shows that total thickness increases toward the southwesternmost part of Kansas, whereas the sand:shale and carbonate:evaporite ratios decrease toward the same area. The canonical trend (fig. 4*A*) fits closely to the total thickness (fig. 4*B*) and slightly to the carbonate:evaporite (fig. 4*C*) and sand:shale ratios (fig. 4*D*).

The second canonical root was 0.8760. The canonical variates were as follows:

$$U = 0.717 \text{ (total thickness)} + 0.663 \text{ (sand : shale)} + 0.217 \text{ (carbonate : evaporite)} \quad (47)$$

and

$$V = -0.545X - 0.803Y - 0.064X^2 - 0.106XY - 0.204Y^2 + 0.004X^3 + 0.029X^2Y + 0.018XY^2 - 0.008Y^3 .$$

The canonical trend (eq. [47] and [48]) yields information contradictory to the first one. This root and other, smaller roots are discarded.

The question is whether it is possible to detect a gross trend representing variations of all or most of the variates. The canonical trends (figs. 2*A*, 2*B*, and 4*A*) answer this question, at least for the first approximation, and also establish the hypothesis expected before the analysis is carried out.

An adequate use of a canonical trend is to test a geologic hypothesis which implies a trend common to all or most variates. The merit of a canonical trend surface is to evaluate a trend common to a set of variates and to condense a set of maps showing areal

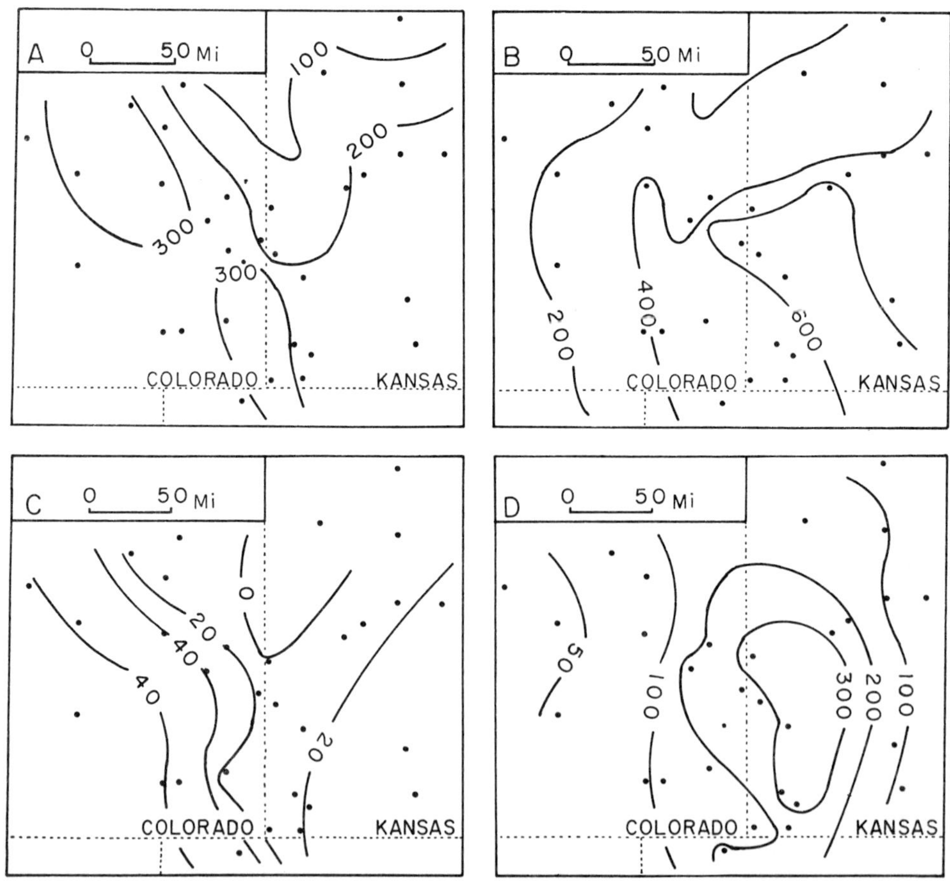

FIG. 3.—Net thickness (in feet) maps of the Permian system of western Kansas and eastern Colorado. *A*, Sand isolith. *B*, Shale isolith. *C*, Carbonate isolith. *D*, Evaporite isolith. (Data after Krumbein, 1962; redrawn by author.)

distributions of geologic variates from a single population. Its maximum correlation property yields a high fidelity between the population response surface and the sample response surface.

FUNCTION OF CANONICAL TREND SURFACES

Stratigraphers and petrologists handle a great number of maps, and usually sort them into groups showing analogous features and groups showing different features. Canonical trend analysis achieves a parsimonious summarization of a set of maps showing distributions of variates from a single population. It is thus a useful technique for screening maps, at least for exploratory studies.

The fundamental principle of canonical trend analysis is to maximize the covariance between a set of geologic variates and (X, Y)-coordinates. This implies that canonical trend surface is a surface most predictable to a particular set of variates. It weighs each variate according to its error variance.

Canonical trend surface will show the nature of variation of any number or any type of geologic variates if they can be amalgamated in a linear function. We can interpret the variations of each variate, even

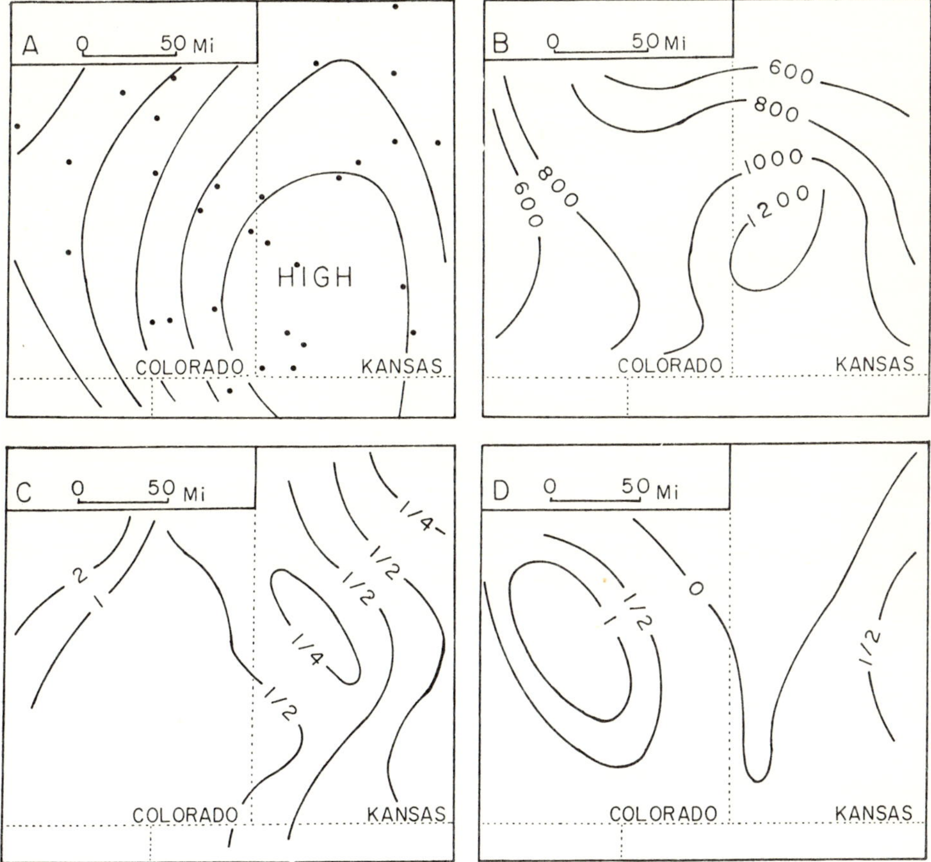

FIG. 4.—Permian system of western Kansas and eastern Colorado. *A*, Third-order canonical trend showing variation in total thickness, sand : shale ratio, and carbonate : evaporite ratio. *B*, Isopach map. *C*, Sand: shale ratio map. *D*, Carbonate : evaporite ratio map. (Data for maps *B*, *C*, and *D* after Krumbein, 1962; redrawn by author.)

though geologic meaning of a particular combination is not yet known. This type of trend surface does not show absolute values at each contour. On the other hand, it shows the general trend of the variates, at least to a first approximation. The coefficients of a particular set of variates would change slightly from low to high order polynomials to have maximum correlation with geographic coordinates, whereas the coefficients of a linear function of a principal component are fixed values for all orders of polynomials.

Theoretically, any number of variates may be mapped at a time. If many variates are analyzed simultaneously, interpretation of the result will not be as obvious as the interpretation for fewer variates. There is no way to decide the appropriate number of variates in advance. Based on the writer's experience of practical problems, a reasonable number is four or fewer. It is not necessary that the geologic variates be uncorrelated. But if they are highly correlated, the correlation matrix will be nearly singular and matrix inversion may yield a meaningless solution as a result of the limited number of digits in the computer. With a suffi ciently large computer, this is not always a serious problem. This problem should be kept in mind, however, when dealing with highly intercorrelated variates. If a number of variates are available for canonical trend analysis, the type of variates chosen for analysis is dependent on which geologically meaningful hypothesis we expect to establish. Poor selection of variates would produce geologically ambiguous and trivial results.

Real linear functions other than the first one cannot be obtained unless equations (5), (6), (13), (14), and (15) are satisfied. Thus, variations obtained from canonical trend analysis exist in nature if they were orthogonal to the previous linear functions; otherwise, variations obtained are mere artifacts brought in by mathematical operations. We should realize, however, that coefficients in the linear function that give maximum correlation may not be suitable as a solution to a practical problem. The linear function associated with the largest canonical root is most likely to have some physical interpretation. If the subsequent functions associated with smaller canonical roots give contradictory combinations, these subsequent canonical roots should be rejected without hesitation, even though these canonical roots are statistically significant at a high probability level.

Coefficients of canonical variates will yield ambiguous signs in the linear function if the sample size is insufficient, if the order of polynomial is too high, if the number of variates is too large, if geologic variates and (X, Y)-coordinates are highly correlated (specimens are taken from a straight line), and if the nature of the problem studied defines its canonical roots poorly (the limiting eigenvalues of eq. [11] are poorly defined). The investigator should be careful that results have a real geologic base. Does this destroy the value of the technique? The answer is, Not at all. In dealing with multivariate statistical analysis such as multiple regression analysis, principal component analysis, and so on, results which cannot be interpreted may be obtained in attempting to solve practical problems. This is a general difficulty in applying multivariate techniques in geology, and it results more from the real complexity of nature than from problems caused by the mathematical analysis.

Examination of the canonical trend for a set of variates frequently leads to an understanding of the intrinsic trend, common to several variates, that would be difficult or impossible to obtain by other means. It is a powerful tool to reconstruct geologic events, such as reef or evaporite patterns of distribution, by considering the variations of many variables simultaneously.

After working a number of practical problems, I am convinced that the canonical trend analysis is an adequate technique to evaluate a trend common to a set of variates in preliminary and concept-formation stages or research as part of a data-reduction scheme. Canonical trend-surface analysis represents a step toward the theoretical development of geomathematics, and its application in geology opens a promising field.

ACKNOWLEDGMENTS.—The writer is deeply indebted to Professor Gerard V. Middleton for his helpful suggestions and valuable comments. The Geological Survey of Canada provided funds for the support of this research.

REFERENCES CITED

ANDERSON, T. W., 1958, An introduction to multivariate statistical analysis: New York, John Wiley & Sons, 374 p.

BARTLETT, M. S., 1938, Further aspects of the theory of multiple regression: Cambridge Philosophy Soc. Proc., v. 34, p. 33–40.

——— 1941, The statistical significance of canonical correlation: Biometrika, v. 32, p. 29–37.

——— 1947*a*, The general canonical correlation distribution: Annals Math. Statistics, v. 18, p. 1–17.

——— 1947*b*, Multivariate analysis: Royal Statistical Soc. Jour., supp. ix, p. 176–197.

——— 1947*c*, The use of transformation: Biometrics, v. 3, p. 39–52.

——— 1948, Internal and external factor analysis: British Jour. Psychology (Statistics Sec.), v. 1, p. 73–81.

——— 1965, Multivariate analysis, *in* WATERMAN, T. H., and MOROWITZ, H. J., eds., Theoretical and mathematical biology: London, Blaisdell Publishing Co., 426 p.

BEECH, H. R., and MAXWELL, A. E., 1958, Differentiation of clinical groups using canonical variates: Jour. Consulting Psychology, v. 22, p. 113–121.

BURT, C., 1948, Factor analysis and canonical correlations: British Jour. Psychology (Statistics Sec.), v. 1, pt. 11, p. 95–106.

BUZAS, MARTIN A., 1966, The discrimination of morphological groups of *Elphidium* (foraminifer) in Long Island Sound through canonical analysis and invariant characters: Jour. Paleontology, v. 40, p. 585–594.

CASSIE, R. M., and MICHAEL, A. D., 1968, Fauna and sediments of an intertidal mud flat: a multivariate analysis: Jour. Expt. Marine Ecology, v. 12, p. 1–23.

COCHRAN, W. G., 1963, Sampling techniques (2d ed.): New York, John Wiley & Sons, 413 p.

CONSTANTINE, A. G., and JAMES, A. T., 1958, On the general canonical correlation distribution: Annals Math. Statistics, v. 29, p. 1146–1166.

COOLEY, W. W., and LOHNES, P. R., 1962, Multivariate procedures for the behavioural science: New York, John Wiley & Sons, 211 p.

DAS, R. S., 1965, An application of factor and canonical analysis to multivariate data: British Jour. Math. Statistics Psychology, v. 18, p. 57–67.

DUNTEMAN, G. H., 1967, A canonical correlational analysis of the strong vocational interest blank and the Minnesota Multiphasic Personality Inventory for a female college population: Educ. Psychology Measurement, v. 27, p. 631–642.

EBERLEIN, P. J., 1962, A Jacobi-like method for the automatic computation of eigenvalues and eigenvectors of an arbitrary matrix: Jour. Social Indus. Appl. Mathematics, v. 10, p. 74–88.

FORGOTSON, J. M., JR., 1960, Review and classification of quantitative mapping techniques: Am. Assoc. Petroleum Geologists Bull., v. 44, p. 83–100.

GLAHN, H. R., 1968, Canonical correlation and its relationship to discriminant analysis and multiple regression: Jour. Atmospheric Sci., v. 25, p. 23–31.

HILL, W. J., and HUNTER, W. G., 1966, a review of response surface methodology: a literature survey: Technometrics, v. 8, p. 571–590.

HORST, P., 1961*a*, Generalized canonical correlation and their application to experimental data: Jour. Clinical Psychology, Mon. Supp. 14, p. 331–347.

——— 1961*b*, Relations among M sets of measures: Psychometrika, v. 26; p. 129–149.

HOTELLING, H., 1936, Relations between two sets of variates: Biometrika, v. 28, p. 321–377.

——— 1957, The relations of the newer multivariate statistical method to factor analysis: British Jour. Statistical Psychology, v. 10, pt. 11, p. 69–79.

HSU, P. L., 1941, On the limiting distribution of the canonical correlations: Biometrika, v. 32, p. 38–45.

IMBRIE, J., 1963, Factor and vector analysis programs for analyzing geological data: Tech. Rept. 6, ONR Task 389–135, Office of Naval Research, Geography Branch, Northwestern Univ., 83 p.

JÖRESKOG, K. G., 1967, Some contributions to maximum likelihood factor analysis: Psychometrika, v. 32, p. 443–482.

KENDALL, M. G., 1961, A course in multivariate analysis: London, Criffin's Statistical Mon. cour., 185 p.

KOONS, P. B., JR., 1962, Canonical analysis, *in* BORKO, H., ed., Computer applications in the behavioral sciences: Englewood Cliffs, N.J., Prentice-Hall, Inc., 633 p.

KRUMBEIN, W. C., 1955, Composite end members in facies mapping: Jour. Sed. Petrology, v. 25, p. 115–122.

——— 1962, Open and closed number systems in stratigraphic mapping: Am. Assoc. Petroleum Geologists Bull., v. 46, p. 2229–2245.

——— and IMBRIE, J., 1963, Stratigraphic factor

maps: Am. Assoc. Petroleum Geologists Bull., v. 47, p. 698–701.

Kshirsagar, A. M., 1962, A note on direction and collinearity factors in canonical analysis: Biometrika, v. 49, p. 255–259.

Lawley, D. N., 1959, Tests of significance in canonical analysis: Biometrika, v. 46, p. 59–66.

——— and Maxwell, A. E., 1963, Factor analysis as a statistical method: London, Butterwork & Co., 117 p.

Lee, P. J., 1968, Applications of canonical correlation in geology: Unpub. Ph.D. thesis, McMaster Univ., Hamilton, Ontario, 124 p.

——— and Middleton, G. V., 1967, Application of canonical correlation to trend analysis, *in* Merriam, D. F., and Cocke, N. C., eds., Colloquium on trend analysis: Kansas Geol. Survey Computer Cont. 12, p. 19–21.

Maxwell, A. E., 1961, Canonical variate analysis when the variables are dichotomous: Educ. Psychology Measurement, v. 21, p. 259–271.

Meredith, W., 1964, Canonical correlations with fallible data: Psychometrika, v. 29, p. 55–65.

Middleton, G. V., and Lee, P. J., 1967, Applications of canonical correlation to sedimentology: Internat. Sedimentological Cong., 7th, England.

Morrison, D. F., 1967, Multivariate statistical methods: New York, McGraw-Hill Book Co., 338 p.

Mudge, M. R., 1967, Paleotectonic investigations of the Permian system in the United States, chap. F: Central midcontinent region: Geol. Survey Prof. Paper 515-F, p. 93–123.

Pearce, S. C., and Holland, D. A., 1960, Some applications of multivariate methods in botany: Appl. Statistics, v. 9, p. 1–7.

Pelto, C. R., 1954, Mapping of multicomponent systems: Jour. Geology, v. 62, p. 501–511.

Rao, C. R., 1955, Estimation and tests of significance in factor analysis: Psychometrika, v. 20, p. 91–111.

Reyment, R. A., 1963, Studies on Nigerian Upper Cretaceous and Lower Tertiary Ostracoda, II: Danian, Paleocene and Eocene Ostracoda: Stockholm Contr. Geology, Acta Universitatis Stockholmiensis, 10, 286 p.

——— 1966, Studies on Nigerian Upper Cretaceous and Lower Tertiary Ostracoda, III: Stratigraphic, paleoecological and biometrical conclusions: Stockholm Contr. Geology, Acta Universitatis Stockholmiensis, v. 14, 151 p.

Roy, S. N., 1957, Some aspects of multivariate analysis: New York, John Wiley & Sons, 214 p.

Seal, H. L., 1964, Multivariate statistical analysis for biologists: London, Methuen & Co., 207 p.

Thomson, G., 1947, The maximum correlation of two weighted batteries: British Jour. Psychology (Statistical Sec.), v. 2, pt. 1, p. 27–34.

Tintner, G., 1946, Some applications of multivariate analysis to economic data: Jour. Am. Statistical Assoc., v. 41, p. 472–500.

Waugh, F. V., 1942, Regressions between sets of variables: Econometrika, v. 10, p. 290–310.

Wilks, S. S., 1963, Mathematical statistics: New York, John Wiley & Sons, 644 p.

AUTHOR CITATION INDEX

Adcock, C. J., 333
Ahmad, M., 314
Ahmavaara, Y., 188
Albert, A. A., 188
Allee, W. C., 412
Allport, G. W., 189
Anderson, E., 101
Anderson, M., 352
Anderson, R. H., 298
Anderson, T. W., 4, 15, 101, 130, 189, 298, 391, 427
Anderson, W. W., 412
Atchley, W. R., 4, 15, 141, 345, 351

Baggaley, A. R., 265
Bailey, D. W., 15, 101
Bargmann, R., 189, 265
Barkham, J. P., 352
Barnard, M., 391
Barrett, M. J., 333
Bartholomew, G. A., 402
Bartlett, M. S., 4, 15, 84, 141, 189, 209, 229, 298, 352, 378, 427
Beech, H. R., 427
Belding, H. S., 402
Blackith, R. E., 4, 101, 345
Blackith, R. M., 345
Blair, W. F., 412
Bock, R. D., 4
Brannstrom, B., 353
Bray, J. R., 130
Brier, G. W., 391
Bronswijk, J. G. M. H., 352
Brown, G. W., 391
Brown, T., 333
Browne, M. W., 141
Bryan, J. G., 391, 412
Bryant, E. H., 4
Burkett, G. R., 298
Burr, E. J., 229
Burt, C., 189, 209, 229, 314, 352, 427
Buzas, M. A., 425

Caffrey, J., 142, 229
Calhoon, R. E., 352, 412
Canadian Department of Transport, 402
Carroll, J. B., 141, 189, 244, 265, 277
Cassie, R. M., 427
Cattell, A. K. S., 189, 265
Cattell, R. B., 130, 141, 189, 209, 265, 266, 345, 412
Chakravarti, I. M., 352
Chebib, F. S., 16, 354
Chen, C. W., 352
Choi, S. C., 101
Coan, R. W., 266
Cochran, W. G., 402, 412, 427
Comrey, A., 237
Constantine, A. G., 427
Cooley, W. W., 4, 10, 130, 412, 427
Cramer, E. M., 4
Cronbach, L. J., 229
Cureton, T. K., 189

Curtis, J. T., 130

Dagnelie, P., 130, 141
Dale, M. B., 131
Daly, H. V., 131
Darrock, J. N., 333
Das, R. S., 352, 427
Davis, F. B., 209
Dawson, W. R., 402
Defrise-Gussenhoven, E., 101
DeGroot, M. H., 352
Dempster, A. P., 4, 84, 352
Dickman, K. W., 189, 266
Dobzhansky, T. H., 412
Driver, H. E., 189
DuBois, P. H., 4
Dunteman, G. H., 427

Eber, H. W., 141
Eberlein, P. J., 427
Edgeworth, F. Y., 352
Elderton, E. M., 378
Eysenck, H. J., 315

Farner, D. S., 402
Ferguson, G. A., 141, 244
Ferrari, Th. J., 130
Fisher, D., 141
Fisher, R. A., 84, 378, 391, 412
Forgotson, J. M., Jr., 427
Foster, M. J., 189
Frisch, R., 84
Fruchter, B., 130, 189

Gabriel, K. R., 10
Galton, E., 352
Garnett, J. C. M., 190
Garwood, R. A., 10
Gauch, H. G., 15
Gittins, R., 130
Glahn, H. R., 427
Gleser, G. C., 229
Goodall, D. W., 15, 130
Gorsuch, R. L., 189
Gould, S. J., 10, 15
Gouldner, A. W., 190
Gower, J. C., 4, 15, 130
Graybill, F. A., 4, 10
Green, B. F., 229
Green, P. E., 352
Greig-Smith, P., 130, 328

Groenewoud, H. van, 131
Grüneberg, H., 333
Guilford, J. P., 298
Gulliksen, H., 298
Guttman, L., 141, 190, 229, 298

Hadley, G., 298
Halbert, M. W., 352
Haldane, J. B. S., 412
Hamilton, T. H., 402, 412
Hamilton, W. J., III, 402
Hammond, W. H., 314
Hannan, G. J., 352
Harman, H. H., 10, 15, 130, 142, 190, 209, 229, 244, 277, 314, 333
Harris, C. W., 229
Harris, E. W., 141
Healy, M. J. R., 352
Heermann, E. F., 298
Hemmerle, W. J., 229
Hendrickson, A. E., 142
Henrysson, S., 190
Heppner, F., 402
Hill, W. J., 427
Hodge, R. W., 353
Holland, D. A., 328, 428
Holzinger, K. J., 142, 190, 209, 244, 314
Hooper, J. W., 352, 391
Horn, J. L., 142
Horst, P., 4, 130, 142, 229, 298, 352, 427
Hotelling, H., 16, 84, 101, 190, 209, 328, 352, 391, 412, 427
Hsu, E. H., 315
Hsu, P. L., 353, 427
Hunter, P. E., 143, 190
Hunter, W. G., 427

Imbrie, J., 427
Ingham, A. E., 378
Ivimey-Cook, R. B., 16, 131

James, A. T., 427
Jameson, D. L., 352, 412
Jardine, N., 10
Jeffers, J. N. R., 16
Jeffery, G. B., 378
Jolicoeur, P., 10, 16, 84, 101, 410
Jones, L. E., 142
Jones, W. H., 141
Joreskog, K. G., 142, 229, 328, 427

Kaiser, H. F., 16, 141, 142, 190, 229, 244, 266, 277, 333
Katz, J. O., 142
Kelley, T. L., 190, 314
Kendall, M. G., 4, 84, 131, 209, 298, 328, 427
Kendall, S. A., 298
Kendeigh, S. C., 402
Kerfoot, W. C., 354
Kestelman, H., 298
King, J. R., 402
Kirby, H., 315
Klatzky, S. R., 353
Klett, C. J., 4
Koons, P. B., Jr., 353, 427
Kraus, B. S., 101
Krumbein, W. C., 427
Kruskal, J. B., 10
Kshirsagar, A. M., 4, 353, 428

Laha, R. G., 353
Lancaster, H. O., 353
Lance, G. N., 131
Landahl, H. D., 266
Lawley, D. N., 16, 131, 142, 190, 209, 229, 298, 328, 353, 426
Lee, P. J., 4, 143, 428
Levene, H., 412
Lewis, C., 230
Lewis, T., 352
Lewontin, R. C., 101, 412
Li, C. C., 352
Lippert, R. H., 402
Littlejohn, M. J., 412
Lohnes, P. R., 4, 10, 130, 412, 427
Lord, F. M., 229
Lorenz, D. N., 391
Lorge, I., 353
Love, W. A., 353, 354
Lund, I. A., 391
Lustick, S., 402

McCloy, C. H., 315
McDonald, R. P., 229, 298, 353
MacEwan, C., 353
McKeon, J. J., 352
Mackey, J. P., 352
Mahalanobis, P. C., 84
Majumdar, D. N., 84
Marriott, F. H. C., 353
Martin, J., 345

Maxwell, A. E., 131, 142, 427, 428
Mefferd, R. B., 190
Meredith, W., 353, 428
Michael, A. D., 425
Michael, W. B., 298
Middleton, G. V., 428
Miller, R. G., 391
Miller, R. L., 101
Misra, R. K., 412
Mittwoch, U., 345
Mook, C. P., 381
Moore, C. S., 131
Moore, T. V., 145, 315
Morishima, H., 142
Morrison, D. F., 4, 10, 131, 402, 428
Mosier, C. J., 229
Mosimann, J. E., 10, 84, 101, 412
Mudge, M. R., 426
Muerle, J. L., 189, 345
Mulaik, S. A., 142

Neuhaus, J. O., 142, 190, 209, 244, 266
Norris, J. M., 352
Novick, M. R., 229, 230

Oka, H., 142
O'Leary, M., 402
Olson, E. C., 101
Ore, O., 10
Orloci, L., 131
Osgood, C. E., 190
Overall, J. E., 4

Pawlik, K., 189
Pearce, S. C., 131, 328, 428
Pearson, K., 17, 25, 84, 101, 353, 378
Peel, E. A., 230
Pelto, C. R., 428
Penrose, L. S., 84
Pijl, H., 130
Pinzka, C., 190, 266
Power, D. M., 16, 142, 402
Powers, D. A., 16
Prim, R. C., 10
Proctor, M. C. F., 16, 131
Prosser, C. L., 101

Quenouille, M. H., 190

Rajaratnam, N., 229

Rao, C. R., 4, 10, 84, 190, 209, 230, 298, 328
Rayner, J. H., 131
Rees, W. L., 315
Reeve, E. C. R., 412
Reyment, R. A., 4, 10, 131, 328, 353, 428
Richmond, R. C., 412
Roberts, M. I., 345
Robinson, P. J., 352
Roe, A., 101
Roff, M., 190
Rohlf, F. J., 10, 16, 131, 142, 143
Romanovsky, V., 378
Roy, J., 84
Roy, S. N., 428
Rozeboom, W. W., 354
Rubin, H., 189
Rulon, P., 412
Rummel, R. J., 142

Salt, G. W., 402
Saunders, D. R., 142, 190, 244, 266
Schmidt-Nielsen, K., 402
Seal, H., 4, 10, 131, 333, 428
Searle, S. R., 4, 10
Shaw, D. C., 84
Sibson, R., 10
Simonds, J. L., 85
Simpson, G. G., 101
Sinha, R. N., 16, 143, 354
Snedecor, G. W., 402
Snyder, W. F., 412
Sokal, R. R., 4, 10, 16, 131, 142, 143, 190
Soulé, M., 354
Spearman, C., 143, 190
Spitz, O. T., 402
Srikantan, K. S., 354
Stewart, D. K., 354
Stone, R., 85
Suci, G. J., 190

Tannenbaum, P. H., 190
Tatsuoka, M. M., 4, 391
Teissier, G., 101
Test, F. C., 412
Thomas, P. A., 143
Thomson, G. H., 143, 190, 209, 230, 244, 298, 315, 354, 428
Thorson, T. B., 412
Thurstone, L. L., 42, 143, 190, 209, 230, 244, 266, 277, 298, 315, 333
Tintner, G., 428
Torgerson, W. S., 85
Tracey, J. G., 131
Tucker, L. R., 266, 315

United States Department of Agriculture, 402

Van Bronswijk, J. G. H., 354
Van de Geer, J. P., 4
Venekamp, J. T. N., 130
Visher, S. S., 402

Wallace, H. A. H., 16, 354
Walshe, B. M., 345
Waugh, F. V., 354, 428
Webb, L. J., 131
Wells, P. H., 143
Wenner, A. M., 143
Wernimont, G., 85
White, P. A., 101
White, P. O., 142
Whittaker, R. H., 15
Whittle, P., 85, 328
Wilks, S. S., 378, 428
Williams, E. G., 85
Williams, G. J., 354
Williams, W. T., 131
Wishart, J., 378
Wright, S., 101, 315, 412
Wrigley, C., 142, 190, 209, 244, 266

Yarranton, G. A., 131
Yule, G. U., 378

Zeuthen, E., 402

SUBJECT INDEX

Allometry, 84, 101, 403, 407, 411
Analysis of covariance, 300
Analysis of variance
  multivariate, 2–3, 7
  univariate, 8, 53, 184, 300, 334, 362, 376

Baldwin effect, 404
Bartlet's test
  for homogeneity of variance-covariance matrices, 7
  for latent roots, 206, 417
Basis
  for factor analysis, 197, 198, 202, 268, 280, 292, 295
  of vector space, 194, 195
Beer's Law, 76, 77, 78
Bergmann's Rule, 396

Canonical analysis (*see* Discriminant analysis)
Canonical correlation
  analysis of, 5, 7, 68, 123, 348, 350, 351, 403, 405, 406, 409, 413, 414, 417, 418
  application of, 351, 352, 355, 387, 404, 408, 410, 411, 412, 427, 428
  applied to trend analysis, 347, 351, 413–428
  coefficient of, 414, 418, 419, 420, 422, 423, 424
  distribution of, 416, 427
  interpretation of, 218, 353
  relation of, to discriminant analysis, 352, 383, 391, 415, 427
  relation of, to factor analysis (*see* Factor analysis, canonical)
  relation of, to multiple regression, 383, 386, 427
  significance tests for, 417, 427
  theory of, 200, 202, 350, 391
Canonical form, 318, 319, 415
Canonical root (*see* Canonical correlation, coefficient of)
Canonical variable (*see* Canonical variate)
Canonical variate
  in canonical correlation, 68, 350, 351, 383, 408, 409, 416, 418, 422, 426, 427
  in discriminant analysis, 8
Canonical vector (*see* Canonical variate)
Center of gravity (*see* Centroid)
Centroid
  approximation to principal axes, 126, 127, 152–159, 185, 249, 250, 266, 302–316
  of a system, 19, 22, 23, 60, 62, 76
Centroid matrix (*see* Factor matrix)
Cephalic index, 73
Characteristic equation, 38, 39, 41–45, 48, 65, 67, 68, 83, 391
Cluster analysis, 3, 5, 8, 9
Coefficient of generalizability, 227
Common factor analysis (*see* Factor analysis)

Common factor space, 135, 177, 181, 194, 224, 237, 238, 267, 280, 290, 292, 294, 302-303
Communality
  definitions relating to, 116, 135, 174, 175, 178, 181, 191, 208, 213, 232, 247, 301, 323, 329
  determination and estimates of, 136, 141, 179, 180, 190, 224, 302, 317, 327
  lower limit to, 176, 219, 222
  use of, 117, 121, 126, 177, 192, 204, 221, 236, 237, 274, 303, 307, 330-332
Component analysis (*see* Principal components)
Component scores (*see* Principal component scores)
Co-relation, 349
Correlation coefficient
  composite, 387
  definitions relating to, 150, 178, 194, 234, 302, 328, 349, 360, 363, 373, 376, 415, 417
  distribution of, 355, 356
  intraclass, 396
  multiple, 3, 176, 177, 187, 208, 347, 349, 351, 355, 362, 365, 376, 380, 381, 384, 388, 392, 395, 397, 409, 427
  partial, 355, 360, 362, 364, 372, 374, 376
  trace, 385
  uses of, 17, 49-50, 155, 305-306, 313, 338, 394-397, 414
  vector, 386
  vector representation of, 107-108, 151
Correlation matrix
  definition relating to, 98, 152, 159, 164-165, 167, 169, 170-172, 179, 192, 194, 216, 218, 220, 259, 280, 417
  determinant of, 27, 38, 39, 356, 380-381
  examples of, 216, 288, 306, 319-321
  geometric structure of, 46-47
  reduced, 136, 171, 173, 177, 197, 239
  tests of hypotheses for, 15, 190, 195, 197, 203-204
Covariance matrix (*see* Variance-covariance matrix)
Covariance structure, 139, 142, 361, 363, 376

Data matrix, 104, 105, 106, 108, 116, 130, 211, 229
Determinant of correlation matrix (*see* Correlation matrix)
Direction cosines
  definitions relating to, 19, 23, 29-30, 43-44, 60, 76-77, 305
  of an ellipse, 89, 91, 94, 96
  examples of, 94-95, 405-408
Discriminant analysis, 3, 5, 8, 84, 98, 217, 350, 383, 391, 412, 428
  applications of, 230, 336, 353, 403-409
  relation of, to canonical correlation (*see* Canonical correlation)
  tests of hypotheses for, 353, 391
Distribution
  Fisher's Z, 362, 365, 375, 376
  multivariate normal, 6, 7, 13, 32, 40, 93, 326, 356
  t distribution, 358, 359, 362, 375
  Wishart, 82, 367, 372

Ecogeographic variation, 404
Equal frequency ellipse, 7, 91, 94, 97, 99, 101
Euclidean space, 65, 75, 76, 81, 105
Experimental design, 168, 349

Factor
  common, 135, 144-145, 174-178, 199, 219, 225, 266, 272, 323
  general (*see* Factor, common)
  specific, 144, 147, 171, 175-178, 193, 212, 282-283
  unique, 212, 281, 301, 329-331
Factor analysis
  alpha, 134, 137, 142, 210, 211, 215, 218, 219, 220, 221, 222, 224, 229
  canonical, 134, 137, 191-192, 202-203, 210-211, 215, 217, 218, 221-224, 278, 280, 290-294, 414 (*see* Factor analysis, maximum likelihood)
  definitions relating to, 3, 7, 10, 134-135, 144, 163, 166-169, 177, 189, 207, 212, 229, 268, 282, 298, 323-324, 328, 329
  image, 136-137, 141
  maximum likelihood, 137, 142, 176, 191, 229, 292, 414, 427
  minimum residuals, 136
  nonlinear, 295, 298

number of factors in, 143, 179-180, 224, 324
principal-factor, 134, 137, 140, 192, 202, 210-211, 215-217, 221-224, 330, 332, 414
Q-mode, 140, 414
relation of, to principal components, 75, 85, 135, 173-175, 181, 192-200, 325
test of significance for, 15, 84, 189-190, 192, 204-207, 324
uses of, 10, 101, 130, 140, 306-313, 325-328, 330-332, 336-339, 352, 414
Factorial analysis (*see* Factor analysis)
Factor loading matrix (*see* Factor matrix)
Factor matrix
definition relating to, 135, 170-173, 176, 194, 211, 267, 269, 272, 279
examples of, 273, 306, 309, 326, 331, 337
Factor rotation
biquartimin, 265, 267, 273, 274, 275, 276, 277
confactor, 184
covarimin, 267, 273
direct oblimin, 138
functionplane, 139, 142
Harris-Kaiser, 139
Kaiser-Dickman, 267
maxplane, 134, 138, 245-265, 336, 345
minimax, 248
MTAM, 137
oblimax, 186, 246, 248, 250, 254, 265, 266, 268
oblimin, 138, 186, 189, 266, 272, 273, 275, 278
procrustes, 139
promax, 139, 142
quadrimax, 195, 209
quartimax, 142, 186, 190, 231, 232, 233, 234, 235, 239, 242, 244, 245, 248, 250, 266
quartimin, 267, 273, 276
varimax, 134, 138, 141, 186, 190, 231-245, 247, 248, 250, 266, 277, 329-333
Factor scores
Bartlet's method for, 280, 296
complete estimation method for, 139
definitions relating to, 134, 139, 142, 177, 214, 229, 277-298
examples of, 339-341
ideal variable method for, 139

Generalized distance, 84
Geographic variation, 10, 101, 140, 142, 335, 392-393, 396, 400-404, 412-413
Goodness of fit, 6, 61, 377
Graph theory, 9
Growth, 87, 95, 98-100, 407
patterns of, 101, 345

Homeostasis, 402, 412
Hyperbolic tangent, 50
Hyperellipsoid, 112, 113, 116, 302

Image analysis (*see* Factor analysis)
Isometry, 405, 408
Isophane, 394

Jacobi method for matrix decomposition, 94, 427

Kurtosis, 6, 232

Lagrange multipliers, 37, 72, 380, 416
Likelihood ratio criterion, 205-206
Loading matrix (*see* Factor matrix)

Major axis (*see* Principal components)
Mahalanobis distance, 81
Minor axis, 111
Moment generating function, 367-369
Multidimensional scaling (*see* Ordination)
Multiple comparison's test, 7-8
Multiple factor analysis (*see* Factor analysis)
Multivariate normal (*see* Distribution)

Network analysis, 5, 8-9
Networks, 9-10
Nonmetric scaling, 3, 6, 8, 10, 14, 75
Numerical taxonomy, 8-9, 16, 130-131, 329

Path coefficients, 299, 315
Principal axes (*see* Principal components)
Principal components
applications of, 89-99, 321-323, 390, 406-408, 413-414, 422
definitions relating to, 3, 6, 8, 12-16, 35-55, 56-57, 66-73, 84, 86, 195-197, 215-216, 316-318

ordination by, 12-16, 102-131
relation of, to eigenvalues and eigenvectors, 6, 12, 44, 57-61, 66-67, 74, 130, 196, 216, 317-322
relation of, to factor analysis (*see* Factor analysis)
tests of hypotheses for, 82-83, 192, 196, 321, 328
Principal component scores, 14-15, 116, 119-121
Principal coordinates, 14

Reference structure, 206, 276-277
Reference vector, 181-186, 246, 252
Regression
coefficient of, 72, 355, 358-359, 367, 390, 397
curvilinear, 323
multiple, 3, 123, 135, 140-141, 202, 278, 351-352, 355, 383, 386, 392, 394-397, 427
parabolic, 374
partial, 355, 361-365, 324, 392-395, 397-405
relation of, to principal axes, 24, 28, 200-201, 207
stepwise, 352, 409
univariate, 85, 348-349, 356, 373, 376-377, 428
Reliability coefficient, 202

Sample space, 104, 106, 109, 110, 114, 116
Scatter diagram, 6, 8, 13, 52, 55, 89-99, 119, 121, 197, 391, 395, 405
Serial correlation, 318
Sex dimorphism, 87, 134, 334-337, 342, 344, 393
Sexual difference (*see* Sex dimorphism)
Similarity
character, 78
coefficient of, 110
ecological, 105, 119
overall, 9
Simple structure
definitions of, 138-139, 183, 232, 234, 239, 302-303
rotation for (*see* Factor rotation)
significance tests for, 187
Size and shape, 2, 10, 57, 71-73, 84-100, 334, 336, 339, 343, 345, 404-405, 412
Skewness, 6, 301

t-test, 373
Table of intercorrelations (*see* Correlation matrix)
Tensor analysis, 32
Time series, 70, 190

Uniquenesses (*see* Variance)

Variance
common, 135-136, 220-221
generalized, 296, 355
unique, 135, 221, 228, 268, 296, 307, 313
Variance-covariance matrix
definitions relating to, 6-8, 100, 110, 212, 268-269, 286, 323, 384, 416, 418
examples of, 89-94
tests of significance for, 15
Vector alienation coefficient, 386